Antibacterial Chemotherapeutic Agents

Antibacterial Chemotherapeutic Agents

Scott L. Dax
The R.W. Johnson Pharmaceutical
Research Institute
Spring House
Pennsylvania
USA

BLACKIE ACADEMIC & PROFESSIONAL
An Imprint of Chapman & Hall
London · Weinheim · New York · Tokyo · Melbourne · Madras

**Published by Blackie Academic & Professional, an imprint of
Chapman & Hall, 2–6 Boundary Row, London SE1 8HN, UK**

Chapman & Hall, 2–6 Boundary Row, London SE1 8HN, UK

Chapman & Hall GmbH, Pappelallee 3, 69469 Weinheim, Germany

Chapman & Hall USA, Fourth Floor, 115 Fifth Avenue, New York NY 10003, USA

Chapman & Hall Japan, ITP-Japan, Kyowa Building, 3F, 2-2-1 Hirakawacho, Chiyoda-ku, Tokyo 102, Japan

DA Book (Aust.) Pty Ltd, 648 Whitehorse Road, Mitcham 3132, Victoria, Australia

Chapman & Hall India, R. Seshadri, 32 Second Main Road, CIT East, Madras 600 035, India

First edition 1997

© 1997 Chapman & Hall

Typeset in 10/12pt Times by Cambrian Typesetters, Frimley, Surrey

Printed in Great Britain by T.J. Press (Padstow) Ltd, Padstow, Cornwall

ISBN 07514 0289 3

A Catalogue record for this book is available from the British Library

Library of Congress Catalog Card Number 96–83762

∞ Printed on acid-free text paper, manufactured in accordance with ANSI/ NISO Z39.48-1992 (Permanence of Paper)

Dedicated in memory of my sister, Francine

Contents

Preface

Antibacterial chemotherapy is a highly valued medical science which has shaped modern humanity in a phenomenal fashion. The practice of administering chemical substances to treat and cure infectious diseases and disorders has been successful on a grand scale. More human lives have been saved by this discipline than any other area of the pharmaceutical sciences. Of the host of microbes that can be pathogenic to humans, particularly rewarding advances have been achieved in eradicating bacterial, fungal and parasitic infections. Many viral infections cannot yet be solely tackled through the use of small molecule drugs; however antiviral chemotherapy along with vaccinations and conferred immunity have had a favorable impact upon some diseases. Despite these accomplishments, more people will continue to die from infectious diseases than from any other unnatural cause worldwide.

Within the past half century, a wide variety of antibacterial substances have been discovered, designed and synthesized; literally hundreds of drugs have been successfully used in some fashion over the years. Today, the worldwide anti-infective market exceeds $20 billion dollars annually and overall, antibacterial agents comprise the bulk of this trade. A number of general classes of antibacterial drugs have emerged as mainstays in modern infectious disease chemotherapy. It is the purpose of this text to address these agents in a format that is useful to the novice as well as to the researcher involved in antibacterial research.

From a personal perspective, upon joining the infectious disease efforts at Hoffmann-La Roche in 1988, I found myself searching for a fundamental text which detailed the issues I believed were pertinent to the outcome of our research program. The universal research mandate remains that new antibacterial compounds must offer some pronounced advantage over existing developmental or marketed agents. Therefore, the successful researcher is absolutely required to have a working knowledge of many other classes of antibacterial agents. Despite some outstanding volumes (which are rightfully recognized within this work), issues of mode of action and resistance are often found separated from purely medicinal chemistry concerns such as synthesis and structure–activity relationships (SAR). In addition, many texts which address these issues on a level useful to the medicinal chemist or biochemist focus upon a single class of agents. In my opinion, this can overwhelm the reader with details that tend to remain unappreciated unless one is intimately involved with that specific area.

Lastly, there is probably some tendency for a trained chemist to shun some key aspects of microbiology, molecular biology and bacteriology as it pertains to the function of antibacterial compounds; I imagine the converse can be true for our colleagues who study the 'bugs'.

My belief remains that any researcher in the field needs to be well-informed on several fronts before wisely envisioning a novel antibacterial agent. It is a tremendous advantage to know of the origin and history of a given class of antibacterials since this lends an appreciation of the evolution and development of landmark agents. In addition, one must understand how the antibacterial drug works at a macromolecular or molecular level (if available). After all, structural modifications of any existing prototype can only be successful if the necessary pharmacophore is left intact. From this knowledge, rational derivatives and novel congeners can be synthesized and a sound structure–activity relationship (SAR) can be gathered. In order to devise new antibacterial regimens, one must also recognize the bacterial enemies and the diseases caused by these pathogens. In this regard, it is important to form a 'therapeutic insight' of how a target molecule would be used and what practical impact such a protocol would have on comparison with existing therapies. The medicinal chemist must identify ways in which structural modifications can lead to superior agents and this is best accomplished by not only tackling the issue of potency, but also addressing absorption, distribution, metabolism and excretion (ADME) early on.

Until rather recently, these combined skills could successfully direct an antibacterial research program. Unfortunately, the most pressing issue at this time, if the others are resolved, is antibacterial resistance, for this phenomena dictates the fate of virtually every agent. Most antibacterial agents in use today can be rendered ineffective due to a variety of bacterial defenses. The emergence of resistance is frightening; some experts have envisioned bacterial strains (so-called 'super bugs') immune to virtually every established chemotherapy. Consequently, antibacterial compounds of the future must be wisely conceived with this issue in mind.

The major classes of antibacterial drugs are presented in this text in keeping with the issues outlined above. Chapter 1 provides a predominantly historical introduction to anti-infective chemotherapy which also includes brief descriptions of traditionally troublesome pathogens and diseases. An emphasis is placed upon general issues which are prevalent in antibacterial research and a basic discussion on resistance is given. Some specific trends in antibacterial chemotherapy are also considered. The sulfa drugs and antibacterial antifolates represent the first general class of antibacterial agents and are described in chapter 2. The bulk of the text is devoted to the β-lactams agents (chapter 3), the tetracyclines (chapter 4), the amino-glycosides including the aminocyclitols (chapter 5), the 'non-peptidic' macrocycles, specifically the macrolides and rifamycins (chapter 6) and the quinolone antibacterials (chapter 7). The obvious attention devoted to these classes of agents, I believe, is justified. These compounds remain the focus of significant pharmaceutical research at this time and are among the most widely used of all modern antibacterial chemotherapeutic agents. Lastly, various antibacterial peptides are also covered (chapter 8) as well as glycopeptidic and small molecule agents (chapter 9). Extensive listings of marketed and clinical candidates are provided in each chapter. By presenting information in this way, I hope this book will find use in pharmaceutical and academic environments.

In closing, I realize that this work would not be possible were it not for the unwavering support of my publisher and colleagues. I wish to thank Dr Barbara Goldman for providing me with this extraordinary opportunity. I sincerely thank all of my reviewers (listed below) for their help; in particular, I remain indebted to Les Mitscher and Herbert Kirst for taking the time to teach a first-time author (standing on the shoulders of giants gives one a much clearer view).

Professor Lester Mitscher (University of Kansas)
Dr Pat N. Confalone (The Dupont Merck Pharmaceutical Company)
Dr Jay Kostman (Temple University Medical Center)
Professor Robert S. Paley (Swarthmore College)
Dr Herbert A. Kirst (Eli Lilly & Co.)
Professor William O. Foye (Massachusetts College of Pharmacy and Allied Health Sciences)
Dr Edgardo Laborde (Ariad Pharmaceuticals)
Dr Norma Dunlap (Vanderbilt University)

To my wife Carrie and my daughter Kristy, I am forever thankful (once more) for their love, support and understanding during this lengthy project.

Scott L. Dax

1 Introduction

1.1 History and overview of anti-infective chemotherapy

Anti-infective chemotherapy is the science of administering chemical agents to treat infectious diseases. This practice has proven to be one of the most successful of all pharmaceutical studies. Historically, the use of anti-infective agents can be credited with saving more human lives than any other area of medicinal therapy discovered to date. Humanity has enjoyed a tremendous increase in life expectancy in the past century; a significant portion of this can be attributed to the control and eradication of infectious pathogens. A better understanding of infectious disease pathogenesis and the importance of sanitation are contributing factors. Regardless, most individuals will become infected with a microbial pathogen many times throughout their lives and in developed countries, anti-infective chemotherapy will be periodically administered. Antibacterials account for the majority of anti-infective agents in comparison to antifungals, antivirals and antiparasitic agents.

For millennia, primitive applications of the principles of anti-infective chemotherapy had been practiced with varying degrees of success. In early times, often a crude extract prepared from a naturally occurring source would be administered to the ill with little or no understanding of its therapeutic action. Uniform protocols or standards were then absent and little regard was paid to the potentially lethal toxicities associated with such reckless use. Consequently it was too common that the 'cure' killed the patient! In addition, some preparations were worthless concoctions often used in conjunction with superstitious or magical rites. Fortunately some of those treated would be cured; some successful medicaments, practices and protocols would be documented, and anti-infective chemotherapy would evolve into a science.

The true origin of anti-infective chemotherapy is open to speculation but curative properties of teas and herbs were documented in early Chinese civilizations (ca 3000 BC). In addition, the use of moldy cheeses to treat infected wounds was a common practice among some peasant populations. Rather remarkably, some of these early remedies have been shown to be efficacious in recent times. In 1632, cinchona bark was imported to Europe for use in treating malaria. The discovery of the antimalarial effect of cinchona bark was accompanied by the tale of a stricken, feverish Indian who drank water from a pond laden with fallen cinchona trees. This warrior made a remarkable recovery and word of the cure was passed on to other natives and eventually to the missionaries in the region. The tree bark responsible for this and subsequent cures, was later named in honor of the wife of the Count of Chinchon. Although the namesake was altered *en route* to Europe, a reliable therapeutic substance was realized and gained widespread use. Nearly two centuries later, in 1820, it was discovered that the alkaloid quinine **1** (Figure 1.1) is responsible for the antiparasitic activity of cinchona. Even today, this saga continues since other anti-malarial agents, which in essence originate from this finding, are commonly used in tropical Third World countries.

In 1683 van Leuwenhoek discovered bacteria and for the first time infectious pathogens could be viewed. In the centuries to follow, an intensive knowledge of microscopic organisms would be assembled and provide a foundation upon which the principles of microbiology and bacteriology would mature. The contributions of van Leuwenhoek are monumental since from this work, researchers would later be able to witness directly the antibacterial effects of certain substances. Interestingly, long before the work of van Leuwenhoek, it was recognized that certain substances were capable of preventing putrefaction, but there was little understanding of the mechanism by which these relatively simple molecules thwarted this process. For example, during the Middle Ages, some physicians used mercuric chloride to retard infection of open wounds. Centuries later, in 1825, Labarraque introduced chlorinated soda for the same purpose, and the use of tincture of iodine followed in 1839. The true establishment of antiseptic protocol into medical practice can be attributed to Lister. (An antiseptic is a substance that prevents the spread of infectious pathogens when applied to living tissue.) In 1865 Lister demonstrated the antiseptic properties of phenol; he used phenolic solutions to treat wound dressings and to disinfect surgical instruments. (A disinfectant is a substance that destroys infectious microorganisms from a non-living surface.) From this work, a family of alkylated and halogenated phenolic compounds would be developed: many are routinely used in hospitals today.

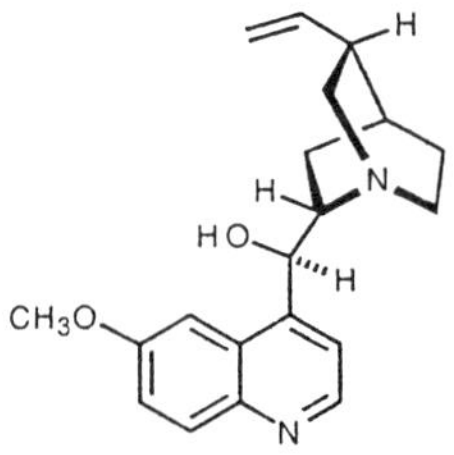

1

Figure 1.1 Quinine.

The works of Pasteur and Koch independently demonstrated micro-organisms to be the cause of infectious disease. In 1876, Koch identified a bacillus as the causative agent of the disease anthrax and in 1882, he described a mycobacterium as the pathogen of tuberculosis. The germ theory of disease was soundly established and ways to destroy infectious pathogens were sought. In 1877, Pasteur demonstrated that microorganisms could be retarded in their growth rate. He and Joubert made the dramatic observation that the anthrax bacillus could be killed by the presence of 'normal common' bacteria. The concept of an antibiotic (against life) substance was put forth by Vuillemin in 1889 but the formal definition, as recognized today, would not be introduced until 1942 by Waksman. An antibiotic is a chemical substance produced by microorganisms that can inhibit the growth of, or kill other microorganisms.

In the latter half of the nineteenth century, exciting new anti-infective compounds were discovered from natural sources. For example, the alkaloid emetine **2** (Figure 1.2), isolated from the Brazilian ipecacuanha root, proved to be effective against cases of amebic dysentery. In 1896, Gosio discovered that microorganisms can biosynthesize substances which possess antibacterial activity. Mycophenolic acid **3** (Figure 1.3) was isolated from a *Penicillium* fungi and found to have the ability to inhibit the growth of the anthrax bacillus. Emmerich and Low isolated an antibiotic product from cultures of the Gram negative bacteria *Pseudomonas aeruginosa*. This product actually contained two antibiotic substances called pyocyanase **4** (Figure 1.4) and pyocyanine, and for a time, the mixture found use in the control of anthrax and the treatment of diphtheria.

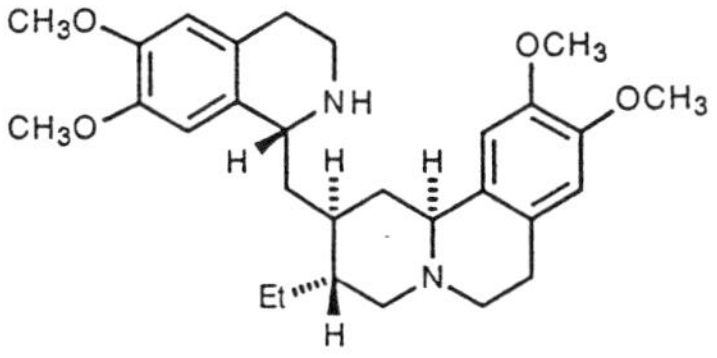

2

Figure 1.2 Emetine.

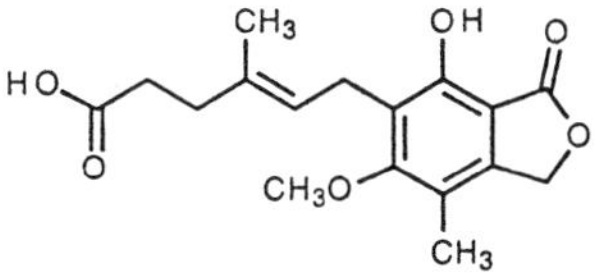

3

Figure 1.3 Mycophenolic acid.

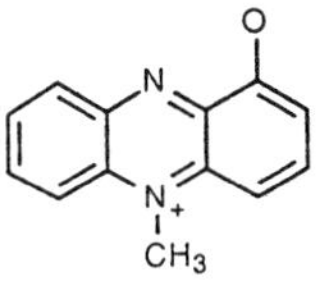

4

Figure 1.4 Pyocyanase.

Towards the end of the nineteenth century, the concept of synthesizing anti-infective agents was realized with the discovery that a synthetic halogenated quinolone called chiniofon **5** (Figure 1.5) possesses antiamebic activity. Following on from this work and the subsequent investigation of other halogenated quinoline derivatives, the earliest synthetic chemotherapeutic agents found use in the treatment of infectious diseases.

Ehrlich investigated the staining of tissues by dyestuffs and observed that certain dyes can be taken up selectively by living tissue. A common dye of the time was Methylene Blue **6** (Figure 1.6), which as Ehrlich noted, was able to accumulate in the animal nervous system. This dye also selectively stained malaria parasites in blood and was subsequently shown to have a beneficial effect against the disease in man. For a time, Methylene Blue was used to treat malaria victims despite some troublesome side effects. Ehrlich also carried out a systematic investigation of hundreds of compounds detailing the effects of structural modifications on anti-microbial activity. This research culminated in the discovery of salvarsan. Salvarsan **7** (Figure 1.7) is an organoarsenic compound that is effective against trypanosomes and found use in the treatment of malaria and syphilis. Other heavy metal compounds such as organoantimonials and organomercurials had been developed during this time but were found to be toxic and their use was limited.

Ehrlich invented the term 'chemotherapy' from his belief that infectious diseases could be successfully treated with synthetic chemicals. He postulated that cells possessed 'receptors' responsible for the uptake of vital nutrients. He envisioned that a chemical agent must similarly be able to undergo binding to some cellular receptor in order to exert activity as a

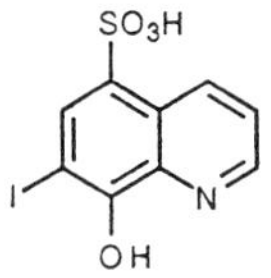

Figure 1.5 Chiniofon.

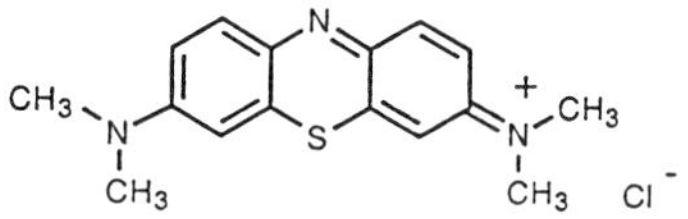

Figure 1.6 Methylene Blue.

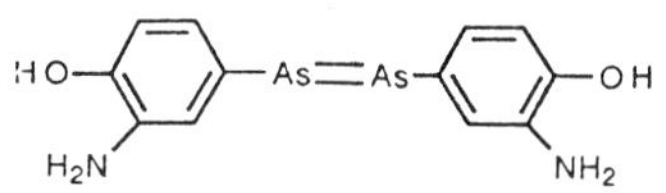

Figure 1.7 Salvarsan.

drug. His hypothesis that a successful anti-infective agent must be selectively toxic to the pathogen in the presence of mammalian cells, is a sound principle that has shaped anti-infective chemotherapy to this day. Ehrlich recognized that an effective antimicrobial agent need not kill all invading pathogens since a normal (immune) host response is often capable of eradicating the organisms if not present in overwhelming numbers. Ehrlich also pioneered the screening of chemicals for biological activity and put forth the concept of metabolic activation of a substance to afford an active drug. Finally, he introduced the basic concept of structure–activity relationships (SARs); a topic which will be addressed throughout this book. As mentioned above, Ehrlich demonstrated that the rational synthesis of structural analogs of a lead compound, which is known to possess a desired biological activity, can afford new agents with predictive properties. Through this exercise, more potent and often less toxic successors can be developed.

From the work of Ehrlich, the search for novel, more potent and less toxic antibacterial substances, which could be derived from dyes, became a common research theme. Soon after the introduction of salvarsan, suramin **8** (Figure 1.8) gained use as an agent against the African sleeping sickness (caused by trypanosomes). Suramin is structurally similar to the dye Trypan Red, but is colorless.

Malaria has been a widespread threat to civilization for a long time but the search for agents to combat the disease became particularly intense as more of the human population traveled to tropical areas. During the early part of the twentieth century, molecules containing a quinoline ring system, reminiscent of some antiparasitic dyes, were investigated for activity against the malarial parasite. Appropriate substitution onto the quinoline nucleus gave rise to derivatives that were more selective for the malarial protozoan than other parasitic pathogens. Antimalarial chemo-

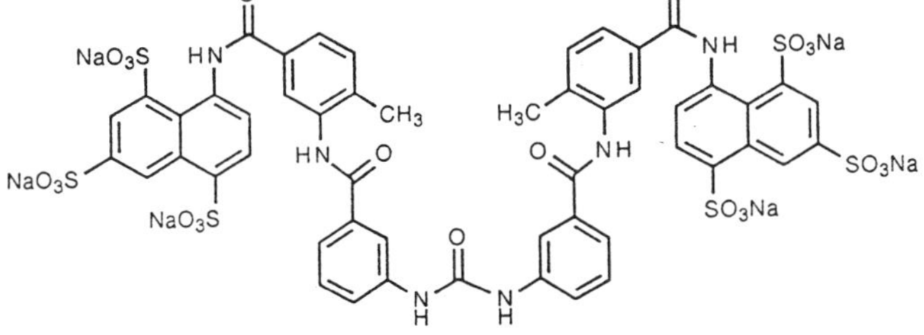

8

Figure 1.8 Suramin.

therapy was successfully established with the introduction of pamaquine **9** (Figure 1.9) in 1924. Quinacrine **10** (Figure 1.10) and chloroquine **11** (Figure 1.11) followed and in 1946, the landmark antimalarial agent primaquine **12** (Figure 1.12) became available.

In 1932, Domagk discovered the antibacterial properties of the azo dye prontosil **13** (Figure 1.13) which would initiate a highly successful area of anti-infective chemotherapy. Trefouel demonstrated that the *in vivo* activity of prontosil was actually due to its metabolite *p*-aminobenzene-sulfonamide **14** (Figure 1.14). By 1938, sulfonamides such as sulfacetamide, sulfapyridine and sulfathiazole were synthesized and some of these new variations possessed superior potency compared to sulfanilamide (from prontosil). Within a decade or so, the synthesis and evaluation of thousands of sulfonamides would be documented; the sulfa drugs became the first major class of antibacterial chemotherapeutic agents (chapter 2). Research in this area also led to the development of antibacterial sulfones (chapter 2) such as the agent dapsone **15** (Figure 1.15), which is highly active against leprosy.

In 1929, Fleming noted that the growth of a staphylococcal culture was inhibited in the presence of a metabolic product of a contaminating mold (*Penicillium notatum*). The remarkable properties of this substance, called penicillin, eventually led to the development of the prototype penicillin G **16** (Figure 1.16). A huge family of penicillin congeners followed and developed into one of the successful classes of antibacterial agents. In 1945, Brotzu discovered the first (cephalosporin C **17** (Figure 1.17)) of another family of *β*-lactam antibacterials, the cephalosporins. Today, the cephalosporins account for the largest share of the antibacterial market

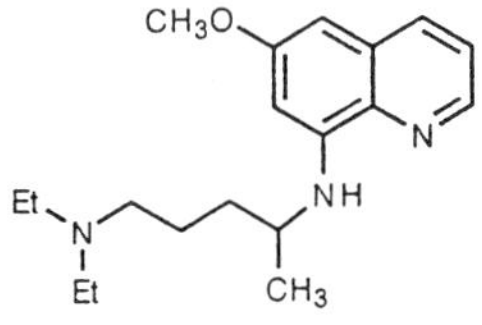

Figure 1.9 Pamaquine.

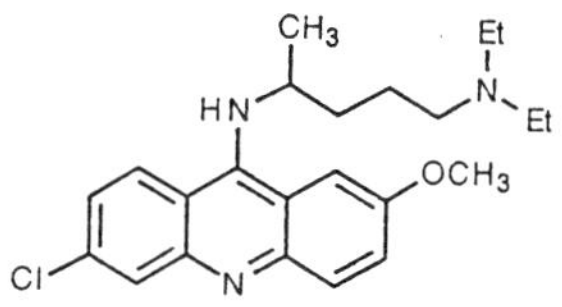

Figure 1.10 Quinacrine.

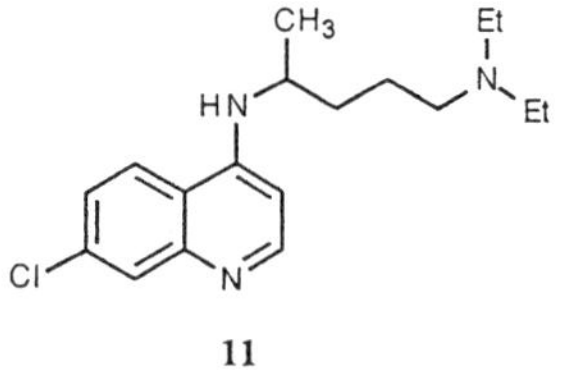

Figure 1.11 Chloroquine.

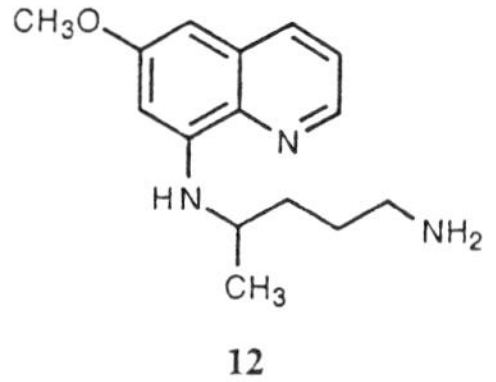

Figure 1.12 Primaquine.

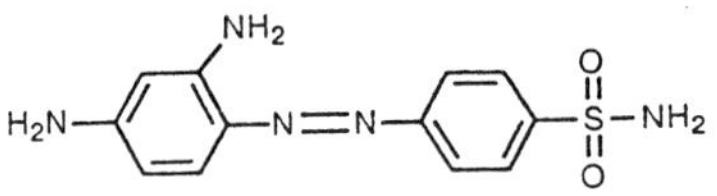

13

Figure 1.13 Prontosil.

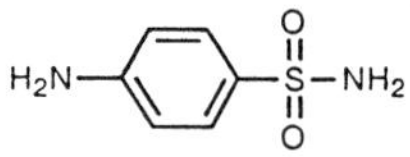

14

Figure 1.14 *p*-Aminobenzene sulfonamide.

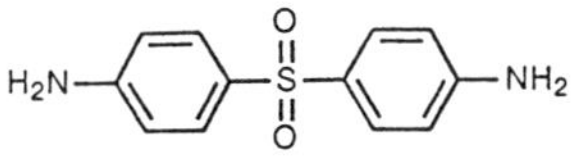

15

Figure 1.15 Dapsone.

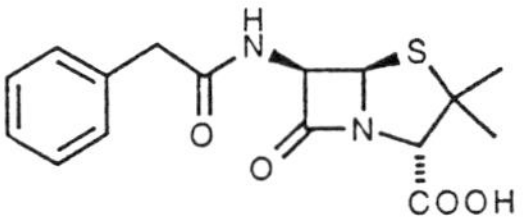

16

Figure 1.16 Penicillin G.

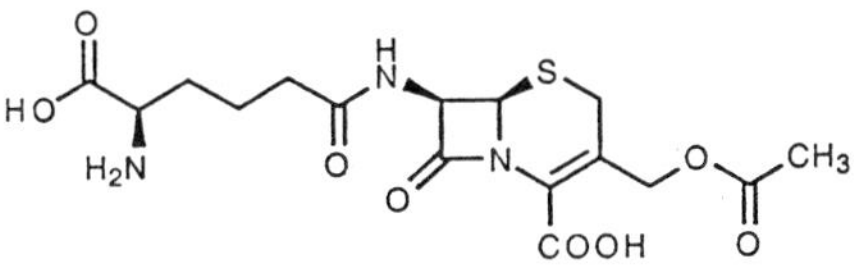

17

Figure 1.17 Cephalosporin.

(in terms of sale dollars). The phenomenal success of the β-lactam anti-bacterial agents (chapter 3) is reflected by the fact that over a hundred agents have been marketed or advanced into clinical studies since the introduction of penicillin G.

The screening of microbial sources began to produce significant contributions in antimicrobial chemotherapy by the early twentieth century. It became possible to assay crude extracts against a number of pathogenic bacteria, and from this practice a number of structurally complex antimicrobial substances were discovered. In 1939, tyrocidine A **18** (Figure 1.18) and gramicidin S **19** (Figure 1.19) (chapter 8) were isolated from a bacillus by Dubos. A year later, Waksman and Woodruff isolated actinomycin **20** (Figure 1.20) from a streptomyces culture; this product would later be recognized as the first cytostatically active antibiotic. Bacitracin **21** (Figure 1.21) (chapter 8) was discovered in 1945 and the polymyxins (e.g. B1 and B2) **22** (Figure 1.22) (chapter 8) followed in 1947.

The *Streptomyces* genus has been a particularly rich source of anti-bacterial substances and has afforded prototypes belonging to the erythromycin A **23** (Figure 1.23), tetracycline **24** (Figure 1.24) and aminoglycoside **25** (e.g. kanamycin A) (Figure 1.25) families. The

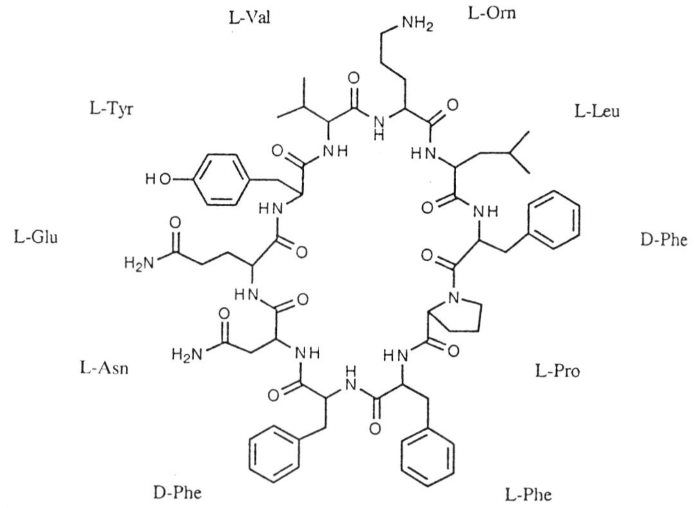

18

Figure 1.18 Tyrocidine A.

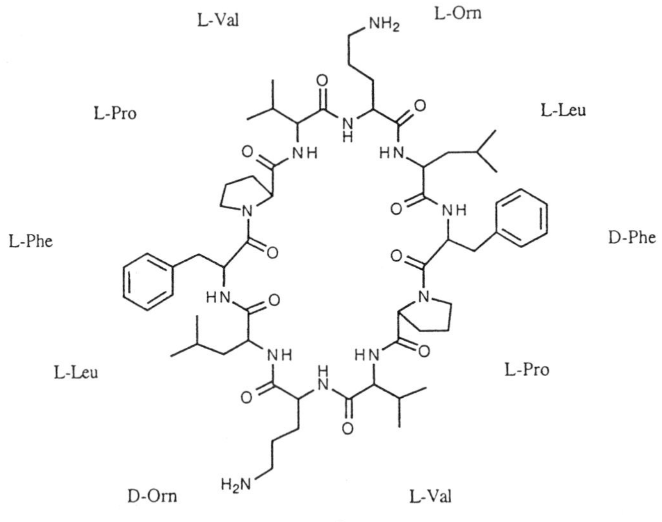

19

Figure 1.19 Gramicidin S.

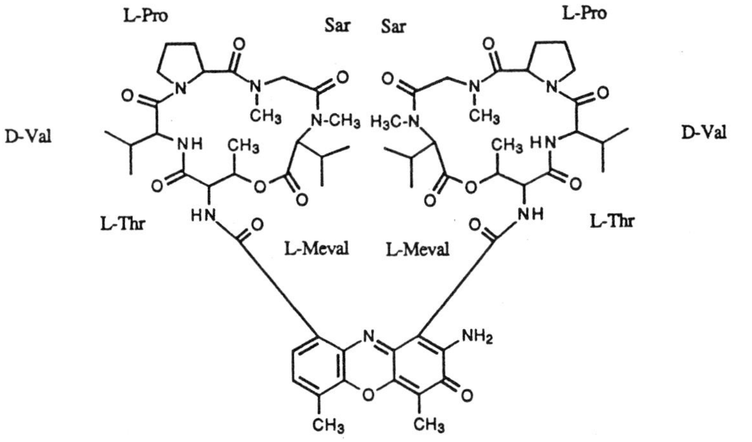

Figure 1.20 Actinomycin.

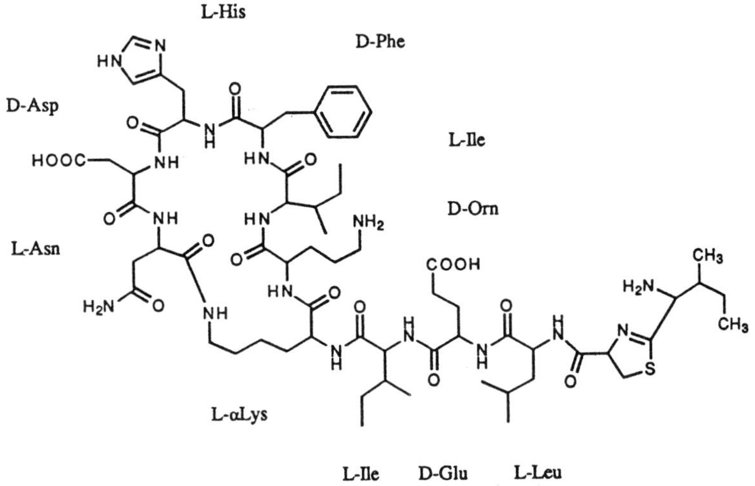

Figure 1.21 Bacitracin.

erythromycins (and macrolides, chapter 6) and tetracyclines (chapter 4) became important antibacterials, due to their oral activity and potency against pathogens not covered by other classes of agents. Members of the macrolide and tetracycline families are widely used today, and like the sulfonamides and β-lactams, a number of congeners have been marketed or have advanced into clinical trials. The aminoglycosides (chapter 5)

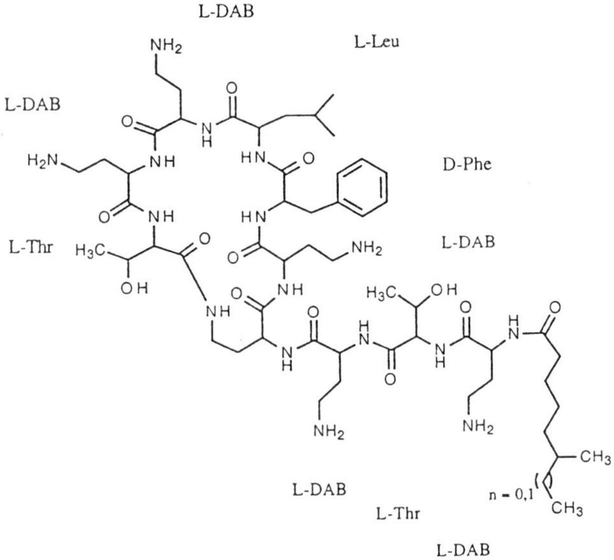

Figure 1.22 Polymixins.

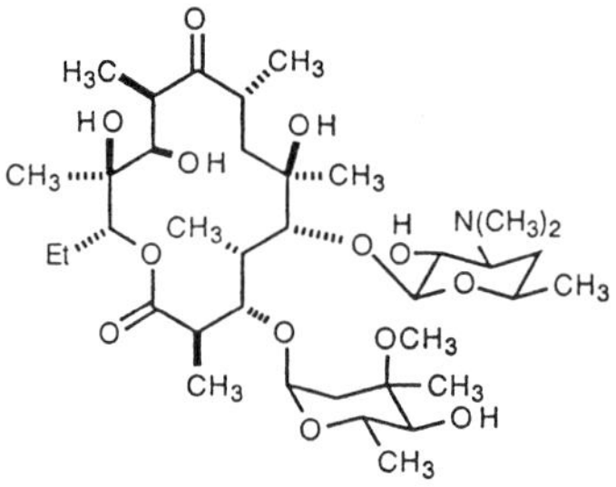

Figure 1.23 Erythromycin A.

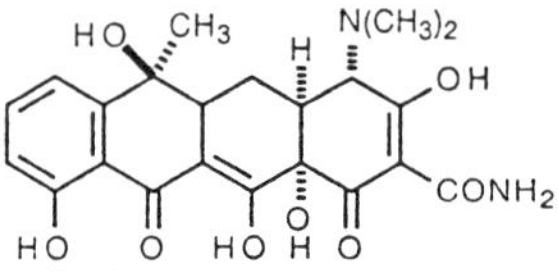

Figure 1.24 Tetracycline.

typically offer good Gram negative potency and are used in some severe infections.

Potent antibacterial activity can also be manifested in a variety of small, synthetic molecules. Isoniazid **26** (Figure 1.26), discovered in 1952, became an important agent to combat tuberculosis (chapter 9). Metronidazole **27** (Figure 1.27), first reported in 1959, possesses particularly potent activity against anaerobic bacteria (chapter 9). A family of nitrofurans have also been developed and marketed (chapter 9).

The discovery of an antibacterial by-product during a synthesis of the

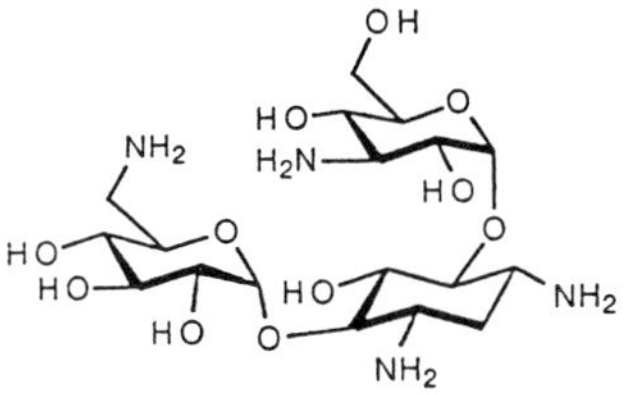

Figure 1.25 Kanamycin A.

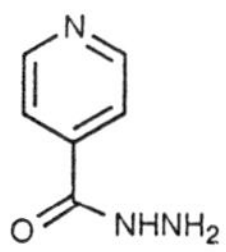

25

Figure 1.26 Isoniazid.

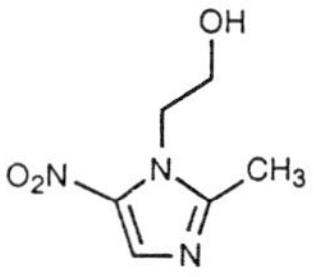

26

Figure 1.27 Metronidazole.

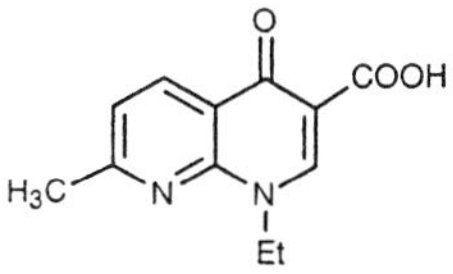

27

Figure 1.28 Nalidixic acid.

28

antimalarial agent chloroquine led to the development of nalidixic acid **28** (Figure 1.28) in 1962. From this finding, a large family of quinolone antibacterials has been advanced, which today are among the most potent and widely used of the synthetic agents (chapter 7).

1.2 Infectious disease outbreaks and plagues

In order to appreciate the impact anti-infective chemotherapy has made on humanity, it is helpful to recognize some of the infectious diseases humanity has endured. The complete global eradication of an infectious pathogen has been achieved only rarely and so most of these dreadful diseases still plague or threaten the human population.

1.2.1 AIDS (Acquired Immune Deficiency Syndrome)

AIDS is a disease characterized by the presence of a retrovirus called human immunodeficiency virus (HIV). The HIV virus, the putative AIDS pathogen, ultimately depletes a specific subset of lymphocytes (CD4 T-cell) which leads to a state of severe immunodeficiency. The virus is transmitted by the intimate exchange of various body fluids such as blood and semen. After initial infection, there is a characteristic false state of latency; recent evidence suggests that the virus is capable of profilic replication during this time and may harbor in lymph nodes. In general,

HIV-positive individuals remain asymptomatic during this time (often years) but for reasons not known, the virus eventually becomes active and exerts its lethal effects on the immune system. As a result, victims succumb to various opportunistic infections and diseases. There is currently no cure for AIDS and to date, AIDS is universally fatal. Estimates place the number of people infected with the HIV virus to be at least 21 000 000 worldwide in 1996.

The chemotherapeutic agents that are available to treat AIDS today can inhibit a specific enzyme (reverse transcriptase) that is required by the virus to express its genomic DNA. Current research is also aimed at inhibiting a different enzyme (HIV protease) that is needed for viral maturation and infectivity. Unfortunately, the virus is able to mutate these enzymes to a state of resistance. Other areas that offer promise are aimed at preventing virus–host cell associations prior to infection, and the development of vaccination protocols is also being explored.

1.2.2   African Trypanosomiasis

Trypanosomiasis (African sleeping sickness) has plagued humanity for a long time. The disease is caused by a family of protozoans of the *Trypanosoma brucei* complex. African trypanosomes have existed for perhaps millennia, but the first reports of sleeping sickness came in the fourteenth century when a sultan was so overtaken by sleep that he could barely be awakened until his death. In the eighteenth century, a 'Negro lethargy' was decribed as a common disorder among the African natives. The development of the Congo region by Stanley and Livingstone occurred during a period when perhaps one-half million people were dying from sleeping sickness every decade.

The African sleeping sickness trypanosomes are transmitted to humans in the saliva of a biting tsetse fly. The severity of the disease can range from a sleeping sickness to incapacitation and death. Once the nervous system becomes involved (meningoencephalitis), prognosis is poor. Today environmental and medical practices have greatly reduced the toll of the sleeping sickness. An understanding of the disease process along with the development and use of effective chemotherapeutic agents to combat trypanosomal infection are responsible for this success.

1.2.3   Bubonic plague

The bubonic plague is the most notorious infectious disease humanity has yet to encounter; the Black Death has claimed many millions of human lives throughout history. There have been three waves of epidemic bubonic plagues and many lesser outbreaks. The first plague occurred in the sixth century and the most recent in the nineteenth century. The most feared was the Black Death that erupted in the fourteenth century and raged for perhaps five hundred years. At its peak, the Black Death claimed several hundred thousand lives per month and eventually killed one-third of Europe's population. There was a much smaller outbreak in 1994 in India, testimony that the plague is still a lethal foe.

The bacterium responsible for the bubonic plague is *Yersinia pestis* which readily thrives in the bloodstream of rodents such as rats. The rodent host will probably die but during its period of infection, the rat will transmit the pathogen to rat fleas and other rats. Even the flea will succumb to the bacteria but not before the reproduction of the bacteria blocks digestion causing the flea to regurgitate or defecate infectious material into a human host. In some cases, the bacteria can enter the lungs via inhalation of infected droplets; the result is pneumonic plague which is highly contagious and extremely fatal. Despite its route of introduction, the plague is quick to kill, often within hours or days. The symptoms are fever, chills and pains in the limbs. The classic painful swelling of the lymph nodes (buboes), delirium and memory loss precede death.

Bubonic and pneumonic plague can be cured if treatment is started in time; for example, tetracycline and streptomycin are effective against *Y. pestis*.

1.2.4 Cholera

Cholera, a disease caused by the bacillus *Vibrio cholerae*, has reached pandemic or epidemic proportions frequently in many different regions. In Russia, over five million were sickened and perhaps two million of those died in the period between the 1820s and the 1920s. For centuries, cholera was sporadically epidemic in India, and parts of Europe, Asia and the Americas suffered outbreaks. The most recent (1994) outbreaks have occurred in refugee camps in Zaire; in some cases death occurred in less than one day. Cholera is transmitted by fecal matter that contaminates drinking water. Once infection is established, the bacteria is able to produce a toxin that disrupts normal intestinal function. Characteristic symptoms are nausea, dizziness, violent vomiting and 'rice water' diarrhea. If fluid loss is not compensated, body temperature drops and death can occur.

Today cholera remains a health problem in populations where poor sanitation is common. Fortunately, cholera can be treated with intensive fluid resuscitation and a variety of antibacterial agents (tetracyclines, sulfamethoxazole–trimethoprim, quinolones).

1.2.5 Diphtheria

Diphtheria is an acute contagious disease, particularly dangerous to children, that is caused by the bacterium *Cornyebacterium diphtheriae*.

Diphtheria is an ancient disease but even relatively recent outbreaks have occurred. A particularly alarming episode that occurred in the late 1880s probably claimed tens of thousands of lives. The infection is primarily transmitted via inhalation of pathogen-containing droplets. Once an infection is established, the generation of a bacterial toxin gives rise to fever, sore throat, malaise and a noticeable cough. The upper respiratory tract becomes inflamed, as can the nasal passages and trachea. Peripheral areas, particularly the heart, can be affected. Death can follow via asphyxiation.

Fortunately diphtheria has been controlled by vaccination which produces a prolonged state of immunity to the toxin although colonization with the bacteria itself is still possible. There are also a number of antibacterial agents (e.g. erythromycin) that are very effective against *Cornyebacterium diphtheriae*.

1.2.6   Dysentery

Dysentery has been a recurring dangerous disease throughout the history of humanity. The disease is caused by a variety of bacteria such as *Shigella* or *Campylobacter*, or can be amebic in nature (due to *Entamoeba histolytic*). In the nineteenth century, soldiers from Europe and the United States, as well as famine victims in Ireland suffered devastating outbreaks. Altogether hundreds of thousands may have died from this disease. Serious outbreaks have also occurred in Africa and underdeveloped areas of the Americas. Today, India and parts of Asia and Central America are routinely plagued by dysentery.

The disease is dependent upon human transmission and so, as with cholera, its incidence coincides with poor hygiene standards. Interestingly, the *Shigella* species behave as strict human pathogens; a relatively small number of organisms can cause the disease in man and yet the bacteria are not harbored in a non-human reservoir (such as an animal or insect). The microorganisms are transmitted through human waste and the contamination of drinking water supplies. Common symptoms of dysentery include abdominal pain and acute diarrhea that can become uncontrollable, frequent and often bloody in composition. Despite effective antibacterial chemotherapy (some cephalosporins and penicillins, quinolone antibacterials, sulfamethoxazole–trimethoprim, for example), dysentery will continue to pose a health problem as long as economic factors prevent the installation of proper sanitary practices in populated areas.

1.2.7   Gonorrhea

Gonorrhea is an old and common foe cause by the bacteria *Neisseria gonorrhoeae*. Gonococci are not normal inhabitants of the genital tract;

transmission by sexual contact results in 'the clap'. It is a venereal disease that effects the genitourinary tract with varying symptoms. In the male, a common symptom is painful urination and sterility is possible. Women are usually susceptible to infection but can often be asymptomatic; damage to the female reproductive system, the risk of blindness and a threat to the newborn can result. Infected men and women are susceptible to serious effects if the organism is able to establish residence in the bloodstream. The joints, the eyes and heart tissue (endocardium) can be targeted.

While antibacterial chemotherapy (penicillins and cephalosporins if susceptible; alternatively, quinolones, tetracyclines and erythromycin) can eradicate this disease, social and sexual practices in many societies allow millions to walk infected and suffer the long term effects of this disease. These victims often act as carriers and infect many others during their lifetime.

1.2.8 *Influenza*

Influenza is an acute respiratory illness cause by a viral pathogen. The infection can often lead to pneumonia and sometimes death. From the sixteenth century, outbreaks have been well documented. In 1917 and 1918, a particularly virulent strain readily caused widespread pneumonia and secondary infections in New York; altogether, tens of thousands of lives were lost. The most likely to succumb are generally the elderly as well as those with chronic respiratory problems, weakened cardiovascular systems or compromised immune responses.

The 'flu' virus can be spread from person to person in the form of respiratory excretions; milder infections can originate directly from an animal reservoir. Several strains of virus are responsible and periodic waves of disease occur as more virulent pathogens spread. Symptoms include fever, aching joints and coughing. These clinical signs are difficult to differentiate from other infectious states, and secondary bacterial infections can complicate diagnosis. Vaccination is an effective measure in high-risk populations and antiviral therapy (amantidine, rimantidine) can shorten the course of the illness.

1.2.9 *Leprosy*

Leprosy, perhaps unlike any other disease until AIDS, has generated such fear and insensitivity that its history describes a disgusting and sorrowful legacy of humanity. The origin of leprosy is difficult to ascertain since perhaps as early as the twentieth century BC afflictions have been documented that may be attributed to the disease. In the Middle Ages, leprosy was considered to be a disease of corruption. Victims were isolated and forgotten; religious beliefs and governmental mandates instilled a widely held fear and misunderstanding of the disease.

The bacillus *Mycobacterium leprae* is the causative agent of leprosy. The disease can take many forms and is slow to progress, often requiring years before many symptoms are obvious. Dimorphous and indeterminate leprosy are relatively mild forms of the disease. The tuberculoid form can affect the nerves and lead to loss of sensation in the limbs which become swollen and deformed. However, lepromatous leprosy is even more serious and can be transmitted through nasal or laryngeal secretions. Progression of the disease produces dramatic and alarming features. The face often becomes swollen and yellowish, and degradation of the nose, throat and nasal passage tissue structure grotesquely disfigure the leper. In the latter stages, blindness can also occur and eventually fever and breathing difficulties can lead to death.

Today, the introduction of antibacterial chemotherapy (the sulfone antibacterial dapsone, often in combination with a rifamycin such as rifampin) has proved successful but requires an extended period of therapy (one-half a year or greater). Unfortunately, tens of millions are currently suffering from some form of leprosy.

1.2.10 Malaria

Malaria is a parasitic disease that has plagued humanity probably since its existence. Outbreaks have been documented from the first century AD, but in the eighteenth and nineteenth centuries, the 'bad air' disease carried incredibly heavy tolls. In the mid 1800s, over 40% of the one-half million British soldiers in India along with the native population were infected, putting the toll in the tens of millions. Regions of China may have been similarly stricken in the nineteenth century. The outbreak during the digging of the Panama Canal peaked with an infection rate of nearly 85% among the workers.

Malaria is most often associated with tropical and subtropical regions where the female anopheles mosquito, the reservoir of the infectious pathogen, is prominent. Four species of the parasite plasmodia are capable of causing malaria in humans but *Plasmodium falciparium* is responsible for the majority of the deaths caused by the disease. The plasmodia undergo development in the stomach of the mosquito and are transmitted to humans through the bite of the insect. The parasites become established in the host bloodstream and lodge in the red blood cells feeding on hemoglobin. This process gives rise to the release of toxins that cause well-recognized clinical features. Headache and dull pains in the limbs and joints occur early. The victim subsequently experiences a shivering chill and fever, caused by red cell rupture, and may possibly convulse and vomit. The fever, sweating and vomiting often stop and the victim will take sleep until the next cycle starts again. Death can occur within weeks or months, sometimes due to secondary infections brought on by the weakened state of the subject and damage to the erythrocytes. In cases involving *P. falciparium*, severe fever may precipitate violent death, due to direct involvement of the central nervous system.

Effective antimalarial treatments (quinine, mefloquine or tetracycline) have become commonplace and morbidity rates have dropped greatly. However, in some areas, malaria remains a serious problem; some strains are resistant to standard therapy.

1.2.11 Measles

Measles (also known as rubcola; German measles (rubella) is another similar RNA virus) is a highly contagious upper respiratory tract viral infection that has proven deadly. Eruptive fevers such as measles may have been prevalent as early as the first century AD. Apparent outbreaks in Greek and Roman civilizations predate similar diseases in Asia. From the sixteenth century, measles afflicted European people claiming numbers of children. Even as the New World came into existence, the Americas suffered outbreaks. In the eighteenth century, epidemic measles infections were common in the colonies. Other areas were similarly affected; an outbreak on Fiji alone may have killed forty thousand natives in 1874.

The measles virus is easily transmitted by nasal and oral secretions. After establishing infection, a fever and a pronounced susceptibility to complicating laryngitis and bronchopneumonia follows and can be fatal. The most recognized symptom is the eruption of gray-white spots encircled by erythema (redness and inflammation of the skin) that appear on the head and in the mouth, and later on the trunk and extremities. (German measles follows a similar pathogenesis, but can appear to be milder.)

Fortunately permanent immunity to the measles is available either by vaccination or through infection. No specific chemotherapy is available to infected individuals other than supportive measures to prevent secondary bacterial infections.

1.2.12 Pneumonia

Pneumonia is an inflammation of the lungs that is often caused by pathogenic microorganisms such as pneumococci. Viruses are more likely to be the cause of the disease in infants. Pneumonia has plagued humankind throughout history. Persons with weakened immune responses or chronic heart conditions are more likely to succumb to the disease. These conditions are prevalent in the elderly and so pneumonia has been often regarded as an 'old man's friend' from the belief that a rather non-violent or painless death can result. However, infants are also particularly susceptible to forms of the disease. Symptoms of the disease include fever and then a characteristic coughing and 'heavy chest' and 'frothy phlegm'. Acute pneumonia can be cause by various bacteria such as *Streptococcus pneumoniae*, *Haemophilus influenzae*, *Klebsiella pneumoniae*, *legionella* species and *Pseudomonas aeruginosa*. Mycoplasma and fungi (e.g. in AIDS patients) can also be responsible.

Many chemotherapeutic agents are available for the treatment of bacterial pneumonia (e.g. the β-lactam antibacterials).

1.2.13   Polio

Polio (poliomyelitis) is a viral infection caused by polioviruses that can be transmitted by ingestion or inhalation. Poliomyelitis may have been a human disease since early times; Hipprocrates described deformities and paralysis apparently caused by polio. In the European Middle Ages, a 'lameless' disease was described and by the nineteenth century, outbreaks were clearly recognized in Europe. In the early twentieth century, thousands fell victim in the United States.

The virus takes hold in the tonsils or lymph nodes and viral RNA is synthesized and expelled into the gut. Many cases can go unnoticed during the early stages of the disease as symptoms can either be extremely mild or wrongfully regarded as an intestinal 'flu'. The real scourge of polio occurs if the virus is able to become established in the bloodstream and subsequently spread to the central nervous system causing paralysis. The degree of paralysis depends upon the specific neuronal region that is affected and the degree of damage. Paralysis can be temporary but if the cranial nerves (bulbar polio) or the spinal cord (spinal polio) are attacked, permanent paralysis often occurs; if extensively involved, polio can be fatal.

An understanding of the disease process emerged during the early part of the twentieth century; different strains of the virus were identified and investigated. In 1953, Salk first used a virus-killed vaccine with success, and an oral vaccine was subsequently developed by Sabin. Due to these advances polio has been brought under control and is now a rather rare disease in most regions of the world.

1.2.14   Rheumatic fever

Rheumatic fever is a condition caused by a type of streptococcus (hemolytic group A). Symptoms include fever, joint pain and swelling. Internal organs, particularly the heart, can be damaged. Rheumatism has been described in the Hippocratic writings and has plagued humanity for ages. The disease can be fatal; in the 1930s the mortality rate approached 20% of the thousands afflicted.

Interestingly, in 1976 a tree bark was found to be effective in treating rheumatism. The bark of willow trees contains salicin which exhibits an aspirin-like activity. (The 'true' salicylate drugs became common therapeutic agents later.) Sulfonamides and β-lactam antibiotics are particularly effective against streptococci. Recent resurgences in rheumatic fever have occurred in defined geographic areas of the United States.

1.2.15 Scarlet fever

Scarlet fever (scarletina) is a disease caused by a subgroup of streptococci bacteria. Scarlet fever was probably first documented in the fourteenth century. Epidemics were common in Europe and the Americas in the eighteenth and nineteenth centuries. Some of these outbreaks claimed thousands of lives.

A rash, of the face first, and then of the chest and trunk accompanies fever and is due to the body's reaction to an erthyrogenic toxin. A severe sore throat, a 'strawberry' tongue, abscesses in the throat and tonsils and inflamed lymph nodes are common symptoms. The disease can be mild, and, infrequently, fatal. Penicillins and other antibiotics have been effective against streptococcal species that cause scarlet fever.

1.2.16 Smallpox

Smallpox is a highly infectious disease caused by the variola virus. Smallpox may date back to Egyptian civilization (1100s BC). Later writings in the sixth century AD clearly delineate the disease process. Europe, Asia and the Americas have, at some point in time, been decimated by smallpox. Ironically some sectors of the population were immune to the disease. For example, smallpox probably claimed one-third of the Indian population in the early-to-mid 1500s. During the same period, the Spanish invaders proved to be immune. Smallpox epidemics were often followed by outbreaks of measles, typhus and influenza that wiped out scores of people with over 90% mortality. It has been estimated that in the eighteenth century, smallpox killed 15 million people every 25 years. Even though less severe in nature, smallpox lay a heavy claim in Europe in the nineteenth century.

Smallpox is spread through nasal and oral secretions and after establishing infection, causes a high fever and characteristic rash. It is this rash that can lead to the dreaded disfigurement associated with smallpox. Rash papules enlarge and fill with pus often causing swelling and scarring. Blindness can occur in the more severe cases. Since a high mortality rate can result, smallpox has historically been one of the most dreaded diseases.

Interestingly, as early as the eighteenth century, it was known that implanting matter from a smallpox pustule could confer immunity. Unfortunately at the time, this practice was also capable of causing the deadly disease. However, Jenner observed that milkmaids would often develop a benign disease similar to smallpox and attributed this phenomenon to their handling of cows. In 1796, he inoculated a boy with matter from the cowpox lesion and noted only a mild reaction. Weeks later, the child was similarly inoculated with smallpox matter and surprisingly, no reaction nor disease was produced. From this work, vaccina (meaning 'cow') or vaccination became widely used to prevent the disease. There is probably no greater triumph in medicinal science today; smallpox disease has officially and literally been wiped from the face of the earth.

The post-smallpox era of the 1990s is now engaged in a debate on whether the remaining two samples (in the USA and the Russian Republic) of the live virus should be destroyed or 'housed' for future study. Opponents to destruction of the virus cite the potential advances that could be gained by studying the virus using advanced technology (of today and the future). The contention is that valuable knowledge of the virus may be beneficial to more contemporary viral disease processes (perhaps even AIDS, hantaviruses, etc.). Proponents of the final demise of smallpox virus cite that an extensive knowledge of the virus is already in hand (the complete viral genome is known), and that terroristic acts of reintroduction into a region could re-establish infection among a population. The latter worst-case scenario would wipe out the global effort that eradicated this pathogen, as well as possibly resulting in loss of human life.

1.2.17 Syphilis

Syphilis is a chronic venereal disease caused by the bacterial spirochete *Treponema pallidum*. The disease has affected millions and has claimed countless lives throughout the past five centuries. Syphilis probably started in Europe in the late fifteenth century and was often deemed a punishment for lack of sexual discrimination, drunkenness and gluttony. In addition, syphilis was often confused with gonorrhea during this time. The issue was settled in 1879 when Neisser isolated the gonococcus bacteria responsible for gonorrhea.

Syphilis is transmitted through sexual contact and can inflict a long period of suffering before death. Soon after the pathogen enters the body, a painless chancre usually forms but can go unnoticed. The primary lesion disappears in weeks and soon after, the organism takes residence in the lymph nodes. Secondary skin lesions can follow, but the real damage of syphilis occurs through its vascular effects. The pathogen generates lesions (gummas) that damage the liver or other internal organs. The heart and the nervous system are particularly susceptible. After decades in a human host, syphilis can attack the brain causing dementia and death. During their lifetime, a person may pass on the disease to many others, and mothers can infect their newborn (congenital syphilis).

At the turn of the century, mercury-containing compounds were used against the syphilis pathogen with some success, but were also toxic. In the early 1890s, bismuth agents proved somewhat safe and the organoarsenic salvarsan, discovered by Ehrlich, gained widespread use at the time. Today, penicillins, tetracylines and erythromycins (macrolides) are used to combat syphilis but unfortunately, the disease is prevalent in many sectors of today's population.

1.2.18 *Tetanus*

Tetanus ('lockjaw') is a dangerous disease caused by the bacteria *Clostridium tetani*. The disease has been recognized for centuries, but in the late 1800s, Nicolaier observed the ability of the bacillus to produce tetanus-like symptoms in laboratory animals. Kitasato hypothesized that it was the production of a toxin rather than the organism itself that caused the disease since the bacillus was only found in local lesions and not systemically, in fatal cases. This toxin, when injected into mice at sublethal concentrations produces a serum with antitoxin properties.

The tetanus bacillus enters the body through a contaminated wound and gives rise to headache, toothache, sweating and eventually, the characteristic muscular spasms in the head and neck ('lockjaw'). At this stage, the toxin has grave consequences for the spinal cord and causes nerve damage. The death rate after infection is established at approximately 50%, usually resulting from respiratory distress or cardiac arrest. The bacterium is common in manure-containing soil and so poor sanitation in areas inhabited by man and animal is a factor leading to the disease. Dirty needles and other puncture wounds can also cause the disease, even in modern society.

Immunization with a (modified) toxoid today has virtually eliminated tetanus in many regions of the world.

1.2.19 *Tuberculosis*

Tuberculosis (also known as consumption) is a disease caused by the bacterial pathogen *Mycobacterium tuberculosis*. Tuberculosis is an ancient disease evident by the description of 'phthisis' (weight loss accompanied by pulmonary symptoms) in the Hippocratic writings. Medieval outbreaks were common and the infected were often isolated and reported to authorities. At its peak, tuberculosis may have brought a premature death to more than 25% of the early nineteenth century European population.

The disease is usually transferred by breathing droplets exhaled by an infected person. Animals can also get tuberculosis from another mycobacterium (*M. bovis*) and can pass another form of the disease to man through their milk. The pathogens can also enter the body through a wound. *Mycobacterium tuberculosis* most often invades the lungs. The bacteria reproduce slowly and give rise to fever, chills, sweats and a productive cough. Physical wasting occurs later in the disease process; if untreated, the weak can succumb as a result.

In 1882, Koch identified the bacillus responsible for the disease; in time, the search for an immunization procedure followed. Tuberculin, an extract of the tubercule bacillus, was eventually developed to diagnose the presence of the tuberculosis bacillus; infected individuals experience a noticeable allergic response to subdermal administration of tuberculin. In the 1930s, a purified protein derivative of the Koch extract was developed which is today used along with a bacilli supplement for the 'tuberculin test'. In 1944, Waksman isolated a new antibiotic streptomycin which was found to be extremely effective against *M. tuberculosis*. However, resistance emerged and so streptomycin was often used in conjunction with *p*-aminosalicylic acid. In the 1950s, isoniazid was developed and in the 1960s, the rifamycins were also introduced for the treatment of tuberculosis. Today, a variety of combinations of these drugs is often used since multi-drug resistant *M. tuberculosis* strains have surfaced. In fact, within the past several years, resistance has become a particularly pressing issue. For example, in New York City there have been cases spread by rather incidental contact, in which only 'cocktails' of agents have proven to be lifesaving.

1.2.20   Typhoid fever

Typhoid fever is a potentially fatal disease caused by salmonella bacteria. Numerous outbreaks have been documented throughout history. The origin of typhoid fever is difficult to trace but some reports from the seventeenth century probably describe the disease. In the early 1600s, as many as six out of every seven new Jamestown colonists may have died from typhoid, although a vitamin deficiency syndrome (beriberi) probably played a role. In general, the disease is problematic during periods where social conditions prevent proper sanitation and hygiene.

This enteric fever is often indistinguishable in symptoms from several unrelated diseases. The *Salmonella typhi* bacteria lives in the digestive tract of humans and can be transmitted via exposure to feces, vomit or urine. Some carriers unknowingly infect others ('Typhoid Mary' was a cook known to have infected dozens and may have been responsible for a subsequent outbreak that affected over a thousand people). The disease is transmitted by the consumption of contaminated food or water. House flies can carry the pathogen and deposit it on food. Floods often spill a source of contamination into drinking water producing typhoid outbreaks. Symptoms include fever, loss of appetite and muscular pains. A distended abdomen, jaundice, rapid respiration and a dark diarrhea usually occur; death can follow.

β-Lactam antibiotics, quinolones and sulfa drugs are useful agents in the treatment of typhoid fever.

1.3   Common bacterial species

Many horrendous epidemics and scores of human casualties have been caused by pathogenic bacteria. It is true that parasites and viruses have been responsible for notable tragedies such as malaria and AIDS respectively; both diseases presently affect millions of people worldwide. However, more people worldwide are affected by bacterial infections and diseases. In order to approach the science of antibacterial chemotherapy, it is helpful to understand what bacteria are and some of the ways that bacterial cells differ from animal cells. After all, these 'differences' can allow chemotherapeutic agents to act as toxins to bacterial cells while being innocuous to the varied mammalian cells in the host animal such as humans. This selectivity is the very 'magic bullet' envisioned by Paul Erhlich in the early part of this century.

Bacteria are single cell microorganisms that are present in all ecosystems on earth. Some species can reside within a plant, animal or human host (or host cell); others are found freely within the biosphere whether the medium be water, air or ground. Although bacteria are rather primitive in composition and function compared to most eukaryotes, bacteria are capable of performing some rather remarkable molecular feats.

Bacteria are essential for the recycling of organic material on the planet Earth. In a general sense, bacteria are responsible for the degradation of complex structures and macromolecules into smaller units that can be used by other organisms whether they be microbe, insect, plant, animal or human. In reality, many aspects of higher animal homeostatis are dependent upon the activity of bacteria. Essential functions such as digestion and waste elimination are carried out by various bacteria that thrive within the host. Under ordinary circumstances these bacteria are kept in check by the body's own regulatory processes such as confinement within an organ, destruction outside a given pH range or elimination via an immune response. However, deviation from this normal state can occur and result in infection and possibly death of the host.

Conversely, bacteria can construct valuable organic (or organic-like) materials from the most elementary of materials. For example, some bacteria are capable of 'fixing' atmospheric nitrogen (gas) and producing nitrogen-containing organic materials that are ultimately elaborated into complex proteins by higher order species such as humans. Somewhat ironically, it is the relatively simple bacterium that can accomplish what no higher order complex creature can — the conversion of atmospheric nitrogen into ammonia.

Bacteria have evolved in a prolific manner in order to arrive at their present-day existence. It is important to realize that these changes are continuing to occur today. For example, a 'megabacterium' of enormous proportions has recently been identified. Some bacteria have been renamed, reclassified or placed in a separate category as new information is gathered using the latest of scientific techniques. In addition, new strains are being isolated at a rapid pace from the clinic or from naturally occurring sources. Also, laboratory-selected mutants are increasingly important for the study of bacteria in greater detail, as well as elucidating the mechanisms by which some species become drug resistant. There are many families of bacteria that consist of dozens of genera and many more individual species. Only a relatively small number of all species of bacteria possess the ability to cause disease under ordinary circumstances. In most ways, a pathogenic bacteria deviates very little from a related benign species of the same genera. However, a minor modification of cell wall composition, the expression or alteration of a bacterial protein or the production of a toxin are just a few ways in which a bacterial species can produce virulent effects.

Bacteria can be distinguished by a number of features but they are most often categorized by cell morphology and cell wall composition. Morphology is readily observed by microscopy that provides a useful, albeit gross, differentiation. Bacteria can be round-shaped organisms which are called cocci, rod-shaped which are termed bacilli or roughly spiral-shaped, which are referred to as spirochetes. The most useful distinction given to bacteria in terms of cell wall composition is based upon the uptakes of dyes. Without doubt, the most recognized and widely employed staining method is the one developed by H. Gram in which bacteria are assayed as being either Gram positive or Gram negative. This differentiation results from the fact that Gram negative bacteria possess a lipopolysaccharide cell wall component that confers an impermeability to many dyes. Gram positive bacteria, on the other hand, possess a peptidoglycan barrier (as do Gram negative organisms) but no lipopoly-saccharide component, and as a result, are permeable to some dyes. Specifically, the Gram method involves first staining the bacterial species with a purple due (crystal violet), followed by treatment with iodine-iodide and then decolorization with alcohol. Gram positive organisms retain the dye–iodine complex; Gram negative bacteria do not. Counter staining with safranin gives Gram negative species a detectable red color whereas

Gram positive bacteria remain purple. There are many other staining techniques; the Gram method is the most widely used and recognized.

The distinction of a bacterial species as either Gram positive or Gram negative also allows for some generalizations pertaining to the cell wall composition. A typical Gram positive bacterium possesses a thick (250 Å) peptidoglycan matrix that covers the cytoplasmic membrane. Peptidoglycan imparts rigidity to the bacterial cell, but is rather porous and does not limit the passage of most small molecule antibacterial drugs. Gram negative bacteria have two sheaths wrapped around the cytoplasmic membrane. The peptidoglycan barrier of a Gram negative bacterium is much thinner (25 Å) than that of a Gram positive organism, but Gram negative species possess a lipopolysaccharide (LPS) layer. The LPS barrier would be impenetrable to most antibacterial agents, as well as nutrients, if not for the presence of certain functional proteins that are lodged within the layer. Porins are proteins that self-associate to form channels or pores of varying size that allow for the influx of nutrients. However, many antibacterial drugs can also cross the LPS layer through porins; in general, porins exclude molecular sizes of greater than approximately 700 amu.

The lipopolysaccharide layer also poses other concerns with regard to antibacterial chemotherapy. Some antibacterial agents ultimately cause lysis of the bacterium and the rupture of the LPS layer. Components of the LPS barrier are extremely potent and highly immunogenic toxins (e.g. lipid A) and can cause septic shock if released within a host. Sepsis and shock are frequently fatal since an uncontrollable overwhelming immune response occurs; current research is aimed at designing agents that block the action of lipid A, or antibacterial drugs that do not cause sudden lysis of Gram negative pathogens.

The barrier(s) that protect the bacterial cell from the hostile conditions of the external environment can also allow the microorganism to resist the antibacterial effects of an antibiotic. Most antibacterial drugs must reach a specific molecular target in order to be effective, although some agents act by randomly disrupting the bacterial cell wall or membranes. At a first glance it may seem that Gram negative bacteria would be the more troublesome and that Gram positive bacteria may dictate less attention in terms of chemotherapy. Unfortunately, Gram positive bacteria can be extremely pathogenic in their own right. Many species are widespread and can easily produce an infection often with severe consequences. Gram positive bacteria, in general, are also well adapted to survive without the added protection of a lipopolysaccharide barrier and are often capable of producing (or overproducing) enzymes which destroy the antibacterial effects of many agents.

With this background in mind, the most relevant bacteria, some of which can prove to be pathogenic, are briefly described below.

1.3.1   Gram positive bacteria

1.3.1.1   Streptococci.   Streptococci are common Gram positive cocci that are often human and animal pathogens. Streptococcal cells associate with each other giving rise to characteristic formations. Two organisms may pair together (diplococci) or dozens of cells (up to about 30) may associate to form a chain. The streptococci are aerobes but an anareobic species, called *Peptostreptococcus*, is now regarded as a member of the genus.

Streptococci can elicit some powerful effcts on mammalian cells and organisms, and it is through these actions that streptococci can be particularly troublesome. Some streptococci are capable of causing hemolysis (the lysis of red blood cells (erythrocytes)). In addition, streptococcal species can produce powerful toxins that can lead to the destruction of other host cells. For example, streptolysin O is an antigenic protein produced by streptococci which can evoke a powerful immune response and cause the lysis of leukocytes (white blood cells) and some tissue cells. Other toxins include streptokinase which affects the action of fibrinogen and blood clotting. In addition, an antigenic nuclease (a deoxyribonuclease (DNA)ase) can also be present.

Streptococci are recognized by a number of their features, the most obvious being shape and staining characteristics. Serious infections caused by streptococci include scarlet fever, rheumatic fever, pneumonia, impetigo pharyngitis and sinusitis. The enterococci (*E. faecalis* and *E. faecium*) are a group of related Gram positive cocci that are particularly troublesome at this time. (Enterococci are included in a new genus *Enterococcus*). Multidrug resistant enterococci have emerged that are not susceptible to commonly used antibacterials; there are concerns that resistance determinants may be passed from streptococci to enterococci, resulting in a true 'super-bug'.

1.3.1.2   Staphylococci.   Staphylococci are round-shaped Gram positive bacteria that are usually larger than the streptococci and which often form clusters resembling grapes. The bacteria can also exist in pairs or remain single. Some species of staphylococci are remarkably hardy and can survive long periods of time in harsh conditions or on environmental surfaces, and with limited nutrition.

There are nearly a dozen species of common staphylococci that can reside in human or animal hosts but only a few behave as pathogens under ordinary circumstances. As with streptococci, some staphylococci produce proteins and toxins that can cause a variety of complicating conditions. For example, *S. aureus* produces an enzyme (a proteinase) which causes coagulation of the blood. α-Hemolysin is a toxin that attacks red blood cells and platelets. There are a host of other toxins and one has recently surfaced after causing the alarming problem of toxic shock in some women wearing tampons. Toxic Shock Syndrome is attributed to the Toxic Shock Syndrome Toxin 1 (TSST-1), an exotoxin produced by a *Staphylococcus* which stimulates cytokine production resulting in organ damage and possible death.

More typically, *S. aureus* causes tissue wound infections and osteomyelitis, but can be responsible for certain types of 'food poisoning'. *Staphylococcus epidermidis* is found on the skin of normal humans but can cause inflammation if it is able to colonize appropriately. *S. epidermidis* can be particularly troublesome to the immunocompromised patient. *Staphylococcus saprophyticus* is another species of this genera that can be pathogenic. It is able to produce urinary tract infections in some women but again this is normally easily remedied.

1.3.1.3 Corynebacterium. Corynebacterium are small Gram positive rods that tend to be flattened at one end giving the microscopic appearance of tiny clubs that cluster to form 'Chinese letters'. The most infamous member of this genus is *Corynebacterium diphtheriae* which has proven throughout the ages to be a threat to mankind in terms of morbidity and mortality. The diphtheria pathogen is potentially fatal due to the toxin it produces, which is among the most toxic of all biological materials known to humanity. The toxin, when present in even minute quantities, is capable of shutting down host protein biosynthesis. This protein can also damage heart tissue and peripheral nerves and is highly immunogenic.

Listeria monocytogenes is another Gram positive bacillus of clinical concern due to its ability to cause meningitis and sepsis particularly in immunocompromised patients.

1.3.1.4 Clostridium. Clostridium are a genus of anaerobic Gram positive rod-shaped bacteria that can vary from relatively harmless to extremely dangerous in terms of pathogenicity. The most notorious species is *Clostridium botulinum* which is the cause of botulism food poisoning. Upon release of the botulism toxin within a human host, death can occur within hours. There are a number of botulism neurotoxins and the most potent of these can be fatal in a single dose of less than 1 μg!

There are other species of clostridium that are also quite pathogenic. *Clostridium tetani* causes tetanus, a disease resulting from the action of a potent neurotoxin. This substance can be treated with formalin to afford a toxoid that is used to protect against the disease. *Clostridium perfringens* is the causative agent of gas gangrene; the *C. perfringens* toxin affects normal phospholipase C activity. Lastly, *Clostridium difficile* is an important member of this family since it is responsible for severe toxic enterocolitis (pseudomembraneous colitis) often associated with antibacterial chemotherapy. Some antibiotics kill off normal bacterial inhabitants of the intestinal tract which allows for super infection of *C. difficile* to occur. The result is a painful diarrhea and fever which will usually resolve if antibacterial treatment is discontinued. In severe cases, vancomycin or metronidazole therapy is effective against *C. difficile.*

1.3.2 Gram negative bacteria and others

1.3.2.1 Enterobacteriaceae. The *Enterobacteriaceae* are a large group of Gram negative bacilli that encompass harmless, even beneficial bacteria as well as pathogenic organisms. For example, *Escherichia coli* (*E. coli*) is present in human and animal intestine as a normal inhabitant. *E. coli* is the most thoroughly studied of all bacteria due to its ease of growth and cultivation as well as its abundance. Likewise, *Klebsiella* species, *Enterobacter* and *Proteus* species are widespread in nature as normal bacterial flora. However, some genera are strictly pathogenic such as *Shigella* which can cause dysentery, and *Salmonella* which can produce severe gastroenteritis. On the far end of this spectrum lies *Yersinia pestis*, the causative agent of the Black Plague, the most deadly pandemic disease of all times. All of these *Enterobacteriaceae* are capable of liberating the lipopolysaccharide endotoxin and some species also produce potent exotoxins.

The majority of the bacterial species belonging to this family are pathogenic only in certain situations. While *E. coli* species are normal and welcomed occupants of the intestinal flora, they can cause urinary tract infections if allowed to spread beyond the confines of the intestines. However, some *E. coli* strains can cause intestinal disorders of varying severity. These intestinal pathogenic species are divided into subclasses (enterotoxigenic, enteroinvasive, enterohemorrhagic and entero-pathogenic) based upon their site of action, the presence of toxins and indirectly by the symptoms which they produce. A particularly frightening *E. coli* infection is neonatal meningitis which is caused by exposure to contaminated amniotic fluid.

The *Klebsiellae* species are not as common as *E. coli* although some species are found in the digestive system of humans and also in nature. *Klebsiella pneumoniae* is a common pathogen belonging to this genera; *Enterobacter* and *Serratia* species are very similar. The *Proteeae* family consists of the *Proteus* and *Morganella* species of bacteria. These organisms can also reside in the normal intestinal flora but *Proteus vulgaris* and *P. mirabilis* are quite opportunistic and are capable of producing serious infection. The *Shigella* and *Salmonella* species are well known pathogens and have produced outbreaks of infectious disease throughout the history of mankind. *Shigella dysenteriae* is the cause of dysentery. Other bacteria belonging to this genera are *Shigella flexneri*, *S. boydii* and *S. somei*. The *Salmonellae* species cause relatively minor intestinal disorders with the exception being *Salmonella typhi* which is the causative agent of typhoid fever.

1.3.2.2 Vibrio and Campylobacter. The *Vibrio* and *Campylobacter* species are small Gram negative rod-shaped bacteria that are primarily pathogenic in nature. The most serious is *Vibrio cholerae* which causes cholera. *Vibrio vulnificus* can produce a devastating sepsis in certain patients and often results from the ingestion of contaminated raw shellfish. Campylobacter are more bothersome to some livestock than humans but can produce a diarrhea upon the consumption of infected poultry. The most disturbing finding concerning these species is that *Campylobacter pylori* (*Helicobacter pylori*) plays a role in the chronic inflammation of the stomach and may be a factor in the development of duodenal ulcers and gastric cancer.

1.3.2.3 Pseudomonas. The pseudomonads are Gram negative rods that are among the most feared of bacterial pathogens. There are many species of this genera and some are exceptionally opportunistic, extremely virulent and resistant to eradication with many antibacterial protocols. *Pseudomonas aeruginosa* is the most dreaded and is capable of expelling dangerous substances into the infected host. Exotoxin A is extremely toxic and other components such as enterotoxins, lipopolysaccharide endotoxin, hemolysins and other proteolytic enzymes contribute to the virulence of this organism. Infections caused by these organisms are extensive; the elderly, immunocompromised and burn and wound victims often fall prey to infection, especially in hospital settings. In other situations, the respiratory tract, urinary tract and even the auditory canal can be targeted. Usually the bacteria enters through a skin injury or an invasive medical procedure. Rather surprisingly, pseudomonads can be found in a significant proportion of the normal population.

1.3.2.4 Neiserria. The neisseria species are Gram negative diplococci that can produce serious infections in humans. Because of the nature of these infections and the severity of their consequences, neisseria remain a great concern. The most pathogenic of the species are the meningococcus (*N. meningitidis*) and the gonococcus (*N. gonorrhoeae*) which are causative agents of meningitis and gonorrhea, respectively.

1.3.2.5 Haemophilus and bordetella. The *Haemophilus* and *Bordetella* species are Gram negative coccobacilli that are often pathogenic. *Haemophilus influenzae* is a common organism that can be found in the nasopharynx and often does not produce a clinical manifestation. However, invasion of this organism beyond its normal confines can produce potentially fatal diseases such as meningitis and serious effects such as arthritis. The most serious species of the *Bordetella* is *Bordetella pertussis* which can attack the tracheobronchial tract producing whooping cough.

1.3.2.6 Legionella. *Legionalla* are Gram negative rods that consist of many species. The most serious outbreak of disease caused by these bacteria was reported in 1976. An illness which manifested in severe pneumonia, fever and shock proved to be fatal to dozens of people gathered for a meeting. The pathogen was identified as belonging to the *Legionella* species, named as *Legionella pneumophila* and dubbed as (the cause of) Legionnaire's disease. Less severe diseases such as Pontiac Fever are caused by *Legionella*.

1.3.2.7 Bacteroides. The *Bacteroides* species are anaerobic Gram negative rods that are found in the upper and lower intestinal tract of man and animals. Some species such as *Bacteroides fragilis* are opportunistic pathogens and can be found in abscesses, surgical wounds and urogenital lesions. *B. fragilis* is usually resistant to many common antibacterial agents but can be treated with suitable antianaerobic agents (e.g. metronidazole).

1.3.2.8 Mycobacteria. The *Mycobacteria* are rod-shaped bacteria that display characteristic staining properties that result from a unique cell wall structure. The most infectious of the species are *M. tuberculosis*, the causative pathogen of tuberculosis and *M. leprae* which produces the disease leprosy. Other mycobacteria can cause severe illness in immuno-compromised people such as AIDS patients. Mycobacterial species can also exist in other mammals and in birds and cause disease in these animals.

1.3.2.9 Chlamydia, rickettsia and coxiella. A number of intracellular bacterial species can produce infectious disease states. *Chlamydia trachomatis* causes an urethral infection that is very common among sexually active individuals today. Certain infections of *C. trachomatis* (trachoma), if not treated, can lead to blindness or severe conjunctivitis. *C. pneumoniae* is an respiratory organism best known as a form of a 'walking pneumonia'. *C. psittaci* can infect those involved in the handling of birds. Members of the *Rickettsia* and *Coxiella* species are predominantly associated with animals. *Rickettsia rickettsii* is the cause of Rocky Mountain Spotted Fever and *Coxiella burnetii* which produces Q fever, can be transmitted from animals to humans.

1.3.2.10 Mycoplasma. The *Mycoplasma* are unique bacterial organisms. They are among the smallest free-living organisms that contain no cell wall and therefore do not undergo conventional staining with the Gram protocol. These bacteria are also unique in possessing sterol components. *Mycoplasma pneumoniae* is a lower respiratory tract pathogen that causes so-called 'walking pneumonia' and other species such as *M. hominis* and *Ureaplasma urealyticum* can produce urinary tract infections.

1.4 Sources of chemotherapeutic antibacterial agents and the impact of organic chemistry

The origin of most pharmaceutical drugs can be traced to naturally occurring sources. Microbial organisms and plants dominate in this respect; a large percentage of compounds and structural leads are derived from these sources. Antibacterial agents are no exception to this trend. The β-lactam, tetracycline, aminoglycoside, erthyromycin and macrolide families of antibacterial agents, as well as a host of anti-infective polypeptides, glucopeptides and polyethers have all resulted from initial discoveries of active, naturally occurring substances.

Even to this day, many antibacterial agents are obtained, directly or indirectly, from a microbial source. Often, these biosynthetic 'factories' have been made more productive by some modification of the fermentation process that either increases the yield of the product, or allows for an improved isolation and/or purification. In some cases, structural manipulation(s) can be performed during fermentation through some intervention of the biosynthetic pathway. For example, all of these advances were monumental in the development of the penicillin antibiotics. Early on, it was the appropriate alteration of the culture broth, and subsequent purification techniques, that allowed for the production of the prototypic penicillin (penicillin G) in a sufficient quantity for study. Subsequently, it was found that additives to the microbial fermentation could be used directly to modify the molecular structure of the antibacterial product. From this, a number of new penicillins were obtained by supplementing the microbial broth with exogenous chemical agents.

However, the majority of antibacterial drugs marketed today are semi-synthetic derivatives that are obtained, at least in part, through chemical modification. Thus while Nature has provided valuable leads and in some cases, early prototypes, humans have been able to elaborate many of these substances to afford more potent and better tolerated chemotherapeutic agents. This will be a recurring theme throughout this text, indicative of the important role organic synthesis has played in the evolution of antibacterial chemotherapy. For example, the quinolone and sulfonamide antibacterials are completely synthetic products.

1.5 Resistance

Resistance is a condition of immunity that a bacterial organism displays in the presence of a specific antibacterial substance. There are varying degrees of resistance and in fact, it is helpful to view antibacterial chemotherapy as being a spectrum that has as opposing endpoints, susceptibility and resistance. Low level resistance (the middle of the spectrum) may be overcome if drug levels can be increased since the bacteria are often susceptible to the agent at higher concentrations. In these instances, the *in vitro* activity of the antibacterial agent against the resistant strain may be several-fold less than that of a sensitive strain. In contrast, high-level resistance (the end of the spectrum) bacteria are typically hundreds- or thousands-fold less sensitive to the antibacterial agent compared to sensitive counterparts.

The degree of resistance is dependent upon several factors such as the specific mechanism that confers resistance and how many modes of resistance are operative in a single bacterium or in a bacterial population. For example, if a given bacterium produces an enzyme that degrades an antibacterial drug, then high levels of the enzyme will favor resistance and so gene expression and copy number (of the enzyme gene) will play a role. In another scenario, a bacterium may be resistant due to a combination of mechanisms such as altered cell wall permeability along with the action of a drug destroying enzyme.

Resistance can be described in a variety of ways depending upon whether one takes a biochemical view or a genetic perspective. In general, resistance can either be intrinsic to the bacteria, or acquired by the microorganism. Intrinsic resistance is inherent in the bacterial species; the phenomenon exists whether the specific antibacterial drug is present or not. Resistance of this type does not depend upon, nor result from exposure to the drug; intrinsic resistance is a stable genetic property that is encoded within the bacterial chromosome. Typical examples are Gram negative bacteria such as *Pseudomonas aeruginosa* that contain a cell wall that is impermeable to many antibacterial agents. Since the drug never enters the cell, nor reaches its molecular target, no antibacterial effect is produced. Intrinsic resistance often dictates which general class of antibacterials is selected for use against a particular pathogen. For example, vancomycin is a valuable agent against certain Gram positive cocci whereas the aminoglycosides are reserved for Gram negative bacilli. The two classes are not interchangeable; vancomycin cannot penetrate Gram negative species and the aminoglycosides are inferior Gram positive agents.

Acquired resistance is the ability of certain strains, belonging to a normally susceptible bacterial species, to obtain and utilize elements (determinants) that result in resistance. Resistance is acquired either by mutation of the chromosome or more commonly, by the acquisition of genetic material that encodes for a resistance function. Chromosomal mutations that produce antibacterial resistance are rather rare; only one-step mutations are statistically feasible (except for evolutionary processes) within a given bacterial population since the frequency of many known mutations ranges from 10^{-6} to 10^{-9}. Single-step mutations are known for several classes of antibacterial agents. For example, resistance to the quinolones, rifamycins and streptomycin arises from single DNA base mutations that result in the production of proteins that do not bind drug but continue to provide native function (e.g. despite a mutation in the gene encoding for the gyrase protein, DNA gyrase activity is maintained in the case of quinolone resistance).

Most acquired resistance results from the expression of DNA fragments that are obtained by the previously susceptible bacteria. The acquired genetic information can be presented to the bacterium in several ways. Most often, the resistance-bearing gene(s) are contained within a circular piece of DNA called a plasmid. Plasmids are self governing, extra-chromosomal loops of DNA that essentially function without regard to normal chromosomal operations. Plasmids that encode for resistance are termed R plasmids or R-factors, but plasmids can also be responsible for the production of toxins and other virulence traits. Many plasmids can be passed on from strain to strain and in some cases, from species to species and so resistance can be spread rapidly among bacterial populations.

The transfer of resistance genes can occur in several ways. Conjugation, or 'mating' involves cell-to-cell contact between two bacterium and the transfer of genetic material from a donor to a recipient. This is accomplished through the action of pili, an appendage that connects the two cells together and provides a route for passage of the DNA. There are even chemical attractants, so-called 'bacterial sex pheromones' that facilitate this process by bringing the donor and recipient cells close to one another. The end result is that a copy of the plasmid is donated to the recipient bacterial cell and resistance is conferred.

Transformation is the process by which bacteria are able to take up freely occurring genetic material and subsequently incorporate these DNA segment(s) into their own chromosomes. This phenomenon is rather uncommon since free DNA undergoes degradation outside the cell. However, there are special means by which either a Gram positive or Gram negative bacterium can make use of the process to achieve resistance. For example, some *Streptococcus pneumoniae* and *Haemophilus influenzae* can use certain proteins that bind to DNA and allow for uptake into the cell.

Transduction is the transfer of genetic material by way of phage infection. Certain bacteriophages can carry plasmids and expel them inside the bacterium once the phage has docked onto the bacterial cell and dissolved away a section of the cell wall. Non-conjugative plasmids, often found in staphylococci can be spread in this fashion.

Transposition involves the movement of DNA segments that encode a resistance function from plasmid to plasmid or chromosome. Transposons consist of one or more resistance genes flanked by insertion sequences, and are usually located within a plasmid piece of DNA. The insertion sequences direct the copying of the resistance gene(s) and relocation of the gene(s) onto new plasmids or the chromosome itself; in the latter case, the genes are inherited. Transposons, like plasmids, are common among both Gram positive and Gram negative bacteria.

The biochemical manifestations of antibacterial resistance, regardless of its genetic origin, can take many forms. Mechanisms that confer resistance include inactivation of the antibacterial drug, decreased permeability of the bacterial cell, alteration of the molecular target and less often, auxotrophy or metabolic by-pass.

Drug inactivation involves structural modification of the antibacterial substance; this phenomenon is common among Gram negative pathogens. For example, the β-lactams are destroyed by bacterial enzymes that cause fragmentation of the drug itself. The aminoglycosides and chloramphenicol are deactivated by enzymes that modify, rather than fragment, the drug. Phosphoryl, acetyl, adenynyl or nucleotidyl moieties are attached to the aminoglycosides through the action of bacterial enzymes, whereas chloramphenicol undergoes acetylation(s). These processes destroy antibacterial activity since the drug conjugates are not recognized by the molecular target (i.e. the bacterial ribosome in the case of the aminoglycosides and chloramphenicol).

Decreased permeability and/or changes in bacteria cell wall composition or cytoplasmic membrane are other ways resistance can occur. In these instances, the antibacterial substance is either prevented from entering the cell, or is quickly pumped out of the cell; in both cases, lethal intracellular drug levels are not reached. For example, tetracycline resistant bacteria possess specific proteins that bind to and remove most tetracyclines from the cytoplasm. Some Gram negative bacteria express mutated proteins in the cell wall that forbid passage of the β-lactams antibacterials to the outer side of the cytoplasmic membrane where peptidoglycan biosynthesis is completed. Lastly, the aminoglycosides must be actively transported across the cell membrane by an energy-dependent process; bacteria such as streptococci that cannot 'drive' this transport mechanism are intrinsically resistant.

Since most antibacterial agents interact with a specific protein or cellular component, modification of the target is a common means by which resistance can be conferred. There are many examples of this phenomenon; as mentioned earlier, a single-step mutation leading to modified DNA gyrase causes quinolone resistance. In a similar fashion, altered DNA-dependent RNA polymerase is responsible for rifamycin resistance, and modification of the bacterial ribosome can cause streptomycin resistance. Some bacteria produce methylase enzymes that add methyl groups onto a specific adenine residue of a ribosomal RNA molecule; this produces resistance to the macrolide antibacterials, the lincosamides and the streptogramins. Finally, changes in penicillin-binding proteins have been shown to be the cause of β-lactam resistance.

The final resistance mechanism that needs to be addressed is that of auxotrophy and metabolic by-pass. If a bacterial species can by-pass a specific metabolic function that is inhibited by an antibacterial agent, then resistance can be conferred. This can be accomplished if the bacterium is able to supply an isoenzyme that is unaffected by the drug, or can circumvent the metabolic transformation all together. Sulfonamide resistance can result from the expression of a new dihydropteroate synthase protein that overrides inhibition of the native enzyme. Likewise trimethoprim resistance can occur through the production of a new dihydrofolate reductase. Lastly, some bacteria can mutate to thymidine auxotrophy, a condition in which the organism is dependent upon thymine, a substrate that is 'downstream' from the metabolic inhibition exerted by the sulfonamides and trimethoprim.

Resistant bacteria have been known virtually since modern antibacterial chemotherapy became an accepted medical practice. Only a few years after the penicillins were introduced, resistant staphylococci (*S. aureus*) were recognized. Since this time (1944), most staphylococci have acquired resistance to conventional penicillins and many streptococci have followed this trend. However, since the 1980s, there has been a complacency regarding antibacterial chemotherapy; many believed that humanity had forever gained 'the upper hand' in the battle against microbes. The past decade alone has been a very alarming 'wake-up call' to battle once again; the emergence of resistant and exceptionally virulent bacterial strains are once again causing outbreaks and deaths, and unfortunately, this trend is likely to continue (Table 1.1).

1.6   Current status and trends in antibacterial chemotherapy

Antibacterial agents dominate the current and projected worldwide market of anti-infective chemotherapy in terms of frequency of use and number of products, as well as earned revenues. The worldwide purse of anti-infective chemotherapy is estimated to be approximately 20 billion dollars annually at this time. Antibacterial products account for the majority of sale dollars, perhaps nearly 70% of the total anti-infective pool. In terms of therapeutic significance, antibacterial agents currently make up 12 to 15% of the total pharmaceutical business; few other areas can seize such a lion's share of the market.

There are apparent trends that partially reflect areas in which notable advances have been realized in terms of potency, spectrum of antibacterial activity and safety profile. For example, the quinolone antibacterials are

Table 1.1 Bacteria that have gained resistance to some drug therapy

Bacteria	Disease/disorder	Date (approx.)
Penicillin resistant		
Pneumococci	Pneumonia, meningitis	mid 1970s–present
Legionella	Legionnaire's disease (pneumonia)	mid 1970s–present
Borrelia burgdorferi	Lyme disease	1980s–present
Salmonella	Gastrointestinal disorders	1980s–present
Staphylococci	Toxic Shock Syndrome	1980s
E. coli O157:H7	Gastrointestinal disorders	mid 1980s–present
Multi-drug resistant		
M. tuberculosis	Tuberculosis	late 1980s–present
Vancomycin resistant		
Enterococci	Wound, blood and enteric infections	late 1980s–present
V. cholerae	Cholera	present
Multi-drug resistant 'super bugs'		?????

Table 1.2 Summary of sales of major chemotherapeutic antibacterial agents

Class	Sales (billion of dollars)	Trend
Cephalosporins	6.0	up
Penicillins	2.5	little change
Quinolones	1.7	strongly up
Macrolides	1.5	slightly up
Tetracyclines	0.5	down
Aminoglycosides	0.5	down
Others	2.0	little change

expected to gain considerable ground in the market in the foreseeable
future. This can be attributed to new agents that possess markedly
improved activity against Gram positive bacteria, superior pharmacokinetic
profiles and favorable dosing regimens. In addition, ciprofloxacin, the
leading quinolone, has recently gained acceptance for serious indications
that demand intravenous administration. The cephalosporins are also
likely to gain ground due to a number of new potent oral agents and the
well-established role of parenteral agents in hospital settings. Rocephin
(ceftriaxone) alone will once again exceed one billion dollars in yearly sales
and is the leading injectable chemotherapeutic antibacterial. The macrolide
family of antibacterials will probably undergo a modest increase in sales
partly due to Biaxin (clarithromycin), Zithromax (azithromycin) and the
continuing success of erythromycin.

Conversely, tetracycline usage should decline as competing agents from
the other classes of chemotherapeutic antibacterials are established.

Aminoglycosides will also decline in terms of sales due to toxicities associated with these compounds which have traditionally limited their use. A brief summary of the sales figures (projected late 1995) and trends of the major classes of chemotherapeutic antibacterial agent is shown in Table 1.2.

Further reading

R.E. McGrew and M.P. McGrew (1985) *Encyclopedia of Medical History*, McGraw-Hill, USA.

G.L. Mandell, R.G. Douglas, Jr. and J.E. Bennett (Eds) (1990) *Principles and Practice of Infectious Diseases*, 3rd edn, Churchill Livingstone, New York.

J.C. Sherris (Ed.) (1990) *Medical Microbiology: An Introduction to Infectious Diseases*, 2nd edn, Appleton & Lange, Norwalk, CT.

P.D. Hoeprich (Ed.) (1983) *Infectious Diseases*, 3rd edn, Harper & Row, Philadelphia, PA.

G.A. Jacoby and G.L. Archer (1991) in 'New mechanisms of bacterial resistance to antimicrobial agents', F.H. Epstein (Ed.) *New England J. Med.*, **324**, 601.

A.S. Evans and P.S. Brachman (Eds) (1991) *Bacterial Infections of Humans: Epidemiology and Control*, 2nd edn, Plenum Medical, New York.

2 Sulfa antibacterials and arylpyrimidine antifolates

2.1 History and overview

The discovery and development of the sulfonamide antibacterial agents in many ways ushered in the modern era of antibacterial chemotherapy. The preceding decades were rich in advances in chemical technology (i.e. synthesis) and in the understanding of many disease processes. At the turn of the century, infectious disease research was evolving from a passive study based primarily upon the power of observation, to a science in which bacterial species could be identified, cultured and examined. The fundamental processes of pathogenesis were recognized and could even be anticipated in some instances. However, prior to the pioneering work of Paul Ehrlich, scientists were essentially incapable of conceiving and designing biologically active substances. Most often chance prevailed and significant advances were slow to come.

Without doubt, it was Ehrlich who lay the groundwork and mechanics that would revolutionize infectious disease study (chapter 1). In retrospect, it was Ehlrich's insight in using dyes to trace biological destinations that directed others to examine properties of dyes on non-living materials and later, on microorganisms and animals. Many dyes of this era contained a sulfamyl group which was presumed to impart a fastness or affinity for wool proteins. While it was not intuitive at the time that antibacterial substances could be derived from dyestuffs, there were good reasons to treat bacteria with dyes; staining methods were demonstrated to be a vital technology for the study of microorganisms. Even in its infancy, Ehrlich's research had established a relationship between selective cell staining by some dyes and antiprotozoal activity. From this finding, research into the action of dyes became commonplace. In 1919, a quinine-derived sulfonamide **1** (Figure 2.1) was prepared and displayed notable antibacterial activity.

However it was a series of major breakthroughs in the early 1930s which boldly reshaped research in this area. The azo sulfonamide dye, prontosil **2** (Figure 2.2) was shown to inhibit the growth of certain bacteria. Domagk demonstrated these exciting properties to *in vivo* disease models by successfully protecting mice from lethal streptococcal infection, and rabbits from a staphylococcus. After an unsuccessful debut in an attempt to treat a case of severe sepsis and meningitis, prontosil was clearly shown to cure an infant from a life-threatening infection. Many sulfonamide antibacterial agents were to follow; Domagk won a Nobel Prize years later.

Curiously, prontosil did not display antibacterial activity *in vitro*. It was suggested that prontosil was undergoing metabolic cleavage of the azo linkage to liberate *p*-aminobenzenesulfonamide (sulfanilamide) which was responsible for the observed bacteriostatic properties. This hypothesis was confirmed in studies in which blood collected from human subjects treated with prontosil was demonstrated to possess antibacterial activity. Subsequently, sulfanilamide was detected in the blood and isolated from the urine of patients receiving prontosil. These examinations established that the therapeutic action of prontosil was indeed due to its metabolism to sulfanilamide. Sulfanilamide **3** (Figure 2.3) became the prototypic sulfonamide antibacterial and initiated research into sulfonamide compounds.

Extensive analog research ensued in the 1930s and continued into the 1940s until another class of antibacterial agents, the β-lactam antibiotics (chapter 3) redirected efforts. During this period of time, a number of sulfonamide antibacterials had reached the market or were undergoing development. Some of the obvious structural modifications of the relatively simple aryl sulfonamide nucleus had been, in principle at least, exhausted. At the same time, the β-lactam antibiotics appeared to be much more challenging in terms of isolation, purification, synthesis and potential structural modification. The penicillins and cephalosporins also offered a much broader spectrum of activity and enhanced potency than most sulfonamides. Accordingly, these new and bizarre antibacterial substances had been given 'great press' both within and outside of the scientific community at the time. Collectively, these issues were reason to invest in

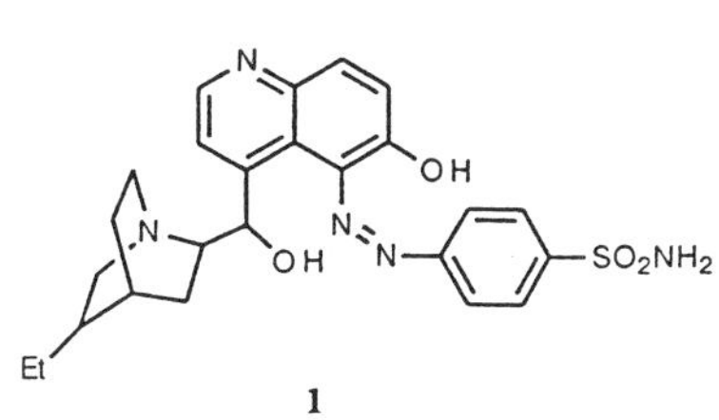

Figure 2.1 A quinine-derived azoarylsulfonamide.

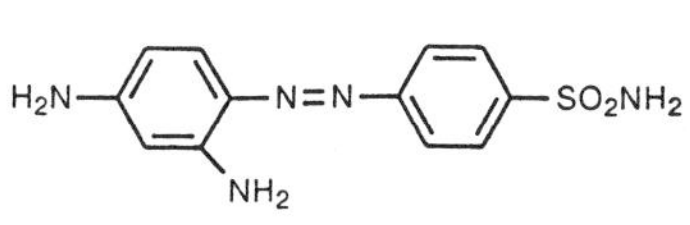

Figure 2.2 Prontosil.

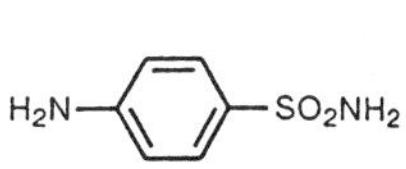

Figure 2.3 Sulfanilamide.

research to study the β-lactam antibiotics at the expense of continuing work on the sulfonamides.

In due course however, the pendulum would swing back the other way. Resistance to the penicillins and cephalosporins emerged and the classical theory of antimetabolites was later advanced. The concept of antagonizing a nutrient or factor needed for bacterial metabolism and growth appeared to present great opportunities for the design and synthesis of novel antibacterial agents. As a result, research on sulfonamides was reinstated and new exciting discoveries followed that not only reshaped antibacterial chemotherapy but some cancer therapy as well. Due to this fragmented history, most sulfonamides originate from two distinct periods. The early agents were produced prior to the β-lactam era (early 1940s). The more contemporary compounds were advanced from the 1950s, from a time at which it was apparent that the penicillins and cephalosporins were not the cure-all for mankind. Over five thousand sulfonamides were prepared within the first decade after the work of Domagk, and thousands more were synthesized and investigated when this area was revisited. Needless to say, precise structure–activity relationships (section 2.3) emerged from this vast work, as well as a number of superb antibacterial agents.

Sulfacetamide **4** (Figure 2.4) followed sulfanilamide to the market; it is still used today, mostly in certain ophthalmic indications. Acylation of the anilino nitrogen gave carboxyacylated sulfacetamides, such as **5** (Figure 2.5), which are an interesting class of congeners due to their pharmacokinetic properties. In general, these agents are extremely poorly absorbed but are able to undergo metabolic deacylation in the intestine to generate the active sulfonamides. In addition, the water solubility of these agents is improved compared to simple sulfacetamides since salt formation is possible.

During the history of development of the sulfonamides, the sulfonamido nitrogen atom has been a major site for structural modification. Some derivatives with good activity contain sulfonamido substituents that are rich in heteroatoms. For example, sulfacarbamide **6** (Figure 2.6) is a carbonic acid derivative and sulfaguanidine **7** (Figure 2.6) is a guanidinyl analog.

The most useful structural modification of the simple benzenesulfonamide nucleus has been the attachment of aromatic heterocyclic substituents

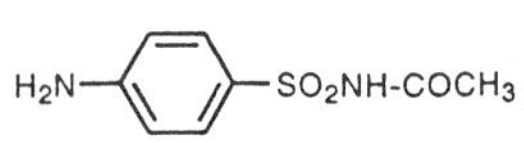

Figure 2.4 Sulfacetamide.

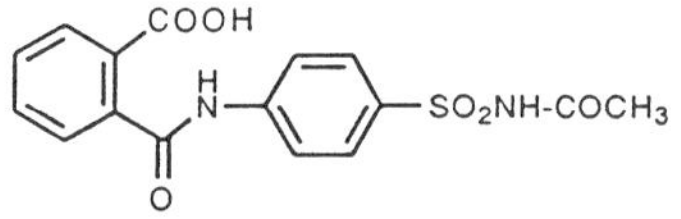

Figure 2.5 A carboxyacylated sulfonamide.

directly on the sulfonamido nitrogen center. Congeners of this type often offer an enhanced spectrum of activity as well as improved pharmacokinetic properties. One prototype, sulfapyridine **8** (Figure 2.7) was the agent of choice against pneumonia for a time.

A variety of other heterocyclic systems were explored and a general trend emerged; the diazine family of heterocycles were found to be particularly attractive substituents. Many marketed agents contain either a pyrimidine, a pyridazine or a pyrazine attached to the sulfonamide. Sulfadiazine **9** (Figure 2.8) is somewhat of a gold standard among the sulfa drugs; the methylated analogs, sulfamerazine **10** (Figure 2.8) and sulfamethazine **11** (Figure 2.8) followed. Sulfisomidine **12** (Figure 2.8) is an isomeric variation on this same theme. In general, these agents complement each other in terms of pharmacokinetic profile and have improved solubility and safety profiles compared to older agents. Sulfadiazine, sulfamerazine and sulfamethazine can be used in combination with each other affording a potent triple sulfa drug. The double sulfa

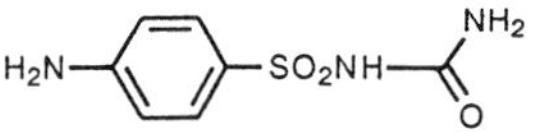

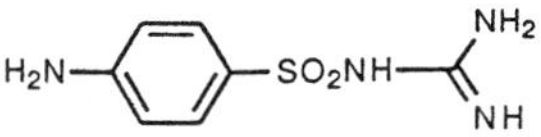

Figure 2.6 Sulfacarbamide and sulfaguanidine.

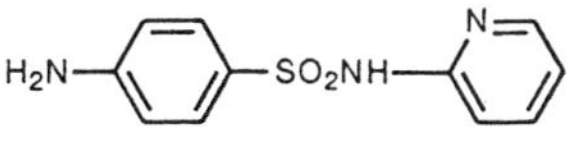

Figure 2.7 Sulfapyridine.

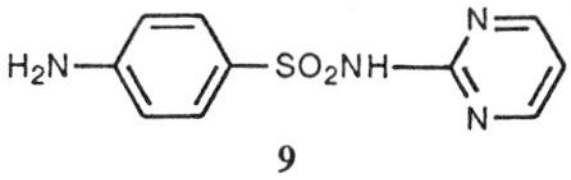

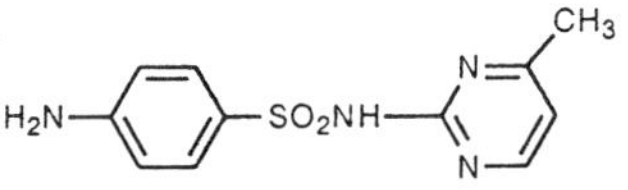

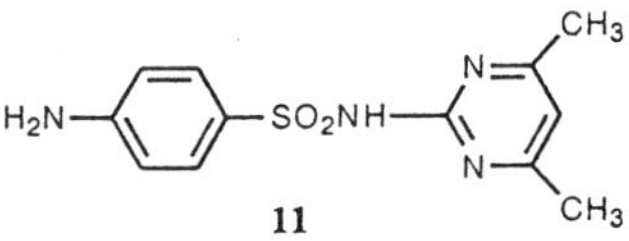

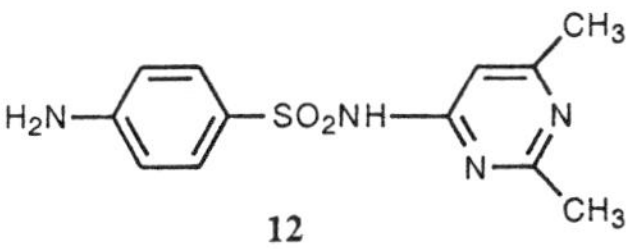

Figure 2.8 Sulfadiazine, sulfamerazine, sulfamethazine and sulfisomidine.

combination of sulfisomidine and sulfadiazine is another useful antibacterial product.

Other heterocyclic systems were investigated and certain azole heterocycles were found to be excellent sulfonamide substituents. For example, the thiazole derivative, sulfathiazole **13** (Figure 2.9) displays good antibacterial activity and is well tolerated; for these reasons, it essentially replaced sulfapyridine. Sulfamethizole **14** (Figure 2.9) contains a thiadiazole heterocycle and sulfaisoxazole **15** (Figure 2.9) has an isoxazole ring. These compounds have been established as marketed products and are often administered for urinary tract infections.

Some sulfonamide antibacterials contain substitution on both the sulfonamido and the anilino nitrogen atoms; the latter position must be acylated rather than alkylated. The best examples are phthalylsulfathiazole **16** (Figure 2.10) and succinylsulfathiazole **17** (Figure 2.10), both of which contain carboxylic acid functionality on the anilino substituent. Interestingly, these compounds remain mostly unabsorbed from the gut without appreciably disrupting the normal intestinal flora. For these reasons, these agents are limited to presurgical sterilization of the gut.

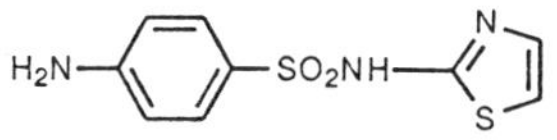

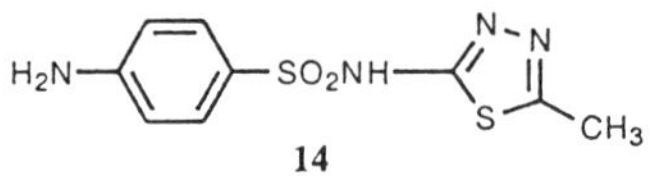

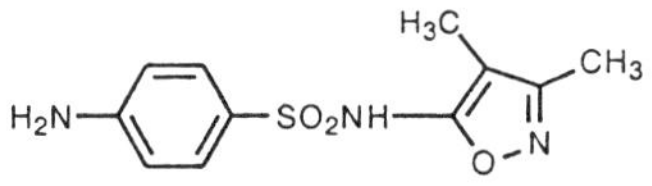

Figure 2.9 Sulfathiazole, sulfamethizole and sulfisoxazole.

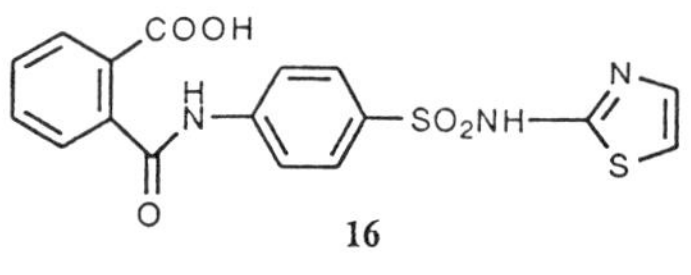

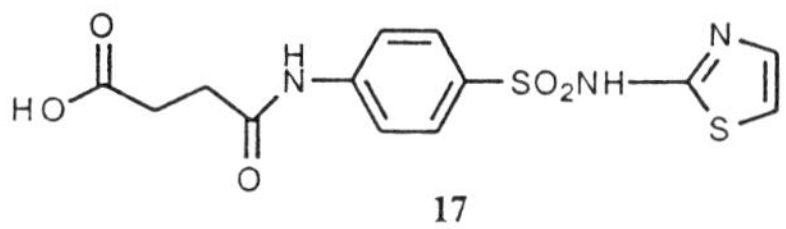

Figure 2.10 Phthalylsulfathiazole and succinylsulfathiazole.

The more recent sulfonamides contain minor variations of the sulfonamido heterocyclic systems, but typically display prolonged levels in humans. For this reason, these compounds can be used for prophylaxis or the treatment of chronic conditions. A classic example is sulfamethoxypyridazine **18** (Figure 2.11) which has a halflife of over one day in humans. The incorporation of methoxy or methyl groups onto the diazine heterocyclic moiety became a recurring theme in sulfonamide research due to the increased binding of such congeners to the protein target. Sulfadimethoxine **19** (Figure 2.12) and sulfamethoxine **20** (Figure 2.12) are both di-methoxylated pyrimidines. Sulfamethoxydiazine **21** (Figure 2.13) and sulfamethyldiazine **22** (Figure 2.13) also follow this trend.

Although pyrimidine is the most frequently encountered heterocycle of the diazine subset of sulfonamides, the pyrazine isomer can be an acceptable modification as evidenced by sulfamethoxypyrazine **23** (Figure 2.14), which also has an exceptionally long halflife in humans. Other

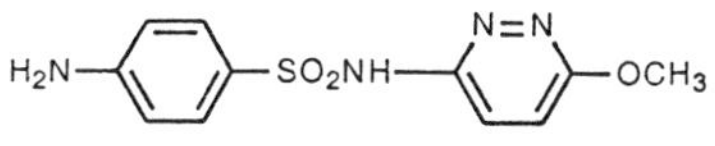

18

Figure 2.11 Sulfamethoxypyridazine.

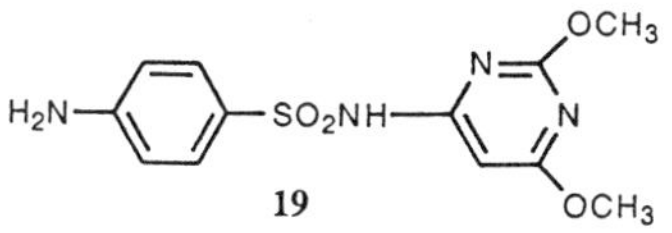

19

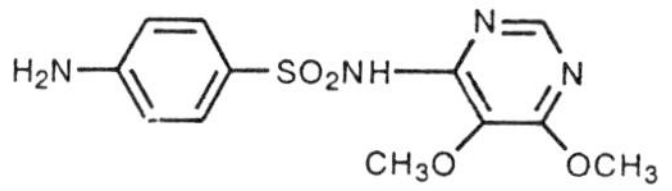

20

Figure 2.12 Sulfadimethoxine and sulfamethoxine.

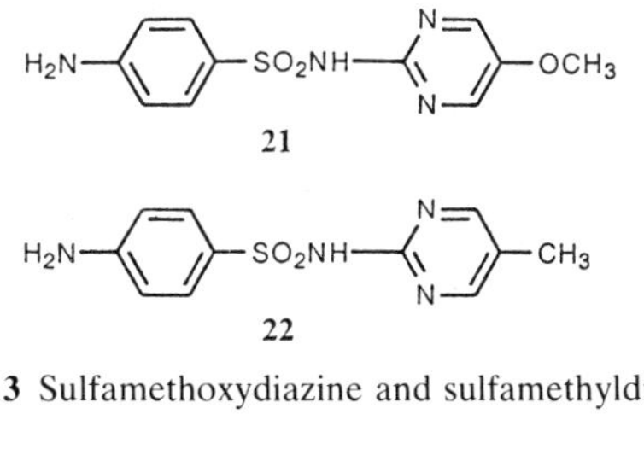

21

22

Figure 2.13 Sulfamethoxydiazine and sulfamethyldiazine.

sulfonamido moieties can be derived from a cytosine substituent such as the sulfa-1-ethylcytosine **24** (Figure 2.15).

Lastly, a number of other heterocyclic systems are attractive alternatives to the azole substituents. The most significant congeners have either a pyrazole, isoxazole and oxazole ring attached to the sulfonamide center; sulfaphenazole **25** (Figure 2.16), sulfamethoxazole **26** (Figure 2.16) and sulfamoxol **27** (Figure 2.16) (respectively) are representative. Sulfamethoxazole is among the best selling of all sulfonamides to date.

In general, the antibacterial properties of the sulfonamides are limited in terms of overall spectrum of activity and potency. By today's standards, the *in vitro* activity of most sulfonamides is often inferior to that of other agents. Regardless, the sulfonamides are quite potent against some staphylococci and streptococci, as well as many common Gram negative bacteria such as *E. coli*, *Proteus* and *Klebsiella*. In addition, some species of *Shigella*, together with *Haemophilus influenzae* and *Neiserria gonorrhoeae* are sensitive to the effects of the sulfonamides. Fortunately, the sulfonamides are easily administered and usually achieve high levels in the urinary tract, and less seldom in the blood and tissues, without serious adverse effects. Consequently, these agents can offer an acceptable success rate for a number of infections. For the most part, the sulfonamide antibacterials have been relegated to the treatment of urinary tract

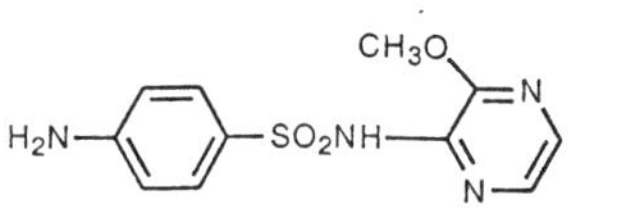

23

Figure 2.14 Sulfamethoxypyrazine.

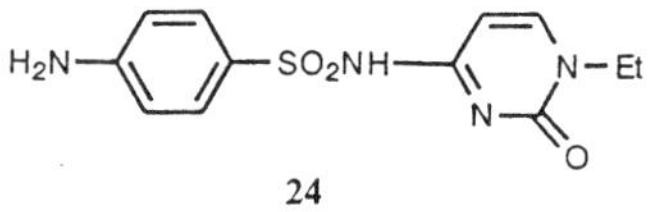

24

Figure 2.15 Sulfaethylcytosine.

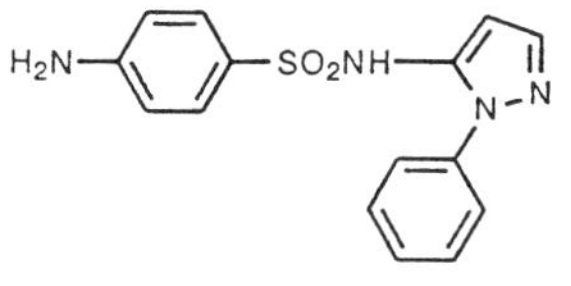

25

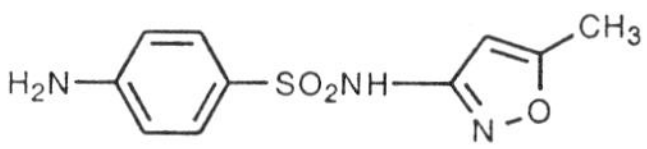

26

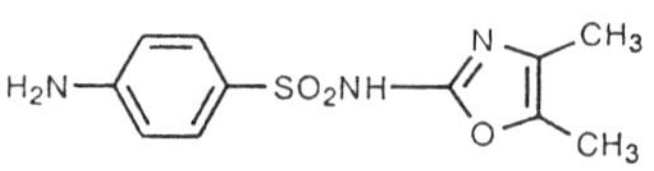

27

Figure 2.16 Sulfaphenazole, sulfamethoxazole and sulfamoxol.

infections. *Chlamydia*, in particular, can be eradicated by a regimen of sulfonamide chemotherapy. Third World countries depend upon sulfonamides to a greater extent than highly developed nations which embrace the use of other, more potent and efficacious agents despite the increased cost of production, administration and therapeutic protocols associated with these other classes of antibacterials.

As research in the sulfonamides progressed, it became apparent that additional opportunities existed for the potentiation of antibacterial activity by inhibiting the same bacterial pathway the sulfonamides affected, but at a different molecular site. It was recognized that sulfonamides and some antifolate drugs were synergistic in their effects upon some species of bacteria. Fildes described antimetabolic antagonism, a concept that was pivotal to the research in this area. From this hypothesis, the development of dihydrofolate reductase inhibitors as antibacterial agents was started. Pyrimethamine **28** (Figure 2.17), a proven (antiplasmodial) antimalarial, was one of the earliest arylpyrimidine antifolates to be examined for antibacterial properties. Certain modifications of the arylpyrimidine nucleus afforded compounds with enhanced antibacterial activity. In 1968, Hitchings described the culmination of this research with the introduction of trimethoprim **29** (Figure 2.18). Trimethoprim is marketed alone or in combination with the sulfonamide sulfamethoxazole **26** (Figure 2.16); bactrim has been one of the most widely used antibacterial therapies in recent history. The closely related congener tetroxoprim **30** (Figure 2.30) has also found use in the antibacterial marketplace.

The era of sulfonamide research also led to the discovery and development of the bis-aryl sulfones, which are a structurally related class of antibacterial agents. In general, the bis-aryl sulfones display outstanding

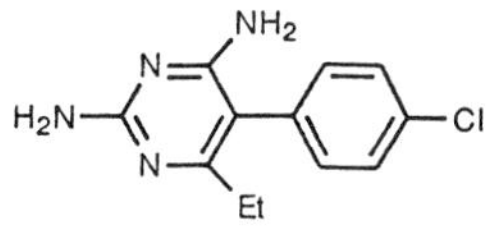

28

Figure 2.17 Pyrimethamine.

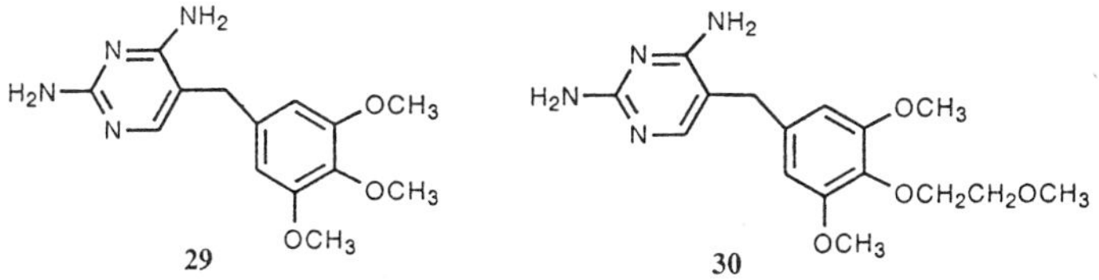

29   **30**

Figure 2.18 Trimethoprim and tetroxoprim.

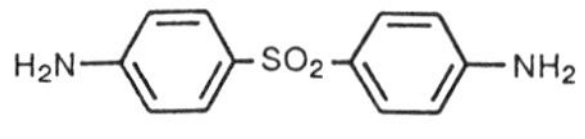

$$H_2N\text{—}\langle\bigcirc\rangle\text{—}SO_2\text{—}\langle\bigcirc\rangle\text{—}NH_2$$

31

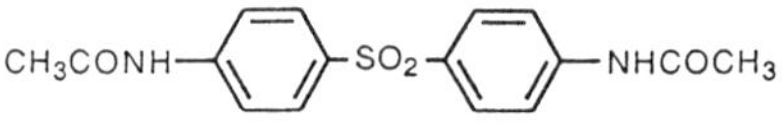

$$CH_3CONH\text{—}\langle\bigcirc\rangle\text{—}SO_2\text{—}\langle\bigcirc\rangle\text{—}NHCOCH_3$$

32

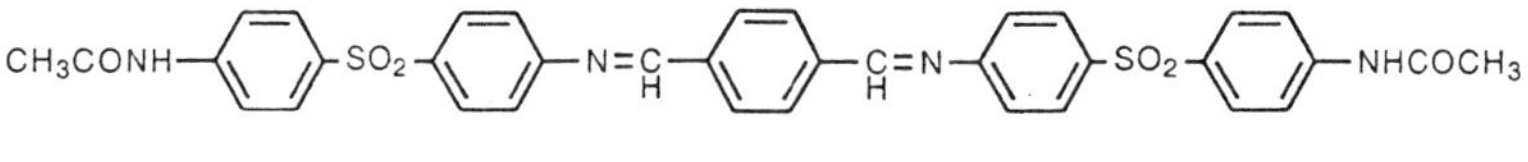

$$CH_3CONH\text{—}\langle\bigcirc\rangle\text{—}SO_2\text{—}\langle\bigcirc\rangle\text{—}N{=}\underset{H}{C}\text{—}\langle\bigcirc\rangle\text{—}\underset{H}{C}{=}N\text{—}\langle\bigcirc\rangle\text{—}SO_2\text{—}\langle\bigcirc\rangle\text{—}NHCOCH_3$$

33

Figure 2.19 Dapsone, acedapsone and DSBA.

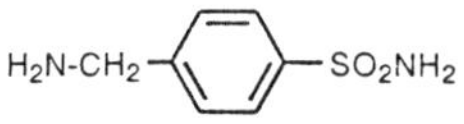

$$H_2N\text{-}CH_2\text{—}\langle\bigcirc\rangle\text{—}SO_2NH_2$$

34

Figure 2.20 Homosulfanilamide.

activity against mycobacteria, and as a result are used almost exclusively as therapy against the leprosy pathogen (*M. leprae*) or for the treatment of tuberculosis (*M. tuberculosis*). Dapsone **31** (Figure 2.19) served as a prototype upon which derivatives such as acedapsone **32** (Figure 2.19) and DSBA **33** (Figure 2.19) were developed. These agents slowly release free dapsone *in vivo* which results in prolonged therapeutic levels of drug and extended dosing requirements (weeks or months). This feature has been important in eradicating mycobacteria, which are typically slow growing, as well as offering an effective therapy in underdeveloped regions where infrequent dosing is more practical.

Finally, it is important to make mention of the antibacterial compound homosulfanilamide **34** (Figure 2.20). Homosulfanilamide is effective against some important anaerobic bacteria, but its mode of action is different than that of the sulfonamides; this agent will not be considered in this text.

2.2 Mode of action: inhibition of bacterial folate metabolism

The modes of action of the sulfonamides, bis-aryl sulfones and aryl-pyrimidine antifolates are well understood; all disrupt specific aspects of normal bacterial folic acid metabolism. From a grander perspective, these agents inhibit the growth of bacteria and exert primarily bacteriostatic effects. This phenomenon occurs at the molecular level in two distinct ways. Sulfonamides inhibit the use of a vital nutrient by interfering with the action of a specific bacterial enzyme needed for the biosynthesis of folic acid. The antifolates target a different enzyme downstream from the aforementioned process. Both processes constitute separate antibacterial mechanisms, although as alluded to earlier, these two antibacterial mechanisms can be simultaneously induced through combination therapy (sulfonamide and arylpyrimidine antifolate) to produce a potent dual mode of antibacterial activity.

It has been well established that the sulfonamides mimic *p*-aminobenzoic acid (PABA) which is a natural substrate required for the biosynthesis of folic acid. Folic acid is a vital nutrient that completes the biosynthesis of *N*-methylated pyrimidines and purines, as well as certain amino acids that are, in turn, necessary for bacterial growth. Bacteria obtain folic acid *de novo*; that is folic acid is biosynthesized from a series of enzymatically driven transformations that utilize PABA. In contrast, mammals obtain folic acid from food consumption (folic acid is a vitamin) and are unable to biosynthesize the same key folic acid precursor(s) as bacteria. This issue is important because adverse effects that would arise from any non-selective action of sulfonamides upon a related mammalian enzyme, are not possible. In this regard, sulfonamides are truly selective antibacterial chemotherapeutic agents although these agents are not free from causing side effects/toxicities to the mammalian host. However, for the most part, the sulfonamides are exceptionally well tolerated and because of this, large amounts of drug can be given in attempts to overwhelm the bacterial pathogen.

Specifically, the antibacterial sulfonamides compete with PABA in the formation of dihydropteroate from dihydropteridine pyrophosphate. In the absence of sulfonamide inhibition, this transformation is carried out by the bacterial enzyme dihydropteroate synthase (DHPS). This enzyme catalyses the displacement of pyrophosphate from an appropriately recognized pteridine substrate. The result is the covalent attachment of the amino moiety of PABA to produce an arylaminomethylpteroate (Figure 2.21). This substrate, dihydropteroate, is further processed by other enzymes eventually to afford dihydro- and tetrahydrofolic acid.

In the presence of sulfonamides, the dihydropteroate synthase enzyme is unable to differentiate between *p*-aminobenzoic acid (the normal substrate) and the sulfonamide agent. In fact, in most cases, the sulfonamide has a greater affinity for the enzyme than does PABA. Consequently, sulfonamide–dihydropteroate synthase binding essentially abolishes (dihydropteroate synthase) enzyme activity. This task is made possible by two distinct binding interactions that exist between the dihydropteroate synthase enzyme and the sulfonamide substrate; these attractive forces mimic those operative in the natural processing of PABA. First of all, the aryl amine portion of the sulfonamide is needed for enzyme recognition. Modification of the anilino moiety is generally not tolerated; enzyme binding is usually completely lost. This feature is similarly present in PABA, and is similarly important for binding of the natural substrate. The second requirement operative in substrate–enzyme or drug–enzyme binding is the presence of an appropriate acidic moiety. In PABA, this group is carboxylic acid (benzoic acid), while in the sulfonamides, the sulfonamido moiety is sufficiently acidic and properly disposed to meet this prerequisite.

Many studies have demonstrated the need of an appropriately acidic sulfonamido proton as a requirement for dihydropteroate enzyme binding. It has been postulated that ionization of this group may produce the (more) active antibacterial species or one which may be (more) readily taken up by the bacterial cell. It has been shown that the pK_a values of antibacterial sulfonamides lie within a somewhat narrow range of values that is close to that of PABA itself ($pK_a \approx 4.7$), and that a pK_a range of 5–8 appears to be favorable, with optimum values being centered about 6.7. However, there are a number of potent agents which fall outside this range (e.g. sulfisoxazole: $pK_a \approx 4.6$; sulfapyridine: $pK_a \approx 8.4$; sulfanilamide: $pK_a \approx 10.4$) indicating that other factors such as electronic charge distribution and lipophilicity are important.

There are other forces that contribute to the overall binding of sulfonamide to DHPS; the aromatic group that extends from the sulfonamido group is obviously important. Certain variations at this position are favorable, evident by the tight binding of agents that incorporate a phenyl ring or some specific heterocyclic systems. As alluded to earlier, the nitrogen-containing heterocycles such as pyrimidines,

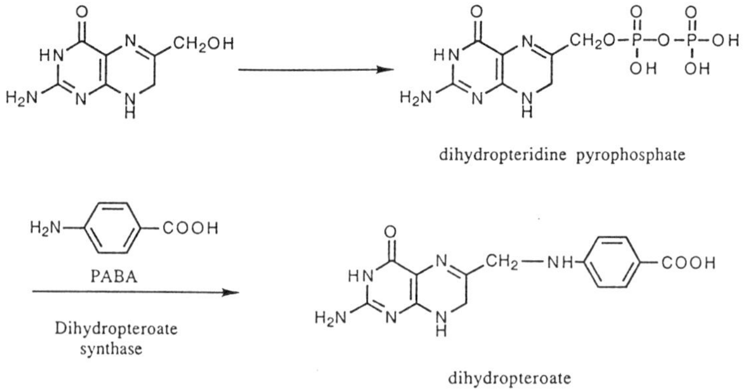

Figure 2.21 Action of dihydropteroate synthase.

pyridines, pyrazoles, oxazoles, isoxazoles and thiazoles are particularly well accepted by the enzyme. Collectively, these observations imply that the essential pharmacophore of a dihydropteroate synthase inhibitor is an aryl- or heteroaryl-substituted *p*-aminobenzenesulfonamide. Of course, congeners which lack a sulfonamido aromatic substituent can also be suitable as demonstrated by sulfanilamide and sulfacetamide.

Usually, the sulfonamides are not structurally modified as a result of binding to the DHPS enzyme; they can be recovered fully intact if removed from the protein. Interestingly, some agents are capable of undergoing direct reaction with dihydropteridine pyrophosphate. Drug–substrate adducts of this type have been obtained; the sulfonamide is covalently bound to the pteridine moiety, the anilino moiety being attached via the displacement of pyrophosphate. The extent to which this conjugation occurs is probably rather minor in terms of antibacterial effect since dihydropteroate synthase activity would be expected to be unaffected (unless the dihydropteroate–sulfonamide hybrid would also serve as an inhibitor of the enzyme and this does not seem to be able to justify the observed effects of sulfonamides on growing bacteria).

In general, the sulfonamides that display the greatest inhibitory activity against dihydropteroate synthase are the most potent antibacterial agents. The inhibitory effect mediated by sulfonamides–dihydropteroate synthase binding is competitive in nature; an increase of PABA is able to diminish the bacteriostatic effects of the sulfonamides. In most scenarios, a bacterial pathogen is unable to acquire, provide or harvest quantities of PABA that are sufficient to thwart the ill effects exerted by the sulfonamide agent. This may seem intuitive but, as is often the case with many antibacterial agents, a drug displaying potent enzyme inhibition may not necessarily exert good antibacterial properties. Frequently, an agent is adversely affected by certain properties of the bacterial pathogen itself or by mechanisms that are operative that do not allow the drug to exert full potency upon its molecular target. In the case of the sulfonamides, a somewhat direct correlation between dihydropteroate synthase inhibition and antibacterial activity does exist. However it is likely that not every issue regarding the mode of action of the sulfonamides has been elucidated. For example, it has been suggested that some sulfonamides may also exert their effects in part by 'down-regulating' the production of dihydropteroate synthase either via a feedback mechanism or by direct stimulation of repressor genes.

The arylpyrimidine antibacterials, also being antimetabolic agents, share some common features with the sulfonamides with regard to antibacterial effects, but target a different enzyme, dihydrofolate reductase (DHFR). Dihydrofolate reductase carries out the (reduction) transformation of dihydrofolic acid to tetrahydrofolic acid (Figure 2.22). Tetrahydrofolic acid is needed for a variety of cellular functions. For example, tetrahydrofolic

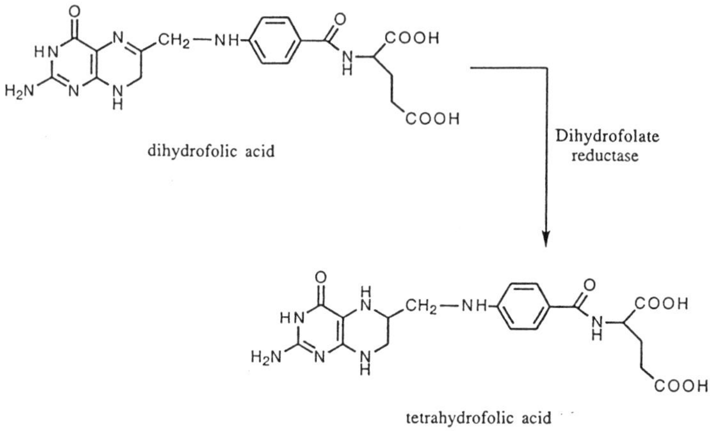

Figure 2.22 Action of dihydrofolate reductase.

acid ultimately serves as a donor of single carbon units in the biosynthesis of some purines, pyrimidines and amino acids.

Both bacteria and mammals are dependent upon the action of endogenous dihydrofolate reductase enzymes. For this reason, it is imperative that antibacterial dihydrofolate reductase inhibitors exert selective activity against the bacterial species without inducing their effects upon the mammalian counterparts. The need of this selectivity has directed most of the research in this area and culminated in the development of trimethoprim **29** (Figue 2.18), which is 50 000–100 000 times more active against most bacterial dihydrofolate reductases than mammalian proteins. To reiterate, this is in marked contrast to the sulfonamides which act upon an enzyme (dihydropteroate synthase) not found in mammals.

Mammalian and bacterial dihydrofolate reductases have been extensively characterized with regard to amino acid sequence, protein shape and functional motifs. Species from both sources are often similar in size; in general, bacterial DHFR consist of 160 amino acids whereas mammalian analogs are slightly larger being about 186 residues in length. Interestingly, mammalian DHFRs are much more homologous (60–70%) than bacterial DHFRs (15–20%), yet some key residues appear to be functionally conserved. One of the best studied dihydrofolate reductase proteins is that present in *E. coli*. The binding of DHFR inhibitors such as trimethoprim has revealed some important molecular interactions. The *E. coli* DHFR assumes a clam-like structure in which five peptide strands of the protein lie in proximity to the bound arylpyrimidine antifolate substrate. On one strand is an aspartic acid residue (Asp-27) that is conserved in bacterial DHFR enzymes. The N-1 nitrogen atom of the antifolate substrate is the most basic site of the agent and probably interacts with this aspartic acid through a conventional acid–base association.

In addition to this site, other specific amino acid residues such as phenylalanine, isoleucine and threonine (Phe-31, Ile-94 and Thr-113 of the (*E. coli*) enzyme) adopt a binding conformation which places these residues in close proximity to the diaminopyrimidine portion of the substrate (trimethoprim). Thus the diaminopyrimidine ring is surrounded by a lipophilic pocket that probably stabilizes the complex through van der Waals interactions. The methoxy groups of trimethoprim are positioned close to similarly lipophilic residues (Phe-31, Ile-50, Leu-28). The shape of

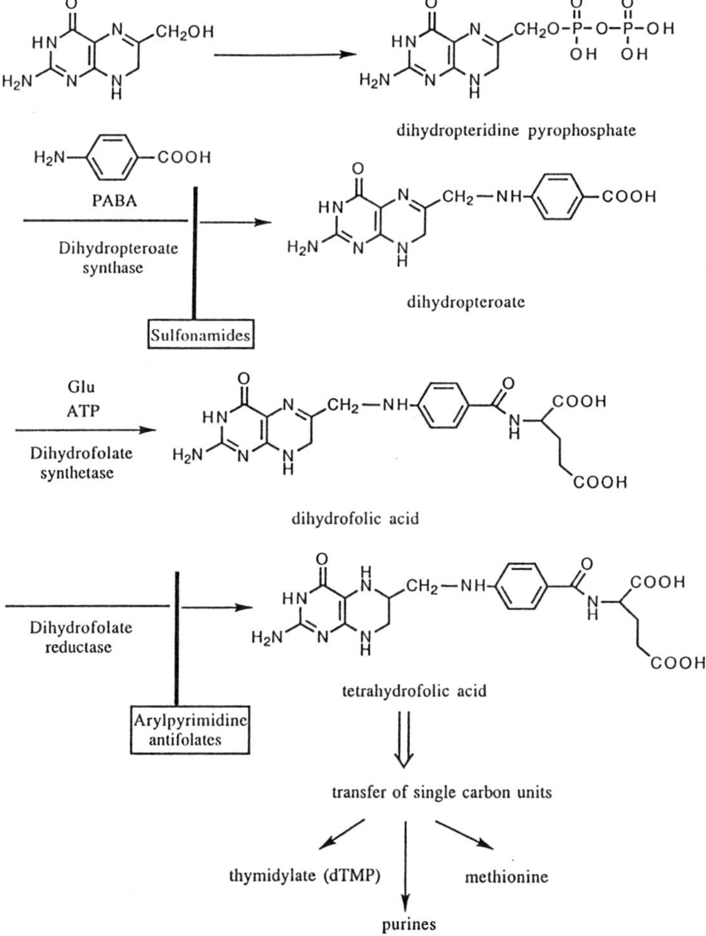

Figure 2.23 Bacterial folate metabolism and sites of action of the sulfonamides and arylpyrimidine antibacterials.

the antifolate molecule is also important in the proper positioning of these moieties within the enzyme. It has been recognized that separation of the aromatic moieties (phenyl and pyrimidinyl) is crucial to DHFR enzyme binding. A single methylene spacer is optimal and this may reflect entropic and steric factors that demand a suitable tilting of the aromatic rings within the enzyme pocket. The spacer is best left unsubstituted which implicates steric as well as rotational factors.

A sound structure–activity relationship between the antibacterial aryl-pyrimidine antifolates has been gathered (section 2.3). During the course of this research, some antifolates have been obtained that are more active against human dihydrofolate reductase (DHFR) than the bacterial proteins. For example, some diaminodihydrotriazines are particularly potent inhibitors of human DHFR (e.g. 1-(*p*-butylphenyl)-2,4-diamino-6,6-dimethyl-5,6-dihydrotriazine is three orders of magnitude more active against human DHFR than bacterial DHFR). This discovery has had profound implications for the treatment of cancer and other hyperprolifer-ative diseases; indeed an intimately related area of research has led to the development of antimetabolites such as methotrexate, which is widely used to treat certain forms of cancer, arthritis and other diseases.

In summary, the sulfonamides and the arylpyrimidine antifolates are potent antibacterial agents due to inhibition of either bacterial di-hydropteroate synthase (DHPS) or dihydrofolate reductase (DHFR) enzymatic activity, respectively. A schematic depiction of the bacterial folic acid metabolic pathway is provided (Figure 2.23) indicating the sites of inhibition of these agents.

The bis-aryl sulfone agents are bacteriostatic agents presumably due to the same mechanism of action as that of the sulfonamides.

2.3 Structural features and structure–activity relationships of the sulfa antibacterials and arylpyrimidine antifolates

The structural features that bestow antibacterial activity to the sulfon-amides, sulfone antibacterials and the bacterial antifolates are well understood. These agents are produced wholly via chemical synthesis and being relatively simple in structure, are easily modified. Although the pharmacophore of each class is somewhat elementary, specific structural changes can produce more desirable properties in terms of antibacterial activity, elimination halflife, solubility and tissue distribution. This is particularly true in the case of the sulfonamides, which when coupled with their ease of synthesis, can be in essence, tailored in terms of the pharmacokinetic properties mentioned above.

In order to serve as an effective antibacterial agent, a sulfonamide agent must be capable of acting as a mimic of the natural substrate PABA which

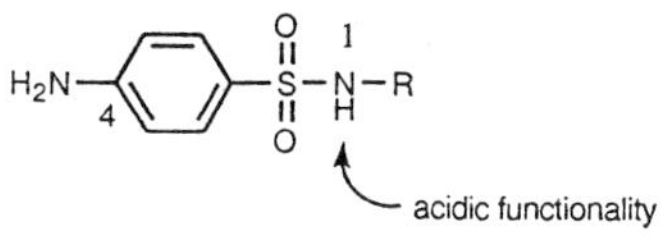

Figure 2.24 Basic sulfonamide SAR; a *p*-aminobenzenesulfonanide containing appropriate acidity.

is normally processed by dihydropteroate synthase. These features are reflected in the sulfonamide structure–activity relationship (SAR). The structural requirements of the sulfonamide antibacterial agents consist merely of a *para*-substituted aniline-derived sulfonamide that contains an appropriate acidic sulfonamido functionality (Figure 2.24). Deviation from this simple construct results in loss of antibacterial activity although it is interesting to note that some diuretic and hypoglycemic agents have originated from this line of research.

With respect to antibacterial effects, *para*-substitution about the benzene ring disposes the anilino group and the acidic sulfonamido moiety to positions that best mimic PABA. This feature is vital; isomeric analogs are useless as antibacterial agents. The replacement of the benzene ring with heterocycles or saturated (hydrocarbon) rings is not tolerated; antibacterial activity is abolished. Alteration of the anilino amine functionality is generally not acceptable except for one modification, that being replacement with a *para*-nitro moiety. In this context, alcohol, ether, halogen, carboxylic acid and simple alkyl groups directly attached at the *para* position of the benzenesulfonamide are inactive. Interestingly, the incorporation of an additional sulfonamido moiety at this position to give a (bis)sulfonamide of *p*-phenyldiamine also results in loss of activity. This finding indicates that a nitrogen atom directly bonded to the aromatic carbon atom is required and that a simple amino group is the best choice. There are some variations; appropriately substituted anilines can behave as active derivatives if the corresponding free amine can be generated *in vivo*. *N*-Alkylation is not acceptable and *N*-acylation is tolerated only if an appreciable rate of deacylation is operative in the body such as in the case of the N-4-carboxyacyl derivatives previously described (section 2.1). When summed, a prerequisite for recognition by the dihydropteroate enzyme is clearly a *p*-aminobenzenesulfonamide; there is only one remaining site of this pharmacophore that has been modified in a productive manner.

2.3.1 Variation of the sulfonamido group: substituents at N-1

Modifications of the sulfonamide itself including varying oxidation states of the sulfur atom have been examined. Attempts to replace the sulfonamide with other acidic groups such as sulfinic, sulfonic, phosphoric or arsenous acids is detrimental and results in a decrease or complete loss of antibacterial activity. Antibacterial activity can be partially retained in some derivatives in which the sulfonyl center is reduced to the sulfoxide oxidation state or exists as a disulfide joined to an additional *para*-substituted aniline ring. Simple *para*-substituted phenylenediamines, congeners in which the sulfonyl center is excised, are inactive. Acetylated amino groups as well as cyano and nitro moieties are not suitable sulfonamide replacements and lead to loss of activity. Overall, the sulfonamide moiety is fairly indispensable with few exceptions, none of which offer any advantage in terms of enzyme binding and antibacterial activity. A simple and general representation of this aspect of sulfonamide SAR is summarized below (Figure 2.25).

The vast majority of synthetic modifications of the sulfonamides has encompassed substitution directly onto the sulfonamido nitrogen center (the N-1 position). In this context, *N*-arylation and *N*-acetylation can be particularly fruitful. Monosubstitution is usually advantageous; disubstitution of the sulfonamido nitrogen atom usually results in loss of antibacterial activity due to the loss of the acidic proton site. Recall that a crucial drug–enzyme association is an electronic interaction that involves the acidic sulfonamide moiety (section 2.2). There are exceptions to this trend; the most notable are some derivatives in which a simple acetyl group is joined to an aryl-substituted sulfonamide at the N-1 position (e.g. **35**, Figure 2.26).

The incorporation of an aromatic group directly onto the sulfonamido N-1 center has been extensively explored. In general, many modifications are tolerated although purely carbocyclic substituents (phenyl groups) afford compounds that do not offer any improvement compared to sulfanilamide.

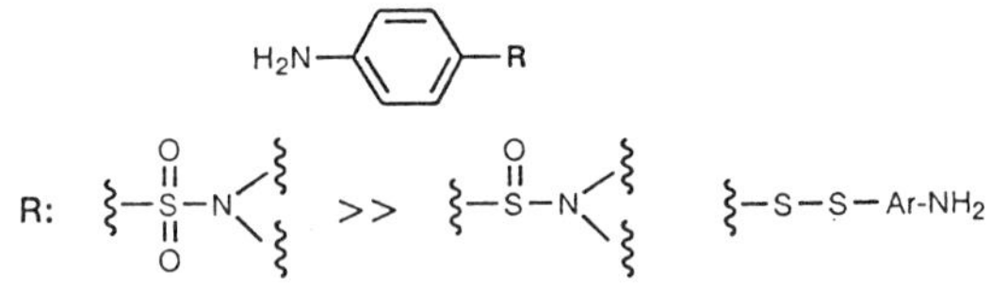

Figure 2.25 The sulfonamide moiety and variations.

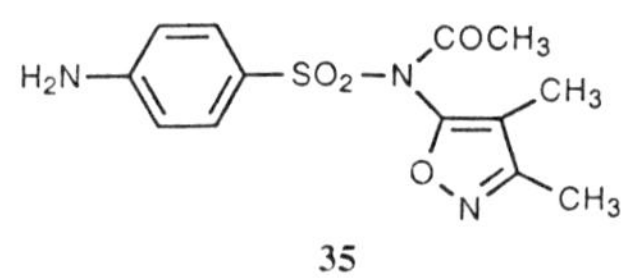

35

Figure 2.26 An N-1 disubstituted sulfonamide.

This is somewhat surprisingly at a first glance since even one heteroatom within the aromatic group can produce superior congeners. In this context, nitrogen-containing heterocycles are particularly attractive substituents in that not only can potency be enhanced, but elimination halflife can often be extended. In addition, tissue distribution and absorption can be greatly altered depending upon the particular heterocycle present as well as substituents emanating from the ring. Other heterocyclic systems that contain oxygen and/or sulfur, but do not contain nitrogen are inferior; a basic nitrogen heterocycle can be regarded as being optimum.

Sulfapyridine **8** (Figure 2.7) contains a 2-pyridinyl group and is a landmark sulfonamide antibacterial agent in several ways. It was the first sulfonamide containing a heterocyclic sulfonamide substituent to achieve widespread use although its once dominant position in antibacterial chemotherapy has since passed. From the vantage point of today, sulfapyridine is somewhat unique in being the only sulfonamide containing a lone nitrogen atom within its heterocyclic array (i.e. a pyridine ring), to achieve widespread use. Its successors today, which comprise the majority of marketed sulfonamide agents, possess at least two heteroatoms. The placement of additional heteroatoms into a sulfonamide N-1 aryl substituent has followed several distinct trends.

A five-membered heterocyclic substituent that contains an additional nitrogen atom, or either a sulfur or oxygen atom, is often a favorable modification. Both oxazole and isoxazole derivatives have been successfully developed as illustrated by sulfisoxazole **15** (Figure 2.9), sulfamethoxazole **26** (Figure 2.16) and sulfamoxole **27** (Figure 2.16). The best oxazole-derived agents have the heterocyclic ring attached at the C-2 position to the sulfonamide N-1 position; isoxazole congeners can be attached at the corresponding C-3 or C-5 positions. The thiazole ring is the most important of the sulfur-containing heterocycles; the most important derivatives are joined at the C-2 position as exemplified by sulfathiazole **13** (Figure 2.9) or any of its *N*-carboxyacyl derivatives. An additional nitrogen atom can be incorporated into these heterocyclic systems, evident by the thiadiazole congener sulfamethizole **14** (Figure 2.9). In general, heterocyclic substituents which contain a third heteroatom offer no advantage over agents which possess the simple oxazole, isoxazole or thiazole moiety. The optimum five-membered heterocycles are depicted below (Figure 2.27).

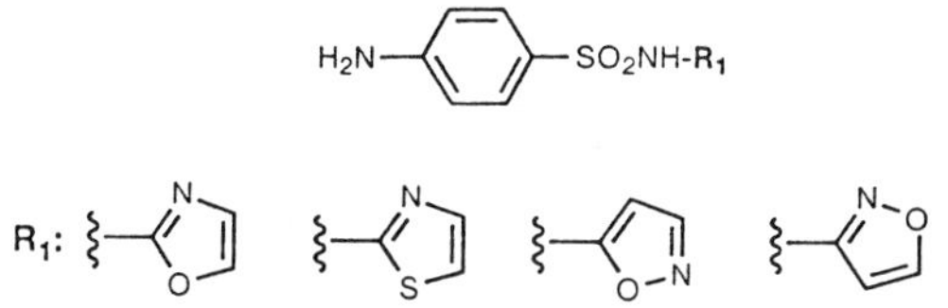

$$H_2N - \text{C}_6H_4 - SO_2NH\text{-}R_1$$

Figure 2.27 N-1 azole substituents.

In addition to the five-membered heterocyclic substituents described above, six-membered rings that exclusively contain two nitrogen atoms, but no endocyclic oxygen or sulfur atoms, have proven to be desirable. Specifically, a pyrimidine or a pyridazine heterocycle joined to the sulfonamido nitrogen can confer superb antibacterial properties. Sulfadiazine **9** (Figure 2.8) is the prototype of the pyrimidine-derived sulfonamides and contains a simple pyrimidine ring joined at the C-2 position to the sulfonamido nitrogen center. The attachment of a pyrimidine substituent at the C-6 position is an important isomeric modification. In addition, the placement of one or two small alkyl substituents onto the heterocycle can be desirable. With pyrimidine congeners, both methyl and methoxy substituents afford agents that display outstanding antibacterial properties such as sulfisomidine **12** (Figure 2.8), sulfamethoxine **20** (Figure 2.12) and sulfadimethoxine **19** (Figure 2.12). The most significant six-membered heterocyclic substituents are summarized below (Figure 2.28).

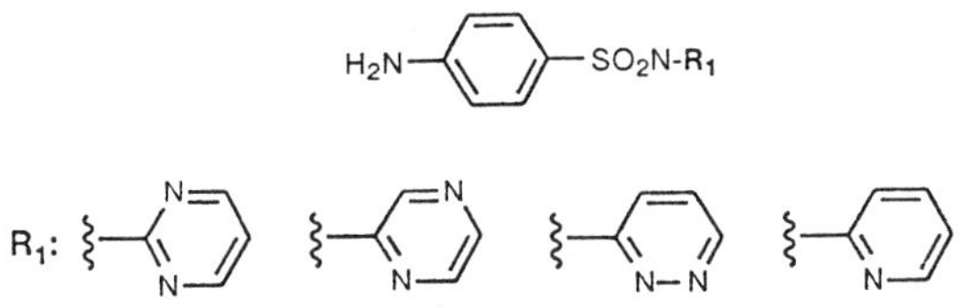

Figure 2.28 N-1 diazine and azine substituents.

2.3.2 Antibacterial dihydrofolate reductase inhibitors

The arylpyrimidine antifolates inhibit the bacterial enzyme dihydrofolate reductase (section 2.2). Dihydrofolate reductase activity is also crucial in mammalian systems and so a high degree of selectivity must be operative in order for an agent to serve as a safe and useful antibacterial agent. There are distinct structural features that produce preferred bacterial enzyme binding without disrupting normal mammalian DHFR function. The most basic requirement is an appropriately substituted aryldiaminopyrimidine (Figure 2.29).

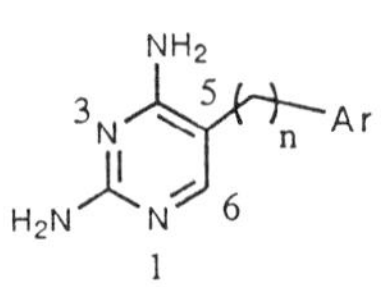

Figure 2.29 Basic arylpyrimidine SAR; a 5-(aryl-substituted)-2,4-diaminopyrimidine.

For all intensive purposes, the pyrimidine nucleus is indispensable for antibacterial activity as are the flanking amino substituents at the C-2 and C-4 centers of the heterocyclic ring. The C-5 substituent is the most critical feature regarding antimicrobial activity. The incorporation of aryl or aryloxy substituents at the C-5 position of the pyrimidine is a common feature among some antimalarial compounds; simple alkyl substitution at C-6 is also prevalent (Figure 2.30). In general, these compounds are limited to antiparasitic applications in humans. Various fused bicyclic congeners, including some pteridines (Figure 2.31), can be active against bacterial dihydrofolate reductases, but usually the therapeutic index is marginal due to effects on the mammalian enzyme(s). This is also true for some corresponding pyridopyrimidine and quinazoline analogs (Figure 2.31). Only specific modifications at the C-5 position are tolerated for noteworthy and selective antibacterial properties.

The incorporation of aromatic groups onto the diaminopyrimidine nucleus at C-5 produces potent antibacterial activity if properly 'spaced' from the heterocyclic nucleus. Only the benzene substituents are suitable C-5 groups in terms of antibacterial potency (Figure 2.29; Ar = benzene), although some pyridine congeners are weakly active. The optimum spacing between the diaminopyrimidine and benzene rings is provided by a simple methylene unit (Figure 2.29; $n = 1$) (section 2.2); longer linkages (ethyl, propyl, etc.) abolish antibacterial activity. Substitution onto the methylene unit is also detrimental, as is the formation of an additional ring from either aromatic ring to this single carbon atom. Lastly, heteroatom (oxygen, nitrogen) spacing units such as ethers, amines, carboxylate esters or amides are not tolerated.

The majority of useful modifications of the aryl moiety itself have been achieved via the placement of groups onto the *meta* (C-3, C-5) and *para*

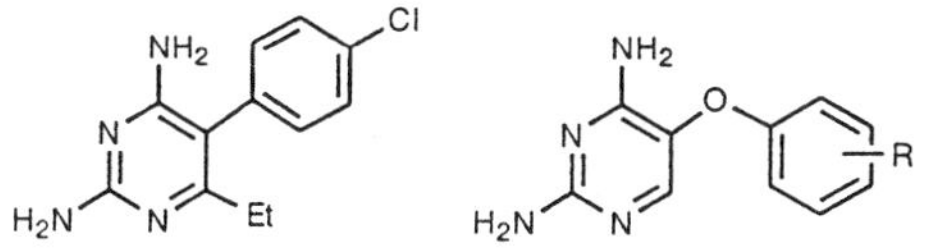

Figure 2.30 Antimalarial diaminopyrimidines.

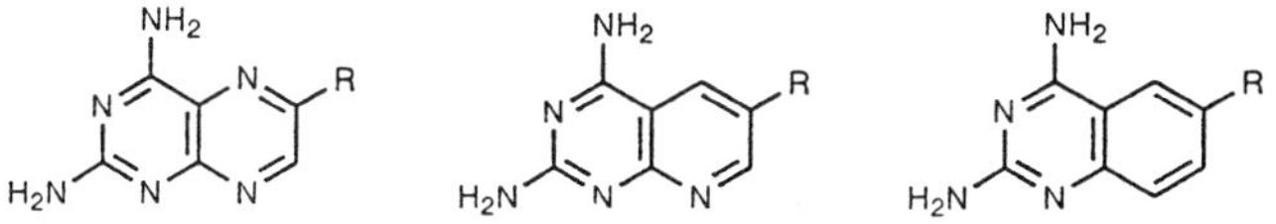

Figure 2.31 Pteridine, pyridopyrimidine and diaminoquinazoline systems.

positions (C-4) of the benzene ring (Figure 2.32). *Ortho*-substitution can afford agents which have good potency against certain protozoan parasites (coccidia) but in general, these compounds do not serve as useful antibacterials. Thus, substitution onto the C-3, C-4 and C-5 benzyl positions has been extensively explored.

Small alkyl and alkyloxy groups at these positions provide maximum enzyme binding and corresponding antibacterial potency. The methoxy group has been the most desirable substituent; ethyl and ethoxy groups can afford compounds with comparable dihydrofolate reductase binding, but frequently with reduced selectivity against the mammalian enzymes. Larger lipophilic moieties as well as simple amine groups and thioethers result in reduced binding to the bacterial enzyme, although some of these derivatives can display notable enzyme binding and antibacterial activity. The C-4 position is the most forgiving of these moieties, as illustrated by the 3,5-dimethoxy-4-(2-propenyl) analog which has an exceptional affinity for some bacterial dihydrofolate reductases. Interestingly, C-4 (free) hydroxyl substituents (phenols) are inferior compared to methoxy moieties which may reflect spacial factors. Hydroxyl substituents can remain essentially co-planar with the benzene ring whereas a trisubstitution pattern that incorporates alkyloxy groups at C-4 presumably forces substituents 'out-of-plane' in order to alleviate buttressing of the adjacent substituents. Thus the degree of substitution of the benzene ring is a crucial issue. Trisubstitution at C-3, C-4 and C-5 is generally optimal and disubstitution is better than monosubstitution. Thus a trisubstituted analog which contains a suitable but inferior group (e.g. small alkyl) to that of methoxy, can be a more potent agent if it is flanked by methoxy groups than a simple monomethoxy congener. This rationale leads to the trimethoxy analog,

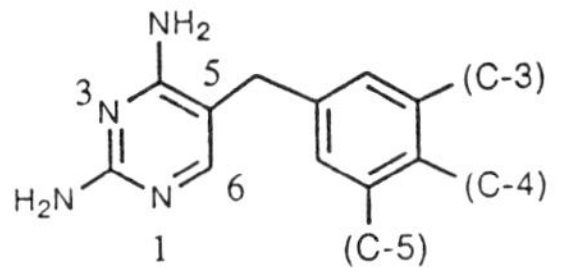

Figure 2.32 *meta* and *para* positions of an arylpyrimidine antifolate.

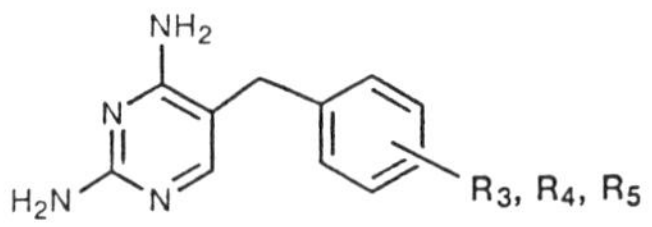

R's: R_3, R_4, R_5 > disubstitution > mono-substitution;

OCH_3 > Et > small alkyl, amino, thioether

Figure 2.33 Preferred aryl substitution.

trimethoprim **29** (Figure 2.18), which remains the premier dihydrofolate reductase inhibitor in use today. A simplistic depiction of the benzene-derived moieties is useful in summarizing this discussion (Figure 2.33).

2.4　Synthetic approaches to sulfa antibacterials and arylpyrimidine antifolate antibacterial agents

2.4.1　Synthesis of sulfonamide antibacterials

There has been little reason to stray from the classic synthetic approaches to the sulfonamides; most variations rely upon two basic themes. The first strategy is centered upon the propensity of *N*-acetylated arylamines to undergo chlorosulfonylation directed to the *para* position. The resultant benzenesulfonyl chloride, being highly electrophilic in nature, undergoes reactions with a host of aryl and heteroaryl amines. Acid or base hydrolytic cleavage of the *N*-acetyl group affords the final sulfonamide product (Figure 2.34). An alternative route involves sulfonylation of nitrobenzene which is similarly amenable to the incorporation of various sulfonamido aryl substituents. Subsequent reduction of the nitro moiety completes the synthesis of the sulfonamide agent (Figure 2.35).

In the case of sulfanilamide **3** (Figure 2.3) and sulfacetamide **4** (Figure 2.4), the elements of ammonia are added to the benzenesulfonyl chloride. Sulfonamides that contain heterocyclic substituents make use of hetero-cyclic amines. The chemistry of thiazoles, oxazoles, isoxazoles, pyrimidines, pyridazines and pyridines provide access to the corresponding

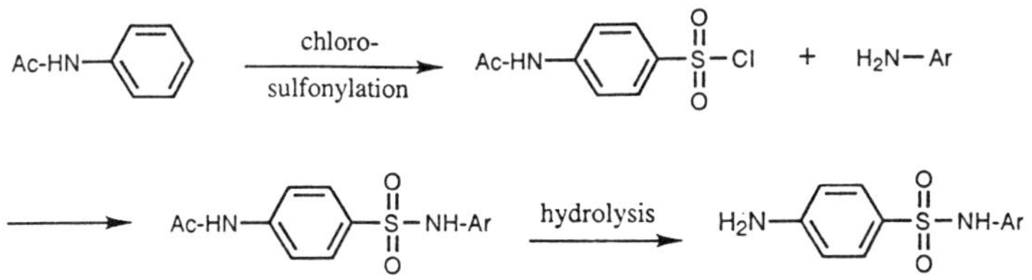

Figure 2.34 Chlorosulfonylation method for the preparation of sulfonamides.

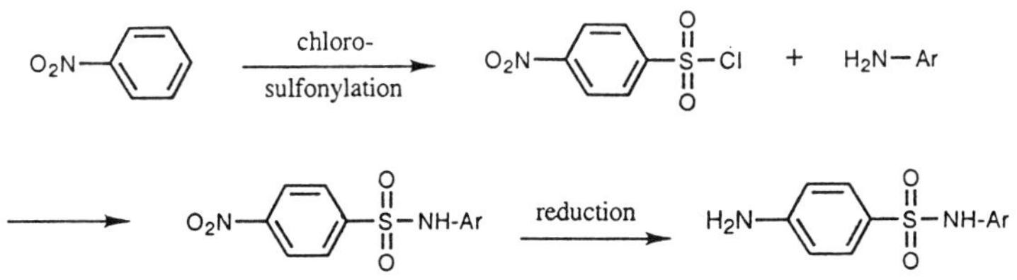

Figure 2.35 Chlorosulfonylation method using nitrobenzene.

amino-substituted heterocycles. These issues are not being considered here, although some details regarding their use in the synthesis of sulfonamide antibacterial agents are fitting.

Thiazoles and oxazoles that contain a free amino substituent at the C-2 position undergo the desired coupling reaction with *para*-acetylated aminobenzenesulfonyl chloride, as do C-2 substituted aminopyrimidines. In these examples, the exocyclic nitrogen atom is the more nucleophilic site; the (remaining) heterocyclic nitrogen atom, being weakly basic due to aromatic resonance stabilization, is ineffective as a competitive nucleophile. Other nitrogen-containing heterocycles such as 5-aminoisoxazoles, 3-aminopyridazines and 5-aminopyrazoles similarly undergo bond formation upon the exocyclic amine functionality for the same reason. In most cases, alkyl or alkyloxy substituents do not significantly alter the reactivity of the arylamine. Therefore, many landmark sulfonamide antibacterials (sulfapyridine **8** (Figure 2.7); sulfadiazine **9**, sulfamerazine **10**, sulfamethazine **11** and sulfisomidine **12** (Figure 2.8); sulfathiazole **13** and sulfaisoxazole **15** (Figure 2.9); sulfamethoxypyridazine **18** (Figure 2.11); sulfadimethoxine **19**, sulfamethoxine **20** (Figure 2.12); sulfamethyldiazine **21** and sulfamethoxydiazine **22** (Figure 2.13); sulfamethoxypyrazine **23** (Figure 2.14);

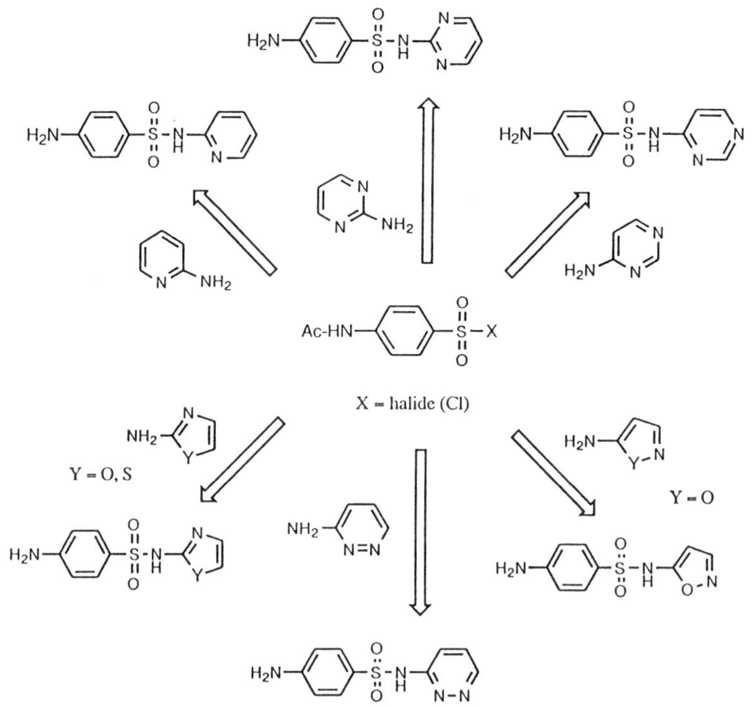

Figure 2.36 Most sulfonamides can be prepared by amine sulfonylation.

sulfaphenazole **25**, sulfamethoxazole **26** and sulfamoxol **27** (Figure 2.16)) are all accessible via this synthetic strategy (Figure 2.36).

There are other strategies that have been used to synthesize sulfonamide antibacterials. It is possible to form the desired heterocyclic substituent after acylation of the benzenesulfonyl chloride although in most cases, this approach does not offer any advantage over the previously described chemistry. For example, the semicarbazides of aldehydes react with benzenesulfonyl chloride to produce intermediates poised for subsequent cyclization to the heterocyclic derivative (Figure 2.37). In some instances, such as the N-1-carbonic acid derived sulfacarbamide **6** (Figure 2.6) and sulfaguanidine **7** (Figure 2.6), simple ureas and guanidines are introduced and the latter cyclization methodology is omitted.

Acetylation of either the anilino nitrogen atom or the sulfonamido nitrogen atom is required in order to synthesize the N-4-carboxyacyl sulfonamides (section 2.1) or the disubstituted sulfonamides (section 2.1), respectively (Figure 2.38). In the first case, phthalic (or succinic) anhydride

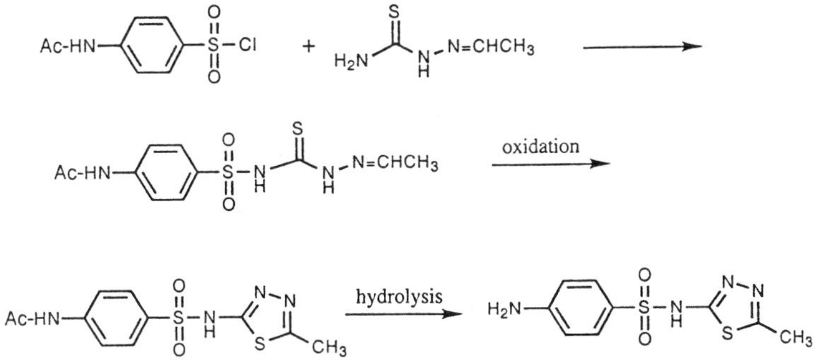

Figure 2.37 Sulfonamide synthesis with subsequent construction of an N-1 heterocyclic group.

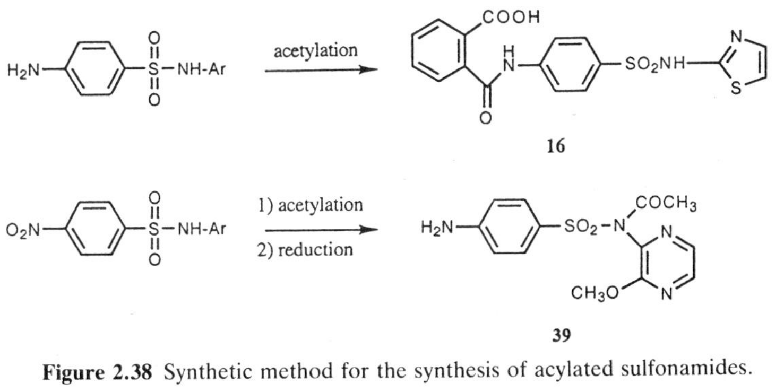

Figure 2.38 Synthetic method for the synthesis of acylated sulfonamides.

is used to acetylate directly the anilino (C-4) position (e.g. phthalylsulfa-thiazole **16** (Figure 2.38)) while in the latter case, a *p*-nitrosulfonamide is reacted with an appropriate acetylating agent and subsequently reduced to the final product (e.g. **39** (Figure 2.38)).

2.4.2 Synthesis of pyrimidine antifolate antibacterial agents

As with the sulfonamide agents, the synthesis of the arylpyrimidine antifolate antibacterials has been successfully developed. The central core of these drugs is a C-5-benzyl-substituted pyrimidine; the synthesis of these substrates have been well documented in the chemistry of the pyrimidine class of heterocycles. There are a number of synthetic approaches to the pyrimidines, but only a few will be considered in the context of producing the antibacterial arylpyrimidine compounds.

An elementary structural dissection of the arylpyrimidine antifolates intuitively focuses upon the construction of the pyrimidine ring in the latter stages of a synthetic approach, or the joining of the carbocyclic benzene nucleus to that of a pyrimidine. Obviously, the formation of an inherently stable benzene ring is an unwise undertaking since either vigorous conditions are needed for this task, and/or the placement of alkyloxy or alkyl substituents would probably be complicated by issues of reactivity and selectivity. For these reasons, the arylpyrimidine antifolates are elaborated from simple benzoic acid derivatives. The synthesis of the prototypic pyrimidine antifolate trimethoprim **29** (Figure 2.18) and ormetoprim, is representative. In the case of trimethoprim, trimethoxy-benzaldehyde is condensed with a propionitrile such as β-ethoxypro-pionitrile to afford the corresponding α,β-unsaturated nitrile. Subsequent reaction with guanidine results in addition to the nitrile with loss of ethanol and aromatization to produce the desired pyrimidine (Figure 2.39). This methodology is a general route to the pyrimidine heterocycles, and other

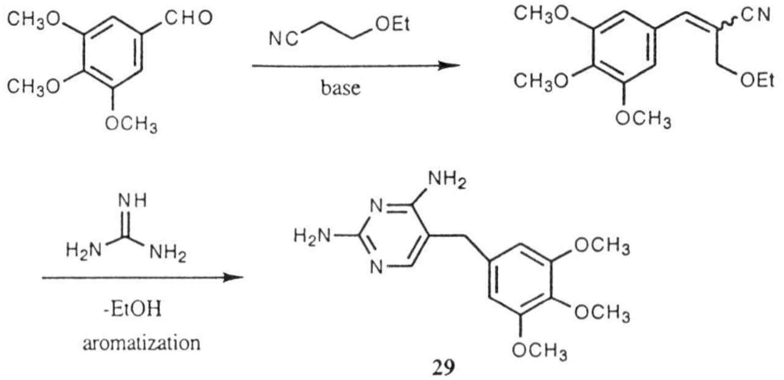

Figure 2.39 Synthesis of trimethoprim with construction of the pyrimidine ring late in the scheme.

aldehydes can be utilized to access mono- and disubstituted arylpyrimidine antibacterials.

In the case of ormetoprim **40** (Figure 2.40), it is possible to join the aryl substituent to the pyrimidine nucleus by the displacement of halide from the corresponding *N*-acetylated bromomethylated pyrimidine. A drawback of this approach is that the amine functionality must be masked and subsequently resurrected.

2.4.3 Synthesis of sulfone antibacterial agents

The bis-aryl sulfone antibacterials are also simple in structure and in synthesis. A general synthetic route is based upon the coupling of *p*-chloronitrobenzene with *p*-acetamidobenzenesulfinic acid, followed by reduction of the nitro moiety to the corresponding amine (Figure 2.41). Dapsone **31** (Figure 2.19) is readily prepared via subsequent

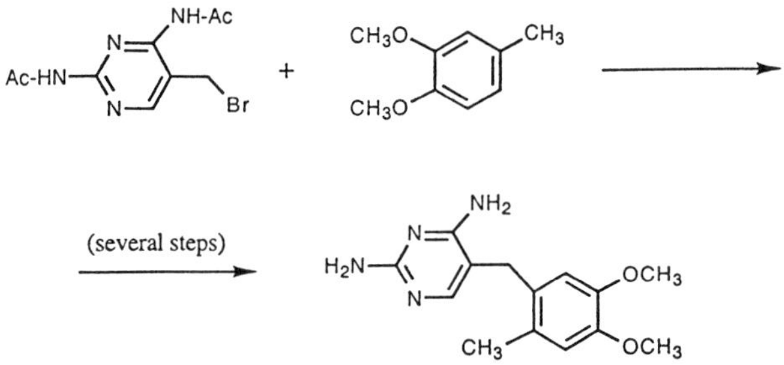

Figure 2.40 Synthesis of ormetoprim by joining a pyrimidine to an aryl substrate.

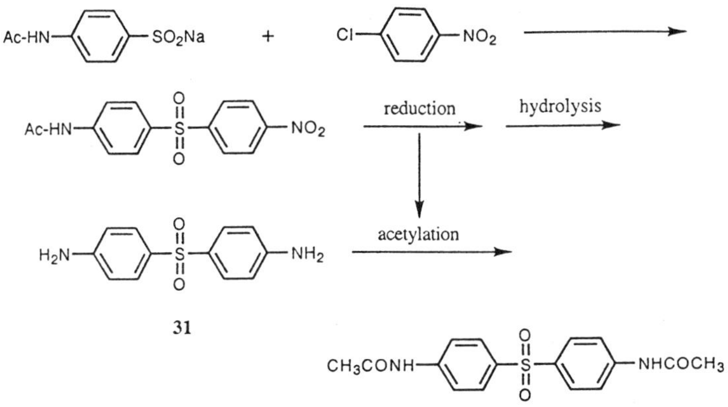

Figure 2.41 Synthesis of bis-aryl sulfones dapsone and acedapsone.

deacylation although water-soluble prodrugs such as the acetylated derivative acedapsone **32** (Figure 2.19) result from further modification (which can include acetylation as shown, or bisulfite adduct formation).

2.5 Bacterial resistance to the sulfa antibacterials and arylpyrimidine antifolates

Bacterial resistance to the sulfonamide, sulfone and arylpyrimidine antibacterial agents is prevalent and continues to restrict the use of many of these agents in current infectious disease chemotherapy. This is particularly evident with the sulfonamides which are, today, often relegated to serve as limited alternatives since more potent antibacterial agents are available that are not hampered by such widespread resistance. While the coadministration of a sulfonamide along with a pyrimidine antifolate remains an effective and widely used regimen, the emergence of trimethoprim-resistant bacteria threatens the continued success of this combination chemotherapy.

A number of mechanisms that are responsible for resistance to the sulfonamides have been elucidated and investigated. Unfortunately many details remain surprisingly unresolved although most of these agents have been in use for many decades. This is due to the difficulty in studying the complex folic acid metabolic pathways present in bacteria, and the formidable challenges that hinder enzyme isolation and purification, as well as the chemical synthesis of viable enzymatic probes. However it is certain that bacteria are capable of conjuring a barrage of defenses that successfully evade the antibacterial effects of the sulfonamides. In fact, there are more mechanisms by which bacterial pathogens express resistance to the sulfonamides and/or trimethoprim and related pyrimidine antifolates, than is observed for most other major classes of antibacterial chemotherapeutic agents.

The inhibition of the intracellular enzyme dihydropteroate synthase is paramount for the antibacterial effects of the sulfonamides and so any bacterial cell that can avoid uptake or accumulation of a sulfonamide agent can display resistance. In some instances, bacterial impermeability is intrinsic, such as in the case of the Gram negative pathogen *Pseudomonas aeruginosa*. Other bacterial species can acquire permeability characteristics that confer sulfonamide resistance. Today, many clinically important *Enterobacteriaceae* are sulfonamide-resistant due to changes in permeability that are encoded for and carried on plasmids. Experimentally, *E. coli* mutant strains have been selected for resistance against virtually any sulfonamide (sulfadiazine, sulfacetamide, sulfanilamide, etc.). Reduced permeability can also be chromosomal and may confer resistance to the sulfamethoxazole–trimethoprim (sulfonamide–arylpyrimidine) combination. Unfortunately, in many cases alteration of bacterial cell permeability may only be a contributing factor to overall antifolate resistance; other mechanisms that produce resistance may be simultaneously in operation.

Even if a sulfonamide is successful in penetrating into the bacterial cell, a number of other mechanisms can render the agent impotent. The sulfonamide must be able to compete effectively with the natural substrate PABA for recognition by dihydropteroate synthase enzyme. In the case of susceptible bacterial organisms, the sulfonamide is able to accomplish this task if sufficient intracellular concentrations of the drug are achieved which overwhelm binding of PABA to enzyme. However, abnormally high levels of PABA can override the antibacterial effects even though the sulfonamide (usually) possesses an increased affinity for the enzyme relative to the natural substrate. Some strains of *Staphylococci* and *Neiserria* species display sulfonamide resistance through the hyperproduction of PABA. The level of PABA production of these bacteria can be greater than one order of magnitude than that of susceptible strains; PABA concentrations of these mutants accumulate in exceedingly high concentrations in the absence of sulfonamides.

Another mechanism of sulfonamide resistance occurs in instances where a bacterial species is able to utilize exogenous folate and thus does not need to rely upon the biosynthesis of this nutrient. A so-called folate auxotroph (see discussion on resistance in chapter 1) such as sulfonamide resistant *Streptococcus faecium*, is inert to the sulfonamides since no antibacterial target site is present. Sulfonamide resistance mediated via the phenomenon of auxotrophy can also be acquired by a number of other potentially pathogenic bacteria. Some species of *E. coli*, *Serratia*, *Proteus*, *Salmonella*, *Staphylococcus* and *Streptococcus* can mutate to thymine auxotrophy, a condition in which a thymidine rather than a folate requirement is operative. This is usually due to the loss of thymidylate synthase activity, an enzyme whose action is downstream from that of dihydropteroate synthase; this, in essence, by-passes the effects of sulfonamide-induced dihydropteroate synthase inhibition.

There are other mechanisms that cause sulfonamide resistance and, as alluded to earlier, several are mediated through the transfer of resistance genes from plasmids to the bacterium. Plasmid-mediated resistance was first observed in sulfonamide-resistant strains decades ago although this phenomenon is prevalent today, and hinders many classes of antibacterial agents. Plasmid-mediated sulfonamide resistance can lead to the production of an altered dihydropteroate synthase enzyme that possesses diminished sensitivity to the sulfonamides. Clinically, some troublesome bacterial pathogens such as *Enterobacteriaceae* (e.g. *E. coli*), *Neiserria*, *Salmonella* and some pneumococci display resistance in this way. In addition, in some cases, cross-resistance to other classes of antibacterials occurs. For example, *Salmonella typhi* that is resistant to sulfonamides can also be inert to chloramphenicol and the tetracyclines, and in some cases, additional resistance to ampicillin and kanamycin was observed. Alternatively, plasmids can pass on mechanisms that lead to the production of mutant enzyme in addition to the wild-type dihydropteroate synthase. Under laboratory conditions, sulfonamide insensitive dihydropteroate synthase can arise as spontaneous temperature sensitive mutations selected for resistance against a number of sulfonamide agents such as sulfathiazole, sulfabenzamide, sulfadiazine, sulfacetamide, sulfamethoxazole and sulfanil-amide (*E. coli*).

Sulfonamide resistance is very common and as a result these agents are now severely limited in use compared to just a couple of decades ago. However, sulfonamide combination therapy, the use of sulfonamides for the treatment of certain urinary tract infections, ophthalmic indications and prophylactic indications, along with their value as therapeutic agents in the poorer regions of the world will probably continue to be significant in infectious disease chemotherapy in the near future.

Interestingly, resistance to the structurally related bis-aryl sulfone antibacterials is relatively rare in comparison to the sulfonamides. The sulfone antibacterials have been primarily relegated to the eradication of mycobacteria which are inherently slow growing microorganisms, and this may be an underlying factor. For the most part, sulfone resistance appears to be related to improper dosage and use, with the important exception of multi-drug resistant *Mycobacterium tuberculosis*.

As is the case with sulfonamide resistance, bacterial resistance to the arylpyrimidine antifolates can occur in several ways. In a general sense, these mechanisms parallel those operative in sulfonamide-resistant strains; cellular impermeability, overproduction of the target enzyme, modification of the target enzyme and auxotrophy are capable of conferring resistance to the arylpyrimidine antifolates; plasmid-mediated resistance is a common occurrence.

In an analogous fashion to sulfonamide chemotherapy, the troublesome Gram negative pathogen, *Pseudomonas aeruginosa*, is inherently resistant to the bactericidal effects of the pyrimidine antifolates. This is due to the inability of the agent to penetrate the sturdy cell wall and gain access to the dihydrofolate reductase protein. Impermeability can be acquired in some species such as *Klebsiella pneumoniae* and *Serratia marcescens* due to alterations of outer membrane proteins. This mechanism of antifolate resistance can also confer multiple resistance to other unrelated classes of antibacterial agents.

Acquired resistance to the arylpyrimidine antifolates is achieved in some mutant bacterial species that lack or have diminished thymidylate synthetase activity and therefore depend upon an exogenous source of thymine or thymidine. Under these conditions, the effects of dihydrofolate reductase inhibition are by-passed leading to resistance. There are varying types of this kind of resistance since some species require only low levels of thymine to acquire resistance whereas other species are absolute in that thymidylate synthetase activity is completely absent. In addition some species contain temperature-sensitive thymidylate synthetase.

Another mechanism of acquired antifolate resistance involves modifications of the dihydrofolate reductase enzyme by mutation of the DHFR (dihydrofolate reductase) gene. The structural elucidation of DHFR genes and the cellular effects mediated upon its expression continue to occupy an important role in both antibacterial and cancer chemotherapy. In the case of *E. coli* and some other bacteria, the DHFR has been sequenced. Resistance is caused by 'promoter' mutations that induce the over-production of dihydrofolate reductase. In some instances, dihydrofolate reductase activity can be increased by several orders of magnitude. Since more enzyme is available, folic acid metabolism remains functional and no antibacterial effects are produced. A variation arises when chromosomal mutation(s) afford bacterial dihydrofolate reductase that displays reduced susceptibility to antifolate antibacterial agents. Mutated enzymes of this type possess a decreased affinity for trimethoprim (or other related agents); DHFR enzyme is not hyperproduced, nor is the protein completely resistant to binding of antifolate.

Plasmids are responsible for the majority of antifolate resistance today. Plasmid-mediated resistance to trimethoprim and other antifolates occurs in *Pseudomonas, Staphylococci, E. coli, Klebsiella, Proteus, Shigella, Salmonella* and *Serratia*. A number of plasmid-mediated trimethoprim-resistant dihydrofolate reductases have been structurally characterized and many of these appear to be related to the wild-type enzyme(s) in terms of amino acid homology and function. As a result, there is speculation that some plasmids originated from a common ancestral chromosomal gene. There are at least eleven distinctive plasmid-borne DHFRs, often disseminated on highly mobile transposons; most affect either *Enterobacteriaceae, Pseudomonas* or *Staphylococci*. The plasmid-encoded dihydrofolate reductase enzymes are often categorized by types and subtypes, but can differ greatly in structure. For example, type I DHFRs exist in monomeric (DHFR type Ib) or dimeric (DHFR type Ia) forms whereas type II DHFRs are tetrameric in nature. DHFR type Ia, caused by a 13.6 kilobase transposon (Tn7), consists of two identical 17.6 kDa subunits that confer cross-resistance to streptomycin. DHFR type Ib originates from a 3 kilobase transposon (Tn 4132) and exists as a 24.5 kDa monomeric protein. Antifolate resistance mediated by these and related transposons are common, particularly the plasmid-borne Tn7 transposon. Type II DHFRs are less common than the type I variants but these proteins (types IIa, IIb and IIc), which are similar in structure (tetrameric) confer resistance to trimethoprim that approaches six orders of magnitude in terms of inhibitory concentrations compared to the wild-type protein(s). In contrast, DHFR type III, a monomeric 16.9 kDa protein found in a *Salmonella* species, is only marginally resistant to trimethoprim. The largest of the plasmid-mediated DHFRs is that of type IV which exists as a 46.7 kDa protein that can be induced upon exposure to trimethoprim and can achieve levels of two orders of magnitude greater than that of the chromosomal-produced enzyme.

Plasmid-mediated dihydrofolate reductase resistance mediated by types I and II is prevalent and particularly troublesome since frequently cross-resistance to other (classes of) antibacterial agents such as streptomycin, spectinomycin, ampicillin, carbenicillin, chloramphenicol, the tetracyclines and the sulfonamides, can result. However, as mentioned earlier, several mechanisms of resistance to the arylpyrimidine antifolates can be operative in the same species; this contributes to the increasing concern surrounding the future of trimethoprim and its congeners in antibacterial chemo-therapy.

One approach to combat resistance to the sulfonamides and the arylpyrimidine antifolates originated from the discovery that these agents, when administered together in combination, could be synergistic in nature. This led to the development of co-trimoxazole, a mixture of trimethoprim and sulfamethoxazole in a 1:5 ratio. The choice of these two particular agents was based upon the individual spectrums of antibacterial activity of each component as well as pharmacokinetic properties, which were favorable in terms of blood levels, elimination halflives and tolerability. Of course, the success of each individual agent was also a crucial factor in the development of co-trimoxazole; both trimethoprim and sulfamethoxazole have occupied a dominant role in the antifolate and sulfonamide markets, respectively. The combination of these agents is also beneficial since bactericidal effects can be produced, whereas sulfamethoxazole, as well as other sulfonamides are bacteriostatic in nature when administered as a sole agent. Trimethoprim, a 'slow' bactericidal by itself against many bacterial species, is also synergized via combination therapy with sulfamethoxazole. In addition to co-trimoxazole, other combinations of a sulfonamide dihydropteroate synthase inhibitor and an arylpyrimidine dihydrofolate reductase inhibitor were also advanced.

However, the main impetus for such combination therapy was actually postulated prior to the development of sulfonamide–antifolate regimens, and focused upon the emergence of resistance to each respective agent. Combination therapy offered the advantage of suppressing resistance to any one particular agent. Despite the overall success of co-trimoxazole, resistance to both components has been documented. Furthermore, in some cases, resistance to either agent greatly diminishes the efficacy of the remaining active substance. There have been some reasons suggested as to why sulfonamide–antifolate combination chemotherapy may be inherently less efficacious than once envisioned. Intuitively, an antifolate such as trimethoprim, in blocking the action of bacterial dihydrofolate reductase, would be expected to increase intracellular concentrations of dihydrofolate. In doing so, the concentration of dihydropteridine is possibly increased and/or dihydropteroate synthase enzyme activity is 'down-regulated'; both events result in decreased susceptibility to sulfonamide. Some advocates point out that trimethoprim use without supporting sulfonamide therapy is preferable for this reason. Trimethoprim is a marketed antibacterial agent in its own right.

2.6   New agents

Recently, there have been only scant efforts to develop novel antibacterial sulfonamide or sulfone agents. The reasons for this trend have already been outlined: the sulfonamides are besieged by bacterial resistance and the sulfone agents remain limited to specific indications. These compounds are seldom regarded as a first-line therapy (the exception is the sulfone treatment/prevention of leprosy).

There has been a considerable research effort into the design and development of novel antibacterial antifolates although mammalian dihydrofolate reductase inhibition remains the most pressing issue (cancer research). A particularly interesting dihydrofolate reductase inhibitor has been described. Epiroprim **41** (Figure 2.42) has been selected for preclinical investigation due to exquisite potency against opportunistic (protozoal and fungal) pathogens such as *Toxoplasma gondii*, *Pneumocystis carinii* as well as *Mycobacterium avium*. Fatal infections of these organisms are frequent among immunocompromised subjects such as AIDS active patients.

Lastly, a new congener brodimoprim **42** (Figure 2.43), has recently been launched although it is not widely marketed at this time. Brodimoprim is a potent inhibitor of bacterial dihydrofolate reductase and displays a fairly broad spectrum of activity; in particular, it is efficacious against amoxicillin-resistant pathogens. Brodimoprim will be primarily used for upper respiratory tract infections.

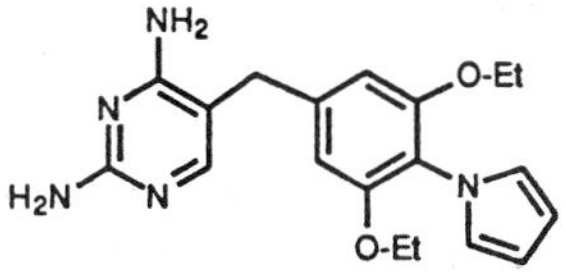

41

Figure 2.42 Epiroprim.

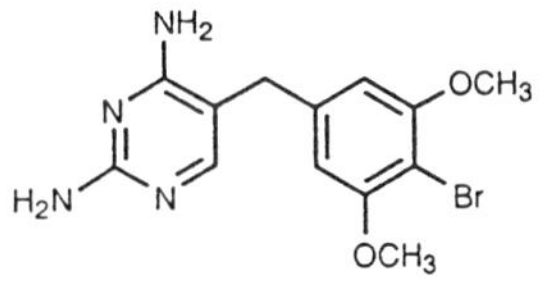

42

Figure 2.43 Brodimoprim.

2.7 Summary

The sulfonamide and sulfone antibacterials as well as the arylpyrimidine antifolates continue to be successful chemotherapeutic agents. However these agents are restricted in their use due to resistance.

1. The sulfonamide agents inhibit the action of bacterial dihydropteroate synthase, an enzyme crucial for the biosynthesis of folic acid. The sulfonamides are well tolerated in the mammalian host since the analogous enzymatic transformation is absent.

2. The sulfonamide antibacterials are primarily used for the treatment of uncomplicated urinary tract infections caused by *E. coli*, *Klebsiella*, *Enterobacter* and *Proteus*, and only seldom for middle ear infections caused by *Haemophilus influenzae*. The sulfonamides are also used for the prophylaxis of recurrent rheumatic fever associated with certain (*β*-hemolytic) streptococcal infections, as well as for prophylaxis for *Neiserria meningitidis*. Among the dozens of agents that have been marketed, the most commonly used sulfonamides are sulfamethoxazole, sulfadiazine and sulfaisoxazole. Some sulfonamides are administered for ophthalmic and topical applications.

3. Sulfonamides can be coadministered with arylpyrimidine antifolate antibacterials, usually trimethoprim, which leads to a powerful synergism that has proven to be efficacious in a number of indications. A consequence of this chemotherapy, along with the issues previously raised, is that sulfonamides are often not used as a single agent.

4. The sulfonamides are totally synthetic substances that are produced by relatively simple chemical synthesis. In general, the cost per dose and per therapy of these agents is advantageous compared to other classes of antibacterial agents. In impoverished regions of the world, sulfonamides continue to offer an economically acceptable and practical means for the treatment and prophylaxis of some infection.

5. The sulfone antibacterials have found an important niche in the antibacterial therapy of specific diseases such as leprosy and

tuberculosis. For these indications, these agents remain as a premier treatment.

6. The arylpyrimidine antifolate antibacterials exert their effects via the inhibition of dihydrofolate reductase. Since the mammalian counterpart of this enzyme is needed for normal eukaryotic (mammalian) cellular metabolism, a highly selective action is dictated in order for an agent to be successful as an antibacterial substance.

7. Trimethoprim is a successful pyrimidine antifolate due to its selective mode of action. Trimethoprim is five to six orders of magnitude more active against many pathogenic bacterial dihydrofolate reductases compared to mammalian counterparts. For this reason, as well as its tolerability and pharmacokinetic properties,

Table 2.1 Major antibacterial chemotherapeutic agents

Generic name	Trade names	Administration*
Sulfonamide		
Sulfamethoxazole	Gantanol	PO
(with phenazopyridine HCl)	Azo-Gantanol	PO
Sulfisoxazole	Gantrisin	PO, IV
(with phenazopyridine HCl)	Azo-Gantrisin	PO
Sulfadiazine	Debenal	PO, IV
(with sulfamerazine, sulfmethazine)	Neotrizine, Sulfonsol, Terfonyl	PO
Sulfamethizole	Thiosulfil	PO
(with phenazopyridine HCl)	Thiosulfil A, Thiosulfil A Forte	PO
Sulfathiazole	Cibazol	PO
Sulfanilamide	Sulfanilamide	PO
Sulfapyridine	Dagenan	PO
Sulfaphenazole	Orisul	PO
Sulfamoxole	Sulfuno	PO
Sulfamerazine	Debenal-M	PO
Sulfamethyldiazine	Pallidin	PO
Sulfadimethoxine	Madribon	PO
Sulfalene	Kelfizina	PO
Sulfone		
Dapsone	Avlosulfon	PO
Antibacterial pyrimidine antifolate		
Trimethoprim	Trimpex	PO
Tetroxoprim		PO
Brodimoprim	Hyprim, Unitrim	PO
Sulfa–pyrimidine antifolate combinations		
Trimethoprim–sulfamethoxazole (co-trimoxazole)	Bactrim, Septra	PO
Sulfadimethoxine–pyrimethamine[†]	Fansidar	PO
Sulfalene–pyrimethamine[†]	Metakelfin	PO
Dapsone–pyrimethamine[†]	Maloprim	PO

* – PO, peroral; IV, intravenous. [†] – Primarily used as an antiprotozoal.

    trimethoprim remains the premier antibacterial antifolate, although other analogs are similarly potent.

8. Trimethoprim is an effective agent against uncomplicated urinary tract infections caused by susceptible *E. coli*, *Proteus*, *Klebsiella*, *Enterobacter* and some *Staphylococci* species. However, trimethoprim is usually more successfully utilized in combination with an appropriate sulfonamide antibacterial agent.

9. Co-trimoxazole is an extremely successful combination of the trimethoprim and the sulfonamide sulfamethoxazole. This agent remains a premier chemotherapy for some urinary tract infections, ear infections and enteric manifestations. Other pyrimidine antifolate–sulfonamide combinations have been demonstrated to be efficacious but co-trimethoxazole remains dominant in this context.

10. The combination therapy operative in co-trimoxazole is intuitively and inherently less susceptible to bacterial resistance since two agents are operative via distinct and different modes of action. However, the synergistic effects of co-trimoxazole can be compromised by resistance to either single component and furthermore, some bacterial species are capable of displaying resistance to both agents. For these reasons, the future of either therapy is in jeopardy.

Table 2.1 shows major sulfa antibacterials and arylpyrimidine antifolates.

Further reading

N. Anand (1975) 'Sulfonamides and sulfones' in *Antibiotics III (Mechanism of Action of Antimicrobial and Antitumor Agents)*, J.W. Concoran and F.E. Hahn (Eds), Springer-Verlag, New York, pp. 668–698.

J.J. Burchall (1975) 'Trimethoprim and pyrimethamine' in *Antibiotics III (Mechanism of Action of Antimicrobial and Antitumor Agents)*, J.W. Concoran and F.E. Hahn (Eds), Springer-Verlag, New York, pp. 304–320.

F.M. Sirotnak, J.J. Burchall, W.B. Erisminger and J.A. Montgomery (Eds) (1984) *Folate Antagonists as Therapeutic Agents; Vol. 1 Biochemistry, Molecular Actions and Synthetic Design*, Academic Press, Orlando, FL.

G.H. Hitching (Ed.) (1983) 'Inhibition of folate metabolism in chemotherapy, the origins and uses of co-trimoxazole', *Handbook of Experimental Pharmacology*, Vol. 64, Springer-Verlag, Germany.

E.H. Northey (1948) *The Sulfonamides and Allied Compounds*, American Chemical Society Monograph Series, Reinhold, New York.

J.K. Seydel (1968) 'Sulfonamides, Structure–Activity Relationship, and Mode of Action', *J. Pharm. Sci.*, **57**, 1455.

M. Kansy, J.K. Seydel, M. Wiese and R. Haller (1992) 'Synthesis of new 2,4-diamino-5-benzylpyrimidines active against bacterial species', *Eur. J. Med. Chem.*, **27**, 237.

J.H. Chan and B. Roth (1991) '2,4-Diamino-5-benzylpyrimidines as antibacterial agents. 14. 2,3-Dihydro-1-(2,4-diamino-5-pyrimidyl)-1H-indenes as conformationally restricted analogues of trimethoprim', *J. Med. Chem.*, **34**, 550.

J.J. Burchall, L.P. Elwell and M.E. Fling (1982) 'Molecular Mechanisms of Resistance to Trimethoprim', *Rev. Infectious Diseases*, **4**, 246.

P. Huovinen (1989) 'Trimethoprim Resistance', *Antimicrob. Agents Chemother.*, **31**, 1451.

M.L. Pato and G.M. Brown (1963) 'Mechanisms of resistance of *Escherichia coli* to sulfonamides', *Arch. Biochem. Biophys.*, **103**, 443.

P.J. Ortiz (1970) 'Dihydrofolate and dihydropteroate synthesis by partially purified enzymes from wild-type and sulfonamide-resistant pneumococcus', *Biochemistry*, **9**, 355.

P.J. White and D.D. Woods (1965) 'The synthesis of *p*-aminobenzoic acid and folic acid by staphylococci sensitive and resistant to sulphonamides', *J. Gen. Microbiol.*, **40**, 255.

B. Wolf and R.D. Hotchkiss (1963) 'Genetically modified folic acid synthesizing enzymes of pneumococcus', *Biochemistry*, **2**, 145–150.

E.M. Wise, Jr. and M.M. Abou-Donia (1975) 'Sulfonamide resistance mechanism in *Escherichia coli*: R plasmids can determine sulfonamide-resistant dihydropteroate synthases', *Proc. Natl. Acad. Sci. USA*, **72**, 2621.

G. Swedberg, S. Castensson and O. Skold (1979) 'Characterization of mutationally altered dihydropteroate synthase and its ability to form a sulfonamide-containing dihydrofolate analog', *J. Bacteriol.*, **137**, 129.

O. Skold (1976) 'R-factor-mediated resistance to sulfonamides by a plasmid-borne, drug-resistant dihydropteroate synthase', *Antimicrobial Agents Chemother.*, **9**, 49.

L. Bock, G.H. Miller, K.-J. Schaper and J.K. Seydel (1974) 'Sulfonamide structure–activity relationships in a cell-free system. 2. Proof for the formation of a sulfonamide-containing folate analog', *J. Med. Chem.*, **17**, 23.

G.E. Dale, R.L. Then and D. Stuber (1993) Characterization of the Gene for Chromosomal Trimethoprom-Sensitive Dihydrofolate Reductase of *Staphylococcus aureus* ATCC, 25923, *Antimicrobial Agents Chemother.*, **37**, 1400–1405.

K.H. Mayer, M.E. Fling, J.D. Hopkins and T.F. O'Brien (1985) 'Trimethoprim resistance in multiple genera of enterobacteriaceae at a US hospital: spread of type II dihydrofolate reductase gene by a single plasmid', *J. Infectious Diseases*, **151**, 783.

L. Sundstrom, T. Vinayagamoorthy and O. Skold (1987) 'Novel type of plasmid-borne resistance to trimethoprim', *Antimicrobial Agents Chemother.*, **31**, 60.

R. Steen and O. Skold (1985) 'Plasmid-borne or chromosomally mediated resistance by Tn7 is the most common response to ubiquitous use of trimethoprim', *Antimicrobial Agents Chemother.*, **27**, 933–937.

3 *ß*-Lactam antibiotics

3.1 History and overview

Serendipity is common to many landmark discoveries that have shaped antibacterial chemotherapy; however the old adage that 'luck favors the prepared mind' has been an integral part of this progress. The development of the *ß*-lactam antibiotics is a testimony in this regard. The true discovery of the *ß*-lactam antibiotics occurred in 1928 when Fleming observed lysis and inhibition of growth of a bacterial culture due to the presence of a contaminating mold. While this finding was indeed fortuitous, it was the keen insight of Florey, Chain and other researchers who subsequently investigated this phenomena, that led to one of the most successful endeavors in all of infectious disease chemotherapy.

The antibacterial effects that Fleming observed were due to a product of a Penicillium mold. Thus this material was named penicillin. Unfortunately isolation of this antibacterial substance was incredibly troublesome and early attempts at purification were unsuccessful. Perhaps because of these undesirable properties, Fleming virtually abandoned his efforts to isolate pure penicillin but his work with crude extracts that contained penicillin did provide some valuable insight into the remarkable properties of this unidentified substance. Crude penicillium culture broths were shown to exhibit extraordinary antibacterial activity against several common and pathogenic Gram positive organisms such as *Staphylococcus aureus* and *Streptococcus pyogenes*. Fleming noted that penicillin selectively exerted its lethal effects on bacteria rather than mammalian cells (leukocytes) and that the substance was not toxic to animals. Since penicillin was non-irritating, Fleming eventually suggested its use as a wound dressing.

It was not until a decade after Fleming's finding that a fruitful era of progress on penicillin began, led by Florey, Chain, Heatley and others. In 1940, the first use of penicillin to treat infection in animals was carried out and shortly thereafter, the first clinical use of penicillin in humans was documented. The early cases are worth reflecting upon and give an appreciation of the progress of antibacterial chemotherapy over the past half-century. The first person to be treated with penicillin was a woman suffering from blood poisoning following a miscarriage. Her prognosis was poor: her fever was 106°F and she was not expected to survive much longer. As a last resort, doctors administered half of the world's supply (at that time) of penicillin (about 5 g) by intravenous injection. Within hours her fever had subsided and she made a rapid recovery to good health. Another case involved a policeman slightly injured during a gardening accident but severely ill with a staphylococcal infection. When treated with penicillin, this patient likewise improved remarkably until the entire supply of drug was consumed. He relapsed after treatment was discontinued and eventually died. A case that soon followed was that of a boy suffering from a streptococcal infection who received some new drug as well as penicillin recovered from the urine of the policeman. The boy dramatically improved while under treatment with penicillin. These accounts are particularly fascinating in that the penicillin administered was likely of poor purity, and yet treatment proved to be, in essence, life saving.

More successful cases followed but therapy would often require tens of grams of material completely to eradicate the infection. A collaborative agreement between the American and the British governments was formed and in a short period of time, the fermentation processes used to produce penicillin were greatly improved. During the campaign to harvest large quantities of penicillin, a strange peculiarity was observed. The American penicillin released phenylacetic acid upon acidic hydrolysis but the British stock did not; the different fermentation protocols had produced two distinct compounds indicating that a family of penicillins existed. The American version was designated as penicillin G (by the Americans; penicillin II by the British). Penicillin G, a benzylpenicillin, demonstrated exciting potency and would eventually become the prototypic penicillin whereas the British compound (F or I), a pentenylpenicillin, was soon forgotten. It would be years before the structural differences between these penicillins and other congeners were fully appreciated in terms of antibacterial activity. However, the most striking structural feature of the penicillins was the presence of a strained and highly reactive β-lactam ring, which was not documented until 1945. In fact, some researchers had speculated that this heterocyclic ring would be too unstable either to exist or to allow for isolation.

By the early 1950s, dozens of penicillins were known. Each congener contained a unique acyl side chain attached to the β-lactam nucleus; these variations resulted from the choice of fermentation conditions and supplements. However these congeners were somewhat overshadowed by the initial success of penicillin G **1** (benzylpenicillin) (Figure 3.1). Penicillin G is extremely potent against Gram positive bacteria, in particular many strains of staphylococci and streptococci. Even to this day, penicillin G remains an agent of choice for the eradication of a number of aerobic Gram positive cocci and bacilli as well as certain anaerobic species. Exquisite Gram positive potency is commonly associated with the β-lactam antibiotics in general although, as will be discussed later, advances of the past decades have afforded semi-synthetic β-lactam antibacterial agents that also possess superb activity against many Gram negative pathogens.

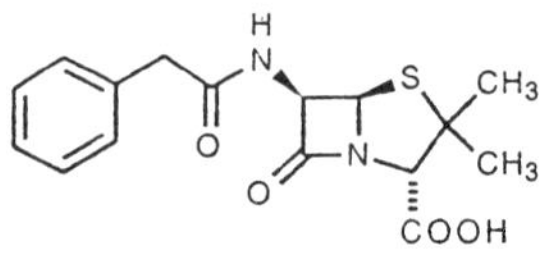

Figure 3.1 Penicillin G.

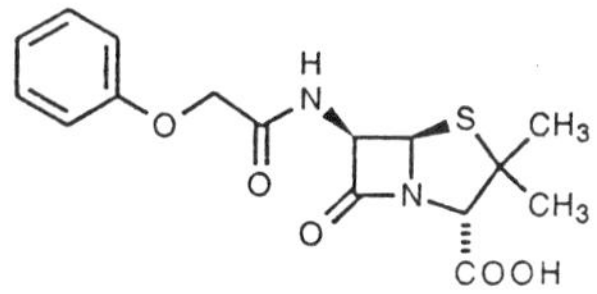

Figure 3.2 Penicillin V.

The early successors to penicillin G were obtained by adding various supplements to the fermentation broths (a practice also known as precursoring). This technology was used to produce penicillin V; phenoxyacetic acid (derived) supplements were added to the penicillin-producing fermentation broths. Penicillin V **2** (phenoxymethylpenicillin) (Figure 3.2) is less potent than penicillin G but displays improved acid stability. Penicillin V has remained a valuable agent since it usually offers the spectrum and potency of penicillin G by oral administration.

Unfortunately, it was soon discovered that precursoring had gross limitations in that only rather simple (essentially acetic acid derived) acyl side chains could be introduced. Frequently, supplements proved to be toxic to the microbial culture. This barrier was successfully tackled in 1956 when Sheehan described the synthesis of 6-aminopenicillanic acid, and demonstrated its utility by converting it to the naturally occurring penicillin V using a chemical acylation method. This work proved to be of extreme importance; penicillins containing novel acyl side chains that could not be introduced by fermentation were now accessible. Methods to produce large quantities of 6-aminopenicillanic acid (6-APA) **3** (Figure 3.3) were eventually developed. The availability of 6-APA led to the synthesis of many new penicillins; this era ushered in the first generation of semi-synthetic β-lactam antibiotics, so called since the β-lactam nucleus (penicillin) was obtained by fermentation but chemical synthesis (usually the introduction of the acyl side chain) was performed to afford the final products. To this day, 6-APA has served as precursor of hundreds of penicillins.

One of the early semi-synthetic agents was phenethicillin **4** (Figure 3.4) which was introduced in 1959 but did not offer much advantage over

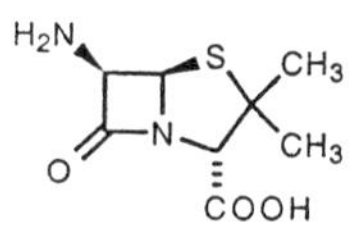

Figure 3.3 6-Aminopenicillanic acid.

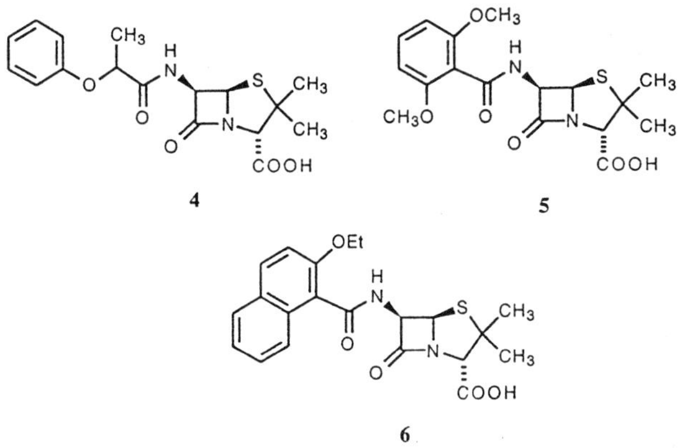

Figure 3.4 Phenethicillin, methicillin and naficillin.

penicillin V. However, a landmark agent was methicillin **5** (Figure 3.4) since it was inert to degradation by certain bacterial enzymes (i.e. penicillinases, sections 3.2 and 3.4). Even at this early stage in the development of the penicillins, bacterial resistance was recognized as a threat; semi-synthetic analogs offered the possibility of reducing susceptibility to enzymatic cleavage by altering the steric and electronic environment around the reactive β-lactam nucleus. Nafcillin **6** (Figure 3.4) is another important semi-synthetic penicillin that also displays a reluctance towards enzymatic degradation.

Oxacillin **7** (Figure 3.5), a successor to methicillin, became a prototype of a new class of penicillins. Oxacillin contains an isoxazolyl side chain that imparts stability to (bacterial) enzymatic degradation, particularly that operative in some staphylococcal pathogens. Although methicillin remains among the most widely used of the early semi-synthetic agents, oxacillin is better tolerated in patients with impaired renal function, and less likely to induce some toxicological side effects. Cloxacillin **8**, dicloxacillin **9** and floxacillin **10** (Figure 3.5) are other isoxazolyl penicillins that are effective in eradicating staphylococci, especially when oral administration is desired.

Aminopenicillins are another important class of penicillin antibiotics and are recognized by the presence of an α-aminoacyl side chain appended to the β-lactam nucleus. These agents possess an extended spectrum of antibacterial activity although their potency against some important Gram positive bacteria, such as staphylococci and streptococci, is often diminished compared to the early penicillins (penicillin G for example). However many of the aminopenicillins possess enhanced activity against certain troublesome pathogens such as *Haemophilus influenzae* and *Neiserria meningitidis*. Ampicillin **11** (Figure 3.6) contains a D-(−)-aminobenzyl side chain and has found widespread use, particularly in pediatric indications.

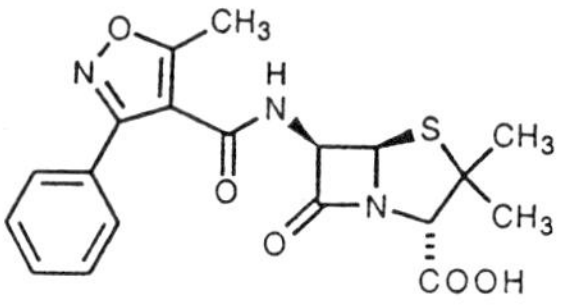

7

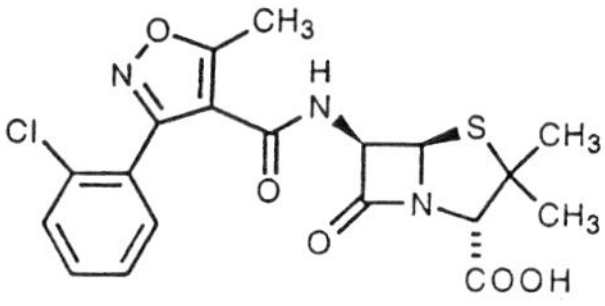

8

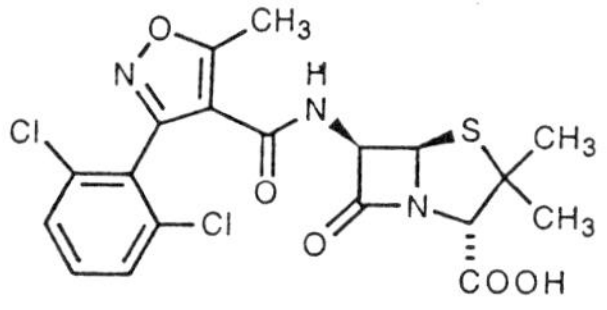

9

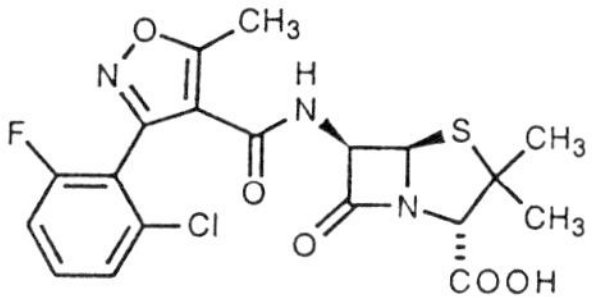

10

Figure 3.5 Oxacillin, cloxacillin, dicloxacillin and floxacillin.

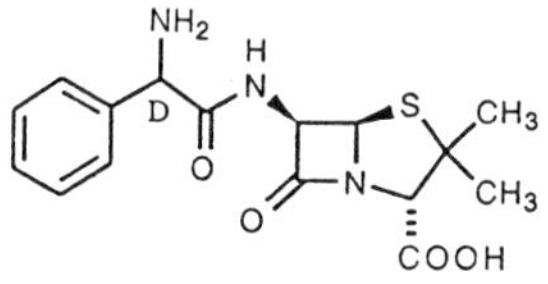

11

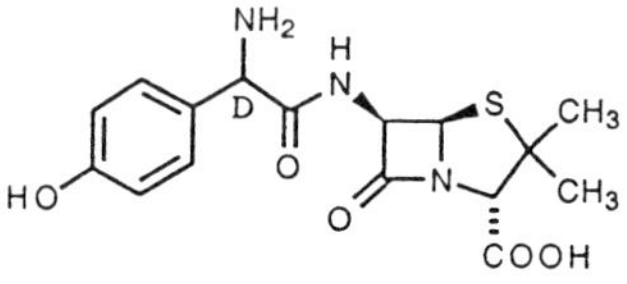

12

Figure 3.6 Ampicillin and amoxicillin.

Amoxicillin **12** (Figure 3.6) is similar in potency to ampicillin but undergoes more facile absorption following oral administration. Ampicillin and amoxicillin are so well recognized today that their names are likely to be household words when a family member is ill.

Research efforts from the 1960s to the present have been primarily focused upon expanding the antibacterial spectrum of the penicillins. The prototypes (e.g. penicillin G, penicillin V) as well as the isoxazolyl penicillins and the aminopenicillins, in general, possess superb potency against aerobic Gram positive bacteria but are often ineffective against many dangerous Gram negative pathogens. The carboxypenicillins were the first class of derivatives to display significant activity against Gram negative bacteria such as *Pseudomonas aeruginosa* and certain *Proteus* species while retaining notable, albeit decreased Gram positive potency. Carbenicillin **13** and ticarcillin **14** (Figure 3.7) are well recognized members of this class of penicillin antibiotics.

Despite these advances, the quest for antipseudomonal agents has continued unabated. The incorporation of certain heterocyclic systems, such as cyclic ureas and piperidines, onto an α-aminoacylated penicillin

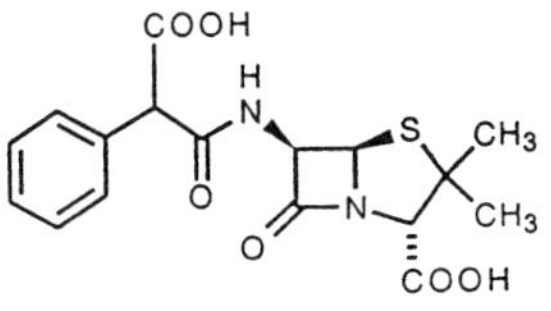

13

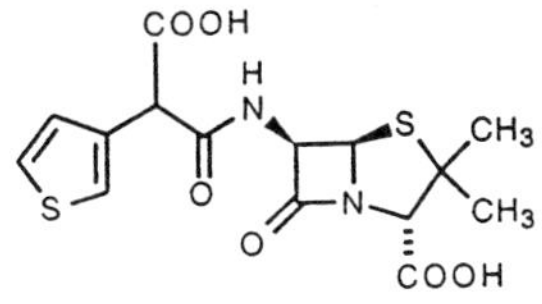

14

Figure 3.7 Carbenicillin and ticarcillin.

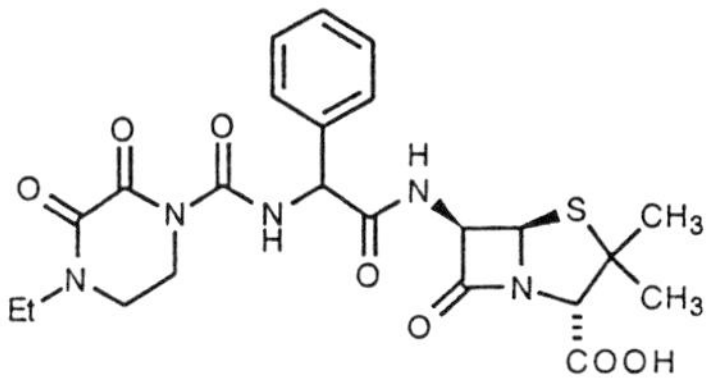

15

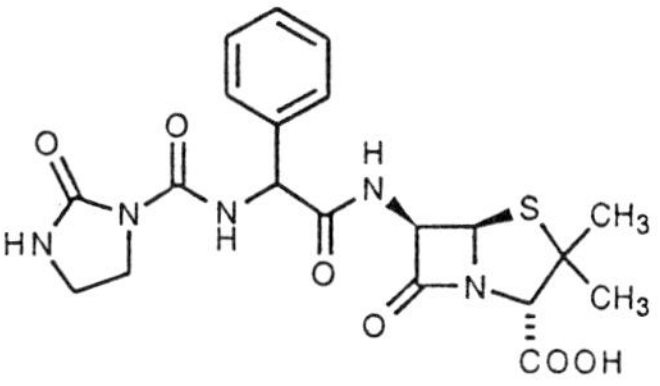

16

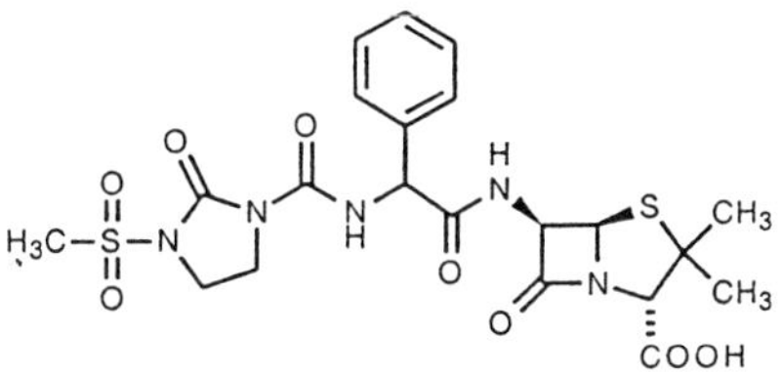

17

Figure 3.8 Piperacillin, azlocillin and mezlocillin.

nucleus, enhanced potency against pseudomonas. These agents contain a characteristic urea linkage in the side chain and are accordingly referred to as ureidopenicillins. Piperacillin **15** (Figure 3.8) is the most widely used of this class of penicillins although azlocillin **16** and mezlocillin **17** (Figure 3.8) are often suitable replacements.

Another significant advance in the development of the penicillin antibiotics are the carboxy-ester prodrugs. Carboxylic acid, carboxylate salts and carboxy-ester forms of penicillins have markedly different solubility properties, and in general the esters offer improved oral efficacy compared to the parent penicillin. Penicillin carboxy-esters are usually inactive substances but can be cleaved by endogenous mammalian enzymes (esterases) to liberate the corresponding free acid (or carboxylate). Most simple alkyl or aryl esters are unable to undergo ester cleavage at an appreciable rate and are therefore ineffective antibacterial agents. However 'doubly activated' penicillin esters such as alkyloxycarboxyalkyl derivatives are extremely labile *in vivo* and readily liberate free penicillin. Pivampicillin

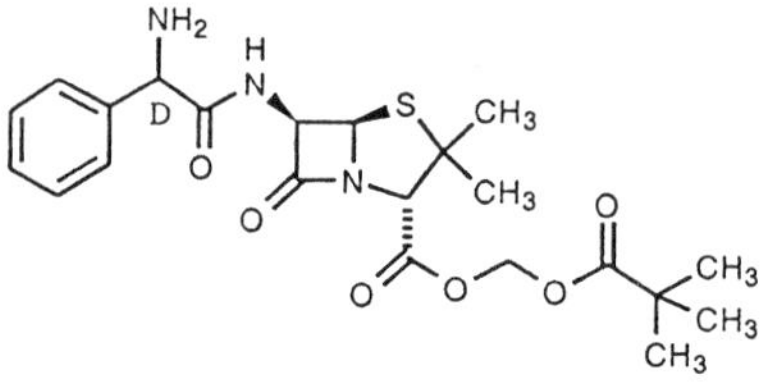

Figure 3.9 Pivampicillin and bacampicillin.

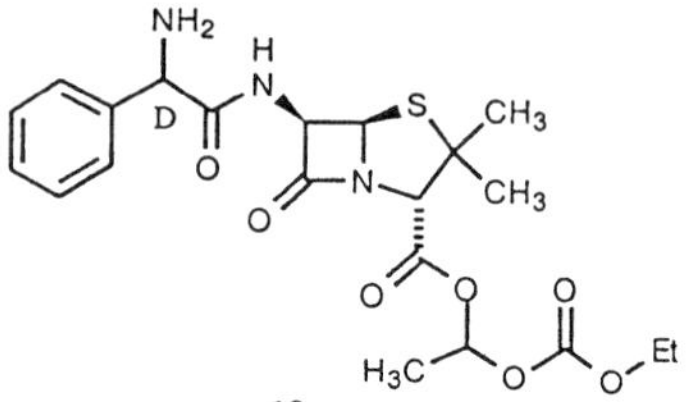

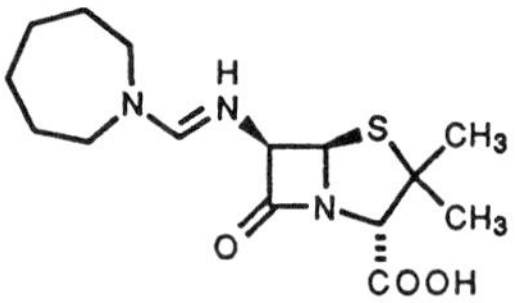

Figure 3.10 Mecillinam.

18 (Figure 3.9), the pivaloyloxymethyl ester of ampicillin and bacampicillin **19** (Figure 3.9), the ethoxycarbonyloxyethyl ester of ampicillin are the most frequently used of the penicillin ester prodrugs. Many ester derivatives of various penicillins have been prepared and studied; the most successful are derived from (and liberate) ampicillin.

There are some unique penicillins that do not formally belong to any of the classes of penicillins described above. Mecillinam **20** (Figure 3.10) is an amdinopenicillin that differs markedly in structure as well as in spectrum of activity compared to the more conventional agents. Mecillinam has a saturated azepine ring attached via an amidine linkage to a penicillin nucleus. Unlike most penicillins, this agent is only weakly active against most Gram positive bacteria. However, the valuable attributes of mecillinam are its potency against *Escherichia coli* and some troublesome pathogens that can be resistant to ampicillin (*Shigella, Salmonella* species). Unlike ampicillin, mecillinam is extremely unstable at low pH, and can only be administered orally as its pivaloyl ester.

Most of the research that culminated in the development of the varied semi-synthetic penicillins originated from attempts to broaden the spectrum of antibacterial activity and potency of the early agents. Some of these trends have carried over to the cephalosporins and other β-lactam antibacterials. While certain historical parallels and structural similarities exist, each subset of the β-lactam family is best addressed separately due to the size of each class and the impact that numerous agents have made on current antibacterial chemotherapy.

It was during the infancy of penicillin research that a related class of β-lactam antibiotics was discovered. In 1945, Brotzu began a search for microorganisms capable of producing novel antibacterial substances. He speculated that the naturally occurring purification processes of water might be due to the action of certain organisms that retard the growth of other microbes. Brotzu isolated a fungus (*Cephalosporium acremonium*) from sea water near a sewage outlet and demonstrated that this organism could produce antibacterial effects. Subsequent studies of fermentation broths of this fungus afforded a filtrate that displayed antibacterial properties in laboratory animals. There were several active components in this broth filtrate; one was subsequently identified as having a β-lactam ring. However, unlike penicillins, this compound, called cephalosporin C **21** (Figure 3.11), possessed a six-membered dihydrothiazine ring fused to the lactam.

These new β-lactam derivatives, the cephalosporins, possessed some desirable qualities that were deficient in penicillins. In general, the cephalosporin nucleus imparted a broader spectrum of antibacterial activity, enhanced stability in acidic media and displayed greater resistance to hydrolytic degradation by certain bacterial enzymes. In addition, cephalosporins were somewhat more accessible by fermentation and available for scrutiny under conditions not tolerated by penicillins. As a result, intense research on cephalosporins was initiated; the groundwork laid during the research on penicillins was a contributing factor to the relatively rapid pace at which new cephalosporins were successfully developed.

Like the penicillins, the cephalosporins can also be categorized in terms of structural modifications that follow a chronological order. The major focus of penicillin research, until recently, has been directed towards the investigation of the acylamino moiety appended to the β-lactam. As described earlier, these groups dictate many important characteristics of the penicillins, the most obvious being spectrum and potency of antibacterial activity. In a general sense, the development of the cephalosporins has followed a similar trend. However by virtue of their structure, cephalosporins offer more opportunity for modification since the dihydrothiazine ring can also be substituted with a host of hydrocarbon, halogen and heterocyclic groups. In many cases, the dihydrothiazine ring substituent

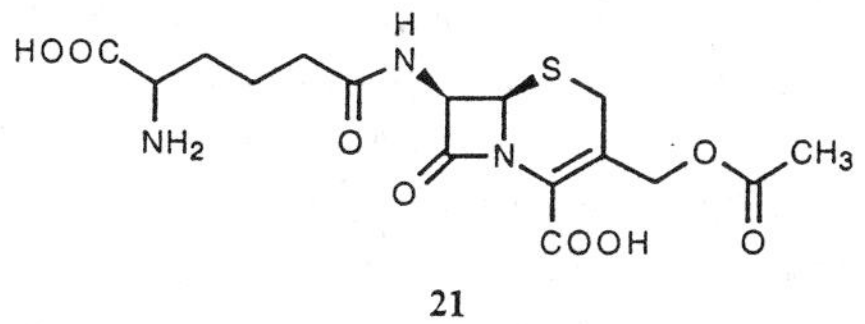

21

Figure 3.11 Cephalosporin C.

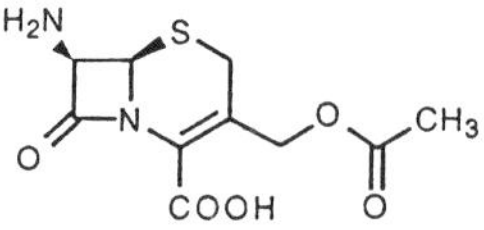

Figure 3.12 7-Aminocephalosporanic acid.

primarily alters the pharmacokinetic properties of the cephalosporin, often with minimal effects on antibacterial activity.

The manufacture of both semi-synthetic penicillins and cephalosporins has relied upon the development and availability of specific precursors that are obtained from fermentation products. In the case of penicillins, it was a practical method of producing 6-aminopenicillanic acid (6-APA) that revolutionized research in this area. In a parallel fashion, cleavage of the 7-acylamino side chain of cephalosporin C to afford 7-aminocephalosporanic acid (7-ACA) **22** (Figure 3.12) had a monumental impact. Thousands of semi-synthetic cephalosporins antibiotics have been prepared from 7-ACA.

As might be expected, the semi-synthetic cephalosporins were designed to contain certain structural features that were present in landmark penicillin analogs. Indeed, many of the early cephalosporins bear striking similarities to successfully developed penicillins with regard to the selection of the acylamino substituent. However, the cephalosporins are more tolerant of structural modifications; a greater variety of acylamino substituents have been successfully developed in conjunction with numerous dihydrothiazine ring substituents, as well as nuclear analogs. As a result, the cephalosporins are usually divided into generations based upon several biological properties; the selection of the acylamino side chain alone does not place a cephalosporin congener in a particular category (as with the penicillins). While it is true that many third generation cephalosporins were introduced at a later date than, say, a number of first generation congeners, it is an extended spectrum of activity and resistance to bacterial enzymatic degradation that actually differentiates a third generation agent from a first generation compound. Nevertheless, it is important to realize that the designation of some cephalosporins to a specific generation remains somewhat arbitrary since several factors come into play.

First generation cephalosporins possess potent activity against many Gram positive bacteria but are only moderately active against common aerobic Gram negative pathogens. These agents lack activity against *Pseudomonas aeruginosa*. This spectrum of activity is similar to many (simple) penicillins and thus some first generation cephalosporins are useful alternatives to penicillin chemotherapy. For example, individuals who have displayed hypersensitivity to penicillin therapy may be treated with a first generation cephalosporin without an adverse event. Cephalexin **23** (Figure 3.13) has an aminobenzyl side chain like that found in ampicillin, whereas cefadroxil **24** (Figure 3.13) has a *para*-hydroxylated side chain analog such as that present in amoxicillin. Both of these agents are widely used and constitute a mainstay in antibacterial chemotherapy due to their potency against troublesome Gram positive pathogens as well as their efficacy following oral administration. Cephradine **25** (Figure 3.13) is another well recognized orally active first generation cephalosporin.

There are also a number of widely used first generation cephalosporins that are administered parenterally. Cephalothin **26**, cephacetrile **27** and cephapirin **28** (Figure 3.14) are illustrative although cephalothin remains the agent of choice in the United States. All of these agents contain an acetoxymethyl group extending from the dihydrothiazine ring but have widely varying 'west-end' moieties.

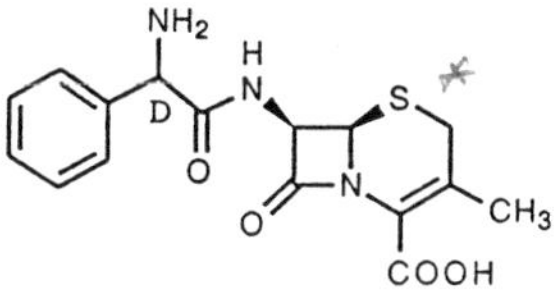

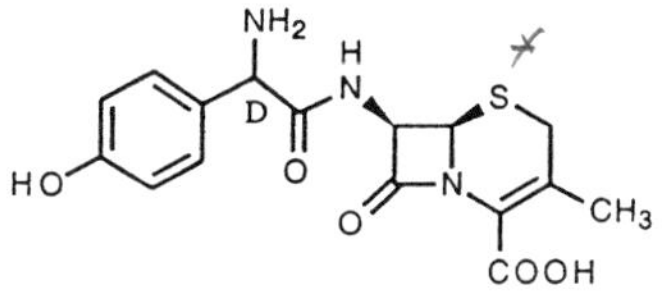

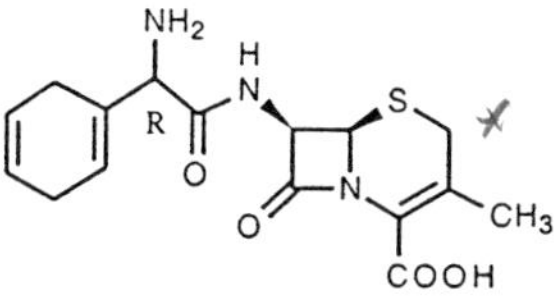

Figure 3.13 Cephalexin, cefadroxil and cephradine.

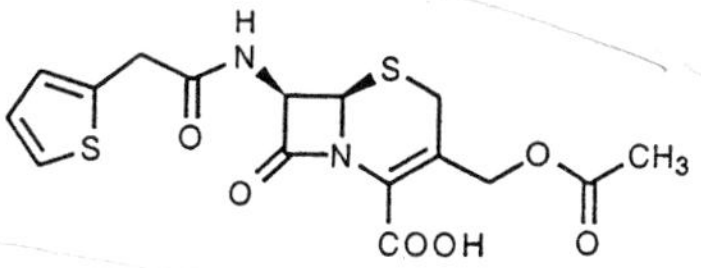

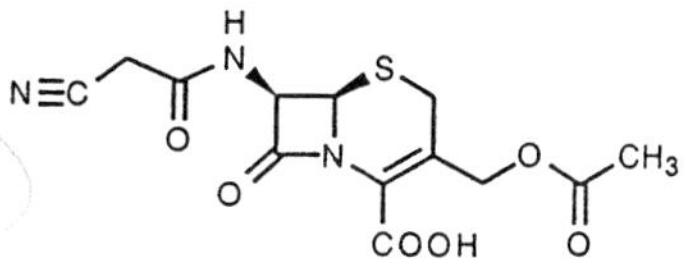

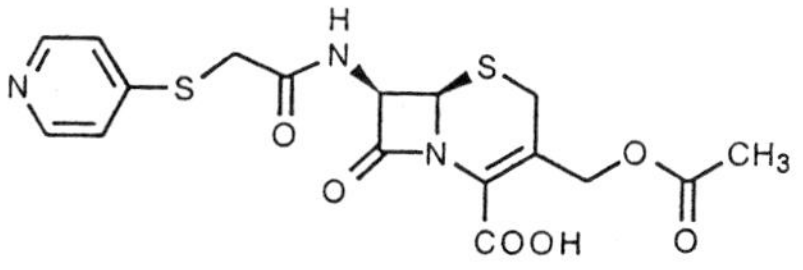

Figure 3.14 Cephalothin, cephacetrile and cephapirin.

The second generation cephalosporins possess an extended spectrum of antibacterial activity compared to first generation agents. In general, these compounds have improved potency against many Gram negative pathogens such as *Escherichia coli* and some *Klebsiella* and *Proteus* species. Some agents are also effective against some *Enterobacter*, *Serratia*, *Neisseria meningitidis* and *Neiserria gonorrhoeae*. As with the first generation congeners, second generation cephalosporins are impotent against *Pseudomonas*. Cefaclor **29** and cefprozil **30** (Figure 3.15) are orally active second generation cephalosporins. However the majority of second generation cephalosporins are parenteral agents. Many of these compounds have a nitrogen-rich heterocyclic moiety extending from the dihydrothiazine ring as evidenced by the tetrazolyl cephalosporins cefamandole **31**, cefonicid **32** and ceforanide **33** (Figure 3.16).

In the 1970s a new type of cephalosporin was isolated from a Streptomyces culture and subsequently investigated. These substances, called cephamycins, typically display broad spectrum antibacterial activity that includes some anaerobic species and enhanced stability to enzymatic degradation by bacterial lactamases. Cephamycins contain a characteristic methoxy substituent on the β-lactam nucleus; the shielding effect of this

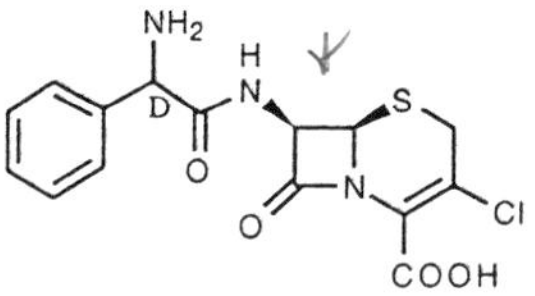

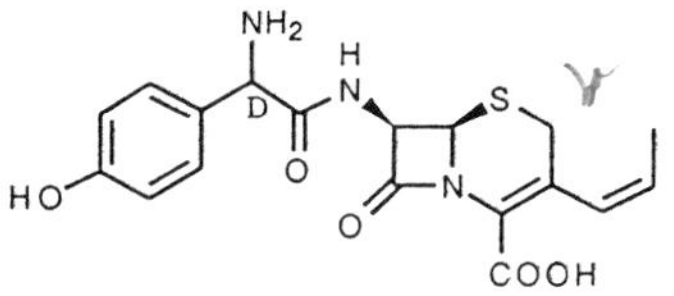

Figure 3.15 Cefaclor and cefprozil.

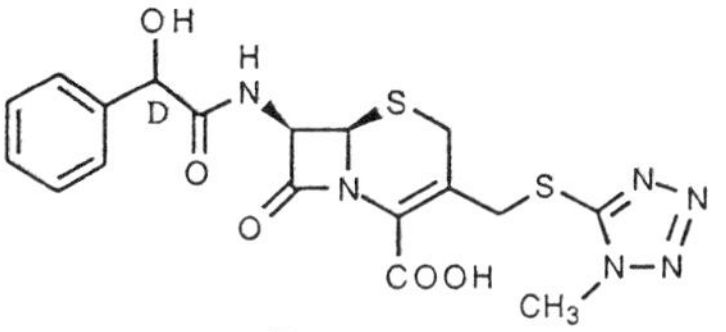

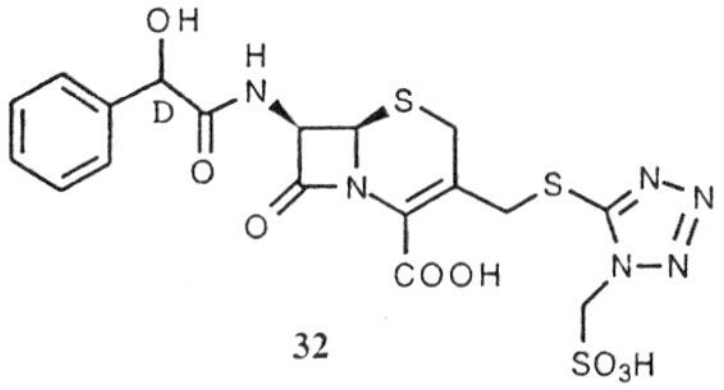

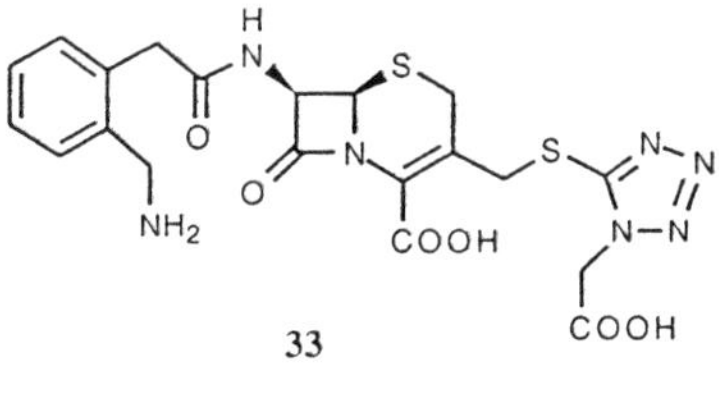

Figure 3.16 Cefamandole, cefonicid and ceforanide.

group is responsible for the increased stability. Successors to the first isolated cephamycins include the semi-synthetic congeners cefoxitin **34**, cefmetazole **35**, cefminox **36** and cefotetan **37** (Figure 3.17).

One of the most recently successful second generation cephalosporins is cefuroxime **38** (Figure 3.18) which can be administered parenterally or given orally as its acetoxyethoxy ester, cefuroxime axetil **39** (Figure 3.18). The latter variation again illustrates the successful application of an activated ester to improve oral efficacy of a parenteral agent.

Third generation cephalosporins display a broader spectrum of anti-bacterial activity compared to the first and second generation agents although potency against Gram positive cocci is only comparable at best.

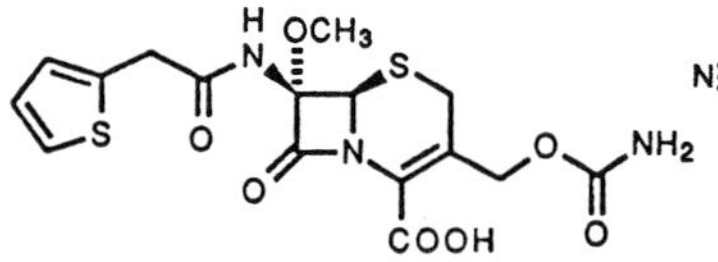

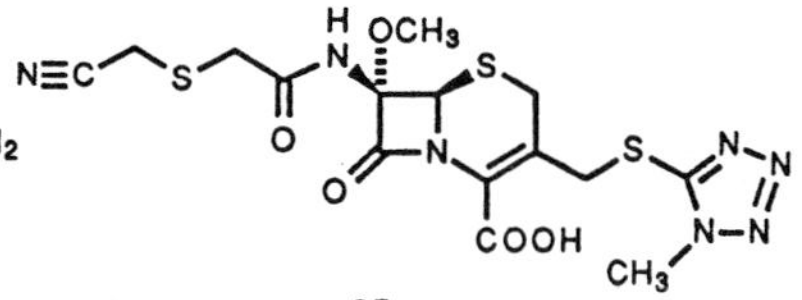

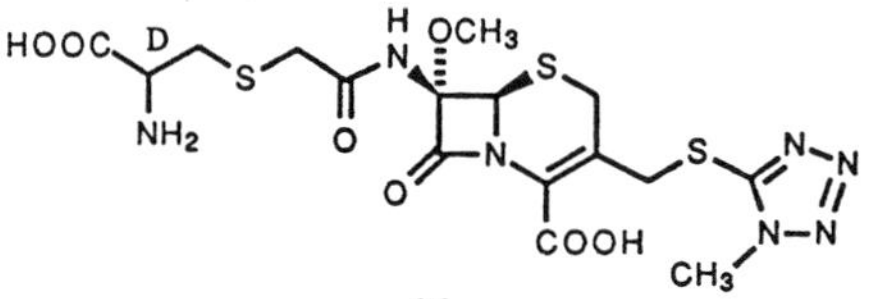

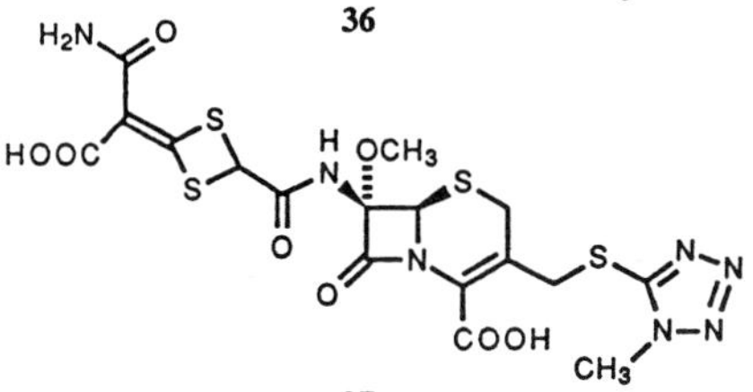

Figure 3.17 The cephamycins: cefoxitin, cefmetazole, cefminox and cefotetan.

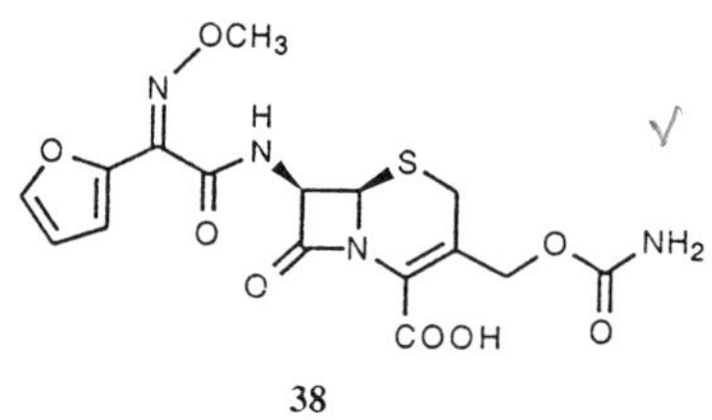

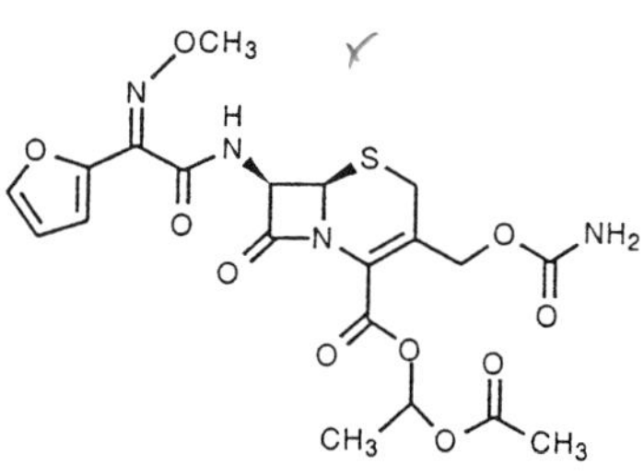

Figure 3.18 Cefuroxime and cefuroxime axetil.

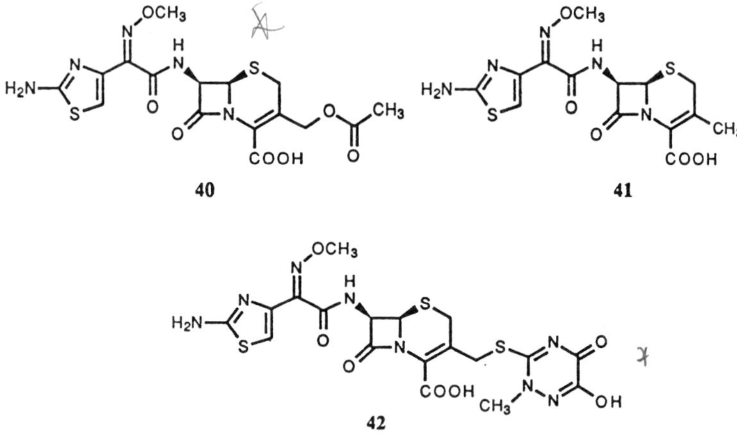

Figure 3.19 Cefotaxime, ceftizoxime and ceftriaxone.

However, enhanced activity against Gram negative pathogens (bacilli, in particular) distinguishes third generation cephalosporins and has allowed many of these agents to compete with other existing Gram negative chemotherapies. In general, the third generation cephalosporins also offer the advantage of β-lactamase resistance, a characteristic primarily attributed to the acylamino side chain. Marketed agents include cefotaxime **40**, ceftizoxime **41** and ceftriaxone **42** (Figure 3.19); the latter is the best-selling parenteral agent due to its long elimination halflife that allows for once-a-day dosing. However, none of these agents are clinically efficacious against pseudomonads.

Several third generation cephalosporins do display good potency against the troublesome Gram negative pathogen *Pseudomonas aeruginosa*. Ceftazidime **43** (Figure 3.20) is a landmark agent in this regard although cefoperazone **44** (Figure 3.20) and some newer agents such as cefepime **45** and cefpirome **46** (Figure 3.21) offer excellent antipseudomonal chemotherapy.

Cefixime **47** and ceftibuten **48** (Figure 3.22) are usually considered as third generation cephalosporin agents and are orally active. Cefixime has been very successful in the marketplace and ceftibuten has been launched in several countries.

Structural modification of the β-lactam nucleus itself has been a subject of intense research for decades. One such class that has been advanced is that of the oxadethiacephalosporins which are often referred to as the oxacephems. These compounds are analogs of cephalosporins in which the sulfur atom of the dihydrothiazine ring is substituted by an oxygen atom. Moxalactam **49** and flomoxef **50** (Figure 3.23) are marketed products best regarded as third generation cephalosporins due to their potent broad

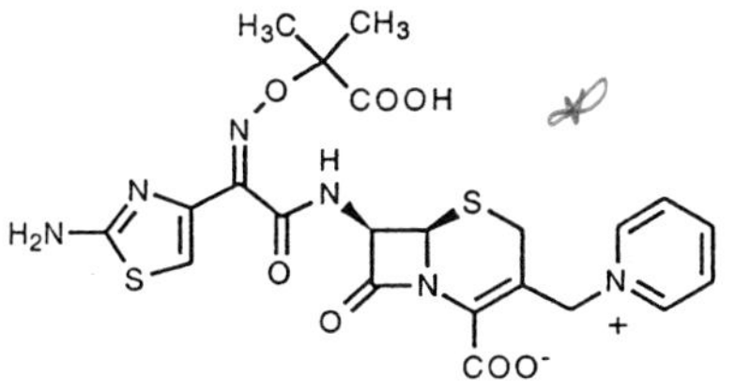

43

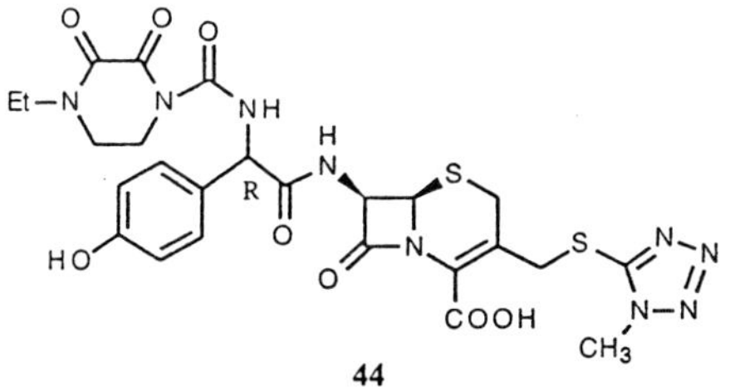

44

Figure 3.20 Ceftazidime and cefoperazone.

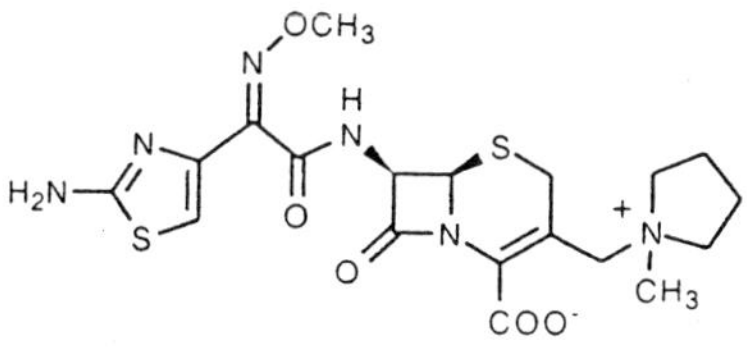

45

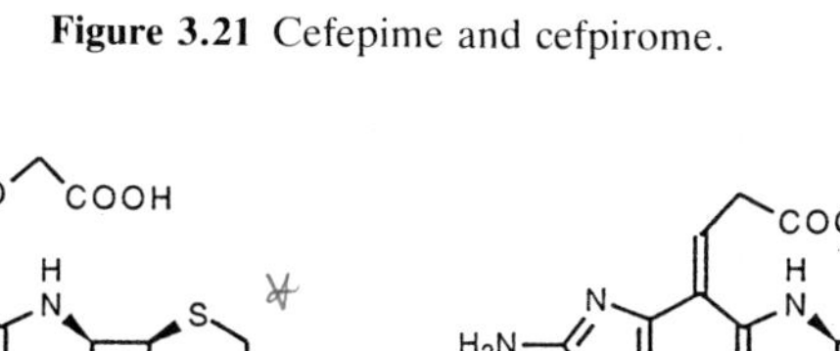

46

Figure 3.21 Cefepime and cefpirome.

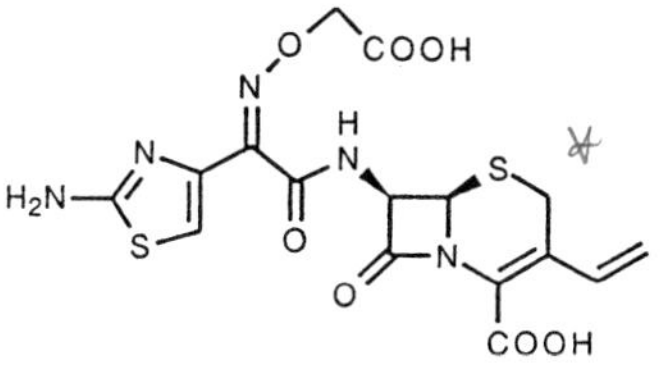

47

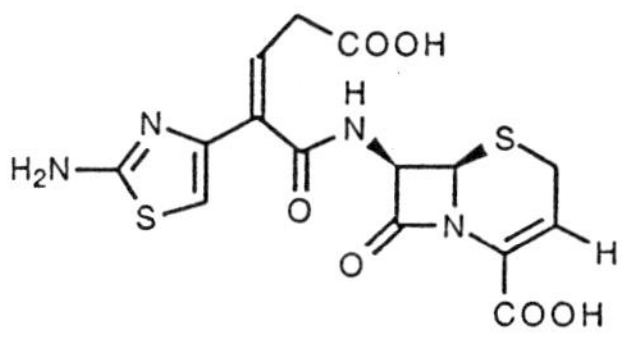

48

Figure 3.22 Cefixime and ceftibuten.

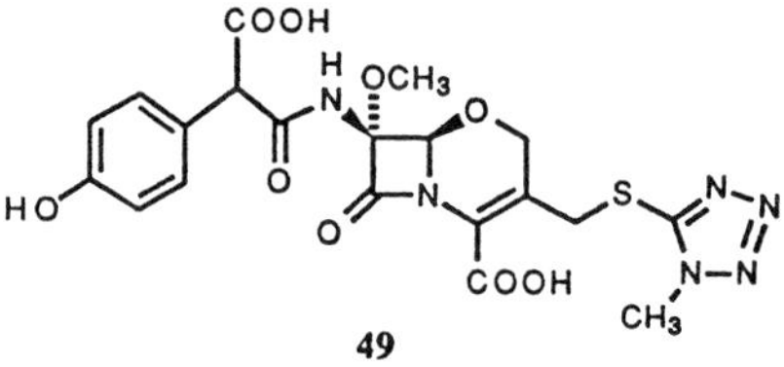

49

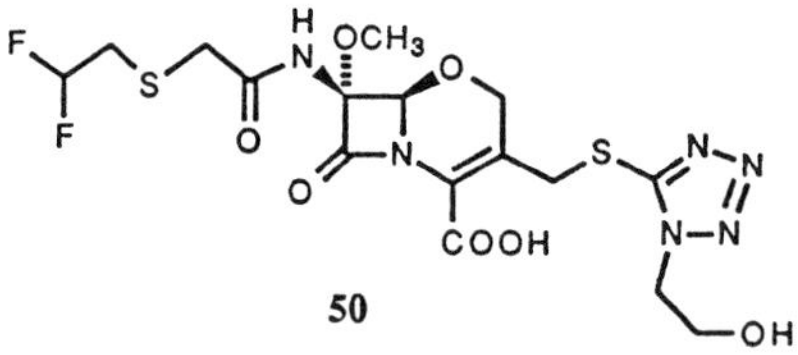

50

Figure 3.23 Moxalactam and flomoxef.

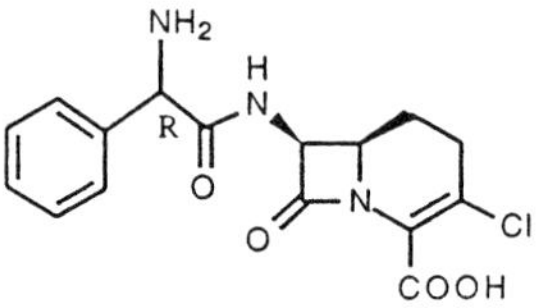

51

Figure 3.24 Loracarbef.

spectrum of activity which encompasses many Gram negative organisms, including *Pseudomonas aeruginosa*.

The replacement of the sulfur atom of a cephalosporin with a carbon unit (dethiacarbacephalosporins, also referred to as carbacephems) has also been explored and recently a prototype, loracarbef **51** (Figure 3.24) has advanced onto the market. This synthetic carbacephem offers greater stability than structurally related cephalosporin analogs and excellent oral efficacy. Although the development of loracarbef was essentially aimed at replacing the oral cephalosporin cefaclor, an agent in decline due to competition from cefixime, loracarbef is the first of what may prove to be a significant new class of *β*-lactam congeners.

Nature has provided its own derivatization of the *β*-lactam nucleus. Thienamycin **52** (Figure 3.25), discovered in the 1970s from a new *Streptomyces* species, was the first known carbapenem antibiotic. Its bicyclic nucleus contains different stereochemical features to those found either in penicillins or cephalosporins. Carbapenems demonstrate superb activity against anaerobic bacteria, an area generally not covered by penicillins or cephalosporins. Reminiscent of early penicillins, the carba-penems are usually quite potent against Gram positive pathogens but are also weak against many Gram negative species. The initial excitement generated by the discovery of thienamycin led to subsequent research that culminated in the synthesis of imipenem **53** (Figure 3.25). Imipenem is considerably more stable to hydrolytic degradation than thienamycin, but like thienamycin, it is susceptible to *in vivo* inactivation by the renal enzyme dehydropeptidase. As a result, secondary research into the synthesis and development of dehydropeptidase inhibitors was undertaken and today, imipenem is coadministered with cilastatin (structure shown in section 3.5, Figure 3.70).

Various monolactams make up yet another subset of *β*-lactam antibiotics. The discovery of a monocyclic *β*-lactam sulfonic acid antibacterial (a so-called monobactam) led to the development of aztreonam **54** (Figure 3.26). Aztreonam not only differs in structure from classical *β*-lactam antibiotics, but also displays a unique spectrum of activity. The potency of this compound is limited to Gram negative aerobic bacteria, in particular, *Enterobacteriaceae*. Other analogs such as carumonam **55** (Figure 3.26), as well as derivatives containing different acidic moieties extending from the *β*-lactam nucleus (e.g. monophosphams, monocarbams, etc.), have also been investigated.

Certain modifications of the *β*-lactam nucleus can afford compounds that are essentially devoid of useful antibacterial activity but are able to inhibit certain bacterial enzymes. *β*-Lactamases, penicillinases and cephalo-sporinases are proteins that are expressed by many bacterial species, often in response to the presence of an antibiotic. These proteins are able to

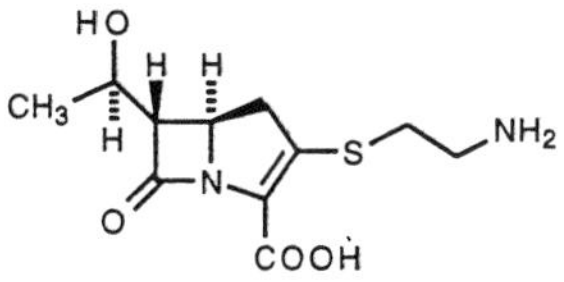

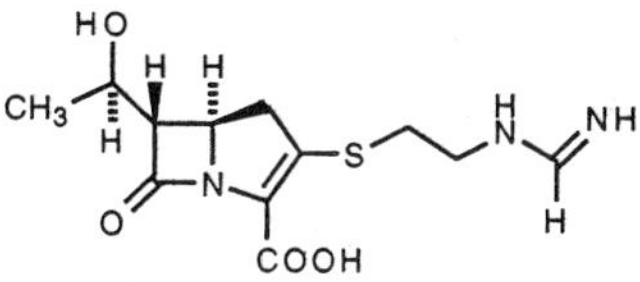

Figure 3.25 Thienamycin and imipenem.

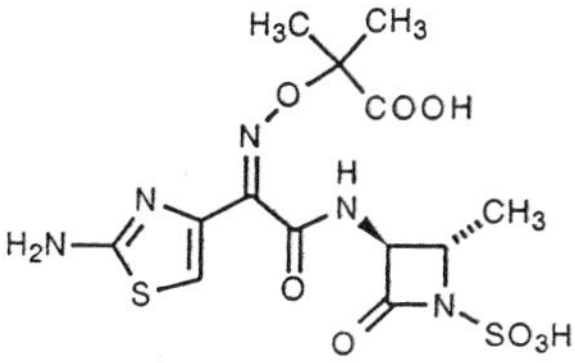

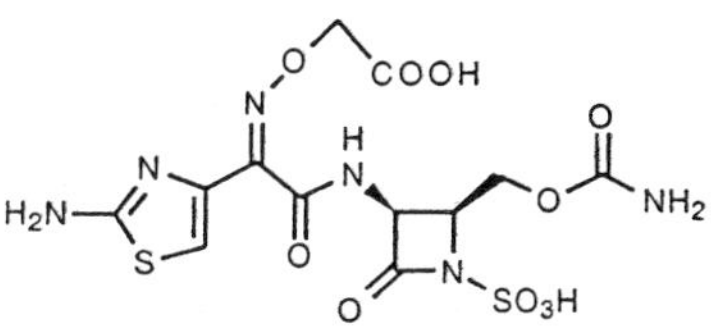

Figure 3.26 Aztreonam and carumonam.

inactivate certain penicillins, cephalosporins and other related agents by hydrolytic cleavage of the reactive β-lactam moiety (sections 3.2 and 3.5). Nature has provided several naturally occurring compounds that act as inhibitors of these bacterial enzymes. Clavulanic acid **56** (Figure 3.27) is produced by a streptomyces and inhibits β-lactamases from a number of Gram positive and Gram negative organisms. Clavulanate is coadministered with several penicillins and can also act in a synergistic fashion with other β-lactam antibiotics.

Sulbactam **57** (Figure 3.28) and its oral ester **58** (Figure 3.28) are desaminopenicillin sulfones that, like clavulanic acid, act as suicide inhibitors of certain β-lactamases. Other β-lactamase inhibitors have been designed and synthesized such as tazobactam **59** and triazolomethyl-idienylpenem **60** (Figure 3.29). Although these compounds structurally resemble a penicillin, their biological action is exerted primarily upon bacterial β-lactamases rather than bacterial transpeptidases (see discussion on mode of action in section 3.2).

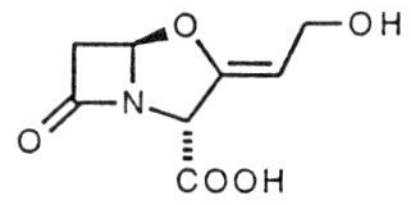

56

Figure 3.27 Clavulanic acid.

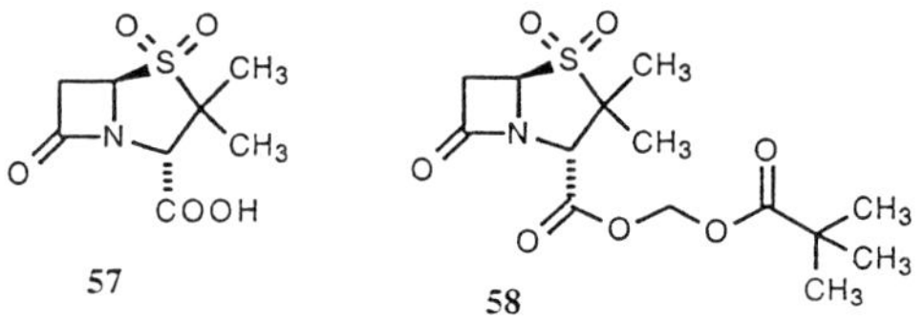

57 **58**

Figure 3.28 Sulbactam and sulbactam pivaloyloxymthyl ester.

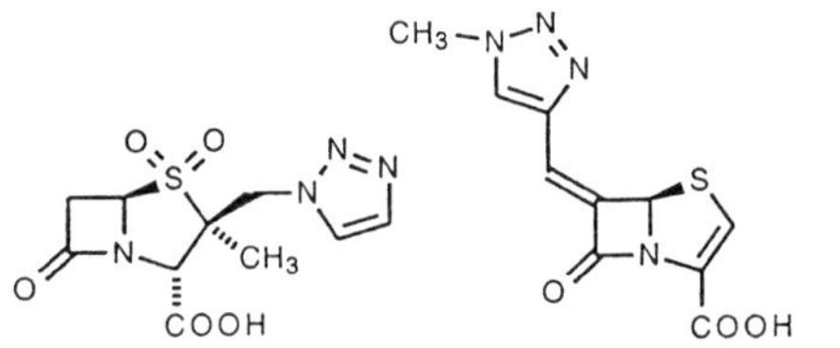

59 **60**

Figure 3.29 Tazobactam and a C-6 triazolomethylidienylpenem β-lactamase inhibitor.

3.2 Mode of action: penicillin-binding proteins and inhibition of transpeptidation

The *β*-lactam antibiotics have been extensively studied and a great knowledge has been amassed regarding the mechanism of action of these agents upon many bacterial species. It is important to address some fundamental issues concerning the composition and biosynthesis of the bacterial cell wall in order fully to appreciate the action of the *β*-lactam antibiotics. Although the cell wall is complex in structure and can vary from species to species, a number of key features are present. First of all, the bacterial cell wall is an absolutely vital 'organelle' for the survival of the bacterial cell. This barrier provides protection from hostile external environments that can cause bacterial cell death. Bacteria are particularly susceptible to even minute changes in pH, ion concentration, osmotic pressure and temperature. In addition, some bacterial components and cellular apparatus can be attacked by a number of widely occurring enzymes and other proteins. The healthy bacterial cell wall is able to withstand the effects of these forces and also prevents toxic substances from penetrating into the cell where the organism's vital genetic material is housed. A concomitant duty of the bacterial cell wall is to allow the uptake of nutrients needed for survival and growth. Most bacteria acquire nutrients from outside the cell by either simple diffusion of the substance across the cell wall, or by some active transport mechanism that ushers in the substance bound to or associated with some carrier molecule. In the healthy bacterium, these mechanisms allow for the degree of selectivity that is needed to nourish the bacterial cell without permitting any poisoning from the outside to occur.

Another function of the bacterial cell wall is to maintain the integrity of the very barrier it has established despite some notably demanding extracellular and intracellular events. This preservation of the bacterial cell wall is a dynamic process. Cell wall material can be marred or shed off, usually as a result of external forces, and new material is constantly required to repair damage. In addition, bacterial growth eventually culminates in the division of the cell and accordingly an adequate supply of cell wall components must be available in order successfully to accomplish this task. In essence, the bacterial cell wall must be yielding and flexible, and yet be able to provide structure to the cell.

A macromolecular component that imparts support and rigidity to the bacterial cell wall is the biopolymeric substance peptidoglycan This substance provides a formidable barrier from the extracellular milieu but it is not the only constituent of the cell wall. Gram negative bacteria also contain a lipopolysaccharide layer and Gram positive organisms utilize other materials to fortify the cell wall. However the peptidoglycan matrix is indeed an integral part of the bacterial cell wall; organisms are prone to undergo death, often by lysis, if the peptidoglycan matrix is disrupted or modified. Peptidoglycan is a polymer of alternating peptide units and glycan subunits. The glycan subunits consist of alternating *N*-acetylglucosamine and *N*-acetylmuramic acid-derived sugars. The peptide chains are ordered and characteristic of the species of bacteria and are attached to the muramic acid glycan (Figure 3.30). A peptidic spacer is present between two linked units such as the Gly-Gly-Gly-Gly-Gly moiety depicted below.

Peptidoglycan biosynthesis occurs via a complex series of intracellular reactions that can be divided into three stages. The first stage occurs in the cytoplasm and encompasses two distinct processes. In the first process, glucosamine and uridine triphosphate are covalently joined and further modified to afford a uridine diphosphate-*N*-acetylmuramic acid conjugate (UDP-MurNAc) (Figure 3.31). In the second process, a pentapeptide chain is assembled onto this 'anchor' by the action of a host of enzymes that

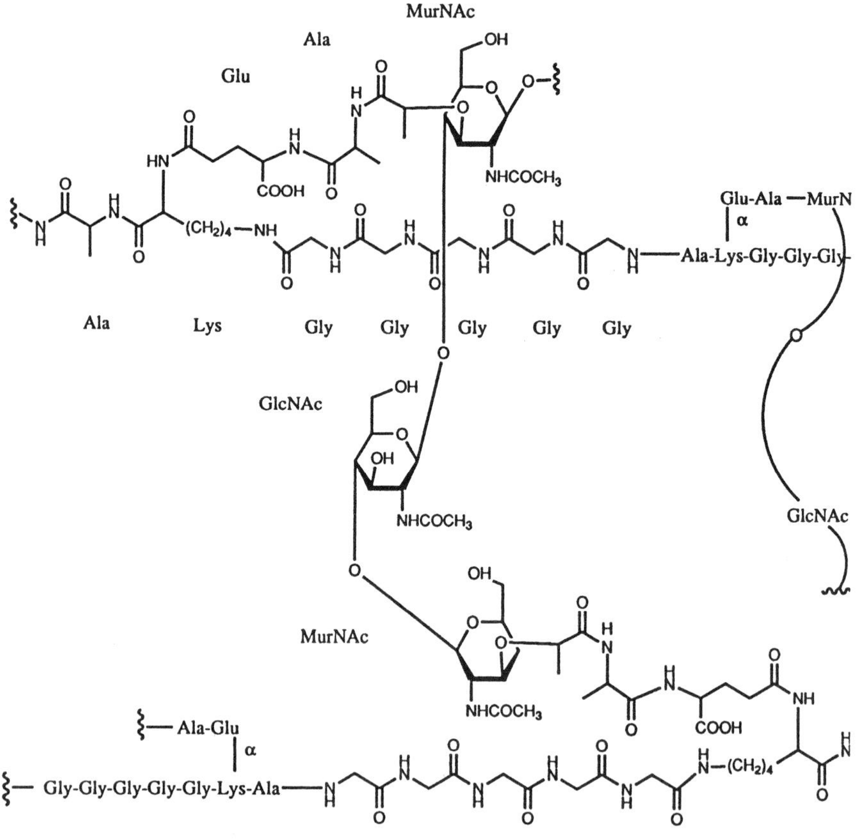

Figure 3.30 Peptidoglycan.

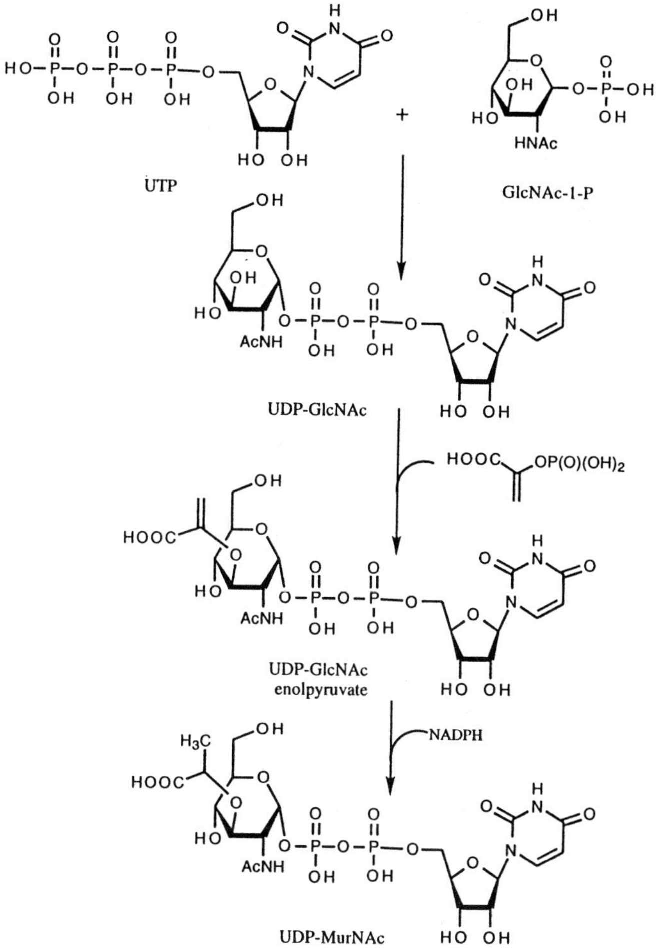

Figure 3.31 Biosynthesis of UDP-MurNAc.

attach individual amino acids onto the growing chain in an orderly and sequential fashion (Figure 3.32). These enzymes are ligases and are species specific. The terminal two amino acids, D-Ala-D-Ala, are joined in a different manner, being attached as a dipeptide unit. This is accomplished by several enzymes, that of a racemase which converts L-alanine to its enantiomer, and two separate ligases responsible for the subsequent dimerization of D-Ala and the attachment of the dipeptide to the UDP-MurNAc-peptide substrate. The culmination of this process (and of the first stage of peptidoglycan biosynthesis) is the formation of the crucial peptidoglycan precursor, UDP-MurNAc-pentapeptide.

Bacterial species vary in the composition of their pentapeptide moiety

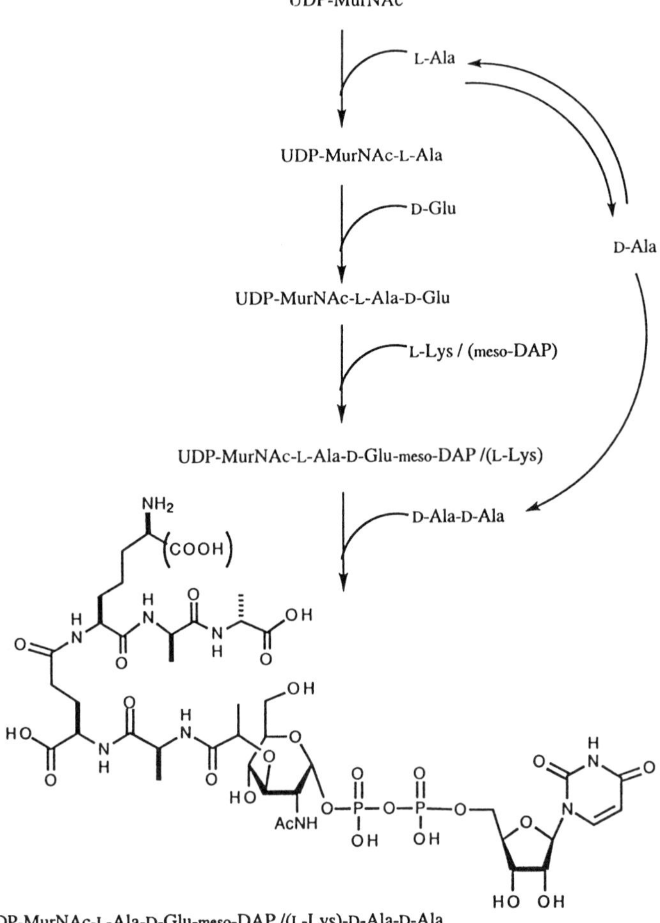

Figure 3.32 Biosynthesis of UDP-MurNAc-pentapeptide.

but one universal feature is the incorporation of an amino acid at the third residue that can accommodate an additional peptide appendage (e.g. Gly-Gly-Gly-Gly-Gly, as shown above). Specifically, this key residue contains a terminal free amine group that is further processed in order to set the stage for the final cross-linking reaction (transpeptidation) that affords peptidoglycan. In Gram negative bacteria, the central amino acid residue of the UDP-MurNAc-pentapeptide is usually *meso*-diaminopimelic acid (DAP). For example, the pentapeptide sequence found in *E. coli* and some other Gram negative bacteria is L-Ala-α-D-Glu-*meso*-DAP-D-Ala-D-Ala (Figure 3.33). The free amine group of the D-center of the diaminopimelic acid residue is the site to which the polyamino acid chain is eventually

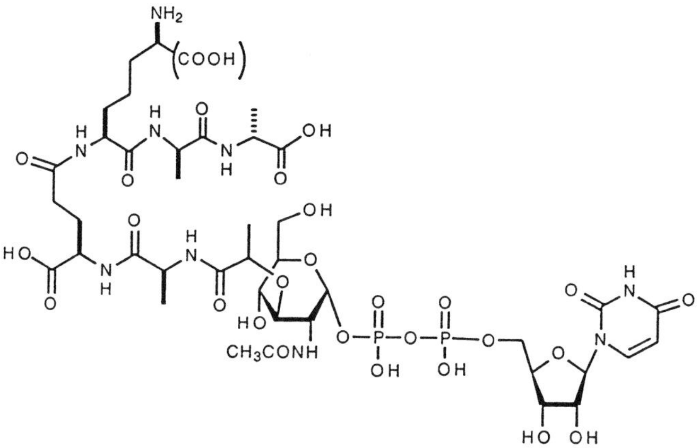

UDP-MurNAc-L-Ala-D-Glu-meso-DAP /(L-Lys)-D-Ala-D-Ala

Figure 3.33 UDP-MurNAc-pentapeptide; central amino acid typically *meso*-DAP(Gram negative) or L-Lys(Gram positive).

attached (Figure 3.35). Gram positive bacteria often incorporate the amino acid lysine (L-Ala-D-Glu-L-Lys-D-Ala-D-Ala) (Figure 3.33) but can have L-ornithine at this position. These residues (*meso*-DAP, L-Lys and L-ornithine) are all capable of coupling to additional amino acids prior to the transpeptidation process.

The second stage (Figure 3.34) of peptidoglycan biosynthesis involves a series of membrane-associated transformations in which the cytoplasmic pentapeptide precursor (UDP-MurNAc-pentapeptide) is modified in a way that allows it to traverse the lipophilic membrane. This process involves the transfer of the pentapeptide substrate onto the lipid undecaprenylphosphate and is called translocation. The undecaprenyl-phosphate lipid needed for the continuation of peptidoglycan biosynthesis also serves as a vital carrier for components that are required for bacterial exocellular polysaccharide components and thus is instrumental in a number of bacterial processes.

The ensuing transformation is the attachment of the other glycan subunit, that of *N*-acetyl-D-glucosamine, which affords the intact glycan–peptide substrate bound to the lipid carrier (GlcNAc-MurNAc(-PP-C$_{55}$)-L-Ala-D-Glu-*meso*-DAP/(L-Lys)-D-Ala-D-Ala). Further modification (Figure 3.35) occurs that is species dependent and which results in the placement of a cross-linking amino acid chain. This is accomplished by the addition of individual amino acids that are activated as t-RNA conjugates. A glycine pentapeptide is frequently encountered although amidated amino acids, D-amino acids and mixtures of these substrates may be incorporated into this peptide chain. The final transformation of this stage of peptidoglycan

UDP-MurNAc-L-Ala-D-Glu-meso-DAP /(L-Lys)-D-Ala-D-Ala

Undecaprenyl (C_{55}) phosphate

UDP-GlcNAc

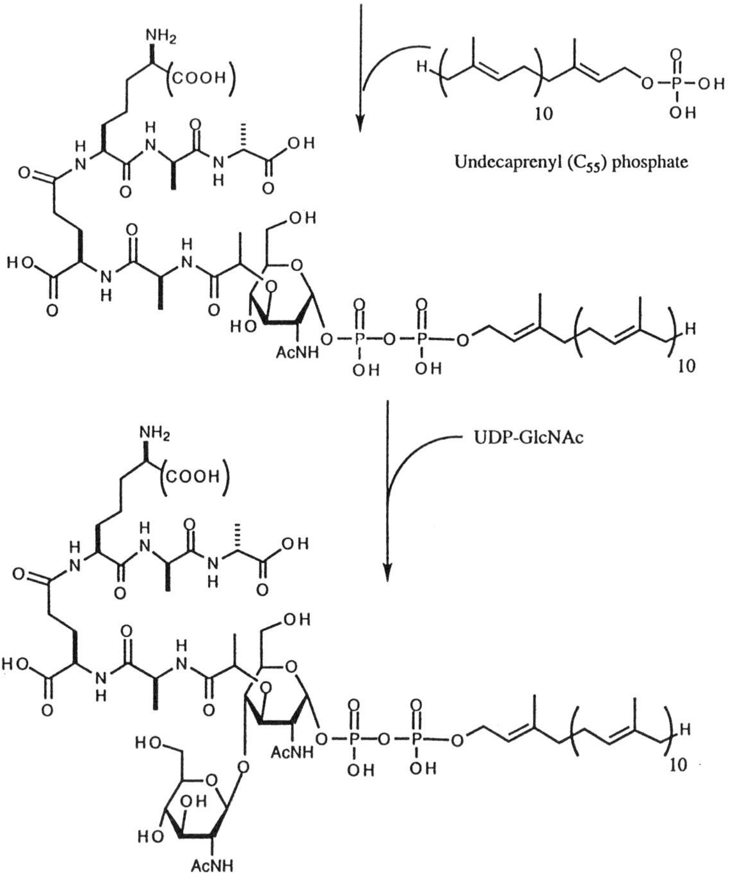

GlcNAc-MurNAc-(PP-C_{55})-L-Ala-D-Glu-meso-DAP /(L-Lys)-D-Ala-D-Ala

Figure 3.34 Second stage of peptidoglycan biosynthesis (lipid bound).

biosynthesis is a transglycosylation reaction that polymerizes disaccharide–peptide units and liberates the lipid carrier. The glycan–peptide polymer is pushed out of the membrane into the periplasm and the next stage of peptidoglycan biosynthesis commences.

The final stage of peptidoglycan biosynthesis is the joining of free peptide strands to one another to rigidify further this bacterial cell wall structure. This transformation is performed by transpeptidase enzymes that catalyse acylation of the terminal amine functionality of the peptide side chain onto the D-alanine residue of an adjacent pentapeptide chain (Figure 3.36). This reaction joins (cross-links) neighboring chains in a

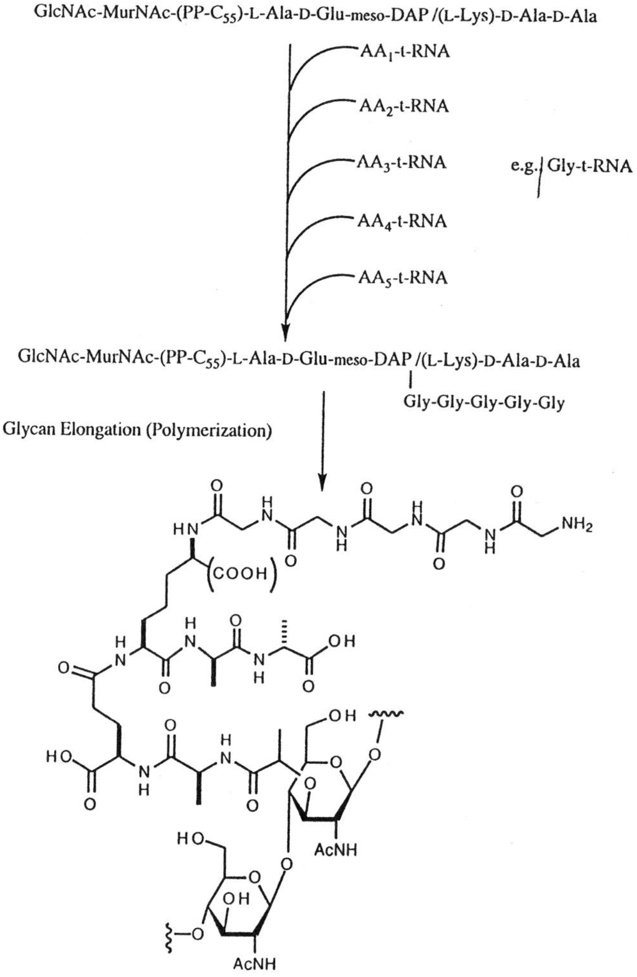

GlcNAc-MurNAc-(PP-C_{55})-L-Ala-D-Glu-meso-DAP /(L-Lys)-D-Ala-D-Ala

AA_1-t-RNA

AA_2-t-RNA

AA_3-t-RNA   e.g. Gly-t-RNA

AA_4-t-RNA

AA_5-t-RNA

GlcNAc-MurNAc-(PP-C_{55})-L-Ala-D-Glu-meso-DAP /(L-Lys)-D-Ala-D-Ala

Gly-Gly-Gly-Gly-Gly

Glycan Elongation (Polymerization)

Figure 3.35 Attachment of amino acid side chain: site of transpeptidation.

covalent manner and completes the formation of the peptidoglycan matrix. Bacterial transpeptidases bind to the nascent peptidoglycan strand whereupon an active site serine residue of the enzyme cleaves the terminal alanine residue of the substrate giving rise to an acylated-D-Ala-enzyme species. An acceptor glycan strand is then able to react at the acyl carbonyl center of this complex and complete transpeptidation, regenerating free enzyme and a catalytic cycle is achieved (Figure 3.37).

The degree of cross-linking varies considerably among bacterial species. In general, transpeptidation may occur to the extent of 75% of the

Cross-Link Formation

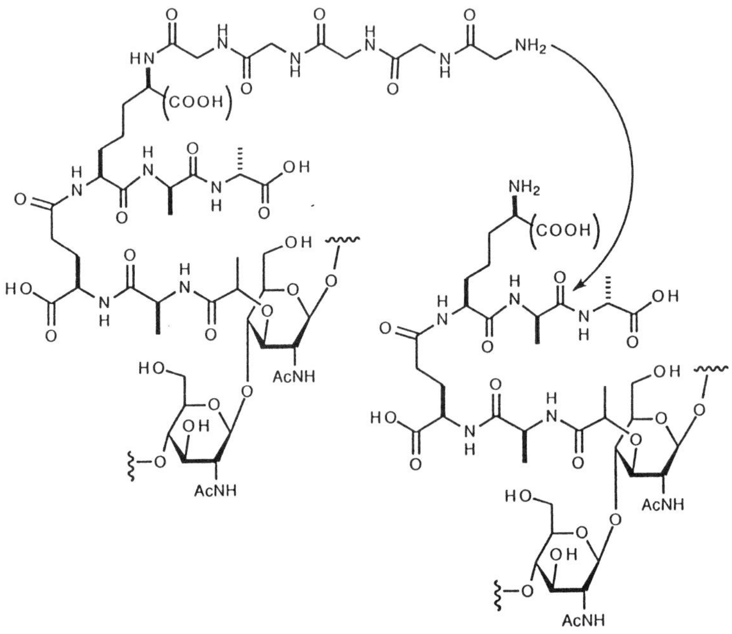

Figure 3.36 Transpeptidation.

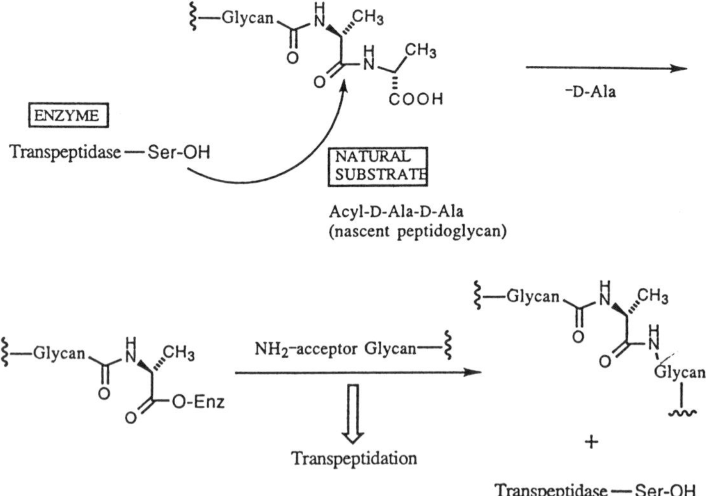

Figure 3.37 Action of transpeptidase.

available sites in some rather prolific species (some *Staphylococci*) but is frequently less in many bacteria (in *E. coli*, for example, cross-linking occurs on approximately one-third of the possible sites). It is not imperative that exhaustive transpeptidation occur in order to furnish a resilient peptidoglycan component. In fact, it is likely that total transpeptidation would afford a matrix unyielding to the dynamics of cell wall maintenance and growth, as well as possibly upsetting processes necessary for nourishment. In this regard, there are opposing enzymatic processes that assure some optimal degree of transpeptidation. Free peptide strands that do not undergo transpeptidation are often rendered unable to do so at a later stage. This operation is accomplished by the removal of the terminal alanine residue from pentapeptide precursor by carboxypeptidase enzymes.

With this brief background in mind, the mode of action of the β-lactam antibiotics can be explored. Due to elegant research by Strominger, Tipper, Ward and others, it has been well established that β-lactam antibiotics induce lethal effects in bacteria by inhibiting the transpeptidation process required to complete peptidoglycan biosynthesis. The first step in this process is the binding of the β-lactam antibiotics to certain bacterial proteins that are generally referred to as penicillin binding proteins (PBPs). Penicillin binding proteins are intimately involved in the construction of peptidoglycan; these proteins can provide either transpeptidase or carboxypeptidase activity. Upon PBP binding, the β-lactam agent is able to adopt a conformation that closely mimics the D-Ala-D-Ala terminus of a peptide–glycan conjugate readied for transpeptidation. The reactive β-lactam moiety is aligned in proximity to the position normally occupied by the scissile D-Ala-D-Ala linkage of the natural substrate. By this association, the β-lactam antibiotics are able to acylate specific serine residues within the active site of transpeptidase enzymes. This renders the enzyme inoperative since the enzyme–(β-lactam)substrate complex is unable to undergo subsequent hydrolysis to regenerate a functional transpeptidase enzyme (Figure 3.38).

A given bacterial species expresses its own unique set of PBPs characteristic of the organism and these proteins are collectively designated based upon molecular size. The highest numerical priority (PBP 1) is given to the PBP of greatest molecular size (weight), the second largest PBP is designated as PBP 2 and so on. While this nomenclature is useful, it is important to realize that PBPs of the same designation from different species are not related. Lastly, the number of penicillin binding proteins differs from species to species (for example, *E. coli* has ten PBPs but many other species contain fewer). In addition, the number of copies of each protein per cell can also vary.

The role of the larger PBPs has been demonstrated in many bacterial species. In general, PBPs 1A, 1B, 2 and 3 of many species are responsible

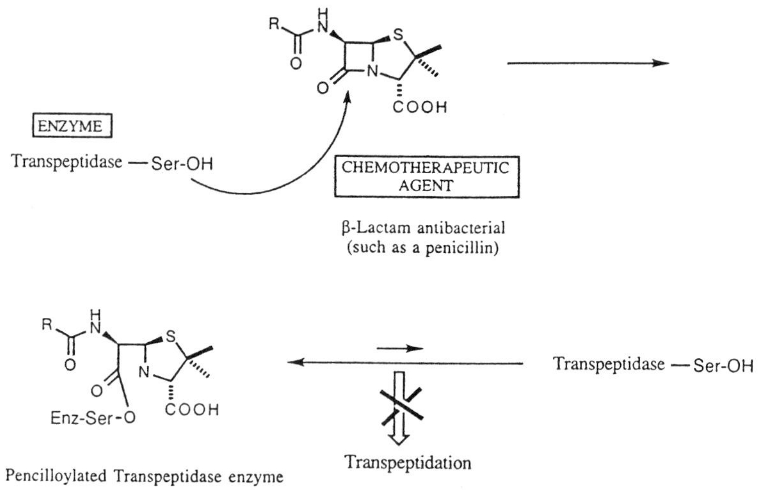

Figure 3.38 Mode of action of β-lactam antibacterial agents.

for a variety of vital cellular operations and most β-lactam antibacterials target one or more of these proteins. Loss of the PBP function has potentially lethal effects on the bacterial cell. For example, in *E. coli*, PBP 1A and 1B play a role in transpeptidation, transglycosylation and cylindrical cell wall synthesis. Inactivation leads to lysis and death of the bacterial cell. The *E. coli* PBP 2 appears to be operational in maintaining cell shape also, and loss of protein activity leads to the formation of spherical cells, an abnormal morphology for the species. The *E. coli* PBP 3 is required for septum cross-wall biosynthesis that is carried out prior to cell division. If this protein is rendered inactive, non-septate cells are formed and filamentation can occur. The PBP 4 of *E. coli* behaves as peptidase whose activity appears to regulate transpeptidation to some degree. The role of the smaller penicillin binding proteins is not fully understood in *E. coli* and many other species. The smaller PBPs may serve some secondary function that has yet to be elucidated or these proteins may just be genetic artifacts that are not needed.

In general, the β-lactam antibiotics display a selectivity in undergoing binding to the PBPs that is unique to each individual agent. For example, cephaloridine undergoes preferential binding with PBP 1 of most suscept-ible bacteria which results in lysis with little change in morphology. In contrast, cefuroxime binds to PBP 3 and obvious morphological changes can occur such as filamentation; the affected bacterium grows but is unable to complete cell division. Some β-lactam antibacterials, such as the cephamycins, bind to PBPs thought to be non-essential (e.g. PBP 5 and PBP 6) but yet are potent agents. In addition, some β-lactam antibiotics

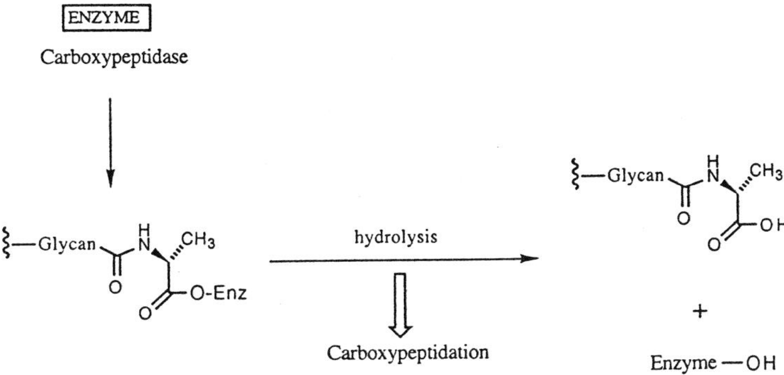

Figure 3.39 Action of carboxypeptidases.

can affect the carboxypeptidase function of a PBP (Figure 3.39) although inhibition of transpeptidase activity is usually regarded as the premier event leading to an antibacterial action. Since the β-lactam antibiotics are able to bind to a number of PBPs with different affinities, and since these proteins are performing various functions, it is not surprising that some mechanistic details remain vague.

3.3 Structural features and structure–activity relationships of the β-lactam antibacterials

The basic structural requirements of the β-lactam agents that are necessary for antibacterial activity are rather simple. However, there are literally thousands of β-lactam congeners and many subtle peripheral modifications are difficult to evaluate within the context of a general overview; therefore the most useful general features are presented here.

The crucial structural feature of the β-lactam family of antibacterials is a highly reactive lactam nucleus which contains a suitably positioned acidic functionality. It is imperative that the lactam carbonyl center be reactive towards nucleophilic attack by the active site serine residue of the bacterial transpeptidase enzyme. Small-ring lactams are more reactive than larger ring analogs due to the strained nature of the amide bond and a four-membered β-lactam nucleus remains optimal. The acidic functionality is important for the proper positioning of the β-lactam substrate within the enzyme site so as to mimic the endogenous peptide substrates (i.e. the D-Ala-D-Ala terminus of the MurNAc-pentapeptide). The simplest structural element containing these features is that of a β-lactam nucleus substituted at the nitrogen atom of an organic acid (Figure 3.40); usually the carboxylic functionality is one carbon center (alpha α) removed from

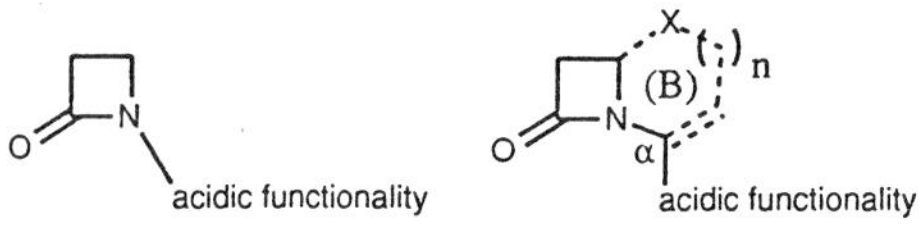

n = 0 (five-membered ring)
1 (six-membered ring)

X = S, CH_2, O

Figure 3.40 Basic *β*-lactam SAR.

the heterocycle. In many cases, the *β*-lactam ring is fused to either a five- or six-membered sulfur-containing heterocyclic ring; the thiazolidine or dihydrothiazine congeners make up the vast penicillin and cephalosporin subfamilies. While the sulfur atom can confer some redeeming properties to the molecule as a whole, the fused (B) ring may be carbocyclic in nature (carbapenems, carbacephalosporins), save the lactam nitrogen atom, or alternatively incorporate the replacement of sulfur with oxygen. However, a bicyclic ring system is not a prerequisite for antibacterial activity; simple monocyclic *β*-lactams constitute yet another subfamily of the *β*-lactam antibacterial agents.

The placement of the heteroatom within the (B) ring of the bicyclic nucleus can vary in the case of the six-membered cephalosporins affording some isomeric analogs that retain potent antibacterial activity. Specifically, the heteroatom (sulfur or oxygen) which normally occupies a position alpha to the *β*-lactam nucleus can be transposed to the beta position thus furnishing the isocephems and iso-oxacephems (respectively). Similar structural alterations are not possible with the five-membered penicillin and carbapenem derivatives due to substitution or saturation at the beta position. However, both saturated penam systems (e.g. penicillins) as well as unsaturated systems (e.g. carbapenems) are found in potent antibacterial compounds.

There have been several variations of the simple *β*-lactam-carboxylic acid nucleus itself. In monocyclic (mono *β*-lactam) congeners, the carboxylic acid functionality can be favorably replaced by other organic acids such as sulfonic and phosphonic acids. This feature is unique to the monolactams; analogous cephalosporanic and penicillanic (carboxylic) acid surrogates are inferior. Lastly, the most distinguishing structural feature of the huge family of *β*-lactam antibacterial compounds has, itself, been found tolerant of certain modifications. While the *β*-lactam nucleus has remained a cornerstone feature, both synthetic endeavors as well as the screening of naturally occurring microbial sources have yielded compounds lacking a *β*-lactam nucleus but capable of displaying antibacterial properties reminiscent of conventional agents. γ-Lactams, lactivicins and pyrazolidinone congeners contain acceptable surrogates for the *β*-lactam nucleus as evidenced by the antibacterial mode of action of these compounds. The five-membered pyrazolidinones continue to demand the greatest attention since a variety of potent compounds have been synthesized.

In addition to structural variations of the *β*-lactam nucleus, the substituents that emanate from the lactam core also play a crucial role with regard to antibacterial activity. There are two key portions of the *β*-lactam molecule that have been extensively modified and remain of paramount importance with respect to chemical, physical and antibacterial properties. The first is the acylamino substituent situated alpha to the lactam carbonyl center, which will be collectively referred to as the 'west-end' substituent. (This terminology has been used by some researchers and merely reflects the positioning of these substituents based upon the common depiction of any *β*-lactam compound as shown throughout this text. In many cases, west-ends will be more specifically addressed by their positioning on a given bicyclic *β*-lactam system). In the penicillin and cephalosporin subfamilies, the west-end chain consists most often of an acylated amino moiety (i.e. a substituted acetamido-derived west-end), while in case of the penems and carbapenems, a hydroxylated alkyl group is usually present. The west-end substituents, being adjacent to the reactive lactam carbonyl center, play a key role in terms of stability and antibacterial activity of any given compound, and are at least partly responsible for penicillin-binding protein activity.

The second type of substitution present in most *β*-lactam antibacterials are groups extending from the five- or six-membered (B) ring (Figure 3.40). These substituents are usually referred to by their positioning along the bicyclic *β*-lactam nucleus. There are an enormous number of variations on this general theme that will be discussed throughout this section by considering the individual subfamilies of compounds. The most significant and extensive modifications have been the host of substituents introduced onto the C-3 (or C-3′) position of the cephalosporin *β*-lactams, although recent advances have allowed for a number of favorable alterations to penem and carbapenem systems.

The most important *β*-lactam ring systems are shown below (Figure 3.41) and grouped by structural similarities without addressing specific bicyclic substituents. Some general issues reserved for later discussion include the stereochemistry of the ring junction, as well as specific substituents extending from the bicyclic nucleus.

3.3.1 The penicillins

The penicillins are formally 6-amino-7-oxo-4-thia-1-azabicyclo[3.2.1]-heptane-2-carboxylic acids, but are commonly named as penams, a

Cephalosporins: I: X = S: cephems; X = O: oxacephems; X = CH$_2$: carbacep
II: X = S: isocephems; X = O: iso-oxacephems

Penems and Carbapenems: III: X = S: penems; X = CH$_2$: carbapenems
Penicillins: IV: X = S

Mono-β-Lactams: V: R = SO$_3$H: monobactams
V: R = PO(OR)OH: monophosphams
V: R = C(O)NHSO$_2$R'R'': monocarbams
V: R = OSO$_3$H: monosulfactams

Norcardicins: VI: R = p-OH-phenyl

Clavams: VII: R = CH$_2$OH

Pyrazolidiones: VIII

Gamma-Lactams: IX: R = acidic functionality   Lactivicin: X

Figure 3.41 Members of the β-lactam antibacterial family.

designation in which the sulfur atom is given top priority (rather than the bridgehead nitrogen atom). Other β-lactam antibacterials are named using similar common designations (e.g. the cephalosporins as cephems). Using this nomenclature, the penicillins have the prerequisite carboxylic acid group placed at the C-3 position. The west-end substituent is joined to the C-6 center which is animated, and is usually substituted via (mono) acylation, thus constituting a variety of C-6 acylamido substituents. The β-lactam carbonyl center is located at position 7 and the C-2 center contains the geminal dimethyl substitution characteristic of the penicillins (Figure 3.42).

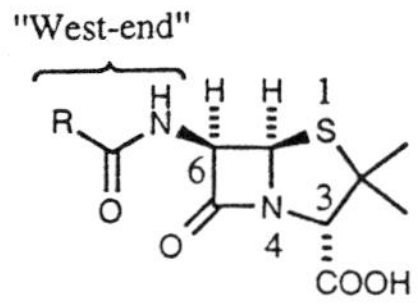

Figure 3.42 Penicillins (penam).

The stereochemistry of the C-3, C-5 and C-6 centers is imperative for the antibacterial activity of the series. The ring fusion (at C-5) is of the (R)-designation, and the C-6 center is also (R). The hydrogen substituents at both of these positions are correspondingly in the alpha (α) orientation. The carboxylic acid must also be properly oriented and must exist in an (S) configuration. All of these stereochemical requirements are imposed by the necessity for the penicillins (and other β-lactam antibacterials) to adopt a spacial arrangement that mimics that of the D-Ala-D-Ala terminus of the UDP-MurNAc-pentapeptide (formal (R) and (S) stereochemical configurations do not inherently correspond to D- and L-amino acid stereocenter assignments).

There are two general modifications to the penicillin nucleus that are important with regard to antibacterial activity. Realistically, there are relatively few positions that are even theoretically available for chemical modification since a small bicyclic nucleus is absolutely required, including all heteroatom substituents and functionality as well as that of the carboxylic acid moiety. In addition, the replacement of the *cis*-α-hydrogen atoms at the C-5 and C-6 centers has been demonstrated to be laborious and usually frustrating in terms of maintaining activity. Most of the useful alterations have been focused upon the introduction of novel west-end substituents and/or the modification of the carboxylic acid to allow for the use of ester prodrugs.

3.3.1.1 C-6 Amino west-end substituents. Most marketed penicillins differ only in the structure of the west-end substituent; these congeners can be segregated according to the type of west-end side chain. Such groupings are based primarily upon the spectrum of antibacterial activity of a compound as a whole, but there are often structural resemblance and similar PBP binding profiles that exist between compounds within any given subset. The west-end substituents encompass an enormous number of acyl groups that are directly attached to the C-6 amine functionality by an amide linkage. Other variations of the C-6 amine moiety (e.g. *N*-alkylation), or the direct substitution of other groups onto the C-6 center of the bicyclic β-lactam itself, are detrimental in terms of antibacterial activity (although some halogenated derivatives (C-6 halopenicillins) do display notable activity against certain β-lactam-destroying enzymes). One exception is the antibacterial (C-6) amidino penicillins which do not contain the typical amide linkage present in virtually all other west-end substituents.

The design and development of the west-end substituents has been aimed at strengthening various weaknesses which have traditionally hampered penicillin agents in terms of activity, stability, resistance and absorption/distribution. Initially, the objective was to discover novel west-ends that could broaden the antibacterial spectrum. Another goal has been to incorporate west-ends that impart improved stability towards bacterial enzymes, or to acidic conditions that are encountered following oral administration. In recent years, the focus has been upon designing new penicillins that are active against some troublesome Gram negative pathogens (e.g. *Pseudomonas aeruginosa*) or Gram positive organisms that are now frequently resistant to earlier agents.

Before considering some specific west-end substituents, it is important to recognize some simple requirements for the west-end itself. First of all, the C-6 amine moiety itself is necessary for appreciable antibacterial activity but, as alluded to earlier, substitution of the amine via monoacylation can afford much more potent congeners. In other words, the west-end side chain can essentially potentiate antibacterial activity. Secondly, alterations that abolish the amide character of the west-end are detrimental (with the exception noted above). Only carboxamido-derived west-end moieties are tolerated; sulfonylation or phosphorylation of the C-6 amine substituent (sulfonamide- or phosphoramide-containing west-end substituents, respectively) are devoid of useful antibacterial activity. Similarly, imide- or carbamate-containing west-ends are inferior. Thus the penicillin west-end substituent is nearly always joined through an amide linkage to the C-6 nitrogen atom off the bicyclic β-lactam nucleus.

The early penicillins contain simple acetylamino west-ends in which the C-6 amine group is formally acylated with an acetic acid derivative that usually contains an aromatic moiety. The most prominent members are benzylpenicillins, such as penicillin G **1** (Figure 3.1) which contains a phenylacetamino west-end, and penicillin V **2** (Figure 3.2) which is the corresponding phenoxyacetylated derivative. These agents are extremely potent against (susceptible) staphylococci and streptococci, but lack activity against many Gram negative species. Accordingly there was an early incentive to develop derivatives that retain this Gram positive potency but extend to Gram negative organisms. Although 'precursoring' fermentation broths (section 3.1) afforded penicillins possessing a variety of west-end substituents, this technique only allows for the incorporation of west-end chains that are similar to those present in the early penicillins.

The second type of penicillin west-end substituents resulted from attempts to develop agents that were inert to specific bacterial enzymes (section 3.5). Resistance caused by penicillin-destroying enzymes was recognized just years after the first penicillins were put to use; staphylococci were particularly predisposed to develop resistance against the early penicillins. Since these enzymes cleave the β-lactam ring, west-end substituents were designed to create a more crowded environment around the β-lactam moiety in hopes of suppressing the action of these enzymes. One way to do this was to move the aromatic ring closer to the β-lactam carbonyl center, and/or to incorporate additional substituents onto the aromatic portion of the west-end.

Two variations upon this theme were advanced. Methicillin **5** (Figure 3.4) contains a well-recognized 2,6-dimethoxybenzamido west-end chain (the penicillin C-6 amino substituent), and has been among the most used of penicillins over the years. The placement of the methoxy groups on the aromatic ring is important; the bis-*ortho* arrangement creates the most effective crowding around the β-lactam carbonyl center, while retaining good activity. Other substituents have been introduced on the phenyl ring and have afforded some potent derivatives, but methicillin remains prominent due to superior pharmacokinetic features that include good *in vivo* efficacy. The oxacillins (Figure 3.5) contain a 5-methyl-3-phenyl-4-isoxazolyl (an isoxazoline-carboxamido) west-end substituent that similarly imposes crowding in proximity to the β-lactam. In these compounds, both the methyl and phenyl substituents are positioned closest to the β-lactam system. Removal of either group increases susceptibility to penicillinases, as does the replacement of the phenyl ring with a smaller alkyl group. However, additional substituents can be favorably placed onto the phenyl substituent and still impart good oral efficacy.

The oxacillins and methicillin reflect some general trends regarding susceptibility towards penicillinase. As described above, it is necessary to increase bulk at the carbon center adjacent to the C-6 amido moiety in order to gain appreciable penicillinase stability. Placing a large group only one carbon unit further out along the west-end chain is ineffectual as evidenced by congeners such as penicillin G and phenethicillin **4** (Figure 3.4). In addition, it is necessary that the alpha carbon center of the west-end amide be fully substituted; most often this position is an sp^2-hybridized carbon center that is part of an aromatic ring. In other words, the aromatic ring, whether phenyl as in the case of methicillin, or isoxazolyl as with the oxacillins, must be directly attached to the C-6 amido substituent in order for the west-end to confer resistance to the action of penicillinases. Finally, the importance of placing additional steric bulk as close to the β-lactam ring as possible is reflected in the substitution patterns of the aryl and heteroaryl portions of the methicillin and oxacillin west-ends. Bis-*ortho*-substitution is optimum; placing even more bulky groups at the other positions is less effective.

Even with the superb Gram positive potency inherent in the early penicillins, as well as the remarkable advantage offered by the penicillinase-resistant derivatives, the need to expand the antibacterial spectrum of the penicillin family became a primary concern decades ago. It was discovered that more hydrophilic west-ends can enhance potency against Gram negative pathogens. Specifically the placement of a basic amine moiety onto the arylacylated west-end is favorable. Ampicillin **11** (Figure 3.6) contains a D-α-aminophenylacetamido west-end and is the most recognized of the aminopenicillins. Even today, ampicillin remains a valuable and widely prescribed chemotherapeutic agent, along with its host of prodrug derivatives (described later). Ampicillin possesses Gram positive potency that nearly rivals that of penicillin G and is markedly more active against a host of Gram negative bacilli. The aminobenzoylated west-end substituent (of ampicillin) is not alone in conferring an extended spectrum of antibacterial activity; the amine functionality can be replaced by other hydrophilic groups. For example, benzylalcohol-derived west-ends can afford congeners comparable to ampicillin *in vitro*.

There have been numerous variations upon the ampicillin west-end substituent in attempts to extend further the spectrum of antibacterial activity or to improve pharmacokinetic properties. In general, substituents on the phenyl ring are detrimental either due to decreased hydrophilicity, or conversely, due to adverse polar effects if an ionizable substituent is present. A notable balance of these opposing forces has been achieved with the placement of a *para*-hydroxyl group onto the phenyl ring. Amoxicillin **12** (Figure 3.6) is essentially comparable to ampicillin in terms of *in vitro* potency, but displays better oral efficacy. Amoxicillin continues to be one of the most widely used oral penicillins for this reason and some interesting amoxicillin prodrug derivatives have been investigated. Other west-end variations include congeners that do not contain an aromatic ring but still bear an amine functionality. Epicillin and cyclacillin possess a cyclohexadiene ring or an aminocyclohexyl ring, respectively.

The penicillin spectrum of activity was further expanded with the discovery of west-ends that imparted useful potency against *Pseudomonas aeruginosa*. In general, these west-ends contain a strongly acidic group at the α-carbon center of the side chain. Specifically, the incorporation of a carboxylic acid or a sulfonic acid is advantageous. Usually the stereo-chemistry at the α-carbon center bearing these groups is rather inconsequential with respect to antibacterial activity and often this center is easily racemized. Many of these congeners display good potency against other Gram negative species such as *Serratia*, which is unaffected by most of the early pencillins. However, in most cases, penicillins containing an acidic west-end display diminished Gram positive potency relative to the premier antistaphylococcal penicillins such as penicillin G.

Carbenicillin **13** (Figure 3.7) possesses an α-carboxyphenylacetamido west-end and is the prototypic penicillin containing an acidic west-end chain. This agent has been extremely valuable for the treatment of potentially deadly infections of *Pseudomonas aeruginosa*. Due to the success of carbenicillin, other antipseudomonal penicillins followed. The phenyl ring of the west-end was replaced by a number of heterocycles; the 3-thienyl congeners are most significant. Ticarcillin **14** (Figure 3.7) is more active than carbenicillin against pseudomonas and is generally more useful. For the most part, heterocyclic variations other than thiophene are not suitable biosteres for the phenyl ring and have afforded inferior analogs. Replacements for the carboxylic acid have been investigated, and a number of potent congeners have resulted although none rival carbenicillin and ticarcillin in terms of use. The most recognized are the sulfonic acid derivative sulbenicillin and the sulfamic acid congener, suncillin (neither is illustrated here). These compounds parallel carbenicillin in terms of spectrum of activity, potency and even pharmacokinetic profile but offer no advantage.

Despite the notable advances ushered in with each new structural breakthrough, the penicillins uniformly demonstrated some glaring weaknesses. To summarize, the early penicillins, although superb against many pathogenic Gram positive bacteria (staphylococci, streptococci), are useless against many common Gram negative (bacilli) organisms, and are readily destroyed by various penicillinases. While the oxacillins and methicillin offer penicillinase resistance, potency against Gram negative microorganisms is lacking. The α-amino penicillin west-ends extend the antibacterial spectrum to many Gram negative pathogens including some *E. coli*, *Shigella* and *Salmonella*, but a number of troublesome bacteria such as *Klebsiella* and *Pseudomonas* species are not covered. Lastly, the antipseudomonal penicillins such as carbenicillin, sunicillin and sulbenicillin suffer from weak Gram positive potency, as well as a number of undesirable physical, chemical and pharmacokinetic properties. These shortcomings have led to other approaches in the search for the ultimate broad spectrum penicillin.

The use of polar west-ends had successfully extended the penicillin spectrum of activity in some ways; ampicillin and carbenicillin are examples in which a basic or acidic side chain (respectively) imparts improved Gram negative potency. Therefore, one approach was to modify these polar groups further, and in particular, the ampicillin west-end was explored. However, as alluded to earlier, the polar nature of the side chain must be maintained in order to keep good Gram negative potency. For example, acylation of the ampicillin amine with lipophilic groups, or esterification of the carbenicillin carboxylic acid is detrimental (although carbenicillin esters can be prodrugs for the parent penicillin). It was discovered that acylation of the ampicillin west-end amine functionality with certain polar groups could afford an exquisite balance of potency. In particular, a number of cyclic urea substituents are exceptional.

Compounds possessing this feature are referred to as ureidopenicillins. The simplest member of the acylureidopenicillins is azlocillin **16** (Figure 3.8) which contains a five-membered cyclic urea system joined via *N*-acylation to the α-amino substituent of ampicillin (an ((imidazolidinyl)-carbonyl)aminophenylacetamido west-end). Many ureidopenicillins are more active against *Pseudomonas aeruginosa* than carbenicillin, and potency against a number of other pathogenic Gram negative species, such as *Klebsiella*, is increased. The presence of the urea group imparts improved penetration into these Gram negative species traditionally resistant to other penicillins. Mezlocillin **17** (Figure 3.8) is an analog that contains a methanesulfonyl group attached to the distal nitrogen atom of the urea ring. Both of these compounds have found use in antibacterial chemotherapy. The most recent addition to the market has been piperacillin **15** (Figure 3.8) which is not formally a ureidopenicillin since it contains a piperidione ring rather than a urea system. However, piperacillin displays a similar spectrum of antibacterial activity, and is in fact, more potent than azlocillin and mezlocillin. Piperacillin is widely prescribed today and is replacing some older β-lactam agents due to its broad spectrum of antibacterial activity.

One apparent requirement of a penicillin west-end seems to be that of an amide-derived west-end; all of the agents considered so far contain this general feature. As noted earlier, most variations that stray from an amide-based west-end chain are detrimental in terms of activity. Surprisingly, a peculiar penicillin west-end has afforded a unique penicillin agent. Mecillinam **20** (Figure 3.10) is a C-6-β-amidinopenicillin that remains somewhat of an anomaly due to its unique spectrum of antibacterial activity, PBP binding profile, as well as the structure of its west-end. Mecillinam contains a hexahydroazepine ring system joined to the penicillin C-6 amine center through an imine bond (a (hexahydro-1*H*-azepinyl)methyleneamino west-end). The discovery of amidinopenicillins was a result of scrupulous testing for antibacterial activity of synthetic intermediates; mecillinam was eventually recognized and advanced onto the market. Unlike most penicillins, mecillinam is only weakly active against Gram positive bacteria and does not undergo appreciable binding to the PBP 1's and/or to PBP 3. Mecillinam exclusively targets PBP 2 of susceptible bacteria and possesses good activity against *E. coli* and other *Enterobacteriaceae* such as *Shigella* and *Salmonella*; it is active against some strains typically resistant to other penicillins, but mecillinam is not active against *Pseudomonas aeruginosa*.

Other amines have been incorporated into amidinopenicillins but few congeners have rivaled the *in vitro* potency of mecillinam. Replacement of the hexahydroazepine ring with other cyclic heterocycles, whether smaller or larger in size, or containing unsaturation within the ring are inferior replacements. Amidines derived from simple dialkylated amines are also less potent overall, as are bicyclic ring systems. As a result, mecillinam is the only amidinopenicillin of significance at this time.

Another variation of the penicillin west-end is the development of carbenicillin prodrugs. Carbenicillin ester prodrugs involve derivatization of the carboxylic acid of the west-end rather than esterification of the penam carboxylic acid which is more common among the penicillin prodrugs. Carbenicillin prodrugs offer the advantage of oral administration and good absorption relative to the parent compound; the active penicillin is liberated as a result of ester hydrolysis *in vivo*. The phenyl ester and the

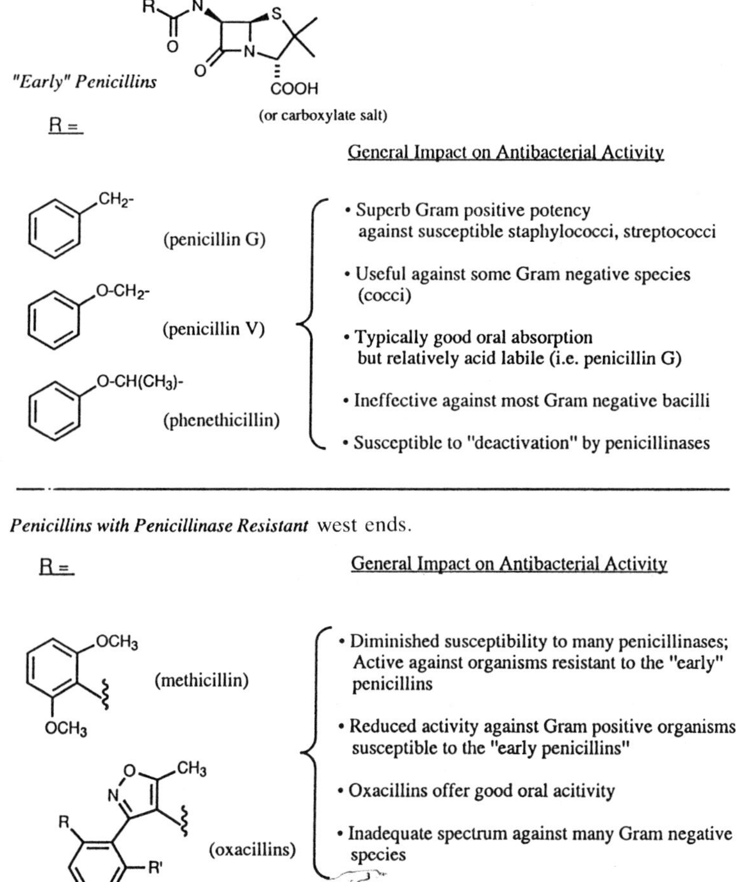

Figure 3.43 Prototypic penicillin west ends.

indanyl ester of carbenicillin (carfecillin and carindacillin, respectively) have been marketed.

It is impossible to present a comprehensive overview of every penicillin that has resulted from modification of the west-end substituent; prototypic west-ends are shown in Figures 3.43 to 3.45.

3.3.1.2 Other variations at C-6. Penicillins naturally contain a C-6-α-hydrogen substituent and in general it is unfavorable to introduce a different group at this position. Substituents at the C-6-α position can

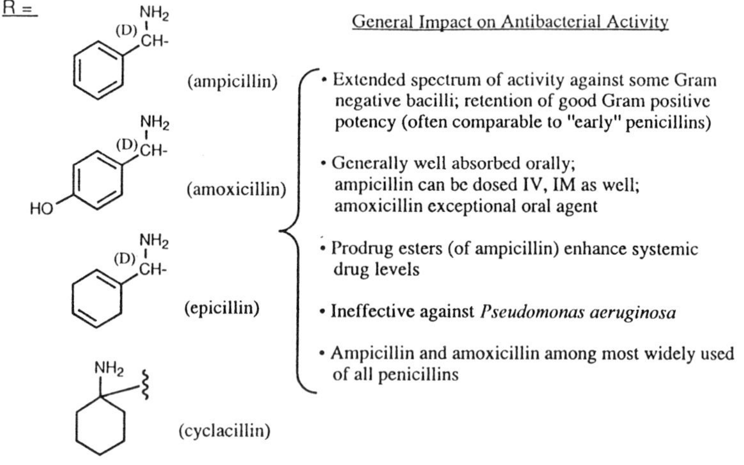

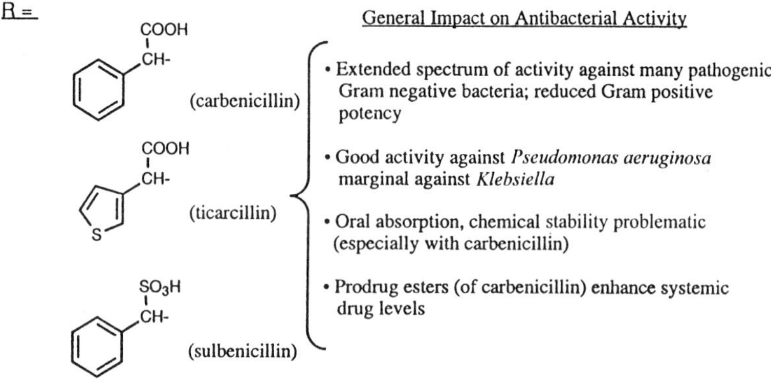

Figure 3.44 Prototypic penicillin west-ends.

Broad Spectrum Uriedo Penicillins (Urea-derived "west-ends" of ampicillin)

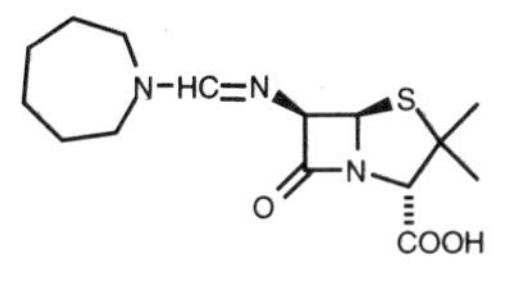

Penicillins with a C-6 Amidino "West-end"

Figure 3.45 Prototypic penicillin west-ends.

crowd the β-lactam center and result in decreased antibacterial activity. Even relatively small groups tend to produce this undesired effect, although there have been some exceptions. The most notable is the introduction of a methoxy group at the C-6-α position which has afforded temocillin (Figure 3.46). Temocillin is resistant to the action of penicillinases, as well as other β-lactamases but unfortunately potency against Gram positive species is diminished relative to its natural counterpart. However, good activity against *Enterobacteriaceae* (e.g. *E. coli*, *Klebsiella*, *Proteus* species) is maintained, thus making temocillin a significant and rather unique penicillin.

3.3.1.3 Substituents at sulfur. Sulfur is the only atom tolerated at position 1 of the penicillins in order to retain appreciable antibacterial

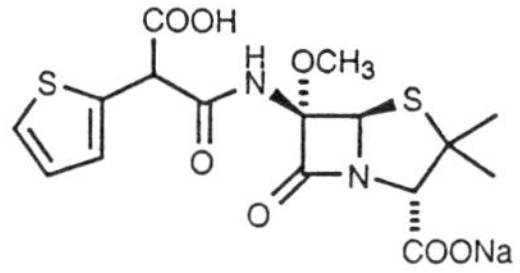

Figure 3.46 Temocillin.

activity. Unlike some other members of the β-lactam family of anti-bacterials, the replacement of sulfur with oxygen, or with carbon, is unacceptable in terms of antibacterial activity. The sulfur atom can be oxidized to the corresponding sulfoxide or sulfone analogs. Derivatives of this type do not rival the antibacterial properties of the naturally occurring penicillins. However, penicillin sulfoxides have found use in the synthesis of new β-lactam antibacterials; the sulfoxide can serve as a means of protecting the sulfur position during chemical manipulations. Furthermore, penicillin sulfoxides are important precursors to cephalosporins via the Morin rearrangement (section 3.4).

3.3.1.4 C-2 substituents.   The geminal dimethyl group at C-2 is character-istic of the penicillins and arises as a consequence of biosynthesis. A number of variations have been obtained by chemical synthesis; in general, no appreciable advantage has been realized. Interesting exceptions are the C-2,C-3-methylenopenams (Figure 3.47) which structurally resemble the penicillins but actually behave more as cephalosporins in terms of PBP binding profiles and spectrum of antibacterial activity.

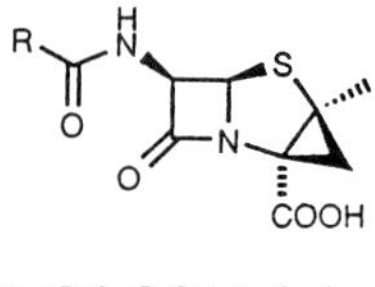

Figure 3.47 C-2,C-3-Methylenopenams.

3.3.1.5 C-3 substituents.   In general, derivatization of the C-3 carboxylic acid functionality is not tolerated unless the free penicillin carboxylic acid can be generated *in vivo*. As mentioned earlier, a number of penicillin carboxylate esters are useful marketed agents but antibacterial activity is attributed to the parent penicillin that arises from ester cleavage. Simple alkyl and aryl penicillin esters are cleaved slowly and are not useful prodrugs. However, 'doubly-activated' penicillin esters, such as alkanoyl-oxyalkyl congeners undergo rapid cleavage *in vivo* to generate active penicillin. Most of the penicillin prodrugs are derived from ampicillin;

pivampicillin **18** and bacampicillin **19** (Figure 3.9) are examples, although ester prodrugs of mecillinam and other compounds are available.

3.3.1.6 Variations at N-4. The nitrogen atom at the ring junction (position 4) is vital for antibacterial activity; the nitrogen atom contributes to the reactivity of the β-lactam carbonyl center. In theory, other atoms may be able to replace nitrogen and keep with the prerequisite of a strained ring system. However, compounds that contain carbon (cyclic ketones) or oxygen (cyclic lactones) in place of the nitrogen, although being part of strained ring systems, are inactive. The nitrogen center is crucial for a number of reasons including its contribution to ring strain, partially via resonance, as well as for factors pertaining to enzyme recognition.

3.3.2 The cephalosporins

The naturally occurring cephalosporins are 7-amino-8-oxo-5-thia-1-azabicyclo[4.2.0]oct-2-ene-2-carboxylic acids commonly described as cephems (Figure 3.48). The cephalosporins are strikingly similar to the penicillins in many aspects including several structural features and mode of action. Like the penicillins, the cephalosporins have an important west-end substituent joined by an amide linkage to the cephem C-7 amine center; the C-8 center contains the β-lactam carbonyl center. The stereochemistry of the cephalosporin C-6 and C-7 centers must be of the (*R*)-designation for antibacterial activity; this spacial disposition is analogous to that of the penicillins. Like the penicillins, the cephalosporins also require an appropriate carboxylic acid group for activity; this group is attached to the C-4 position, placing it in a similar orientation along the β-lactam system as the penicillin carboxylic acid.

The cephalosporins differ from the penicillins in several ways. The naturally occurring cephalosporins contain a dihydrothiazine (B) ring fused to the β-lactam unit; in contrast, the penicillins have a thiazolidine (B) ring (Figure 3.40). The cephalosporin dihydrothiazine ring is not saturated nor does it contain geminal dimethyl substitution; both are characteristic of the

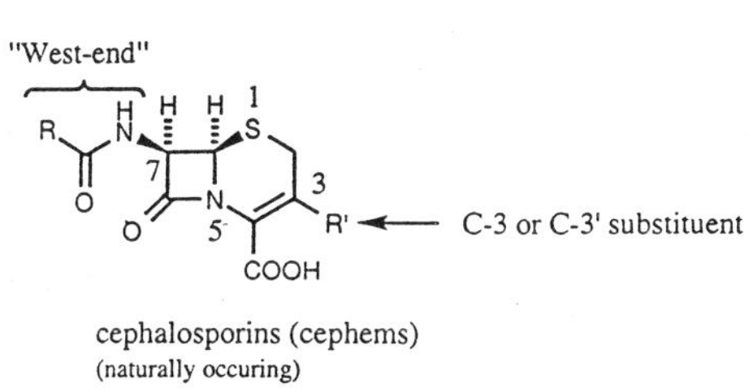

cephalosporins (cephems)
(naturally occuring)

Figure 3.48 Basic cephalosporin SAR.

penicillins. Furthermore, in the case of the cephems, a sulfur atom at position 1 is not an absolute requirement for antibacterial activity; carbacephems and (dethia)oxacephems are potent analogs. In addition, a variety of isomeric derivatives in which the heteroatom is transposed with the C-2 carbon center (i.e. isocephems, iso-oxacephems), have been shown to possess good antibacterial properties. The cephalosporins contain an endocyclic carbon–carbon double bond conjugated to the carboxylic acid functionality, therefore there is no stereocenter within the dihydrothiazine ring. The cephalosporins also differ from the penicillins in having a host of substituents that extend from the C-3 position; penicillins contain their characteristic 'rabbit ear' methyl groups at C-2.

The most common structural variations of the cephalosporin family involve the west-end chains at the C-7 position and the C-3 substituents; both groups are important in conferring various properties to the molecule as a whole. Most early cephalosporins, being active against some Gram negative bacteria as well as Gram positive organisms, inherently offer a more balanced antibacterial specturm than the prototypic penicillins. A number of direct comparisons between cephalosporins and penicillins containing identical west-ends have shown this to be a general trend. Unfortunately the cephalosporins, as a family, are innately less potent than their penicillin counterparts. For example, susceptible staphylococci are inhibited by penicillins (e.g. penicillin G) at much lower concentrations than those needed with the analogous cephalosporin (the congener containing the identical phenylacetyl west-end).

3.3.2.1 C-7 Amino (west-end) substituents. The cephalosporin west-ends consist of a variety of acyl groups attached to the C-7 amine functionality through an amide linkage. As in the case of the penicillins, the west-end substituent can impact greatly on antibacterial activity. The cephalosporins are typically placed into 'generations' based upon their spectrum of activity but this is not only dictated by the west-end; the C-3 substituent can influence the potency or spectrum of action as can nuclear variations.

The first cephalosporin (cephalosporin C) contains a rather unique amino acid-derived (5-amino-5-carboxy-1-oxopentyl) west-end. Cephalosporin C **21** (Figure 3.11) is not clinically significant but it is important; cephalosporin C can be converted to the key synthetic intermediate, 7-aminocephalosporanic acid (7-ACA), precursor to thousands of cephalosporins (section 3.4).

The first generation cephalosporins resemble the early penicillins in spectrum of antibacterial activity, but do not contain the same west-end substituents. These compounds are highly active against Gram positive cocci (staphylococci and streptococci), but have a limited Gram negative spectrum. The first generation cephalosporins are usually active against common Gram negative bacilli such as *E. coli*, and some *Klebsiella* and selected *Proteus* species, but are inactive against *Serratia*, *Haemophilus* and *Pseudomonas aeruginosa*. Since this activity closely parallels that of the early penicillins, the first generation cephalosporins have been used as alternatives; the cephalosporins do offer the advantage of being unaffected by penicillinases that destroy their penicillin counterparts. However, cephalosporins are susceptible to other *β*-lactamases.

A number of first generation cephalosporins have excellent oral activity although this property can not be solely attributed to the west-end substituents. Other cephalosporins containing the same west-ends can have poor oral absorption and may be reserved for parenteral use. As will be discussed later, the cephem C-3 substituent can also have a profound influence on oral absorption and other pharmacokinetic properties. Nevertheless, the D-α-aminobenzyl side chain is exceptional in that many of the oral cephalosporins contain this west-end. Indeed, the most successful first generation cephalosporins, cephalexin **23** (Figure 3.13), cefaclor **29** (Figure 3.15) and cephaloglycin (not illustrated), all contain the D-α-aminobenzyl side chain identical to that found in ampicillin. The trend of incorporating penicillin west-ends has carried over to the use of the amoxicillin west-end in cefadroxil **24** (Figure 3.13) and cefprozil **30** (Figure 3.15). These cephalosporins are also effective oral agents; cefprozil is normally considered to be a second generation agent due to its expanded spectrum of activity.

A number of variations to the α-aminobenzyl-derived west-ends have been investigated, and like the penicillins, reduction of the phenyl ring is tolerated in the cephalosporin series. For example, the 1,4-cyclohexadienyl west-end analogous to that present in epicillin, affords cephalosporins with good activity. Cephradine **25** (Figure 3.13) is the most significant, although it is somewhat weaker in potency than cephalexin. However, additional saturation of the phenyl ring is generally unfavorable. Lastly, substitution onto the aromatic ring itself generally offers no advantage, and in many cases, antibacterial activity may actually be diminished, probably as a result of increased steric bulk about the *β*-lactam center.

Interestingly, the incorporation of simple phenylacetylated west-ends, such as those present in penicillin G and penicillin V, produces cephalosporins with only marginal antibacterial activity. But in contrast to the penicillins, the replacement of the phenyl ring with various heterocyclic systems can be quite favorable. This is exemplified by cephalothin **26** (Figure 3.14) which has found significant use as a parenteral first generation cephalosporin. Cephalothin contains a characteristic thienyl-acetylated west-end chain, and is one of a handful of *β*-lactam agents to have a thiophene ring as a phenyl ring biostere (this west-end will be revisited briefly in the discussion on the cephamycins). Cephapirin **28** (Figure 3.14) contains a rather unique pyridinylthioacetylated substituent;

this parenteral agent has been replaced by newer agents. Even a tetrazole can serve as the aromatic moiety, as evident by cefazolin (see the summary in Figure 3.49) which was introduced on the parenteral market about the same time as cephapirin. In some cases, a west-end lacking an aromatic moiety can afford notable cephalosporin antibacterials; cephacetrile **27** (Figure 3.14) has a simple cyanoacetylated west-end substituent.

The second generation cephalosporins have an extended spectrum of activity that includes *Serratia*, *Neiserria* and *Haemophilus* species but not *Pseudomonas aeruginosa*. A wide variety of west-ends are included in this category and some impart β-lactamase stability by introducing steric protection around the β-lactam carbonyl center. The D-α-aminobenzyl west-end, reminiscent of many first generation cephalosporins, is present in several orally active second generation agents; cefaclor **29** (Figure 3.15) has been a prolific agent for over a decade. As mentioned earlier, cefprozil, another second generation oral cephalosporin, possess the *para*-hydroxylated (amoxicillin) version of this west-end. The most recent addition to the oral cephalosporin arsenal has been loracarbef **51** (Figure 3.24) which is the carbacephem analog of cefaclor (section 3.6).

Replacement of the amine functionality of the D-α-aminobenzyl (ampicillin) west-end with a hydroxy group (with retention of stereo-chemistry; mandelic acid derived) is a favorable modification. A number of second generation cephalosporins contain the D-α-hydroxybenzyl west-end. The most significant of these compounds also contain a characteristic tetrazolylthiomethyl C-3' substituent (which will be addressed later). Cefamandole **31** and cefonicid **32** (Figure 3.16) are second generation parenteral cephalosporins with these features. Ceforanide **33** (Figure 3.16) is also included in this group because of its C-3' substituent, as well as its similar spectrum of activity and pharmacokinetic properties. This compound possesses a rather unusual (aminomethyl)phenacetyl west-end. Cefamandole, cefonicid and ceforanide all accumulate to high concentrations in the serum, often for long periods of time and for this reason these agents are well suited for certain prophylactic uses.

Bacterial resistance towards the cephalosporins did not emerge as quickly as with the early penicillins, but nevertheless most of the agents described above are now susceptible to deactivation by a variety of β-lactamases. In response, the cephalosporin west-ends were designed to encumber the approach of β-lactam-destroying enzymes (β-lactamases) by introducing steric bulk around the β-lactam center. Unlike the penicillins, the introduction of additional substituents onto the carbon center adjacent to the C-7 amido moiety (also known as the α-carbon atom of the west-end chain) abolishes useful antibacterial activity. As a result, a more subtle way in which to introduce steric bulk, in order to shield the β-lactam center, was needed. The most significant advance to date has been the use of heteroaryl(methoxyimino)acetamido west-ends; a number of potent cephalosporins with β-lactamase stability have been developed from this theme. Cefuroxime **38** (Figure 3.18) contains the furanyl(methoxyimino) acetamido west-end and is a widely used cephalosporin due to its broad spectrum of activity (second generation agent) as well as its stability towards many lactamase-producing species. In addition, it can be administered parenterally, or can be given orally as a prodrug ester.

In addition to the furanyl(methoxyimino)acetamido west-end present in cefuroxime, some other structurally related west-ends have been advanced. The most significant variation has a characteristic aminothiazole ring joined to the oximino carbon atom. This aminothiazolyl(methoxyimino) acetamido west-end is present in many third generation cephalosporins such as the parenteral agents cefotaxime **40**, ceftizoxime **41** and ceftriaxone **42** (Figure 3.19). These compounds are potent against most Gram positive and Gram negative bacteria, with the notable exception of pseudomonas, and are usually not susceptible to the action of many β-lactamases. Aminothiazolyl(methoxyimino)acetamido west-ends are also present in the latest cephalosporins which are being touted as fourth generation agents due to exceptional antipseudomonal activity without compromised Gram positive potency. The most advanced agents at the present moment are cefepime **45** and cefpirome **46** (Figure 3.21) although other compounds (section 3.6) with similar oximino west-ends are in development.

The 2-amino-4-thiazolyl moiety is the preferred heterocyclic system of the heteroaryl(methoxyimino)acetamido west-ends investigated to date, although some 2-furanyl derivatives (i.e. cefuroxime) are important. Other heterocycles have been incorporated and have afforded cephalosporins with potent antibacterial activity and good β-lactamase stability. The corresponding thiophene congeners and some simple phenyl derivatives are noteworthy but these compounds generally offer no advantage. The point of attachment of the heterocycle onto the oxime center is important; the 2-furanyl, 2-thienyl, 4-thiazolyl systems are preferred. The regio-chemistry about the oxime moiety is crucial; only the *syn* regiomers impart good antibacterial potency. Oxime functionality is preferred over other sp^2-centers such as ketones. Most oximino west-ends are *O*-alkylated; many are derived from methoxylamine although a number of new compounds (section 3.6) are substituted with more complex groups. The oxime need not be *O*-alkylated; the clinical candidate cefdinir contains a hydroxyimino west-end.

Other modifications have included elaboration of the imine oxygen substituent as illustrated by ceftazidime **43** (Figure 3.20), which is a premier antipseudomonal (parenteral) cephalosporin. The presence of an acidic group on the ceftazidime west-end substituent enhances Gram negative potency but reduces Gram positive activity. Recall that a similar trend was observed with the penicillins carbenicillin, sulbenicillin and sunicillin. Cefixime **47** (Figure 3.22) also has a carboxylic acid-substituted oximino west-end, and has good Gram negative activity and acceptable Gram positive potency. Cefixime, generally considered as a third generation agent, may be destined to become the preferred oral cephalosporin.

One of the more interesting modifications has been the removal of the heteroatoms of the oxime system. It turns out that the oxime itself is not necessary for antibacterial activity and the corresponding alkenyl (carboxyethylidenyl) west-end, complete with aminothiazole portion, can be a suitable modification. The alkenyl system acts as an oxime mimetic and properly disposes steric bulk about the β-lactam system, resulting in notable β-lactamase stability and good antibacterial activity. Ceftibuten **48** (Figure 3.22) is the archetype and has recently been introduced onto the chemotherapeutic market. Its spectrum of antibacterial activity is similar to that of most third generation cephalosporins.

The final cephalosporin west-end modification to be addressed can be traced from the development of the ureidoacylpenicillins such as piperacillin. It was natural that the piperacillin west-end be extended to the cephalosporin family in an attempt to produce potent broad spectrum agents whose coverage included the pseudomonads, along with good β-lactamase stability. As with the ureidoacylpenicillins, the strategy was centered around modification of the ampicillin and amoxicillin west-ends via acylation of the α-amine functionality, and a number of exciting cephalosporins have recently been synthesized using this approach. The first marketed ureidoacyl-derived cephalosporin is cefoperazone **44** (Figure 3.20), a parenteral agent. Some prodrug forms of cefoperazone have also been prepared and are promising in terms of oral activity. There are a variety of other *N*-acylated congeners such as cefpiramide that are in early stages of development (section 3.6).

The cephalosporin west-end substituent plays a premier role in the manifestation of antibacterial activity; the general trends are presented below (Figures 3.49 to 3.51).

3.3.2.2   C-3 (C-3′) substituents.   The C-3 substituents of the cephalosporins can have profound influence on the molecule as a whole with respect to pharmacokinetic properties and metabolism, but usually contribute little in terms of spectrum of antibacterial action or potency. From the previous discussion, it is the C-7 west-end substituent that primarily dictates potentiation of antibacterial activity and/or the expansion of spectrum of action. One such anomaly involves cephalosporins that contain tetrazoylthiomethyl substituents at the C-3 position, as will be described below. Nevertheless the significance of the cephalosporin C-3 (or C-3′) substituent has stimulated an enormous amount of research. For example, it may be desirable to attempt to improve upon the *in vivo* efficacy of a particular cephalosporin that possesses exceptional *in vitro* potency. In such cases, variations of the C-3 substituents could be helpful

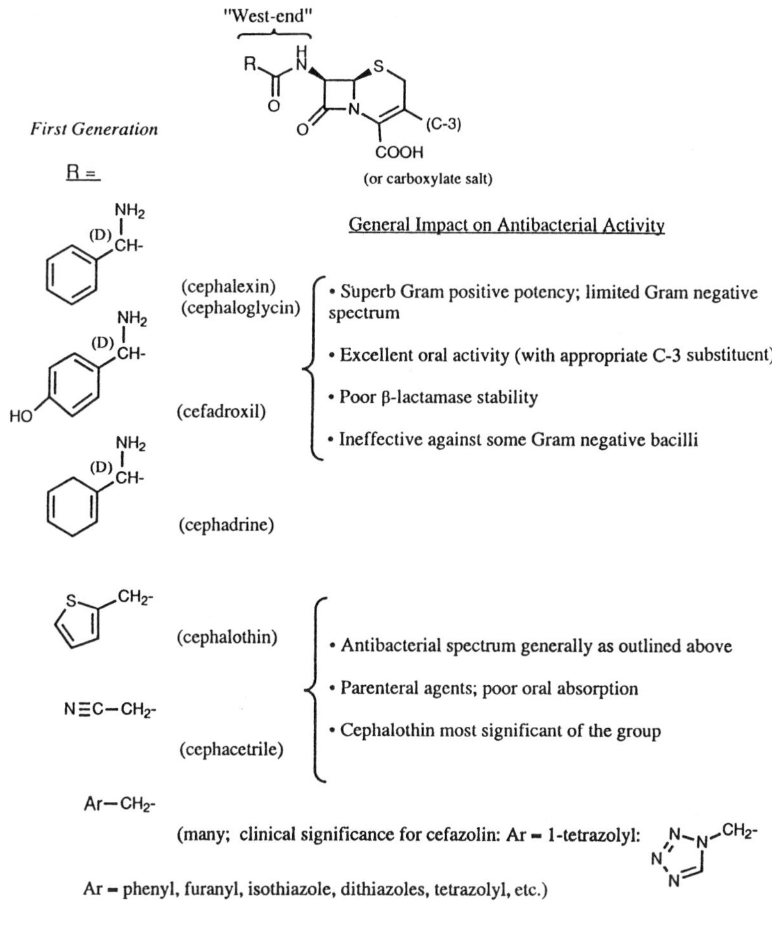

Figure 3.49 Prototypic cephalosporin west-ends.

since absorption, distribution or metabolism may be favorably altered. In another scenario, certain C-3 substituents may be employed in the hope of conferring oral activity.

The most common C-3 substituent of the cephalosporins family is a simple acetoxymethyl group, produced naturally by microbial biogenesis. This substituent is sometimes referred to as a C-3′ acetoxy moiety; this terminology is used more in conjunction with the chemical reactivity of this vinylogous acetate. The most common transformations of the C-3′ acetoxy group are centered around the facile displacement of acetate and this chemistry has allowed for the introduction of a host of nucleophiles onto the C-3′ carbon center (section 3.4).

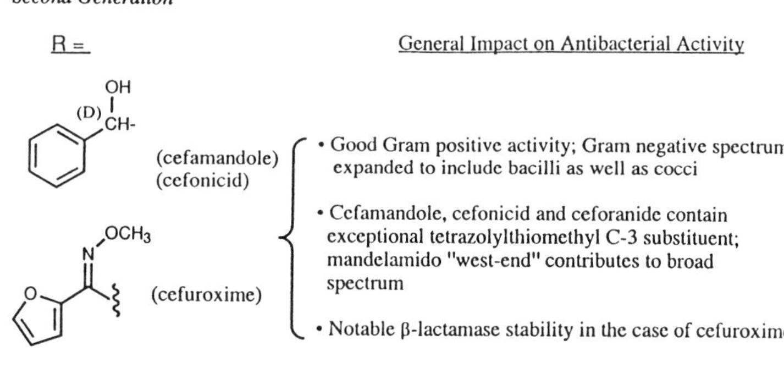

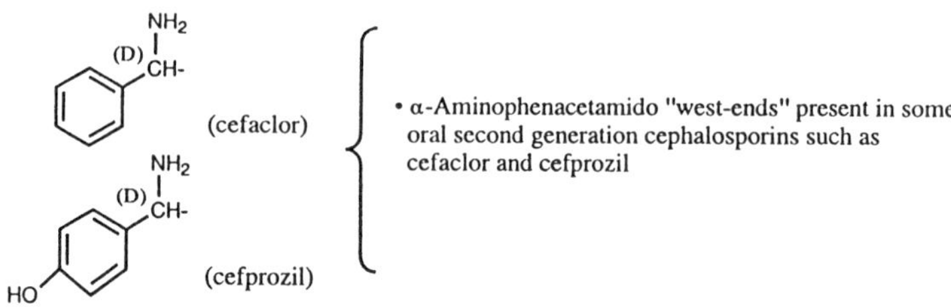

Figure 3.50 Prototypic cephalosporin west-ends.

Rather surprisingly, the C-3 acetoxymethyl group spans the 'generations' of cephalosporins even though there are inherent stability problems associated with this moiety. The first generation cephalosporins cephalothin **26** and cephacetrile **27** (Figure 3.14), as well as the third generation agent cefotaxime **40** (Figure 3.19) all contain this moiety. There is a pronounced tendency for the C-3′ acetoxy group either to undergo displacement as a consequence of β-lactam cleavage, or simple deacylation. These processes are most prevalent upon exposure of the cephalosporin to an acidic or basic pH, or by the action of certain enzymes. Deacetylation to the corresponding cephalosporin alcohol is detrimental since antibacterial activity is significantly diminished. Although extremes in pH can lead to deacetylation, the phenomenon is commonly caused by esterases that target this position without cleaving the β-lactam system.

The C-3′ acetoxy group is one of several C-3′ substituents that are expelled as a consequence of β-lactam ring opening. This can be caused by a bacterial enzyme such as a transpeptidase and lead to antibacterial activity, or it can be mediated by the destructive action of β-lactamases. There has been some detailed investigation into the sequence of events that leads to expulsion of the C-3′ substituent. In most cases, acylation of

Third Generation

R =

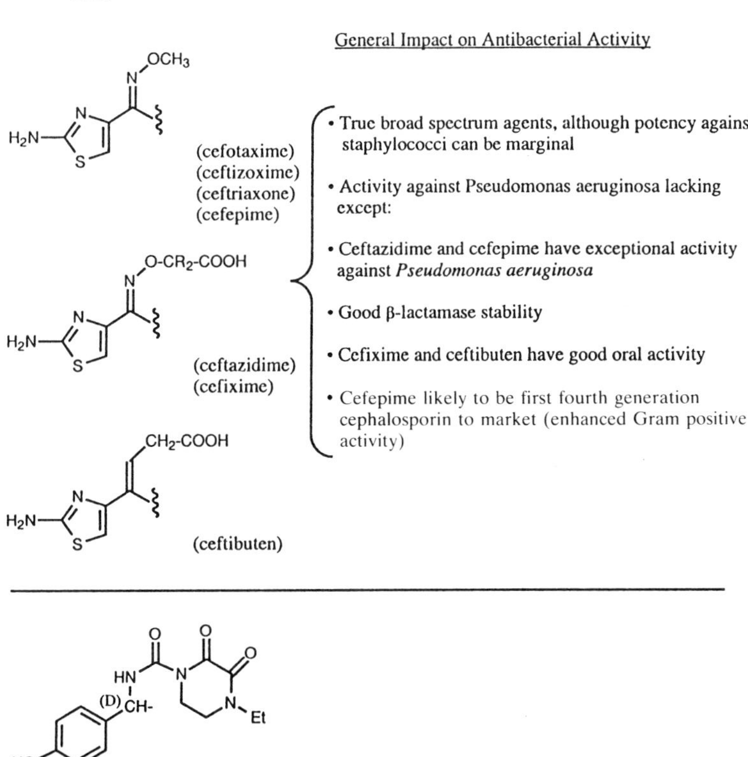

Figure 3.51 Prototypic cephalosporin west-ends.

the bacterial enzyme with commitment *β*-lactam cleavage occurs prior to the loss of this group. This implies that *β*-lactam reactivity is probably not enhanced by the C-3′ substituent, as could be inferred if a concerted process was operative. Furthermore, the C-3′ substituent does not necessarily need to be expelled from the molecule (see below) for antibacterial activity to result.

Orally active cephalosporins contain distinctive C-3 substituents, although it should be recalled that many also contain the α-amino-phenacetamido west-end (another structural feature implicated in conferring oral activity). Oral cephalosporins almost always contain a small uncharged group joined directly to the C-3 carbon center. Cephalexin **23**, cephradine **25** and cefadroxil **24** (Figure 3.13) all have a C-3 methyl group; cefaclor **29** (Figure 3.15) and loracarbef **51** (Figure 3.24) possess a chlorine atom bonded to the C-3 carbon center. Small hydrocarbon substituents are acceptable, evidenced by cefprozil **30** (Figure 3.15) and cefixime **47** (Figure 3.22) which have a propenyl and ethenyl moiety at C-3, respectively. Lastly, the C-3 position can be left unsubstituted; ceftibuten **48** (Figure 3.22) has a hydrogen atom directly attached to C-3 carbon center. Cephalosporins with other C-3 substituents can possess oral activity but certainly there is an obvious trend in the preference for small, uncharged groups.

In contrast, most parenteral cephalosporins contain C-3 substituents that are large in size and rich in heteroatoms and sometimes contain a charged species. Unlike the oral cephalosporins, these groups typically have a sp^3-hybridized carbon atom at the C-3' position that is joined to a polar group such as a heterocycle, a carboxy-derived functionality or a charged group. Thiolated nitrogen heterocycles are common C-3' substituents; the tetrazoylthiomethyl analogs cefamandole **31**, cefonicid **32** and ceforanide **33** (Figure 3.16) are illustrative. An exceptional variation on this theme is the triazinylthiomethyl C-3 substituent found in ceftriaxone **42** (Figure 3.19). This group, (a (1,2,5,6-tetrahydro-2-methyl-5,6-dioxo-1,2,4-triazinyl-3-yl)thiomethyl C-3 substituent) is unique; isomeric variations and other minor structural changes in the C-3' heterocyclic system afford potent derivatives but without the advantages that have made ceftriaxone the leading injectable cephalosporin agent.

Among the other polar C-3' substituents, a carbamate group can be considered to be advantageous compared to the acetoxy moiety since it is inherently more stable. The carbamoyloxymethyl C-3 substituent is found in cefuroxime **38** (and its ester prodrug **39**) (Figure 3.18), as well as the cephamycin cefoxitin **34** (Figure 3.17) (see below). The remaining 'type' of cephalosporin C-3 substituents are those that contain a (charged) quaternary nitrogen atom generally associated with good antipseudomonal activity. Most of these cephalosporins are third generation or even fourth generation agents since good Gram positive potency can be kept in some cases. The prototype is ceftazidime **43** (Figure 3.20) which has a simple pyridinium C-3' substituent; cefpirome **46** (Figure 3.21) is an advanced version in which an additional ring is present. The ring system need not be aromatic as evidenced by cefepime **45** (Figure 3.21) which possesses a methylated pyrrolidine C-3' substituent.

There have been other C-3' substituents, designed to take advantage of the fact that the C-3' substituent can be expelled upon β-lactam cleavage and exert its own distinctive biological properties (section 3.6). Certainly, modification of the cephalosporin C-3 (or C-3') substituent remains a promising area of research. A brief overview of the common cephalosporin C-3 substituents is shown in Figures 3.52 and 3.53.

3.3.2.3 C-7-α-substituents: the cephamycins. The cephamycins are cephalosporin analogs that contain a C-7-α-methoxy substituent. The

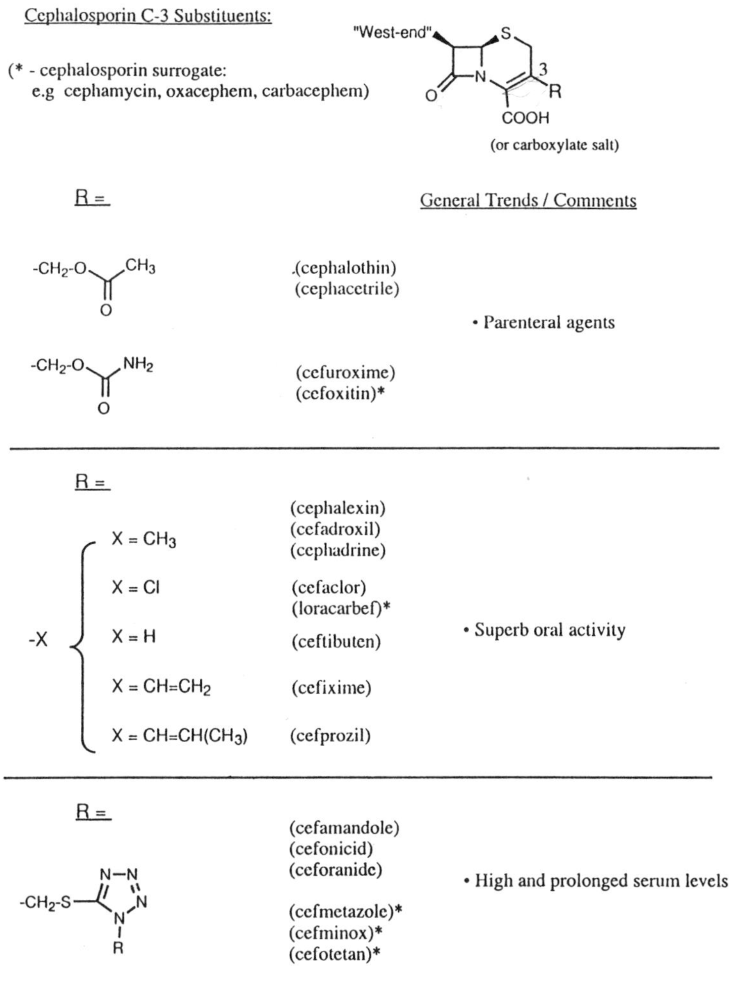

Figure 3.52 Cephalosporin C-3 substituents.

cephamycins typically display a spectrum of activity comparable to the second generation cephalosporins. However, the cephamycins offer the advantage of good activity against some anaerobes and remarkable stability towards *β*-lactamases due to the presence of the methoxy group. Many structural features of the cephalosporins carry over to the cephamycins. Some cephamycins (cefmetazole **35**, cefminox **36** and cefotetan **37** (Figure 3.17)) contain a C-3 tetrazolylthiomethyl substituent identical to that present in cefamandole **31**. Cefoxitin **34** has a carbamoyloxymethyl

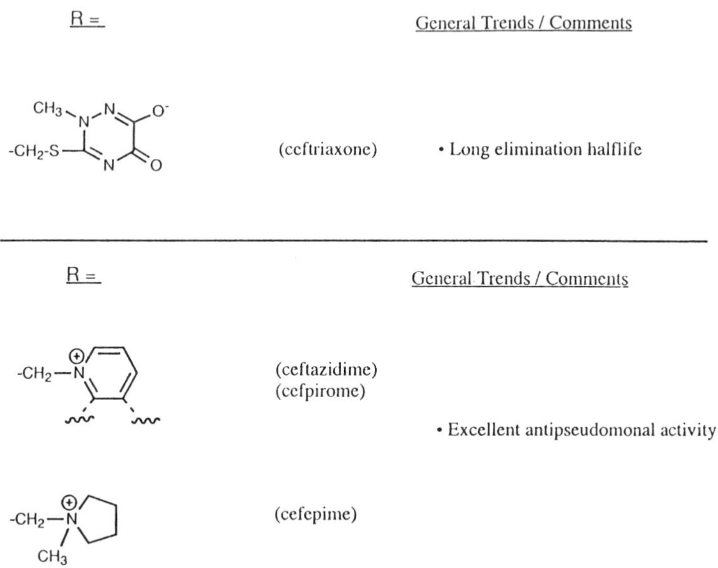

Figure 3.53 Cephalosporin C-3 substituents.

C-3 substituent like that found in cefuroxime **38**, and cefoxitin has a 2-thienylacetamido west-end identical to cephalothin **26**.

3.3.2.4 Nuclear variations. There are various nuclear analogs of the naturally occurring cephalosporins that are good antibacterials. The sulfur atom can be replaced by either oxygen or carbon yielding (1-dethia)oxacephems and (1-dethia)carbacephems, respectively. Alternatively, the sulfur or oxygen heteroatom can be transposed to the adjacent (C-2) position affording isocephems (1-dethia-2-thiacephalosporins) and iso-oxacephems (1-dethia-2-oxacephalosporins) (see Figure 3.41). Less promising variations include the 1-dethia-2-thiacephalosporins, (dethia)silacephalosporins and various azacephalosporins, to name a few.

Two nuclear variants of the cephalosporins have impacted upon antibacterial chemotherapy: the oxacephems and the carbacephems. The oxacephems have been shown to be generally as active *in vitro* as the corresponding cephem counterparts and to possess attractive tolerability and pharmacokinetic features. Two oxacephems have been advanced and are comparable to the third generation cephalosporins regarding breadth and potency of antibacterial activity. Moxalactam **49** and flomoxef **50** (Figure 3.23) more closely resemble a cephamycin in having a C-7-α-methoxy group; both compounds have a tetrazolythiomethyl C-3 substituent. The other nuclear variation is that of the carbacephems. Loracarbef **51** (Figure 3.24) is the first marketed carbacephem and has comparable antibacterial activity but improved stability as an oral agent compared to its cephem counterpart cefaclor.

3.3.3 The carbapenems and the penems

The carbapenems are officially 7-oxo-1-azabicyclo[3.2.0]hept-2-ene-2-carboxylic acids (Figure 3.54). The penems are structurally related analogs that possess a sulfur atom at the position adjacent to the ring fusion (i.e. 1-thiacarbapenems). Although both the carbapenems and the penems may appear at first glance to be structural hybrids of the cephalosporins and the penicillins, they are quite different in several ways. First, the west-end substituents associated with the cephalosporins and penicillins are not applicable for the design of carbapenem or penem agents. The carbapenems and penems, whether by virtue of microbial biosynthesis or chemical synthesis (only the carbapenems are accessible from natural sources), do not have an amine functionality present at the C-6 position. As a consequence, the cephalosporin and penicillin west-ends are not suitable for incorporation into a carbapenem or penem. In fact, some carbapenems or penems that contain (portions of) classic cephalosporin west-ends have been prepared and are essentially devoid of antibacterial activity. Secondly, the carbapenems and penems have different stereochemical features to the cephalosporins and penicillins. The penem series must contain a *trans* configuration about the *β*-lactam edge at the C-5 and C-6 positions, and only the C-5(*R*):C-6(*S*) diastereomers are potent antibacterials. In addition, the carbapenems and penems both have an important stereocenter adjacent to the C-6 center which is usually a hydroxyalkyl substituent bearing the (*R*) designation.

The carbapenem and penem antibacterials also stray from the general spectrum of antibacterial activity that is characteristic of the classic penicillins and cephalosporins. The carbapenems and penems usually offer good potency against certain anaerobic (bacteroides) organisms and often display noteworthy potency against pseudomonas (*P. aeruginosa*). In addition, these compounds offer the potential for oral activity, a different profile against resistance-conferring bacterial enzymes and unique pharmacokinetic profiles. The carbapenem and penem antibacterials

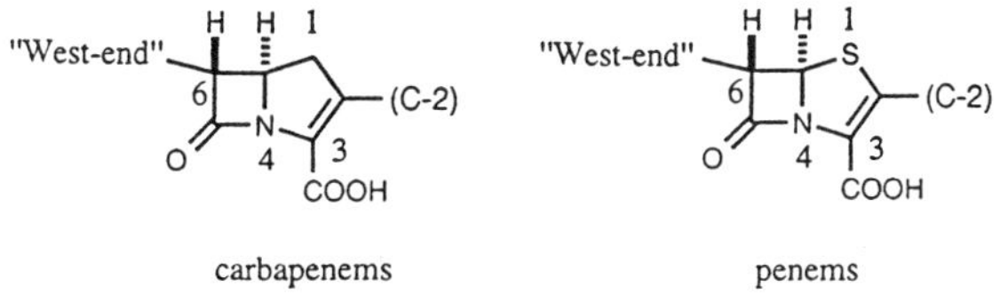

Figure 3.54 Stereochemistry of the carbapenems and penems.

represent a relatively new generation of β-lactam agents and research efforts in the synthesis, investigation and development of new congeners is ongoing.

3.3.3.1 C-6 west-end substituents. The most promising carbapenem and penem antibacterial agents contain a hydroxyalkyl C-6 substituent. In particular, a (*R*)-hydroxyethyl west-end has become something of a standard upon which to judge novel structural modifications. This west-end is present in the prototypic carbapenems thienamycin **52** and imipenem **53** (Figure 3.25). Alkylation and acylation of the hydroxyl moiety is detrimental to activity and the replacement of the alcohol by several other functional groups has been mostly unrewarding. The lone exception is a series of aminoalkyl west-ends (Figure 3.55) which, in the carbapenem series, may offer some advantages despite an inherent propensity towards intramolecular β-lactam cleavage. Interestingly, the stereochemistry of the carbon bearing the amine moiety plays a rather insignificant role in terms of *in vitro* activity. Lastly, simple alkyl substituents such as an ethyl group can suffice as carbapenem or penem west-ends; extension of the hydrocarbon chain is permitted but not advantageous.

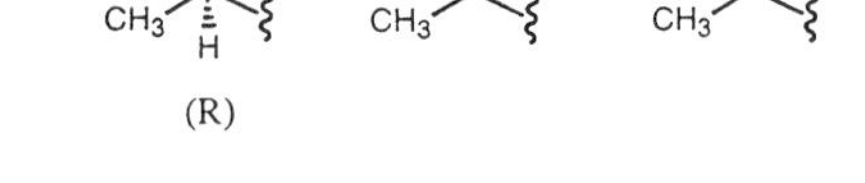

Figure 3.55 Carbapenem and penem 'west-ends'.

3.3.3.2 C-2 substituents. The C-2 substituent of the carbapenems and penems is still at an early stage of development. It is apparent that alkylthioethers are appropriate C-2 substituents, again evidenced by thienamycin and imipenem and many other small alkylthio (alkylmercapto) substituents which have afforded good antibacterial carbapenems and penems (Figure 3.56). The most recent advance has been the incorporation of thioaryl and thioheterocyclic substituents at C-2. Thus while no penem antibacterial agent has been marketed to date, some new congeners in both the carbapenem and penem series have been favorably evaluated in preliminary clinical testing.

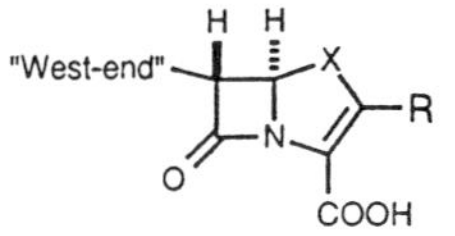

X = CH$_2$, S

R =   -SCH$_2$CH$_2$NH$_2$ (thienamycin)

       -SCH$_2$CH$_2$NHC(H)=NH   (imipenem)

       -SEt

       -SAr

Figure 3.56 Carbapenem and penem C-2 substituents.

3.3.3.3 C-1 carbapenem substituents. The naturally occurring carbapenem thienamycin and many other derivatives (e.g. imipenem) are unsubstituted at the C-1 position (bear a methyleno carbon center). Despite possessing excellent activity, these agents are prone to degradation (section 3.5) by the mammalian enzyme, renal dehydropeptidase-I (DHP-I). The introduction of small alkyl substituents at the C-1-*β* position of the carbapenem nucleus has been shown greatly to decrease susceptibility towards DHP-I without compromising antibacterial efficacy. A simple methyl group is optimal at this time; larger C-1-*β* substituents confer DHP resistance but at the expense of antibacterial potency. Small alkyloxy and alkylthioxy moieties are also suitable C-1 substituents (Figure 3.57). A number of exciting new carbapenems have been described in which the C-1 alkyl group is joined to the C-2 position by a hydrocarbon tether; the tricyclic carbapenems (so-called tribactams) and some tetracyclic congeners are the latest advances in carbapenem research (section 3.6).

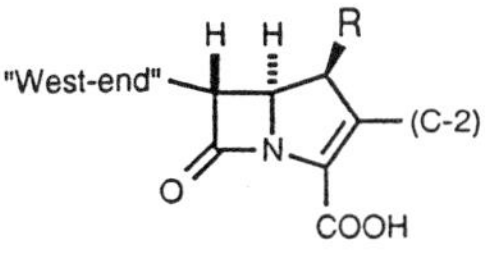

R =   -CH$_3$,

       -OCH$_3$, - SCH$_3$

Figure 3.57 Carbapenem C-1 substituents.

3.3.4 The mono-*β*-lactams

The structural features of the mono-*β*-lactams has been outlined previously (Figure 3.41). The most prevalent monobactams, aztreonam **54** and carumonam **55** (Figure 3.26), are representative. Both compounds have a sulfonic acid moiety joined to the lactam nitrogen atom and both possess an oximino west-end reminiscent of the third generation cephalosporins. Other variations in which the sulfonic acid is replaced by other acidic moieties constitute the entire family of mono-β-lactam antibacterial compounds. However, to date, only the monobactams (aztreonam) have been of chemotherapeutic significance.

3.4 Synthesis of β-lactam antibacterials

3.4.1 Classic approaches

Most of the significant β-lactam antibacterials are semi-synthetic derivatives of compounds obtained from naturally occurring microbial sources. This is particularly true in the case of the penicillins and cephalosporins. Both 6-APA (6-aminopenicillanic acid **3**) and 7-ACA (7-aminocephalosporanic acid **22**) (Figures 3.3 and 3.12, respectively; see section 3.1) are made directly from fermentation products. Over the years, 6-APA and 7-APA have served as precursors for literally thousands of compounds, and in retrospect both the penicillin and cephalosporin families remained largely underdeveloped until these key synthetic intermediates were made available.

Prior to this era, research was intensely focused upon the total synthesis of the penicillins and later the cephalosporins in an attempt to understand the structural requirements for antibacterial activity and to obtain novel compounds for biological testing. The successful strategies needed to address a number of issues. First of all, the β-lactam ring proved to be extremely challenging in developing a synthetic route. This moiety can be labile under relatively innocuous reaction conditions; it is reactive towards nucleophilic species and decomposes in acidic or basic media. For these reasons, most of the early synthetic approaches were designed with the intention of forming the β-lactam ring at a late stage of the scheme. The second issue that needed to be tackled was the relative and absolute stereochemistry of the penicillins and cephalosporins. Only one diastereomer in each family possesses useful antibacterial activity. Any promising route necessitated that ring-forming processes to the β-lactam ring occur with highly selectivity and that an optically pure starting material could be used to 'set' the stereocenter(s). Lastly, separation techniques needed to be developed for the isolation of pure β-lactam products in order to determine precise antibacterial activity. Some of the early, classic work elegantly addresses these issues.

The Sheehan research group is credited with the first synthesis of a penicillin as well as for an ingenious approach which was centered around the formation of the C-5–C-6 bond with concomitant thiazolidine ring cyclization. Sheehan focused upon establishing the correct oxidation state of the C-5 carbon center early in the synthesis and eventually completed construction of the bicyclic system via carbon–nitrogen bond formation. Specifically, dimethylcysteine was reacted with a phenacetylglycine-derived oxazolone (A), or alternatively with an aminomalonate mono-aldehyde (B) to afford the corresponding thiazolidine adducts, (C) and (D), respectively. Cyclization to the bicyclic *β*-lactam products was extremely inefficient, but did allow for the isolation and identification of small amounts of penicillin G thus constituting the first total synthesis of a penicillin (Figure 3.58). In conjunction with the latter approach, Sheehan probably described the first use of a dicarbodiimide to induce cyclization via a formal dehydration process. This route culminated in the total synthesis of penicillin V despite a number of problems pertaining to

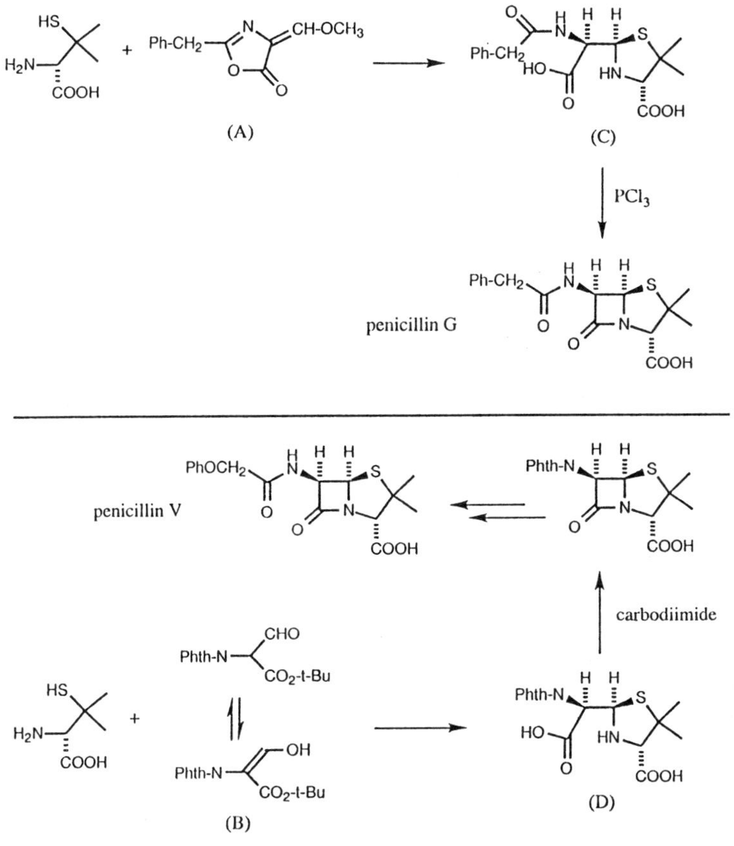

Figure 3.58 Sheehan method for the total synthesis of penicillins.

protecting group manipulations. Sheehan was able to overcome these shortcomings; he recognized the compatibility of the trityl protecting group and benzyl ester to mask the C-6 amine and C-3 carboxylic acid moieties, respectively. This strategy climaxed in the synthesis of 6-aminopenicillanic acid (6-APA).

Application of the Sheehan methodology to the total synthesis of the cephalosporins proved to be difficult and impractical. The corresponding amino acid needed to employ the Sheehan route was neither readily available, nor stable enough to be useful. Several research groups found ways around this problem by modifying the amino acid. For example, lactonization and S-protection did confer some stability and the synthesis of a cephalosporin was achieved. However, unlike the penicillins, the cephalosporins do not contain an asymmetric C-3 center and so precursors (E) of this type led to a racemic synthesis (Figure 3.59).

The most recognized total synthesis of the cephalosporins is the work by the Woodward research group. This strategy uses natural L-cysteine as a starting material; this allows for optical activity throughout the synthetic scheme as well as high diastereoselectivity during a key cyclization step (Figure 3.60). After protection of the amino acid (F), a dialkylazodicarboxylate was added with high selectivity to the carbon center adjacent to the sulfur atom (G). This transformation presumably occurs by addition of the electrophile onto the sulfur atom followed by a 1,2 shift that directed to the alpha face of the thiazolidine. Oxidation resulted in acetate formation with stereospecificity, regardless of small amounts of the *cis* hydrazide starting material. Alternatively, a related alcohol (H) was formed directly from the

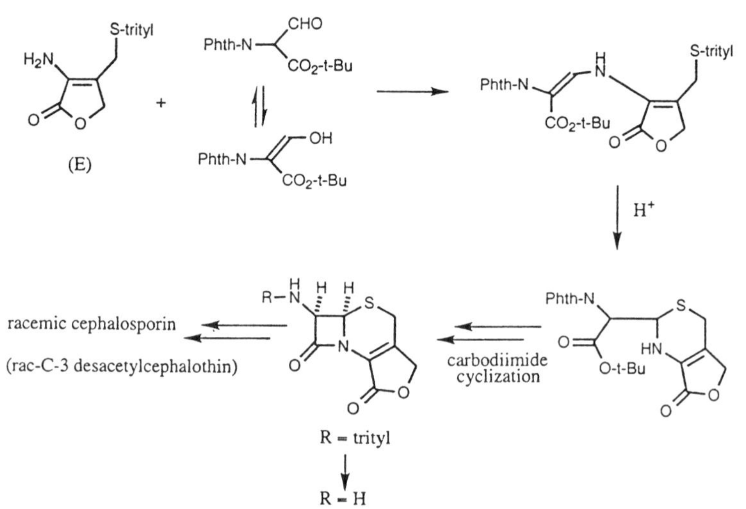

Figure 3.59 Modification of Sheehan approach for the total synthesis of cephalosporins.

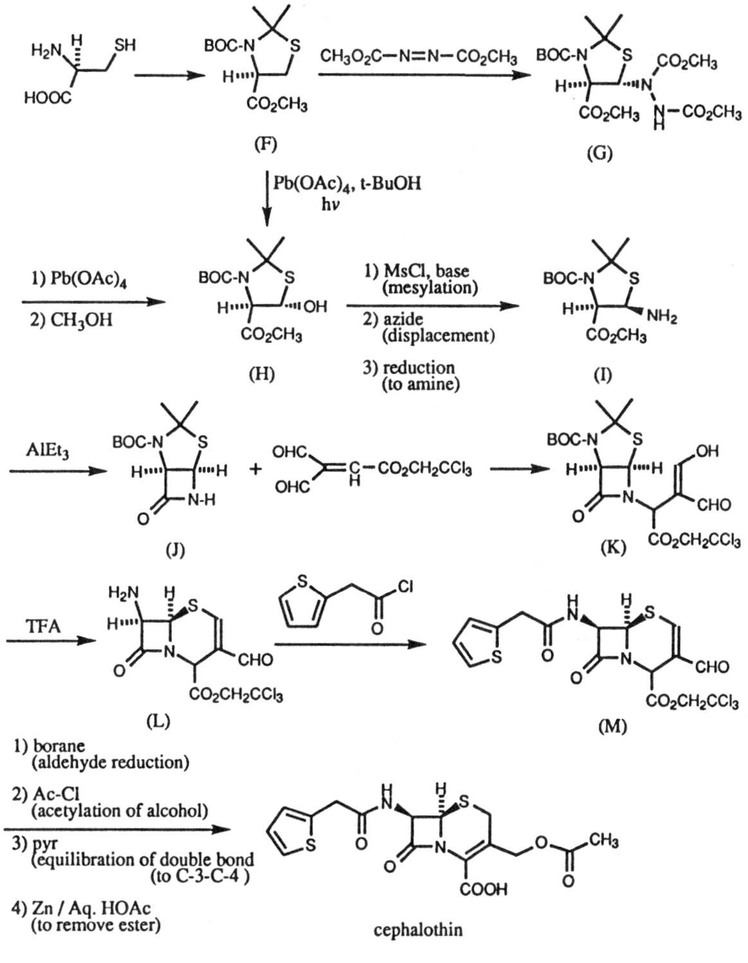

Figure 3.60 Woodward method: total synthesis of cephalosporins.

protected cysteine (F) using an irradiation protocol in the presence of *t*-butanol and an oxidant. The alcohol (H), whether obtained directly as described above, or by acetate hydrolysis, was activated towards displacement (mesylation), reacted with azide and the corresponding adduct reduced to the amine (I) with overall inversion. Cyclization to form the *β*-lactam ring (J) was achieved using a trialkylaluminum compound to activate the ester towards alkoxide displacement. The bicyclic adduct (J) underwent Michael-type addition to a highly activated α,*β*-unsaturated dialdehyde to afford the cephalosporin precursor (K). Treatment with acid induced cyclization to the Δ-2 (C-2, C-3 double bond) cephalosporin and removed the Boc (*t*-butoxycarbonyl) protecting group (L). Subsequent *N*-acylation, under standard conditions, attached the cephalothin west-end side chain (M). Reduction of the aldehyde followed by acetylation produced the corresponding C-3′ acetoxy congener. Equilibration with a mild base isomerized the double bond into conjugation with the ester giving the desired Δ-3 cephalosporin ester. Lastly, deesterification was accomplished under mild conditions which did not disrupt the bicyclic structure nor isomerize the double bond. Both cephalothin and cephalosporin C were synthesized by the Woodward group using this ingenious method.

The Woodward group also developed a premier route to the penem framework, a system not obtainable from natural sources. This approach is based upon the use of an intramolecular Wittig-type cyclization to construct the penem bicyclic nucleus. The Woodward group used the azetidinone thioester (N) with the intention that the thioester could strategically serve as a Wittig-acceptor at a later stage of the synthesis. In order to construct the complementary phosphorylidene tether, the azetidinone center was elaborated via a series of reactions. Condensation with a glyoxylate ester yielded the corresponding aminal (O′) which was immediately converted to the α-chloroazetidinyl ester (O″) via activation and displacement of the oxygen functionality. Lastly, treatment with triphenylphosphine resulted in displacement of chloride and the formation of the ester-stabilized phosphorylidene (P). Heating in an inert solvent smoothly afforded the penem via Wittig cyclization and subsequent ester cleavage was readily effected under mild conditions so as not to disrupt the bicyclic system (Q) (Figure 3.61).

A particularly attractive related protocol is that of the oxalimide method which often reduces competing side reactions (e.g. C-5 epimerization), shortens the synthesis and allows for the introduction of C-2 thioalkyl and thioaryl substituents. In this approach, an oxalimide thiocarbonate (R) is reacted with a trialkoxyphosphorane to induce ring closure to the penem system (S) (Figure 3.62). Formally, this process probably involves the formation of an incipient carbenoid species since a number of other products have been identified that arise from carbon–carbon bond insertion reactions. Nevertheless, some potent penem antibacterial agents have been made by this route including one (shown) which has undergone clinical evaluation. Another deviation from the Woodward precedent is the use of an allyl ester to protect the carboxylic functionality; this modification is noteworthy since mild conditions (palladium (0)) are used to obtain the final product (T). Lastly, the hydroxyethyl west-end, which is among the most potent of the penem (and carbapenem) series, has been amenable to this method (albeit in a protected form as a trichloroethyl carbonate (TCE)).

The carbapenem nucleus (V) has also been prepared using similar cyclization methodology (Figure 3.63). In these instances, a simple ketone serves as the acceptor; hydroxylation alpha to the carbonyl center provides

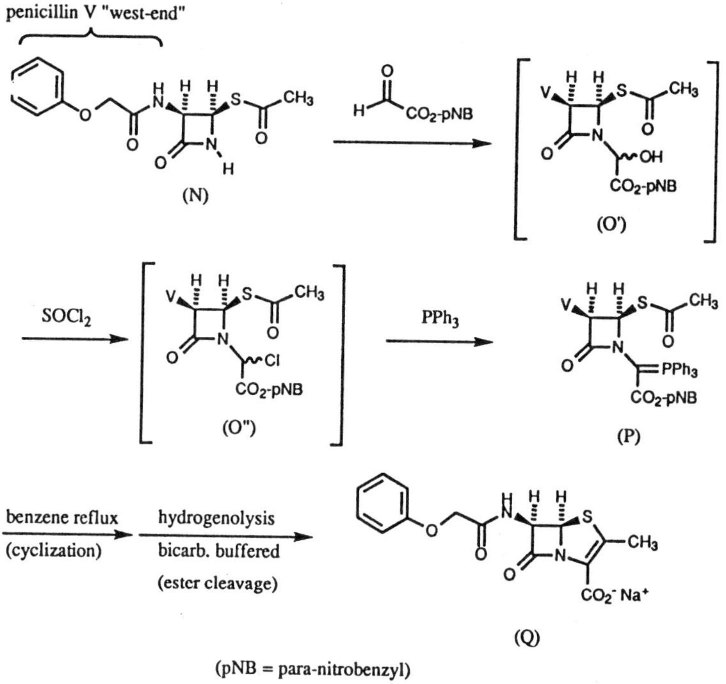

Figure 3.61 Woodward method: synthesis of penems.

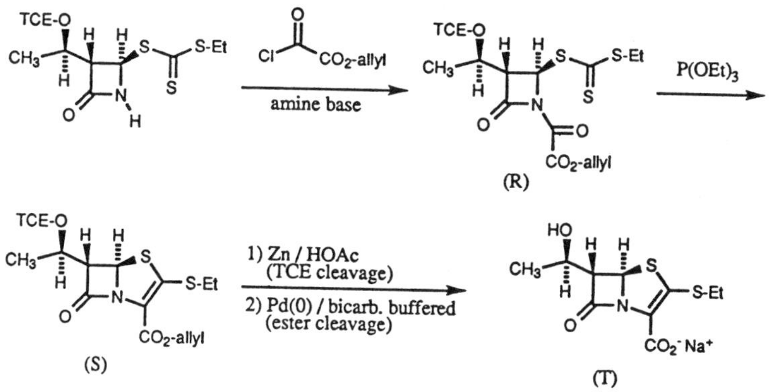

Figure 3.62 Oxalimide cyclization for the synthesis of penems.

a synthetic handle for the introduction of substituents at the C-2 position. There has been a desire to introduce small alkyl substituents at the carbapenem C-1 position in order to reduce susceptibility towards dehydropeptidase degradation. This task is readily accomplished using this methodology although a complex azetidinone precursor (U) containing four contiguous asymmetric centers is required.

There are other methods of constructing β-lactams, many relying on intramolecular cyclizations. Another attractive route involves a carbene insertion directly onto the azetidinone nitrogen center (Figure 3.64). A

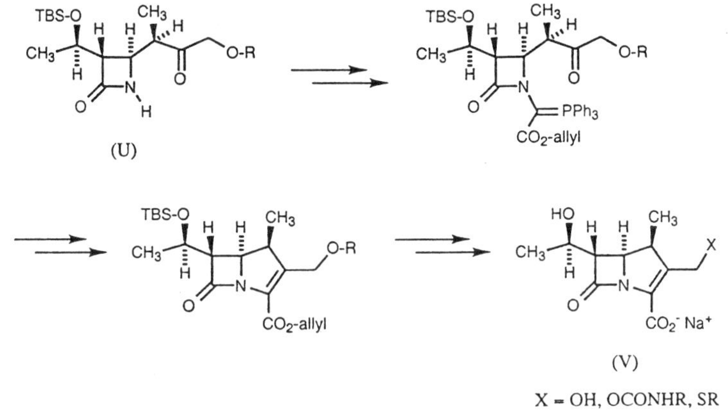

Figure 3.63 Method for the synthesis of carbapenems.

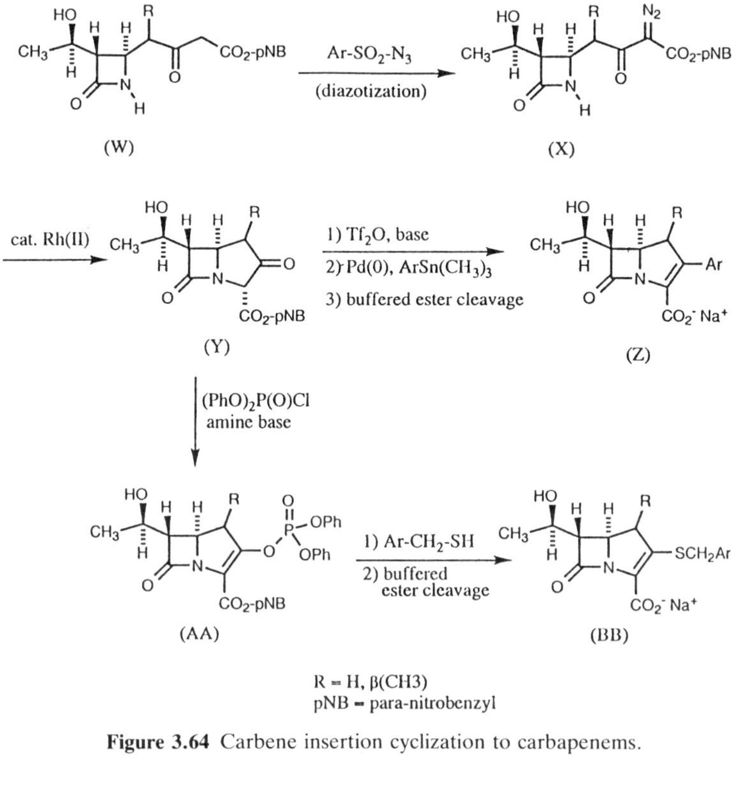

Figure 3.64 Carbene insertion cyclization to carbapenems.

(carbapenem) C-2 ketone results; this can be further elaborated in a number of ways so that the need to protect the hydroxy group of the west-end is avoided. For example, the azetidinone keto-ester (W) undergoes diazotization to the corresponding diazo-azetidinone (X). Treatment with catalytic rhodium (II) species induces cyclization to carbapenam (Y). A variety of potent C-2 aryl-carbapenems (Z) have been synthesized by subsequent enol triflate formation, followed by palladium(0)-mediated coupling to arylstannanes. Alternatively, the carbapenam (Y) can be converted to the corresponding phosphonate (AA) and displaced by a mercaptide (BB).

From a practical viewpoint, the most valuable synthetic approaches are those that construct penems and carbapenems; most penicillins and cephalosporins are obtained from 6-APA and 7-ACA, respectively. Acylation of aminopenicillanic acid and aminocephalosporanic acid or ester derivatives, proceeds using standard protocols developed for amino acid acylation (i.e. peptide synthesis). Among the most common routes are the reactions of a β-lactam derivative with an acyl chloride or anhydride under (Schotten–Baumman) buffered conditions (bicarbonate). Alternatively, carbodiimide-mediated coupling has also been used to incorporate

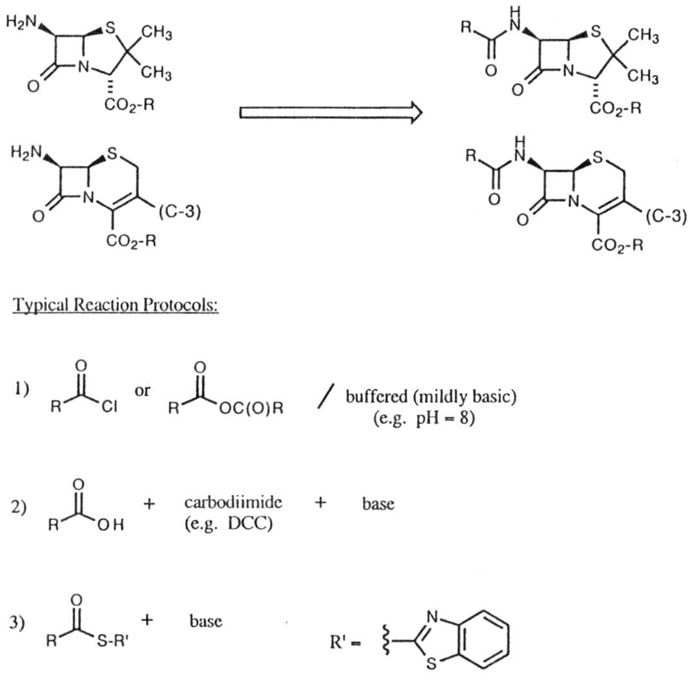

Figure 3.65 General method for the synthesis of penicillins and cephalosporins from 6-APA and 7-ACA.

highly functionalized west-ends that are too reactive for acyl chloride or anhydride formation. Lastly, activated thioesters have been used to attach certain west-ends (Figure 3.65).

Nuclear analogs of the β-lactam antibiotics present formidable synthetic challenges and specific methodologies have been designed for each agent. The most important cephalosporin analogs are the oxacephems and the carbacephems. The synthesis of moxalactam **49** (Figure 3.23) is indicative of the complexity of introducing a heteroatom into the cephalosporin nucleus. Likewise, the synthesis of the carbacephems such as loracarbef **51** (Figure 3.24) shows that removing a heteroatom (sulfur) from the cephalosporin nucleus is equally challenging.

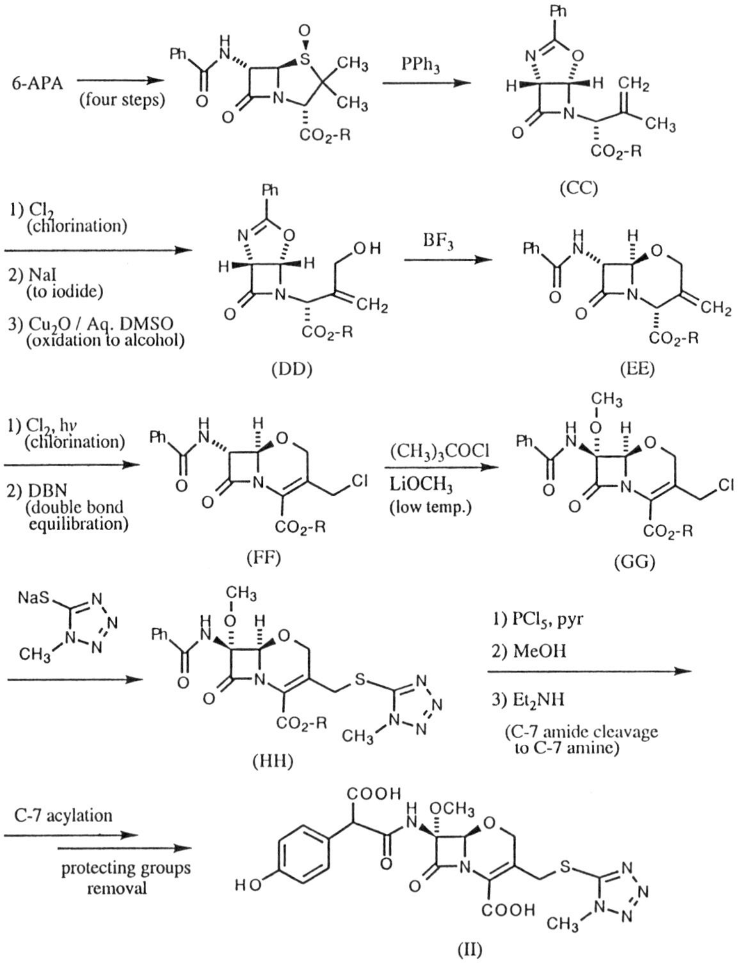

Figure 3.66 Synthesis of the oxacephem moxalactam.

Moxalactam is obtained from 6-aminopenicillanic acid as shown below (Figure 3.66). The penicillin starting material is oxidized to the sulfoxide, which allows for epimerization at C-6 and subsequent ring opening upon treatment with a phosphine. Sulfur extrusion and cyclization gives the bicyclic oxazoline (CC). A series of reactions are used to introduce the prerequisite oxygen functionality. The resultant allylic alcohol (DD) undergoes Lewis acid cyclization to the methyleno–oxacepham system (EE) with concomitant opening of the oxazoline ring. Functionalization of the C-3 position is achieved via chlorination and treatment with mild base equilibrates the double bond into conjugation with the C-4 ester thereby establishing the oxacephem nucleus (FF). The next critical operation is the introduction of the C-7-α-methoxy group which is achieved by treatment with lithium methoxide and pivaloyl chloride and results in overall inversion of the stereocenter (GG). At this stage the C-3 substituent is incorporated via tetrazoylthiolate displacement of chloride (HH). Lastly, the west-end group is introduced by amide cleavage to the free C-7 amino congener followed by acylation, and the protecting groups are then removed to afford the final product (II).

The carbacephem nucleus has been made in several different ways; in general, an azetidinone is joined to a tether and cyclization of the six-membered ring is induced. The crucial bond-forming process can occur via Michael addition, carbene insertion, epoxide opening, Dieckman cyclization or radical addition, depending upon what type of C-3 functionality is desired. All of these approaches require the construction of fairly elaborate (optically pure and highly functionalized) azetidinone precursors in order to be useful for the generation of potent antibacterials. A summary of some of these strategies is shown in Figure 3.67.

3.5 Bacterial resistance to the β-lactam antibacterial agents

As a family, the β-lactams continue to be the most widely used of all systemic agents, but not because bacteria have been slow to evolve efficient mechanisms of resistance. To the contrary, many Gram negative and Gram positive bacteria have developed remarkably effective means by which to debilitate β-lactam antibacterials. To date, the battle between susceptibility and resistance has been favorably controlled by the regular and timely introduction of new agents; often these new compounds are effective against organisms that are resistant to older drugs. Although the therapeutic success of the penicillins and cephalosporins continues, the emergence of bacterial resistance is alarming. There is concern that in the near future, infections caused by certain species will no longer be treatable by β-lactam therapy. An imposing example is certain streptococcal species resistant to most penicillins. This is striking since the penicillins have

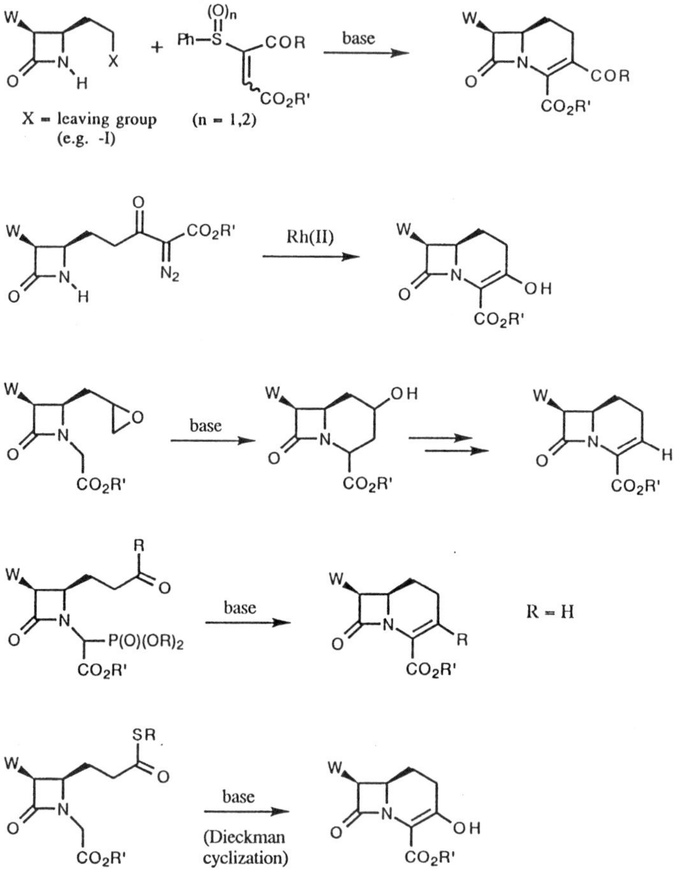

Figure 3.67 Cyclization strategies for the carbacephem nucleus.

historically offered exquisitely superb potency against Gram positive species.

Resistance to the β-lactam antibacterial agents epitomizes the ability of bacteria to mount defenses in order to secure survival of the species. Since their introduction into antibacterial chemotherapy, the course of β-lactam development has followed a sort of cycle that has been revisited many times to date. Usually, shortly after a new cogener or new 'generation' is put to practice, resistance emerges which in turn inspires the research and development of an improved follow-up compound to tackle such problematic pathogens. Unfortunately, with each advance and every new wave of agents, some response of bacterial resistance has eventually surfaced. Even more troubling, many forms of resistance can be spread to otherwise susceptible bacteria and consequently disarm even the most promising β-lactam antibacterial.

There has been ample warning that bacterial resistance would pose a threat to the success of β-lactam chemotherapy. Bacterial resistance to some of the early β-lactam antibiotics was noted before the use of penicillins became a well-established practice in antibacterial chemotherapy. Probably the first documentation of this phenomena was an experimental demonstration that certain cell-ruptured bacterial strains (*E. coli*) were capable of abolishing the antibacterial effects of penicillin. Only a few years later, some clinical isolates of *Staphylococci aureus* were shown to exert a similar resistance to penicillin. In both cases, bacterial resistance could be traced to the action of an enzyme that caused cleavage of the β-lactam antibiotic to inactive penicillanoic acid. These bacterial enzymes were first referred to as penicillinases, a proper designation based upon their substrate target. As the cephalosporins and other β-lactam antibiotics were introduced, it became apparent that other bacterial enzymes could confer resistance in the same way. These enzymes belong to the β-lactamase family which collectively represents the major cause of bacterial resistance to the β-lactam agents.

β-Lactamases can be found in most bacteria including both Gram negative and Gram positive species, as well as some other microorganisms. However, the mere presence of β-lactamases does not necessarily guarantee resistance. These enzymes must directly encounter the β-lactam agent in order for inhibition of antibacterial activity to occur. For many β-lactams, most of which are relatively hydrophilic, drug entry into a (Gram negative) periplasm occurs by passage through porins. The β-lactam agents are usually below the molecular size restrictions of most porin protein channels (an approximate molecular weight of seven hundred is often recognized as the cut-off point). Hydrophobic congeners usually diffuse across the lipid-rich (e.g. lipopolysaccharide (LPS)) outer membrane in order to enter the cytoplasm. (See chapter 1 for a general discussion on bacterial cell composition and mechanisms of antibacterial resistance.)

Once the β-lactam reaches the periplasm, it can be recognized by a β-lactamase enzyme, undergo binding and be rapidly hydrolysed to the corresponding inactive product (e.g. penicillanoic acid from a penicillin). This is easily accomplished in many Gram negative organisms since the β-lactamase is contained inside the outer membrane and can achieve high concentrations within the periplasmic space. One inherent drawback to the microorganisms is that the antibacterial substance must actually approach the bacterial cell cytoplasm, so if β-lactamase activity is inhibited, antibacterial activity can be restored.

β-Lactamase mediated resistance in Gram positive bacteria is less common (especially from a clinical viewpoint) than that of the Gram negative species, but not necessarily because Gram positive organisms are inferior β-lactamase producers. Despite having a thick peptidoglycan sheath, Gram positive bacteria do not have an outer membrane, nor a periplasmic space (chapter 1). As a consequence, the majority of β-lactamase enzyme is not contained within the cell (although most Gram positive species actually do have a very small intracellular supply of β-lactamase). The supply of β-lactamase is therefore primarily exocellular and subject to external conditions which can alter or destroy the activity of the enzyme. In order for resistance to be conferred, β-lactamase concentrations must reach sufficient levels in the vicinity of the bacterial cell so as to deactivate any β-lactam antibacterial. However, dilution or simple diffusion of β-lactamase away from the bacterial cell is essentially an unavoidable process, and so resistant Gram positive bacteria need to be capable of overwhelming the local environment with β-lactamase. This can be accomplished in two ways. The bacterial cell may be capable of expressing high levels of β-lactamase for extended periods of time to ensure that adequate amounts of enzymes are available. Alternatively, bacteria can display resistance if external β-lactamase concentrations are sufficient, regardless of the origin of the enzyme. In other words, Gram positive β-lactamase resistance can be controlled by the population rather than each specific bacterium, and so susceptible cells within a neighborhood of prolific β-lactamase producers can resist the antibacterial effects of a β-lactam antibiotic.

Most β-lactamases are extremely efficient at destroying β-lactam agents assuming that the antibacterial is recognized by the enzyme and undergoes binding. In fact, some β-lactamases are capable of hydrolysing hundreds of drug molecules per minute. β-Lactamases probably serve no function in the construction of peptidoglycan although they embody a similar active site. Like most PBPs, the β-lactamases in general are able to accommodate the β-lactam into their reactive pocket through a number of attractive interactions that include hydrogen bonding and lipophilic associations. For example, typical β-lactam–β-lactamase binding motifs entail complementary charge association of the penicillin carboxylate moiety with specific arginine and lysine residues of the enzyme. A serine or histidine residue in proximity to the β-lactam nitrogen center serves to stabilize the resultant acyl–enzyme complex. On the other side of the enzyme pocket, an aromatic amino acid residue such as a tryptophan can associate with the lipophilic aromatic group of the β-lactam west-end. These interactions place the reactive β-lactam center of the drug molecule near the serine residue of the enzyme (Ser-70 in a number of β-lactamase species). This culminates in acylation of the serine with concomitant cleavage of the β-lactam ring (Figure 3.68). The ring-cleaved substrate is released from the enzyme accounting for the catalytic activity of the β-lactamases. Since regeneration of the substrate β-lactam ring is energetically prohibited, antibacterial activity is lost.

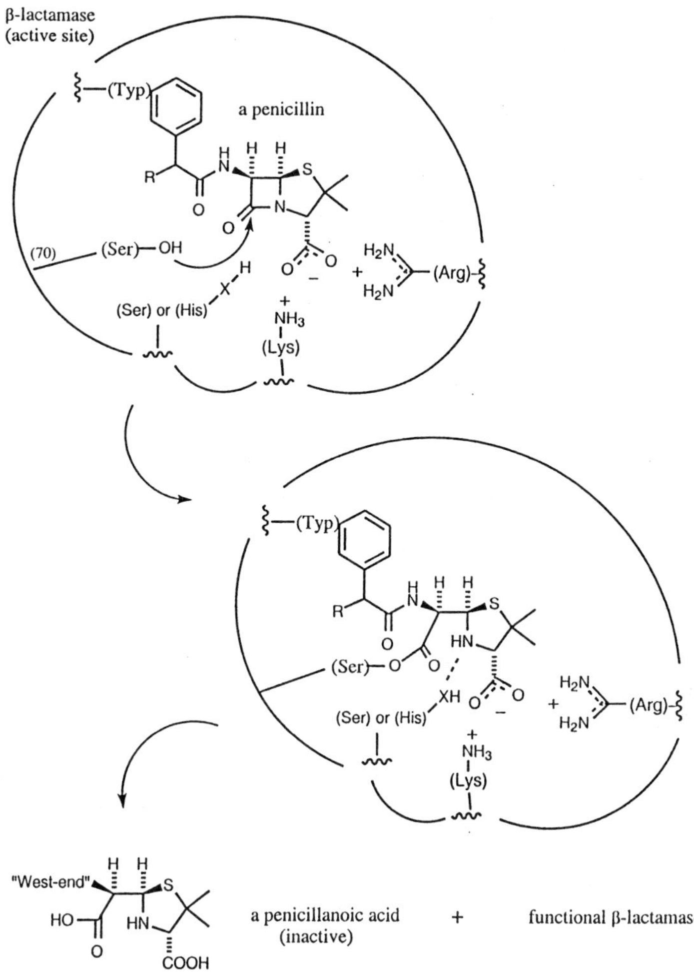

Figure 3.68 Action of β-lactamases.

There are two distinct types of β-lactamase production in bacteria. Inducible β-lactamase formation results from exposure to a stimulus that causes the expression of genetic material encoding for β-lactamase enzyme(s). Gram positive bacteria, in particular, can be induced to synthesize β-lactamase. Ironically, the resultant enzyme can, in turn, destroy the very agent that induced its formation. Even more troubling, multiple forms of the β-lactamase enzyme can result and so other β-lactam agents are often susceptible to inducibly expressed β-lactamases and can follow a similar demise regardless of the specific inducer. Both bacterial chromosomal genes as well as nucleotide sequences of plasmids can express inducible β-lactamases. Typical examples include a variety of Gram positive bacteria, most notably staphylococci, as well as certain Gram negative *Pseudomonas* and *Enterobacteriaceae* species.

Constitutive β-lactamase production is prevalent among resistant Gram negative bacteria and results from the regular (and inherent) expression of genetic material that encodes for β-lactamase proteins. In some cases, the enzymes originate from chromosomal genes, although the β-lactamases are more often acquired on plasmids (see chapter 1 for a discussion on plasmid-mediated resistance). Constitutive plasmid-mediated β-lactamases account for the majority of β-lactam resistance; the transfer of plasmids is a facile process and widespread among bacterial populations. However, the presence of constitutive β-lactamase production does not necessarily lead to bacterial resistance. A number of factors come into play such as the expression of chromosomal β-lactamase genes, the frequency by which plasmids encode for β-lactamase protein or the copy number for β-lactamase present within a transposon. In other words, since β-lactamase production is unaffected by outside forces (not induced), the level of β-lactamase production inherent in a given bacteria strain will determine whether resistance is conferred to a particular β-lactam agent.

There are many β-lactamases and historically a variety of schemes have been devised to differentiate this large family of proteins. Some approaches were based upon substrate profiles, others by molecular weight or size, isoelectric point (pI), genetic origin or homology. Perhaps the most useful system is that introduced by Richmond and Sykes in the early 1970s (Figure 3.69), which essentially distinguishes the Gram negative β-lactamases by their general substrate profiles. Using this classification, Gram negative β-lactamases are divided into five categories. Class I β-lactamases are cephalosporinases, while class II enzymes are penicillinases. Class III enzymes are truly 'β-lactamase' in character being active against a host of both penicillins and cephalosporins. In addition, these proteins are resistant to inhibition by p-chloromercuribenzoate (this reagent selectively reacts with cysteine amino acid residues), but are sensitive towards the penicillin cloxacillin. Class IV β-lactamases likewise generally behave as both penicillinases and cephalosporinases, but are sensitive to p-chloromercuribenzoate and resistant to cloxacillin. The class V enzymes are penicillinases with an extended activity against carbenicillin and other oxacillin penicillins. Using this scheme, the Gram negative β-lactamases can also be divided into two broad categories based upon origin: chromosomal or plasmid (Figure 3.69).

Plasmid-mediated β-lactamases are particularly troublesome since these determinants can often transfer resistance to other bacteria. The temocillin (TEM) β-lactamases are the most commonly encountered class III broad-spectrum penicillinases although the SHV and HMS enzymes can also be problematic. For example, the TEM proteins are often found in *Haemophilus*

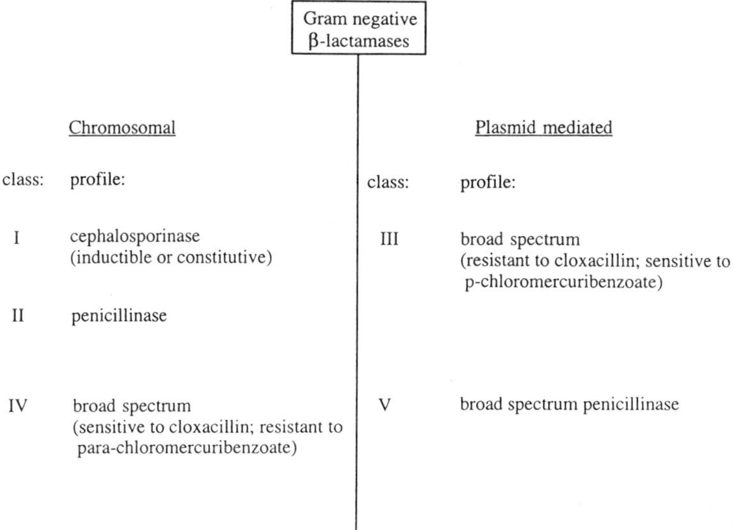

Figure 3.69 β-lactamases.

and *Neisseria* species resistant to ampicillin and cephaloridine. The class V β-lactamases include the oxacillin-hydrolysing enzymes designated as OXA, as well as the carbenicillin-hydrolyzing proteins belonging to the *Pseudomonas aeruginosa* (PSE) subfamily. Certain *Klebsiella* and *Proteus* species, as well as *Pseudomonas aeruginosa* can display resistance to isoxazolyl penicillins such as oxacillin due to the action of the OXA enzymes. PSE (and some *E. coli*) can achieve resistance to the antipseudomonal penicillin carbenicillin as a result of PSE β-lactamases.

β-Lactam resistance is prevalent even without the plasmid transfer. Chromosomally controlled class I β-lactamases are frequent in *Proteus*, *Pseudomonas*, *Serratia*, *Shigella*, *Salmonella* and other *Enterobacteriaceae*. In addition, class II β-lactamases are responsible for the penicillin resistance of certain *Proteus* species. With so many origins, β-lactamases can cause some level of resistance in almost any bacterial species.

Fortunately, for the most part, the development of new β-lactam antibacterials has been able to keep up with β-lactamase resistance and some landmark advances have been realized as a result of this research (section 3.1). For example, the discovery of the cephamycins demonstrates that β-lactamase susceptibility can be reduced via the introduction of a methoxy group at the C-7α position (7α-methoxycephalosporins). In parallel with this finding, C-6-α-formylamidino-penicillins have been investigated. Certainly, it has been demonstrated that β-lactamase-resistant cephalosporins and penicillins can be successfully developed. For instance, semi-synthetic cephalosporins possessing oxime-containing west-end substituents (in particular 2-amino-4-thiazolyliminoacetyl) generally possess good stability against many Gram negative β-lactamases (section 3.3).

Another rewarding area of research directed towards combating β-lactamase-producing bacteria has been the design and development of β-lactamase inhibitors. To date, most β-lactamase inhibitors are structurally related to penicillins or penems. Modification of the penicillin nucleus has been shown to be particular well suited for the construction of β-lactamase inhibitors. For example, penicillin-derived sulfones and C-6-halopenicillins typically offer good activity against a variety of penicillinases, and recently some C-6-vinylidienylpenems (e.g. **60** (Figure 3.29)) are being pursued. Some β-lactamase inhibitors have been advanced onto the market but must be administered along with an active β-lactam agent. Clavulanic acid **56** (Figure 3.27), sulbactam **57** (Figure 3.28) or its pivaloyloxymethyl ester prodrug **58** (Figure 3.28) are currently in use, although several newer compounds are undergoing development.

As with efforts to combat bacterial β-lactamase resistance, dehydro-peptidase inhibitors have been designed, synthesized and developed for administration with susceptible carbapenem agents. Cilastatin **61** (Figure 3.70) is coadministered with imipenem as a marketed product.

Another mechanism of acquired resistance that does not rely upon the action of β-lactamases has recently generated concern for the future of β-lactam antibacterial chemotherapy. As early as the 1970s, it was demonstrated, under laboratory conditions, that mutant bacterial strains could display resistance due to the presence of altered PBPs (penicillin-binding protein). Such species were reluctant to undergo binding to a variety of penicillins, and as a result inhibition of transpeptidation and the resultant loss of peptidoglycan integrity were avoided. Bacteria possessing altered PBPs have been rare under clinical conditions until recently.

Resistance caused by PBP alteration has probably remained relatively obscure for a variety of reasons. First of all, most β-lactam agents do not target one specific PBP and so even a modification of a given PBP may be ineffective in bestowing a pronounced resistance. Secondly, it seems likely that PBP alteration must be carefully orchestrated by the bacterium since

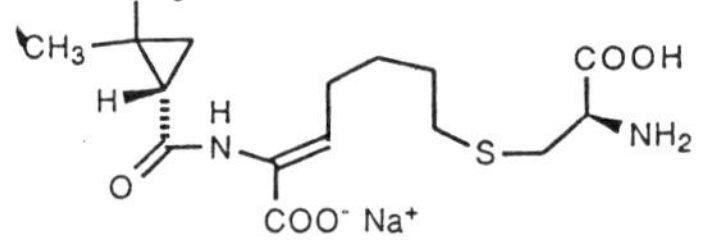

61

Figure 3.70 Cilastatin.

mutations within the active site could be detrimental to the normal construction and regulation of peptidoglycan biosynthesis. It is conceivable that mutations of a PBP may create a faulty peptidoglycan sheath possibly leading to cell death, so perhaps such mutants are easily eliminated. Furthermore, unlike many bacterial enzymes that can often be over-expressed as a means to confer resistance, the hyperproduction of many PBPs would be deleterious to the bacterium since peptidoglycan regulation would be altered. Excess functional PBPs would be likely to result in extensive cross-linking and a peptidoglycan barrier perhaps impenetrable to vital nutrients. Lastly, it is important to realize (section 3.2) that the higher molecular weight PBPs (e.g. PBP 1, PBP 2, PBP 3 proteins) are those which can be successfully targeted by the β-lactam agents for the manifestation of a bactericidal effect. Conversely, lower molecular weight PBPs often provide (D-alanine) carboxypeptidase function which plays a role in 'pruning' the peptidoglycan matrix so that exaggerated cross-linking cannot occur. Inhibition of this process, that of the function of the 'lesser' PBPs, is usually not a lethal event to the bacteria organism, since several PBPs can supply this action.

Despite these inherent obstacles, altered PBP composition has been well documented and the emergence of resistant bacteria within the past years is troubling. Bacteria often acquire resistant PBPs by evolution, but in some alarming cases, resistance is transferred from other strains. Multiple steps/mutations must be achieved before a bacterium can display β-lactam resistance by altered PBP composition. Replacement of the site active serine amino acid residue can contribute to this type of resistance but other amino acids may be altered.

Gram positive species such as streptococci and staphylococci, as well as Gram negative *Haemophilus influenzae*, *Neisseria gonorrhoeae* and *N. meningitidis* have acquired resistance via PBP alteration. Resistant strains of *Neisseria* have been shown to contain altered PBP 2; resistant *Haemophilus influenzae* can possess altered PBP 3a and 3b proteins. Among the streptococci (*S. pneumoniae*), PBP 1a, PBP 1b and PBP 2a are modified, while staphylococci (*S. aureus*) have modified PBP 2, responsible for PBP-altered resistance. The phenomenon of PBP-altered bacterial resistance has now been observed almost worldwide and predictions are that this type of resistance will increase in the near future.

Lastly, there are certain bacteria which contain feature(s) that impart an intrinsic resistance to the β-lactam agents. Upon encountering a Gram negative bacteria, the β-lactam antibacterials need to cross the outer membrane in order to reach the PBP proteins and produce antibacterial effects. Gram negative bacteria can exclude passage of β-lactam compounds into the periplasmic space thereby eliciting an intrinsic resistance. Susceptible bacteria normally contain protein channels, porins (section 3.2), within their outer membrane that allow most β-lactam agents to enter into the periplasm. However porin proteins are not created equal among bacterial species, nor are these channels continually accessible or available. For example, the outer membrane of most *Neisseria* species typically offer little hindrance to say, a variety of penicillins, and so these bacteria are quite susceptible to the actions of penicillins, provided that β-lactamase- or altered PBP-mediated resistance is not operative. At the other end of the spectrum, *Pseudomonas aeruginosa* species are usually quite adept at avoiding β-lactam uptake due to a hardy outer membrane. For these reasons, antipseudomonal β-lactams have been traditionally prized. The most recent advances have made use of an iron-mediated uptake mechanism (e.g. catechol and hydroxylated pyridone west-ends (section 3.6)).

3.6 Recent advances

3.6.1 New β-lactam agents

A host of new β-lactam agents have recently entered clinical trials; many appear poised to enter the market within the near future. Some of the more advanced congeners are shown below (Figures 3.71–3.73). A glance at these compounds indicates several notable trends. There is a continued interest in utilizing oxime-containing west-ends to impart an acceptable level of β-lactamase durability to broad spectrum cephalosporins. This feature has carried over to oral as well as parenteral agents. There remains a need for antipseudomonal cephalosporin agents and several candidates possess a positively charged nitrogen heterocycle as the C-3 substituent, reminiscent of the prototype ceftazidime. Cefclidin, cefozopran, FK-037 and its related congener FK-518 are illustrative; these parenteral agents are truly broad spectrum and possess good stability against many β-lactamases. Compound E-1077 is an interesting variation and demonstrates excellent activity against animal models of urinary tract pseudomonal infections (Figure 3.71).

The incorporation of iron-chelating groups such as catechols and hydroxylated pyridones within the cephalosporin west-end (Figure 3.72) can increase drug penetration into Gram negative bacteria by an iron-uptake mechanism and potentiate antipseudomonal activity. Cefetecol, KP-736 and BOF-12013 are among the most promising of these new congeners. In some cases, these parenteral cephalosporins are even more active than classic antipseudomonal agents such as ceftazidime in addition to prototypic fourth generation cephalosporins such as cefpirome. Furthermore, superb activity is carried over to many other Gram negative species as well as Gram positive pathogens.

The use of carboxylic ester prodrugs has enhanced oral activity of some

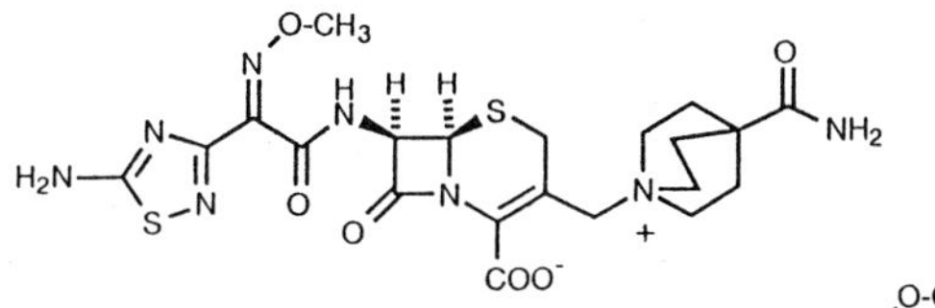

Cefclidin

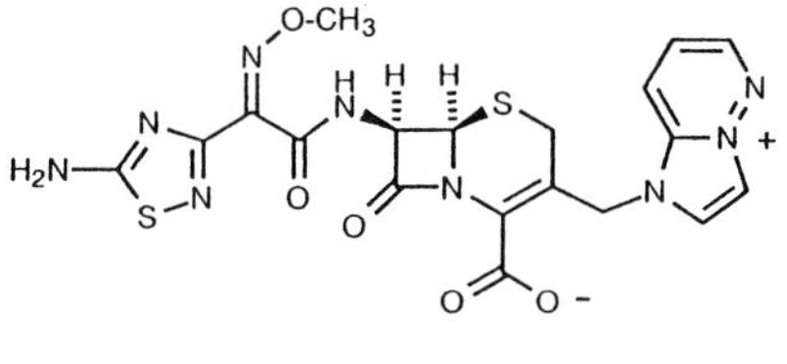

Cefozopran

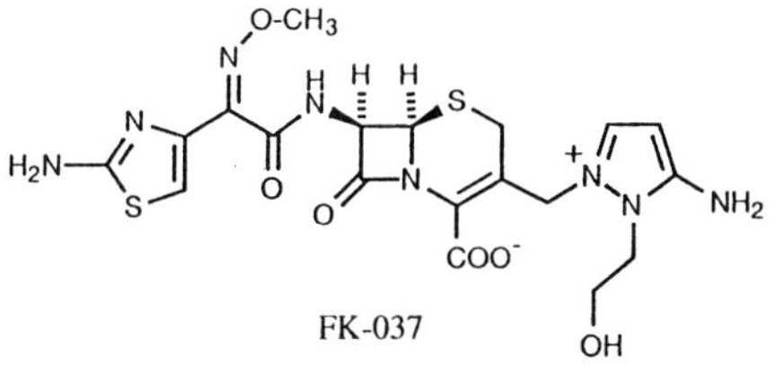

FK-037

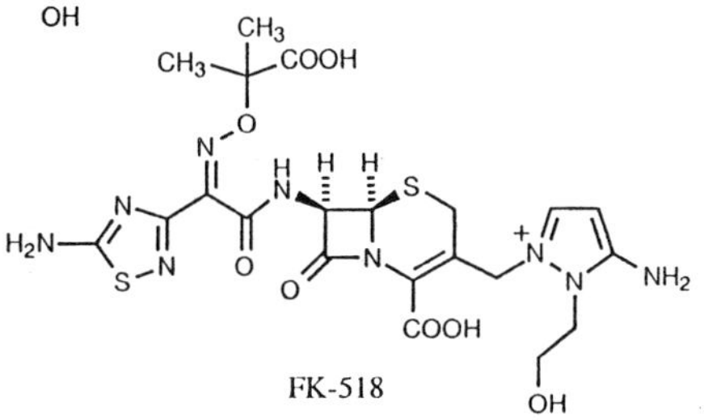

FK-518

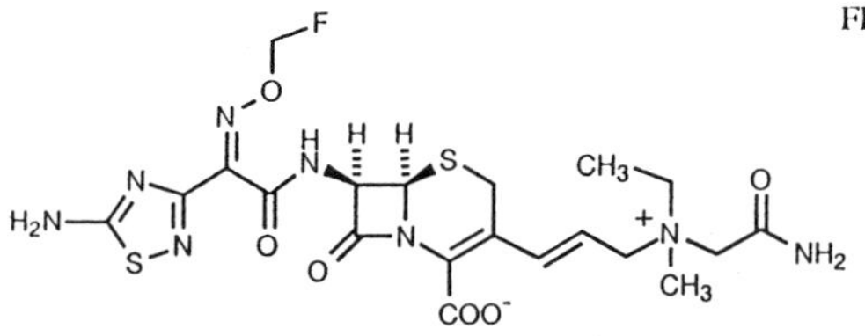

E-1077

Figure 3.71 Parenteral cephalosporins under development.

cephalosporins that contain small C-3 substituents and an extended spectrum of antibacterial activity. Cefetamet pivoxil, cefdaloxime pentexil, cefcapene pivoxil, cefditoren pivoxil and cefcanel daloxate are advanced orally active cephalosporins that should enter the market shortly (Figure 3.73). In general, these compounds offer a broader spectrum of antibacterial activity than third generation oral agents and greater potency against many Gram negative pathogens. The use of cephalosporin C-3 carboxylic acid ester prodrugs has also extended the use of some older proven cephalosporins. Both cefpodoxime proxetil and cefotiam hexetil have been

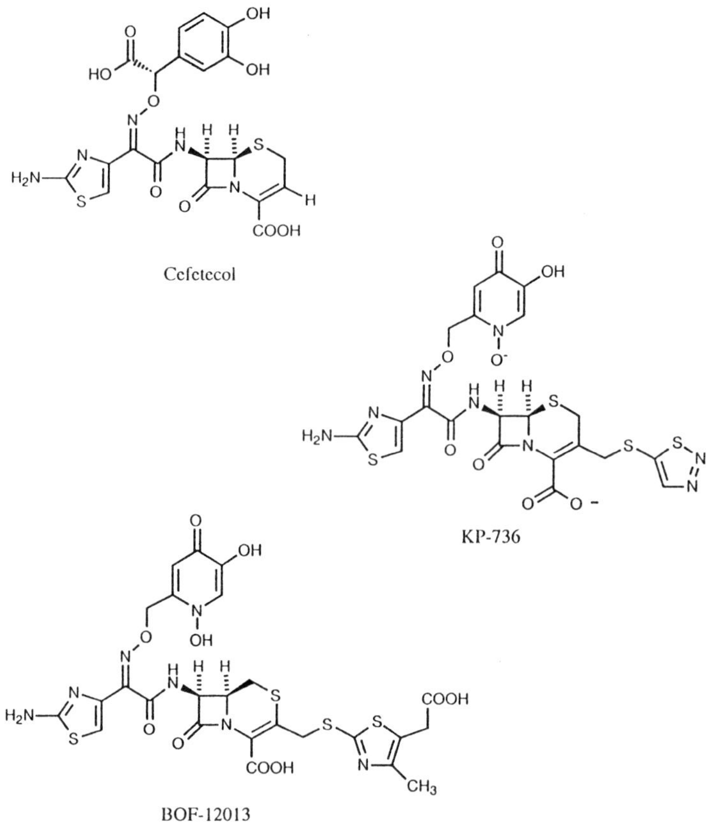

Cefetecol

KP-736

BOF-12013

Figure 3.72 Other promising cephems and isocephem.

launched recently and appear to be quite effective; cefetamet pivoxil appears also destined for the market (Figure 3.74).

In recent years, a series of potent dual-action β-lactam agents have been described and several of these compounds have or are currently undergoing clinical evaluation. Dual-action cephalosporins (DACs) are molecular hybrids of cephalosporins and quinolone antibacterials (chapter 7) constructed in a way that allows both components to exert their individual antibacterial properties (Figure 3.75). The dual-action mechanism of action has been demonstrated: upon PBP binding, the β-lactam center undergoes acylation of bacterial transpeptidases which results in subsequent disruption of peptidoglycan biosynthesis (section 3.2). As a consequence of β-lactam ring cleavage, free quinolone is released at the cellular level which allows for inhibition of DNA gyrase activity (chapter 7, section 2).

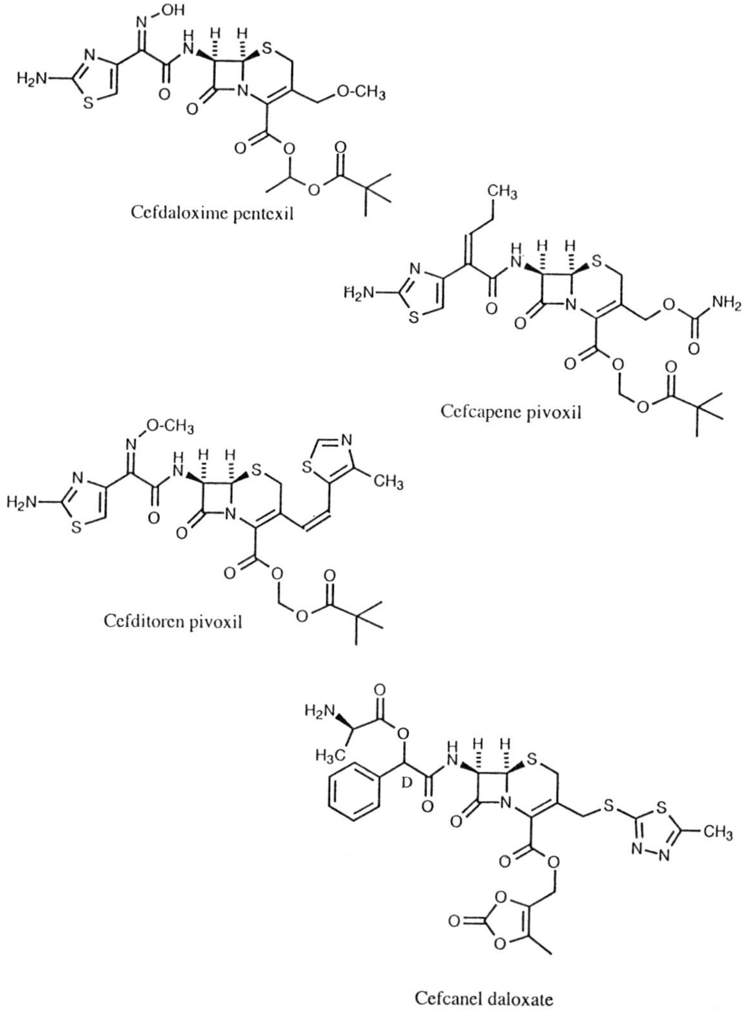

Cefdaloxime pentexil

Cefcapene pivoxil

Cefditoren pivoxil

Cefcanel daloxate

Figure 3.73 Orally active cephalosporins under development.

Carbapenem- and penem-based dual-action agents have also been reported. Many other variations on this theme are being pursued; some include the linking group between the quinolone and the *β*-lactam (ester, carbamate, amine, ammonium, etc.). Dual-action *β*-lactam antibacterials characteristically display a composite spectrum of antibacterial activity indicating contributions from both agents. This can be advantageous since *β*-lactam (cephalosporin) activity can typically bestow potent activity against Gram positive pathogens such as streptococci and staphylococci while quinolone

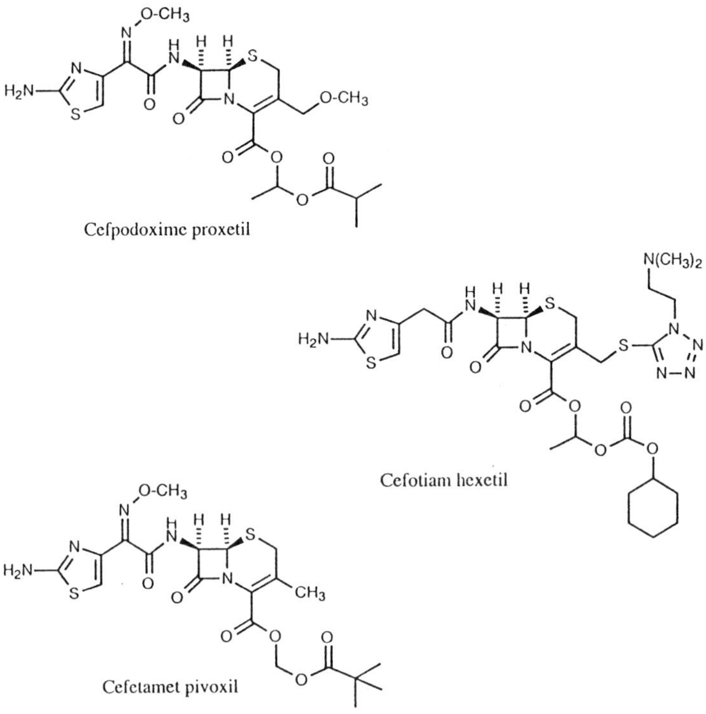

Cefpodoxime proxetil

Cefotiam hexetil

Cefetamet pivoxil

Figure 3.74 Cephalosporin prodrugs.

potency is often primarily manifested against Gram negative species including *Pseudomonas aeruginosa*. In addition, anaerobic organisms can be covered if a penem or carbapenem is incorporated into a dual-action agent. Lastly, some of the undesirable properties of each component can be notably reduced by combining the two antibacterial agents as a dual-action hybrid.

Although a penem-derived antibacterial agent has yet to reach the market, several congeners have displayed favorable clinical results. Among the most promising compounds are ritipenem and its prodrug derivative (FCE 22891) and faropenem (Figure 3.76). Furthermore, the success of imipenem has led to a second generation of carbapenems as exemplified by panipenem, biapenem and meropenem (Figure 3.76); meropenem was launched in 1995. Like imipenem, these penem and carbapenem agents generally offer a broad spectrum of activity that includes useful potency against *Pseudomonas aeruginosa*. Meropenem and biapenem contain a C-1β-methyl substituent responsible for a pronounced resilience to mammalian dehydropeptidase degradation.

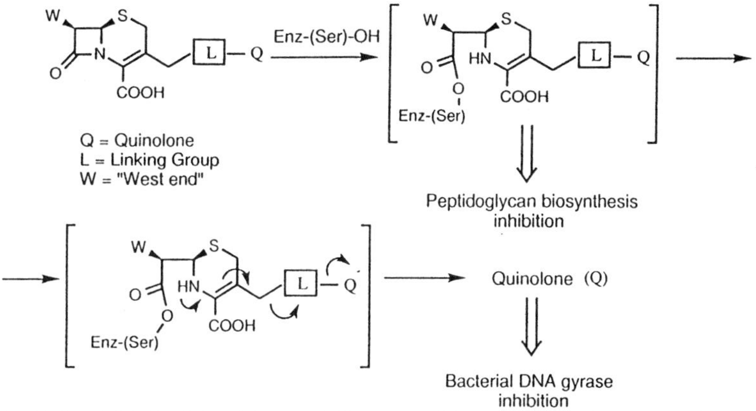

Ro 23-9424
(desacetyl-cefotaxime ester-linked fleroxacin)

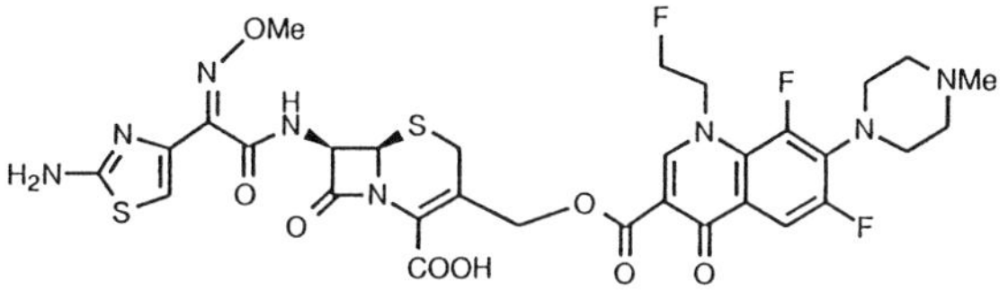

cephalosporin quinolone

Ro 25-0993
(carbapenem carbamate-linked ciprofloxacin)

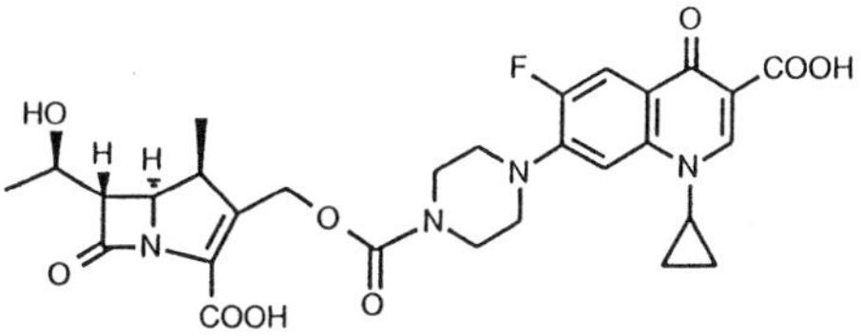

Figure 3.75 Dual action β-lactams.

3.7 Summary

1. The β-lactams are potent, broad spectrum antibacterial agents. Prototypic congeners such as the classic penicillins are limited in activity to certain Gram positive bacteria such as streptococci and staphylococci. Cephalosporins typically offer an extended spectrum of activity against some Gram negative species. Today, the modern generation of semi-synthetic cephalosporins are truly broad spectrum and extremely potent. In addition, numerous other β-lactams such as monobactams, carbapenems, carbacephems and oxacephems have

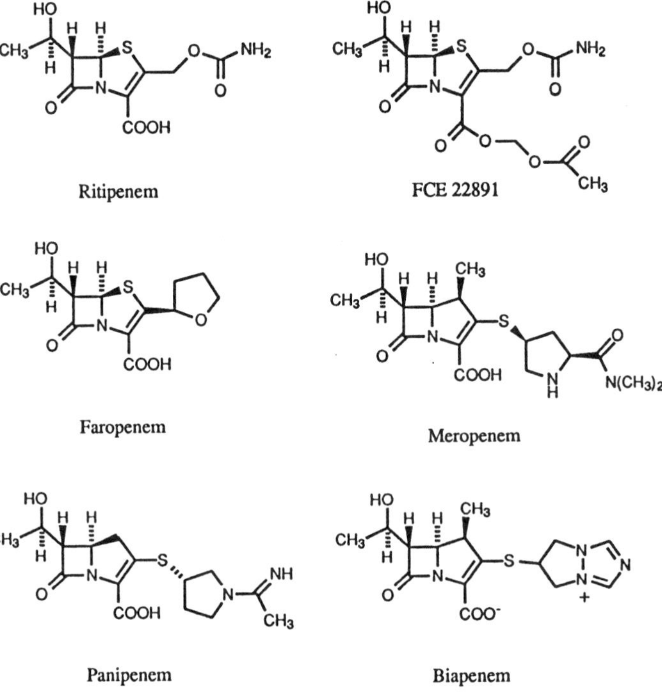

Figure 3.76 Penems and carbapenems under development.

been developed. The penems and various β-lactam-quinolone hybrids may be the next generation of β-lactam antibacterial agents to reach the market.

2. The β-lactam antibacterial agents are usually well tolerated. Some individuals are prone to potentially serious hypersensitivity which can result in anaphylactic shock, particularly upon exposure to the penicillins. Despite this problem, there are literally dozens of marketed β-lactams at this time; several congeners are preferred or alternative agents for many indications thus making the β-lactams the most useful of all families of antibacterials. In addition, the β-lactams offer a variety of dosing regimens; depending upon the congener, therapy can be administered via both parenteral and oral routes. The cost per dose of most β-lactam agents is competitive with most other classes of therapeutic antibacterial agents.

3. The β-lactam antibacterial agents exert bactericidal effects by inhibiting bacterial transpeptidase enzymes essential for the construction of peptidoglycan. This sheath is needed to protect both Gram negative and Gram positive bacteria from external forces and

Table 3.1 Major β-lactam antibacterial chemotherapeutic agents

Generic name	Common trade names	Administration*
Penicillins		
Amdinocillin	Coactin	IM, IV
Amoxicillin	Amoxil, Larotid, Polymox	PO
(with clavulanic acid)	Augmentin	PO
Ampicillin	Amcil, Alpen, Omipen, Pen A/N, Pensyn, Principen, Polycillin, Totacillin, Supen, Probampcin, Penbritin, S-K ampicillin (and others)	IM, IV
(with sulbactam)	Unasyn	IM, IV
Azlocillin	Azlin	IM, IV
Bacampicillin	Spectrobid	PO
Carbenicillin	Pyopen, Geopen	IM, IV
Cloxacillin	Tegopen	PO
Cyclacillin	Cyclapen	PO
Dicloxacillin	Veracillin, Pathocil, Dynapen	PO
Epicillin	Dexacillin, Spectacillin, Omnisan	PO
Floxacillin (flucloxacillin)	Floxapen	PO
Hetacillin	Versapen	PO
Indanylcarbencillin	Geocillin	PO
Methicillin	Staphicillin, Celbenin	IM, IV
Mezlocillin	Mezlin	IM, IV
Nafcillin	Unipen, Nafcil, Nallpen	IM, IV
Oxacillin	Prostaphilin, Bactocil	PO, IM, IV
Penicillin G	Pfizerpen, Pentids, Kesso-Pen	PO
Penicillin G Potassium		IV
(with benzathine)	Bicillin, Permapen	IM
(with procaine)	Wycillin, Duracillin, Crysticillin (and others)	IM
Penicillin V	Pen Vee K, V-cillin, Ledercillin, Veetids, Betapen, Uticillin, Penapar, Compocillin, Robicillin (and others)	PO
Phenethicillin	Broxil, Syncillin, Maxipen, Pensig	PO
Piperacillin	Pipral, Pipracil	IM, IV
(with tazobactam)	Zosyn	IV
Pivamdinocillin	(Pivmecillinam)	PO
Pivampicillin	Pivatil, Pondocillin, Alphacillin	PO
Sunbencillin	Sulfocillin	
Talampicillin	Talpen	PO
Temocillin		IM, IV
Ticarcillin	Ticar	IM, IV
(with clavulanate)	Timentin	PO
Cephalosporins		
Cefaclor	Ceclor	PO
Cefadroxil	Duricef, Ultracef	PO
Cefamandole	Mandol	IM, IV
Cefazolin	Ancef, Kefzol	IM, IV
Cefixime	Suprax	PO
Cefepime	Maxipime	IM, IV
Cefmenoxime	Cefmax, Tacef	IV
Cefmetazole	Zefazone, Cefmetazon	IV
Cefminox	Meicelin	IV
Cefonicid	Monicid	IM, IV
Cefoperazone	Cefobid	IM, IV

Table 3.1 *Continued*

Generic name	Common trade names	Administration*
Ceforanide	Precef	IM, IV
Cefotaxime	Claforan	IM, IV
Cefotetan	Ceftan, Apacef	IV
Cefotiam	Halospor, Pansporin	IV
Cefoxitin	Mefoxin	IV
Cefpiramide	Cefpiran, Sepatren, Suncefal	IM
Cefpirome	Cefrom	Parenteral
Cefpodoxime proxetil	Vantin	PO
Cefprozil	Cefzil	PO
Cefsulodin	Cefomonil, Monaspor, Pseudocef	IM
Ceftazidime	Fortaz, Pentacef, Tazicef, Tazidime	IM, IV
Ceftezole	Celosin	IM, IV
Ceftibuten	Cedax	PO
Ceftizoxime	Cefizox	IM, IV
Ceftriaxone	Rocephin	IM, IV
Cefuroxime	Zinacef, Kefurox	IM, IV
Cefuroxime axetil	Ceftin	PO
Cephacetrile	Celospor	IM, IV
Cephradrine	Anspor, Velosef	PO, IM
Cephalexin	Keflex	PO
Cephaloglycin	Kafocin	PO
Cephalothin	Keflin	IV
Cephapirin	Cefadyl	IM, IV
Cephradine	Forticef	PO
Others		
Aztreonam	Azactam	IM, IV
Flomoxef	Flumarin	Parenteral
Imipenem (with cilastatin)	Primaxin	IM, IV
Loracarbef	Lorabid	PO
Moxalactam (latamoxef)	Moxam	Parenteral

*PO, peroral; IM, intramuscular; IV, intravenous.

toxins; as a result, transpeptidation inhibition can be lethal to nearly all bacterial species. This molecular target is absent in humans and therefore potential adverse effects are minimal.

4. Resistance to the β-lactam antibacterials is widespread and threatens the future of these agents. The major cause of resistance is the production of bacterial β-lactamase enzymes which deactivate the antibacterial by cleaving the β-lactam center. Most often, β-lactamases are encoded on plasmids (or transposons) which can be readily transferred to susceptible bacteria. Nevertheless, chromosomal β-lactamases are also problematic. In addition, the emergence of bacteria that have achieved resistance via the expression of altered PBPs is particularly alarming at this time. Such organisms are capable of undergoing multi-step transformations that allow for functional (transpeptidase) PBPs which do not bind β-lactam agents. Lastly,

some resistant Gram negative bacteria, such as pseudomonas species, possess outer membranes impervious to the β-lactams.

5. Marketed β-lactam antibacterial agents include oxacephem, carbacephem, carbapenem and monobactam congeners in addition to the numerous penicillins and cephalosporins. Several penems have or are currently undergoing clinical evaluation. Various β-lactamase inhibitors are now marketed for use with select penicillins and cephalosporins.

Table 3.1 shows major β-lactam antibacterials.

Further reading

R.B. Morin and M. Gorman (Eds) (1982) *Chemistry and Biology of β-Lactam Antibiotics*, Academic Press, New York, Vols 1–3.

E.H. Flynn (1972) *Cephalosporins and Penicillins; Chemistry and Biology*, Academic Press, New York.

A.I. Demain and N.A. Soloman (1983) *Antibiotics Containing the Beta-Lactam Structure*, Part I and Part II, Springer-Verlag, New York.

E.P. Abraham (1971) 'Penicillins and cephalosporins' in *Symposium on Antibiotics*, Organic Chemistry Division Section on Medicinal Chemistry in conjunction with the Medical Chemistry Division of the Chemical Institute of Canada and the Canadian Society for Microbiologists, March 1–3, 1971, Ste. Marguerite, Quebec, pp. 399–412.

A.G. Brown and S.M. Roberts (Eds) (1984) *Recent Advances in the Chemistry of β-Lactam Antibiotics, Third International Symposium*, Special Publication No. 52, The Royal Society of Chemistry, London.

P.H. Bentley and R. Southgate (Eds) (1989) *Recent Advances in the Chemistry of β-Lactam Antibiotics, Four International Symposium*, Special Publication No. 70, The Royal Society of Chemistry, London.

E. Squires and R. Cleeland (1985) *Microbiology and Pharmacokinetic of Parenteral Cephalosporins*, Hoffman-La Roche, Nutley, NJ.

E. Squires and R. Cleeland (1986) *The Aminothiazolyl Methoxyimino Cephalosporins*, Hoffmann-La Roche, Nutley, NJ.

E. Squires and R. Cleeland (1987) *Microbiology and Pharmacokinetic of Parenteral Penicillins*, Hoffmann-La Roche, Nutley, NJ.

S.F. Queener, J.A. Webber and S.W. Queener (Eds) (1986) *Beta-Lactam Antibiotics for Clinical Use*, Marcel Dekker, New York.

D.J. Waxman and J.L. Strominger (1983) 'Penicillin-binding proteins and the mechanism of action of β-lactam antibiotics, *Ann. Rev. Biochem.*, **52**, 825.

J.B. Ward (1985) 'Biosynthesis of peptidoglycan: points of attack by wall inhibitors', *Pharm. Ther.*, **25**, 327.

D.J. Tipper (1985) 'Mode of action of β-lactam antibiotics', *Pharm. Ther.*, **27**, 1.

M.H. Richmond and R.B. Sykes (1973) 'The β-lactamases of Gram negative bacteria and their possible physiological role', *Adv. Microb. Physiol.*, **9**, 31.

H. Nikaido (1985) 'Role of permeability barriers in resistance to β-lactam antibiotics', *Pharm. Ther.*, **27**, 197.

R. Sutherland (1991) 'β-Lactamase inhibitors and reversal of antibiotic resistance', *Trends Pharm. Sci.*, **12**, 227.

H.C. Neu (1985) 'Relation of structural properties of beta-lactam antibiotics to antibacterial activity'. *Am. J. Med.*, **79** (suppl. 2A), 2.

A.G. Brown (1987) 'Discovery and development of new β-lactam antibiotics', *Pure Appl. Chem.*, **59**, 475.

T. Kametani (1982) 'Synthesis of carbapenem antibiotics', *Heterocycles*, **17**, 463.

G. Franceschi, M. Alpegiani, C. Battistini, A. Bedeschi, E. Perrone and F. Zarini (1987) 'Penems: some recent advances', *Pure Appl. Chem.*, **59**, 467.

S.W. McCombie and A.K. Ganguly (1988) 'Synthesis and *in vitro* activity of the penem antibiotics', *Med. Res. Rev.*, **8**, 393.

J. Roberts (1992) 'Cephalosporins', in *Kirk-Othmer Encyclopedia of Chemical Technology*, 4th edn, Vol. 3, John Wiley, New York, pp. 28–82.

M. Nasirada (1987) 'Structure-activity relationships of cephem analogs', *Pure Appl. Chem.*,**59**, 459.

R.D.G. Cooper (1992) 'Novel β-lactam structures – the carbacephems' in *The Chemistry of Beta-Lactams*, M.I. Page (Ed.) Chapman & Hall, London, pp. 272–305.

G.I. Georg (Ed.) (1993) 'Recent advances in the chemistry and biology of β-lactams and β-lactam antibiotics', *Bioorg. Med. Chem. Lett.*, Symposia-In-Print Number 8.

L. Jungheim, R.J. Ternansky and R.E. Holmes (1990) 'Bicyclic pyrazolidinone antibacterial agents', *Drugs Future*, **15**, 149.

M. Menard, J. Banville J., A. Martel, J. Desiderio, J. Fung-Tomc and R.A. Partyka (1993) '1-β-Aminoalkyl versus 6-aminoalkyl carbapenems: synthesis and SAR', in *Recent Advances in the Chemistry of Anti-Infective Agents*, P.H. Bentley and R. Southgate (Eds), The Royal Society of Chemistry, Cambridge, UK, pp. 3–20.

A. Perboni, B. Tamburini, T. Rossi, D. Donati, G. Tarzia and G. Gaviraghi (1993) 'Tribactams: a novel class of β-lactam antibiotics' in *Recent Advances in the Chemistry of Anti-Infective Agents*, P.H. Bentley and R. Southgate (Eds), The Royal Society of Chemistry, Cambridge, UK, pp. 21–66.

S.L. Dax (1992) 'Dual-action β-lactam antibacterials', *Current Opinion in Therapeutic Patents* (Current Drugs, Ltd.), **2**, 1375.

D.D. Keith, H.A. Albrecht, G. Beskid, K.K. Chan, J.G. Christenson, R. Cleeland, K. Deitcher, M. Delaney, N.H. Georgopapadakou, F. Konzelmann, M. Obake, D. Pruess, P. Rossman, A. Specian, R. Then, C.-C. Wei and M. Weigele (1993) 'Mechanism-based dual-action cephalosporins', in *Recent Advances in the Chemistry of Anti-Infective Agents*, P.H. Bentley and R. Southgate (Eds), The Royal Society of Chemistry, Cambridge, UK, pp. 79–92.

4 Tetracycline antibiotics

4.1 History and overview

For ages, humanity has recognized that Nature can provide biologically active concoctions, extracts and substances that are effective in the treatment of a variety of disorders and diseases. Probably no other area of pharmaceutical science has benefited more from this phenomenon than that of infectious disease chemotherapy. The screening of naturally occurring sources in search of novel antibacterial substances became a firmly established practice following the successful saga of the penicillins. By the 1930s and 1940s, the technology had evolved to allow even crude fermentation mixtures to be assayed for antibacterial properties.

There has been no steadfast rule, even to this day, as to what kinds of naturally occurring sources are most likely to elaborate antibacterial materials. Both animal and plant life have evolved elegant defense mechanisms against pathogenic bacteria that include substances with superb antibacterial properties. Higher life forms, such as plants and animals, often utilize large proteins that can recognize intricate molecular features on the surface of an invading pathogen in order to mount an effective antibacterial defense. A complex interaction between highly specialized cells ensues and, in man and animals, an immune response is induced. In addition, these life forms possess organs that can retard the spread of a pathogen; in some cases, host cells may be sacrificed in order to limit the extent of infection.

In contrast, lower life forms such as single-celled microbes have been forced to evolve more fundamental means to maintain cell survival. Usually microbes mount a defense by producing selectively toxic, small molecules that fend off other microorganisms that are in competition for nourishment or a host. Often these substances are also noxious to human and animal pathogens. For this reason, microbes have been a rich source of novel antibacterial compounds; some of these substances have been marketed without modification, but the majority have served as valuable chemical leads or prototypes for semi-synthetic congeners. Like the penicillins and cephalosporins, the tetracycline antibiotics originated from this practice.

In 1948, a potent broad spectrum antibacterial substance was obtained from a culture of *Streptomyces aureofaciens*. Chlortetracycline **1** (Figure 4.1) was exciting due to its oral activity and broad spectrum of antibacterial activity. In general, the tetracyclines display good activity against many Gram negative bacteria such as *Haemophilus influenzae*, pneumococci and *Escherichia coli* and are effective for sinusitis and bronchitis, as well as urinary tract infections caused by these organisms (respectively). Very importantly, the tetracyclines, being potent against gonococci and *Chlamydia*, are a choice therapy against many sexually transmitted diseases and pelvic and urethral infections. Lastly, these antibacterials are also effective against Lymes disease (*Borrelia burgdorferi*), cholera (*Vibrio cholerae*) and rickettsial infections.

Within a period of only several years following the discovery of chlortetracycline, oxytetracycline **2** (Figure 4.2) and tetracycline **3** (Figure 4.3) were isolated from related *Streptomyces* fermentations. Shortly thereafter, demeclocycline **4** (Figure 4.4) was discovered (1957). After elegant structural elucidation that relied mostly upon degradation studies (nuclear magnetic resonance spectroscopy and mass spectrometry were not available), the 'tetracyclines' were recognized as a class of new antibacterial agents. The tetracyclines were soon established as landmark antibiotics due to superb oral efficacy, broad spectrum antibacterial activity and exceptional tolerability in man. Furthermore, the tetracyclines were effective against some pathogens resistant to existing β-lactam antibiotics at the time.

The excellent antibacterial properties of the naturally occurring tetracyclines enticed many laboratories to engage in attempts to produce semi-synthetic derivatives. The parent compound, tetracycline, was obtained directly from chlortetracycline by chemical reduction (hydrogenolysis), indicating that semi-synthetic modifications were possible. Some academic

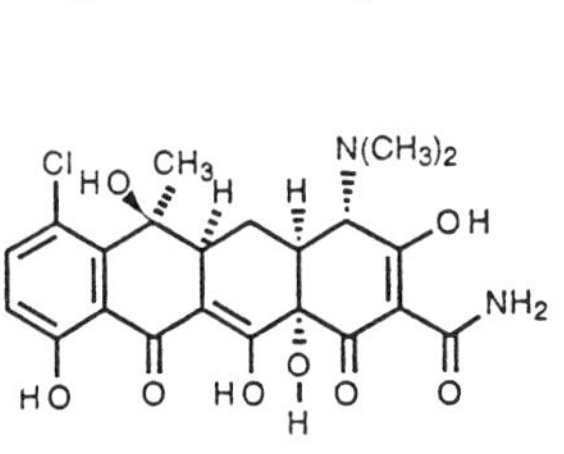

1

Figure 4.1 Chlortetracycline.

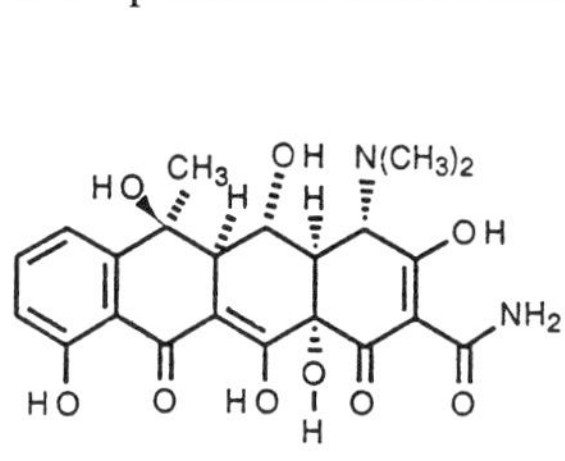

2

Figure 4.2 Oxytetracycline.

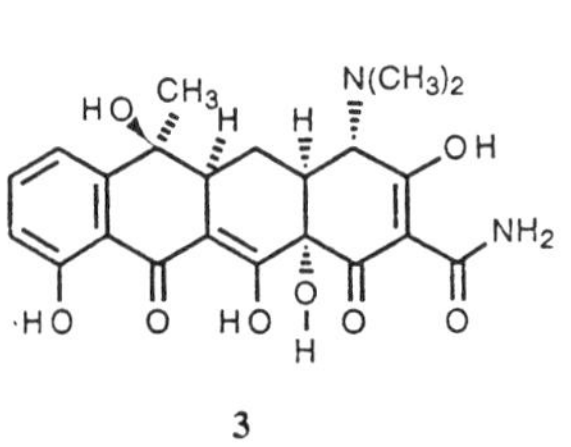

3

Figure 4.3 Tetracycline.

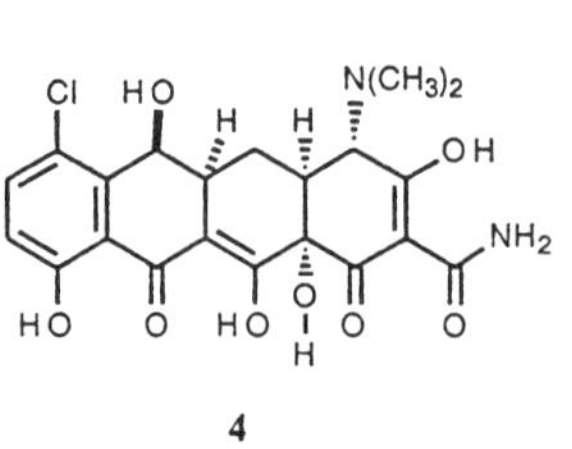

4

Figure 4.4 Demeclocycline.

groups devised approaches for the total synthesis of tetracyclines; some efforts were aimed at building on each ring in a sequential fashion. Another concern was the manner in which the complex array of functionality could be introduced onto the tetracyclic nucleus. Despite many innovative approaches, only a few total syntheses of prototypic tetracyclines would be realized (section 4.4). Most disheartening, it became apparent that wholly synthetic tetracyclines would not be practical due to the sheer number of chemical reactions and troublesome chromatographic purifications. However, these efforts were not without merit since along the way, several structural features were identified as being crucial for antibacterial activity. In addition, some chemical transformations would later prove useful for the construction of potent semi-synthetic tetracycline derivatives.

Indeed the next generation of tetracyclines to be developed were obtained by partial synthesis. The marketed semi-synthetic congeners include methacycline **5** (Figure 4.5), doxycycline **6** (Figure 4.6) and minocycline **7** (Figure 4.7). These derivatives differ from the older agents in the substitution pattern on the tetracycle nucleus; the tetracycle itself is not changed. Not surprisingly, a considerable number of additional analogs have been described that are variations on the same theme (i.e. substituent changes at the same positions), indicative of the fact that these positions are amenable to synthetic manipulations (section 4.4) without the destruction of antibacterial activity.

Despite the successes of the early tetracyclines, congeners with enhanced water solubility were sought in order to improve absorption and offer different delivery routes. For these reasons, a variety of tetracycline

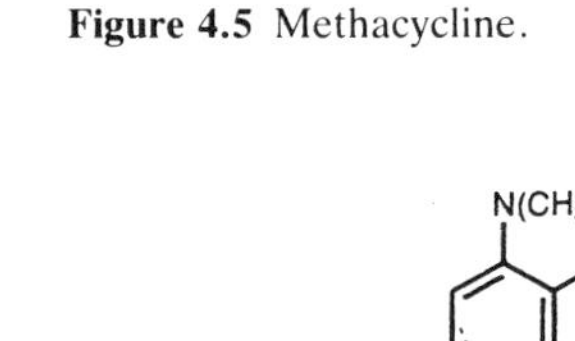

Figure 4.5 Methacycline.

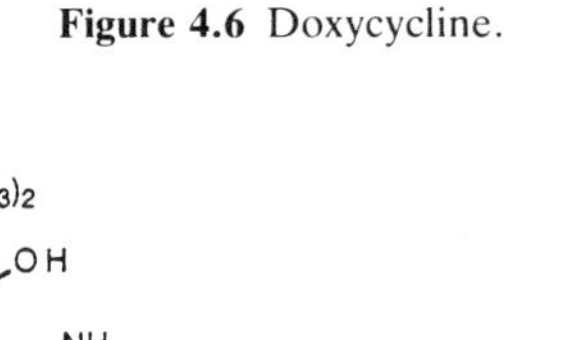

Figure 4.6 Doxycycline.

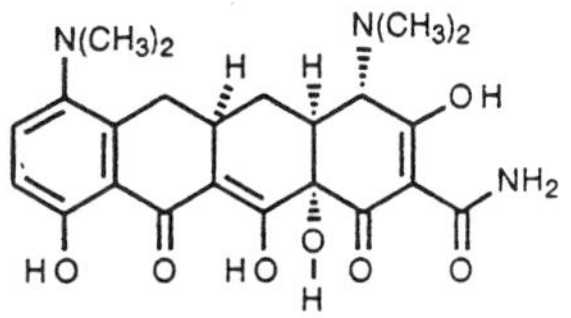

Figure 4.7 Minocycline.

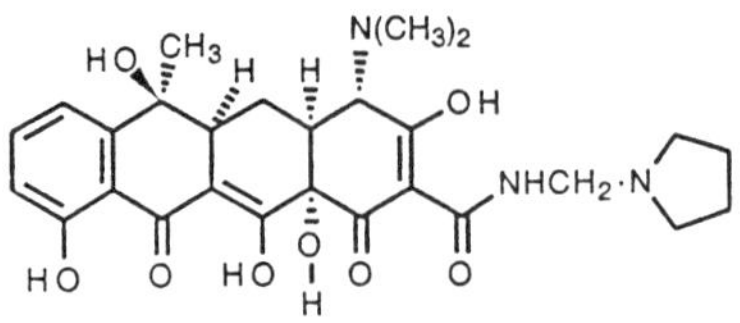

Figure 4.8 Rolitetracycline.

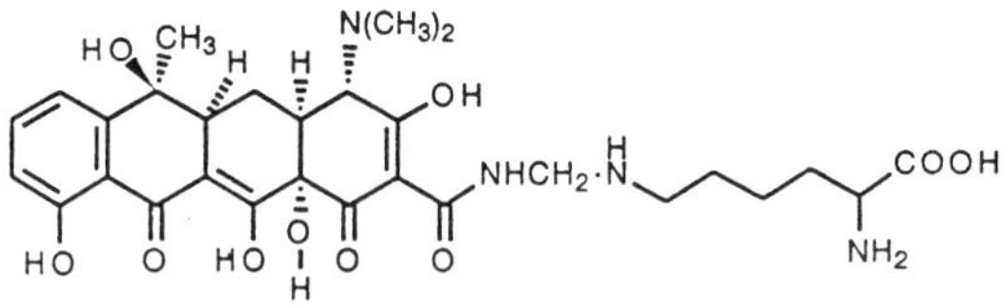

Figure 4.9 A lysinomethyltetracycline.

prodrugs were designed and synthesized. Most approaches were centered around modifications of the carboxamide functionality and some compounds of this type have been developed. For example, the pyrrolidinyl-methyl derivative rolitetracycline **8** (Figure 4.8) readily liberates the free tetracycline but is several thousand times more water soluble. Unlike the conventional tetracyclines, this compound can be administered parenterally. A number of amino acid congeners, such as the lysinomethyltetracycline **9** (Figure 4.9) are similarly water soluble.

At this time, the future of tetracycline antibacterial chemotherapy is in doubt for several reasons. First of all, tetracycline-resistant organisms are common today and often dictate that an alternative therapy be administered. Secondly, structural modifications to the tetracycline nucleus have been limited and total synthetic approaches are not practical. Lastly, many laboratories have abandoned tetracycline research in order to investigate newer and more promising antibacterial agents. However, there have been some recent findings that are likely to renew interest in the tetracyclines (section 4.6).

4.2 Mode of action: bacterial protein synthesis inhibition

Tetracyclines accumulate in sensitive bacteria and are primarily bacteriostatic in nature. However, tetracycline-induced inhibition of bacterial cell growth can be overcome, indicating that a reversible process is operative at some point of drug interaction with the bacterial cell. At the macromolecular level, the tetracyclines are inhibitors of bacterial protein biosynthesis; this phenomenon is responsible for the observed bacteriostatic effects.

The inhibition of protein biosynthesis presents grave consequences to both the prokaryotic and eukaryotic cell. Many proteins are susceptible to degradation by proteolytic enzymes, toxins and external forces such as pH, temperature and ion concentrations. Therefore most proteins have a limited functional existence and so the construction of new material is vital for cell survival. In addition, proteins provide highly specialized functions that are necessary for cell homeostasis such as enzymatic activity and the construction of various cellular components and structures. Since most bacteria are rapidly growing organisms that usually exist within a hostile environment (host or host cell), efficient and often prolific protein biosynthesis is needed in order to repair damage and maintain optimum metabolic processes. The importance of this process is reflected by the fact that a number of antibacterial agents target cellular components (e.g. the bacterial ribosome) which are directly involved in the construction of protein. For example, in addition to the tetracyclines, erythromycins, aminoglycosides, chloramphenicol and clindamycin, despite being structurally unrelated, all exert antibacterial properties as a consequence of protein biosynthesis inhibition. Since this issue will need to be revisited several times throughout this text, it is appropriate to provide a working outline here on the mechanics by which proteins are constructed.

4.2.1 Overview of bacterial protein biosynthesis

The amount of genetic material contained within a bacterium that needs to be correctly processed for cell survival and growth is considerably less in quantity (number of nucleotide bases) and variety (number of genes) than that of a eukaryotic cell. Nevertheless, the production of bacterial protein from DNA is an involved and highly specialized process; an understanding of the functions of the bacterial ribosome and its proteins, along with messenger RNA (m-RNA) and transfer RNA (t-RNA) is needed. For these reasons, an overview on protein biosynthesis is provided (details of gene expression, DNA–protein and DNA–RNA interactions are omitted).

Upon the completion of DNA gene transcription, an intact messenger RNA molecule encoding for a specific protein results and is eventually processed by the bacterial ribosome. The ribosome itself is a highly specialized organelle whose primary role is carrying out the construction of peptides that ultimately become functional proteins. The ribosome is a complex entity and many details pertaining to its interactions with RNA molecules and the role of its protein components are not completely understood. Fortunately, a great amount of information has been gathered about the molecular operations by which the ribosome and RNA molecules work together to synthesize proteins.

Since eukaryotic cells must also make proteins to survive, a bacterial protein biosynthesis inhibitor must be selectively toxic to be useful. Fortunately many antibacterial agents that inhibit protein biosynthesis preferentially bind to the bacterial ribosome or are unable to affect mammalian ribosomal function. In these cases, the antibacterial drug proves to be toxic and frequently lethal to the bacterium cell while being relatively innocuous to the mammalian cells which are present in far greater numbers in the host. This is not to suggest that all antibacterials that inhibit protein synthesis specifically target the bacterial ribosome. In fact, the tetracyclines do exert a pronounced effect on mammalian ribosomal preparations; however, in general tetracyclines are unable to accumulate within most mammalian cells and therefore, mammalian protein biosynthesis is not disrupted. Other types of antibacterial drugs are more selective than the tetracyclines against bacterial ribosomes, and for these reasons the inhibition of bacterial protein biosynthesis is a powerful chemotherapeutic action.

Prokaryotic and eukaryotic ribosomes differ in several ways; the intact bacterial ribosome has a sedimentation coefficient of 70S (Svedberg units) while the mammalian ribosome has a value of 80S. The bacterial ribosome consists of two subunits also distinguished by their sedimentation co-efficients. The subunits, designated as 30S and 50S, are made up of various proteins and RNA molecules. The best studied ribosome is that of the *Escherichia coli* bacteria but there is a great degree of functional similarity among bacterial ribosomes in terms of the subunits and accompanying proteins. The *E. coli* 30S ribosomal subunit consists of one 16S 55 kDa RNA molecule and 21 proteins that are termed as S proteins. These proteins are designated with an accompanying numerical value which reflects their relative molecular weight; accordingly S1 is the 'largest' protein while S21 is the 'smallest'. The 50S subunit consists of a 5S 40 kDa RNA molecule along with a 23S 1100 kDa RNA molecule and 32 proteins designated as L proteins in keeping with the protocol described above. (The S proteins belong to the smaller subunit and the L proteins are part of the larger subunit). Many of the ribosomal proteins have been investigated in greater detail; only components that have been implicated as serving a functional role in protein synthesis will be considered here.

The ribosome orchestrates protein biosynthesis; however, the function of other specialized 'factors' helps initiate peptide formation, elongate the peptide chain and terminate the process. The major events of protein biosynthesis involve association of the ribosomal subunits, their interaction with the m-RNA template and the procession of a variety of aminoacylated t-RNA molecules to the ribosome. A molecular channel results upon formation of the intact ribosomal complex and this groove allows the ribosome to move along the m-RNA template uninterrupted and to 'read' each 'bit' of genetic information. The m-RNA molecule contains a continuous series of nucleotide triplets, or codons, that encode for each and every amino acid of the protein product. The ribosomal complex has two sites termed P (peptidyl) and A (acceptor); the acceptor site accommodates individual t-RNA molecules which each supply one specific amino acid. Individual amino acids are sequentially transferred from these t-RNA molecules to the growing peptide chain which is 'stored' within the peptidyl site. (A third site termed E, for exit, has been recognized in some ribosomes). When translation (the reading of the m-RNA molecule) is completed, the elaborated peptide is released from the ribosome.

The ribosome carries out many necessary functions in addition to reading the m-RNA template and accepting the aminoacylated t-RNA molecules. Some events in the process are energetically demanding and depend upon GTPase activity (hydrolysis of GTP (guanosine 5′-phosphate) to GDP (guanosine diphosphate), energy and phosphate); this energy source is supplied by the ribosome itself. In addition, the formation of each peptide bond is carried out enzymatically by peptidyl transferase activity furnished by the ribosomal complex.

Ribosomes are not constantly engaged in protein biosynthesis. In fact, the ribosomal subunits can disassociate to a 'resting state' and even an intact ribosome needs to encounter a signal from the m-RNA template before protein biosynthesis will begin. This occurs with the recognition of a specific start codon (AUG) that initiates peptide formation and hence protein biosynthesis. This start codon, that for the amino acid methionine, universally leads to the production of methionyl-capped peptides. Of course, few proteins are 'capped'; most peptides are modified by some (often post-translational) process that cleaves off or alters the methionine terminus. An inherent difference between mammalian cells and most bacterial species exists at this early stage of protein biosynthesis. In bacteria, the initiation codon gives rise to a formylated methionine residue as the first amino acid of the peptide chain whereas mammalian proteins are capped with a simple (non-formylated) methionine.

With these issues in mind, the typical synthesis of a bacterial protein can be described (Figure 4.10). The 30S bacterial ribosomal subunit binds to the m-RNA molecule region which contains the start codon. Formyl-methionyl-t-RNA is then joined to the 30S m-RNA complex occupying the P site of the ribosomal unit. The 50S subunit is attached forming the 70S initiation complex which, governed by the action of initiation factors, accepts the appropriate aminoacylated t-RNA molecule. This interaction is specific and only an aminoacylated t-RNA molecule (AA-t-RNA) that encodes for the proper amino acid is able to undergo binding and eventually provide the next amino acid to the peptide chain precursor (discussed in more detail below). This event, whether initiating peptide formation or elongating the chain, is not only due to an exquisite binding specificity between an aminoacylated t-RNA molecule and the ribosome binding site itself; proteins called initiation and elongation factors govern this process.

The AA-t-RNA molecule binds to the acceptor site of the ribosomal complex. Ribosomal peptidyl transferase catalyses the construction of the first peptidic bond giving rise to a formylated methionyl-aminoacylated t-RNA species bound in the A site. Movement of this species to the P site occurs with translocation of the ribosome complex along the m-RNA template and the next codon is read. This event also frees the acceptor site of the ribosome allowing for the binding of the next AA-t-RNA molecule which donates its amino acid conjugate and the peptide chain is lengthened by a single proper amino acid residue. This cyclic process is repeated until a

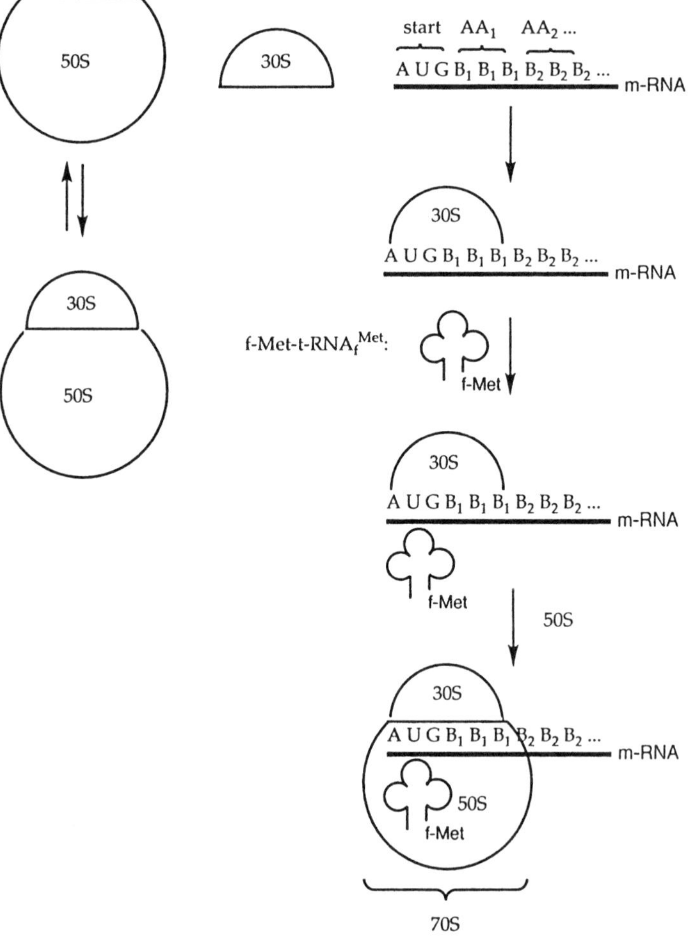

Figure 4.10 Ribosome-directed protein biosynthesis (part 1).

stop codon is reached; this signals the intervention of termination and release factors that halt the process and liberate the intact peptide from the ribosome. Post-translational modifications and protein folding usually complete the construction of the functional protein destined for metabolic, enzymatic or structural duties.

As shown in the preceding scheme, a key feature of protein synthesis is centered around the interaction of AA-t-RNA molecules with the ribosomal complex. In order for protein biosynthesis to be successful, hundreds or thousands of amino acids must be covalently joined in proper sequence (often, several ribosomes will process a lone m-RNA substrate). Only the correct aminoacylated t-RNA molecule must be allowed to bind to the A site of the ribosomal complex and ultimately donate its amino acid

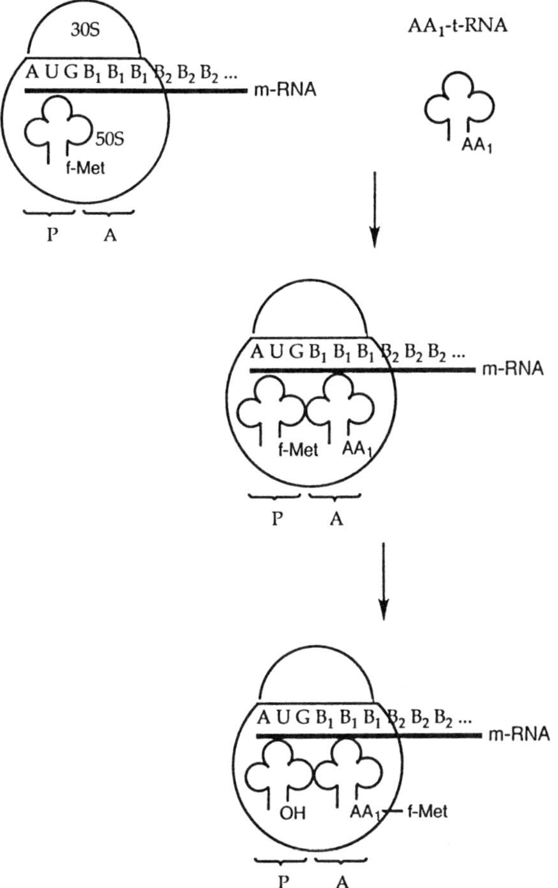

Figure 4.10 Ribosome-directed protein biosynthesis (part 2).

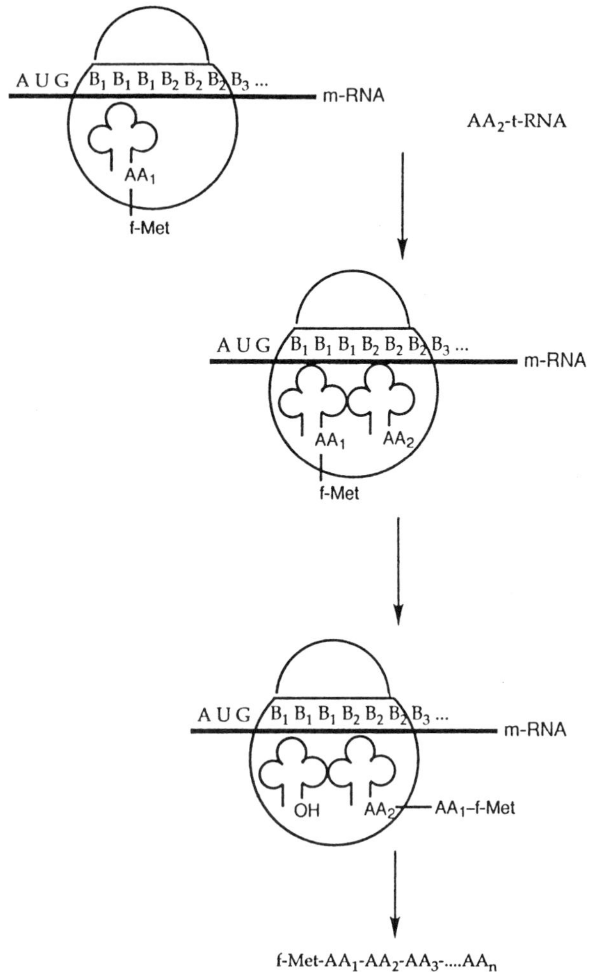

Figure 4.10 Ribosome-directed protein biosynthesis (part 3).

to the peptide chain. An incorrect reading of just one amino acid can produce a non-functional faulty protein, and cell homeostasis and survival may be jeopardized. Under normal situations, the fidelity of this process is remarkable and a wide variety of diverse proteins are correctly elaborated.

The codon presented by the m-RNA molecule remains ribosomally bound during the binding of the AA-t-RNA and is able to recognize a portion of the AA-t-RNA molecule called the anticodon. Codon–anticodon association is paramount to the fidelity of protein biosynthesis. The structure of the t-RNA molecule has been well established and a clover leaf shape is often used to describe its two-dimensional shape although the molecule is actually L-shaped in three dimensions. The anticodon occupies the central loop or leaf of the three-leafed clover. Although t-RNA molecules can vary considerably in composition, the anticodon appears to be a universal feature and one that is functionally conserved from species to species. At the terminus of the t-RNA is an acceptor stem that can accommodate an amino acid giving rise to the AA-t-RNA (aminoacylated-t-RNA) species. The amino acid is joined to the t-RNA molecule terminus as an adenosylmonophosphate conjugate and interestingly, the point of attachment onto the ribose ring of the terminal adenosine residue can vary (e.g. C-2' or C-3') and retain the ability to serve as an amino acid donor (Figure 4.11). The formation of this amino acid-t-RNA conjugate essentially activates the amino acid for incorporation into the growing peptide chain. The carboxy ester linkage of the aminoacylated t-RNA molecule occupying the peptidic site is cleaved and a new peptide bond is formed with the amino acid of the aminoacylated t-RNA species; in other words, the protein grows out from the N-terminus (Figure 4.11).

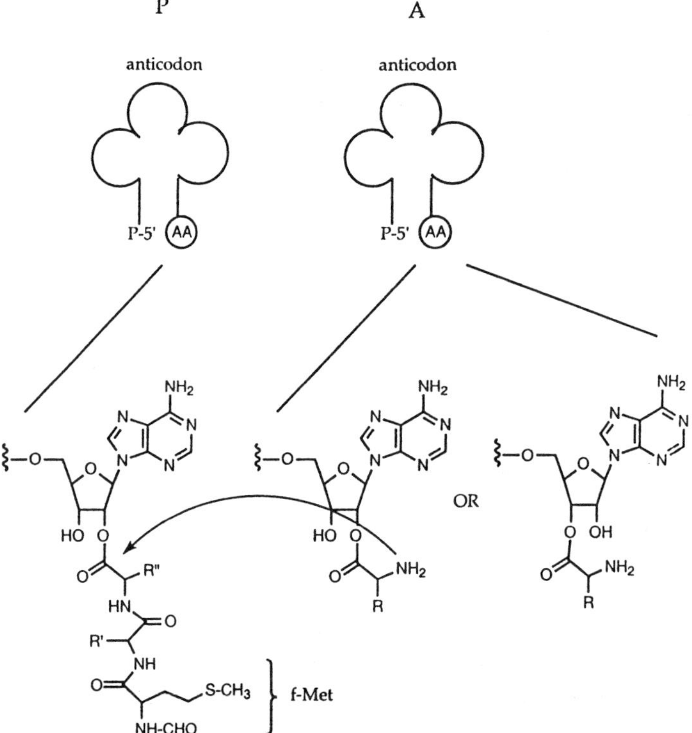

Figure 4.11 Terminus of AA-t-RNA conjugates and peptidyl transferase function.

With this background in mind, the mechanism of action of the tetracyclines can be addressed. The tetracyclines exhibit antibacterial effects in sensitive organisms by binding to the bacteria ribosome. Several hundred molecules of tetracycline can associate with a ribosome and both ribosomal subunits are able to bind tetracycline if the drug is in excess. This phenomenon can be induced under laboratory conditions and is not likely to reflect the real scenario under normal chemotherapeutic situations. Nevertheless, most of these tetracycline–ribosome associations are weak and 'cell-washing' techniques remove most tetracycline from the ribosome. However, a single tetracycline molecule remains tightly bound to the bacterial ribosome in (sensitive) whole-cell systems as well as ribosomal preparations; this tetracycline-binding site is located in a specific region of the 30S subunit. Neighboring ribosomal proteins play a role in accommodating the tetracycline substrate; tetracycline binding effects specific ribosomal proteins residing on the smaller (30S) subunit (S5, S18, and to a lesser degree, S7, S13 and S14). The same proteins are not present in eukaryotic systems and this phenomenon is probably responsible for the selective action of the tetracyclines.

The binding of tetracycline to the 30S ribosomal subunit prevents the subsequent attachment of aminoacylated t-RNA molecules. Specifically, the ribosomal acceptor (A) site is blocked; amino acid conjugates are unavailable for the continuation of peptide formation and protein biosynthesis is halted (Figure 4.12). Ribosomally bound tetracycline essentially interferes with the establishment of normal codon–anticodon interactions but this may not be due to a single direct effect on the acceptor site. It is more likely that the binding affinity of the ribosomal acceptor site is compromised since it shares a proximal or overlapping region of the ribosome with the tetracycline binding region. Tetracycline–ribosome complexes are stabilized relative to AA-t-RNA-ribosome complexes and therefore aminoacyl-t-RNA substrates are unable to compete with or disrupt drug binding. Interestingly, the binding constant of tetracyclines to the 30S ribosome subunit is approximately 10^{-5} M, a marginal difference compared with normal substrates (10^{-6} M). This raises the questions as to whether other factors contribute to tetracycline-induced inhibition of protein biosynthesis; even after nearly four decades of study, this problem has not been resolved decisively.

Tetracyclines can also exert some secondary effects upon ribosomal protein biosynthesis but it is not clear how significant these events are compared to the phenomenon described above. Under high concentrations of magnesium cation, tetracyclines can associate with both the P site and the A site of the ribosome complex. In addition, tetracyclines may also be able to inhibit chain termination by altering the normal function of release factors. Until these actions can be fully appreciated, the mode of action of the tetracyclines remains open for modification.

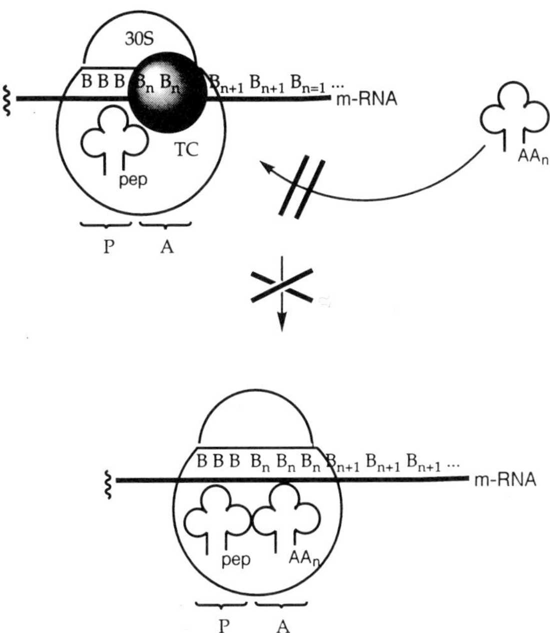

Figure 4.12 Ribosomal binding of tetracycline and inhibition of bacterial protein bio-synthesis.

Despite these minor reservations, it is clear that the tetracyclines primarily target the bacterial ribosome. The tetracyclines do not affect the availability of AA-t-RNA molecules and so protein biosynthesis is not compromised as a result of deprivation of any amino acid source. The tetracyclines display no inhibitory effects towards the numerous aminoacyl-t-RNA synthetase enzymes that catalyse the covalent bonding of amino acids to t-RNA molecules. The recognition of the (m-RNA) start codon by the ribosome and the subsequent binding of the 30S ribosomal subunit to m-RNA substrate are also not targeted. The binding of the formyl-methionyl-t-RNA molecule (f-Met-t-RNA) to the ribosomal site is unimpeded and tetracyclines do not inhibit the action of initiation factors. It is significant that tetracyclines do not inhibit the function of elongation factor(s) since peptide elongation comprises the majority of protein biosynthesis in terms of number of events (compared to initiation and termination). Tetracyclines do not interfere with the transfer of amino acid units onto the growing peptide; peptidyl transferase activity is not a target. Rather peptide bond formation is precluded since delivery of AA-t-RNA to the acceptor site is thwarted. The movement of the ribosomal complex along the m-RNA template is not affected by the tetracyclines, nor is the fidelity of the 'reading' process. Tetracyclines do not cause faulty proteins to be produced, rather protein biosynthesis is halted. Furthermore, the GTPase activity of the ribosome (50S subunit) is inert to the tetracyclines. When totaled, the mechanisms of peptide formation remain operative except for docking of aminoacylated t-RNA substrate.

Interestingly, the tetracyclines are predisposed to a variety of weak binding interactions with many biologically important entities. This phenomenon is probably due to the diverse array of functionality present in these molecules in conjunction with a flat region; these characteristics are likely to establish hydrogen bonding and lipophilic associations and allow for insertion into various molecular environments. Under appropriate conditions, tetracyclines are able to bind to DNA, t-RNA and poly-nucleotides in addition to ribosomes. However, only tetracycline binding to bacterial ribosomes produces antibacterial effects.

4.3   Structural features and structure–activity relationships of the tetracycline antibacterials

The structural features that confer antibacterial activity to the tetracyclines have been well established. The key feature is a linear fused tetracyclic nucleus (rings designated as A, B, C and D as shown in Figure 4.13) from which a variety of functional group substituents emanate. The most important of these are an array of conjugated hydroxyl and ketone groups and the octahydrodioxonaphthacenecarboxyamide core itself. The simplest tetracycline to display detectable antibacterial activity is 1,4,4a,5,5a,6, 11,12a-octahydro-3,10,12,12a-tetrahydroxy-1,11-dioxo-2-naphthacenecar-boxyamide (Figure 4.13) and so this compound may well represent a minimum pharmacophore. The absolute stereochemistry is crucial; only compounds that possess the naturally occurring configurations at the 4a and 12a positions (A-B ring junction) are active as antibacterials. (No unambiguously proven C-4a-β analogs are known, so the stereochemical requirement at this position is inferred).

Despite decades of research, many positions of the tetracycle nucleus have remained essentially unexplored due to the lack of successful or practical synthetic methods which introduce unnatural substituents. The same is true for almost any alteration in the tetracyclic nucleus itself; alteration in ring size or the incorporation of heteroatoms within the tetracycle have remained mostly unexplored. Consequently, a great deal of conventional wisdom has been essentially implicit and there are certainly gaps in current tetracycline SAR that wait to be resolved. This is not to say that a sound working SAR is not in place today. For example, a series of new and exciting tetracyclines (section 4.6) have recently been reported partly as a response to traditional tetracycline SAR.

It is the tetracyclic nucleus from which the tetracyclines derive their

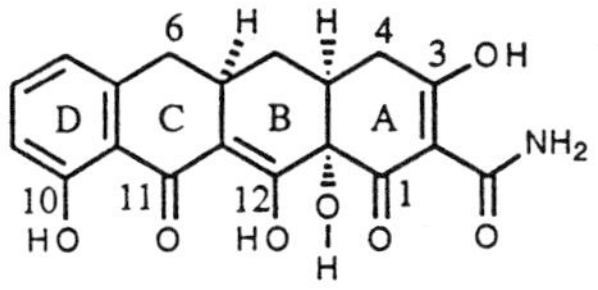

1,4,4a,5,5a,6,11,12a-octahydro-3,10,12,12a-tetrahydroxy-
1,11-dioxo-2-naphthacenecarboxyamide

Figure 4.13 Basic tetracycline SAR.

namesake. In retrospect, this naming was somewhat prophetic in that the tetracycle, to this day, remains one element virtually intolerant of modification. There are unique chemophysical properties of this structural core; the tetracyclines are highly chromophoric compounds and structural modifications that reduce the magnitude of this phenomena generally afford inferior compounds. For example, the change of oxidation states of tetracyclic carbon atoms, or the removal of certain key functional moieties, mostly hydroxyl or carbonyl moieties, is detrimental. The linear array of the fused tetracycle is absolutely vital; any bond cleavage that destroys a ring or frees a ring from the confines of the tetracyclic nucleus abolishes activity. Each ring needs to be six-membered and purely carbocyclic; the introduction of sulfur at the C-6 position is an exception but compounds of this type are not significant at this time. Nortetracyclines, derivatives in which the B ring consists of a five-membered carbocycle, display only a hint of antibacterial activity. The A ring and the D ring are generally regarded as unchangeable; the D ring needs to be aromatic and the A ring must be appropriately substituted at each of its carbon atoms for notable activity. However, the D ring has been amenable to a number of substituent changes and some promising congeners contain novel substituents on this ring.

The tetracyclines are elaborated through a polyketide biosynthetic pathway and this origin is reflected by the conjugation of oxygen-based functionality along the periphery of the tetracyclic ring system. The B ring and the C ring tolerate certain substituent changes so long as the keto–enol system (at C-11, C-12, C-12a) remains intact and conjugated to the phenolic D ring. Aromatization of either the B ring or the C ring is detrimental. Anhydrotetracyclines, derivatives in which the C ring is aromatized are ubiquitous by-products under acidic reaction conditions and are essentially inactive. One notable exception is the naturally occurring chelocardins that structurally resemble anhydrotetracyclines but are best regarded as curiosities in their own right; their substituent pattern and spectrum of activity is markedly different from the tetracyclines and their biosynthetic origin is different. (These compounds, along with the anhydrotetracyclines and the nortetracyclines will be revisited later). The

D-C-B-ring phenol–keto–enol system is imperative and the A ring must also contain a conjugated keto–enol system. Specifically, the A ring contains a tricarbonyl-derived keto–enol array at positions C-1, C-2 and C-3. It should be noted that in highly conjugated systems such as these, tautomeric forms can exist and in the case of the tetracyclines, this appears to be an important feature for antibacterial activity.

Other structural requirements for good antibacterial activity include a basic amine function at the C-4 position of the A ring. Derivatives (Figure 4.13) unsubstituted at the C-4 center are regarded as antibacterials in an academic sense only. The simplest tetracycline that displays a broad spectrum of antibacterial activity is the C-4 (α)-dimethylamino congener sancycline (Figure 4.14) which has not been developed as a marketed product.

4.3.1 C-1 substituents

The keto–enol system of the A ring is indispensable for antibacterial activity. No variation at the C-1 position has been successful (or reported) and this position is regarded as intolerable to modification.

4.3.2 C-2 substituents

The carboxyamide moiety is present in most naturally occurring tetracyclines and this group is crucial for antibacterial activity. The amide is best left unsubstituted at the nitrogen atom but monosubstitution is acceptable if in the form of activated alkylaminomethyl amides (often referred to as Mannich (condensation) derivatives). Examples include rolitetracycline **8** and the lysinomethyltetracycline **9** (Figures 4.8 and 4.9) although there are many other analogs including those derived from the Mannich condensation of a tetracycline with various piperazines, pyrrolidines, morpholines, amino acids, alkyloxyamines and alkylamines. In these cases, the free (C-2 carboxyamido) parent tetracycline is responsible for the antibacterial activity. However, simple monoalkylated C-2 carboxyamide derivatives can retain notable activity providing the alkyl group is small, such as congeners containing a nonmethylated carboxyamide. Larger alkyl substituents on the amide nitrogen atom are likely to be too lipophilic to

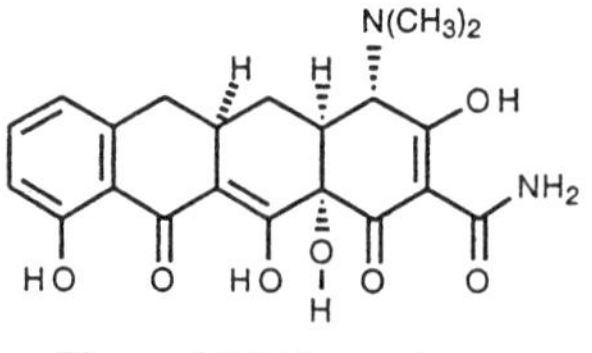

Figure 4.14 Sancycline.

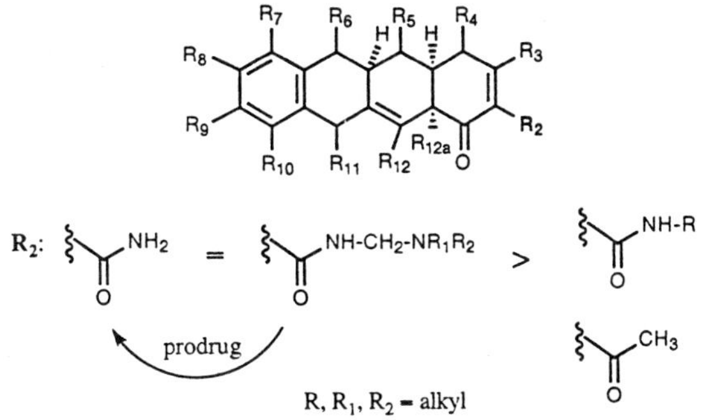

Figure 4.15 Tetracycline SAR: C-2 substituents.

penetrate into Gram negative bacteria; the monosubstituted *t*-butyl amide displays notable Gram positive activity but is essentially inactive against Gram negative organisms. Large alkyl groups on the carboxyamide may alter the normal keto–enol equilibrium of the C-1,C-2,C-3 conjugated system and thus diminishes inherent antibacterial activity.

The replacement of the C-2 carboxyamido group with that of a simple methyl ketone retains some antibacterial activity, but this is weak in comparison to the parent tetracycline. Dehydration of the carboxyamide to the corresponding nitrile results in loss of activity and again electronic factors, as well as the loss of the carbonyl moiety, are likely to be responsible for this phenomenon. The tetracycline C-2 position is limited with regard to SAR (Figure 4.15).

4.3.3 C-3 substituents

In conjuction with the C-1 position, the keto–enol conjugated system is imperative for antibacterial activity and no variations at the C-3 center have been reported.

4.3.4 C-4 substituents

The naturally occurring tetracyclines contain an α-C-4-dimethylamino substituent that favorably contributes to the keto–enolic character of the A ring; β-C-4 epimers are inferior congeners. Epimerization at the C-4 position is problematic at acidic pH, particularly with C-5 unsubstituted tetracyclines; oxytetracyclines (C-5 hydroxylated) are more resistant due to intramolecular hydrogen bonding. Chelocardin **10** (Figure 4.16) possesses a (β)-C-4 amino substituent that is of an inverted configuration relative to the tetracyclines. Chelocardin is obtained from a different

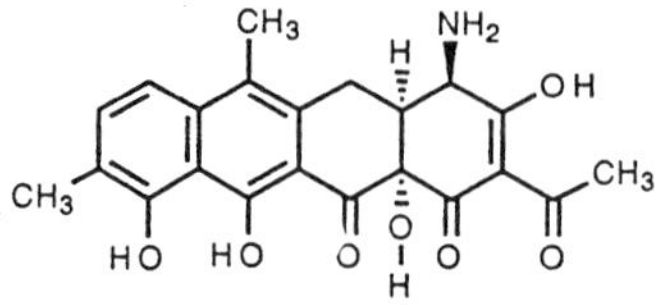

10

Figure 4.16 Chelocardin.

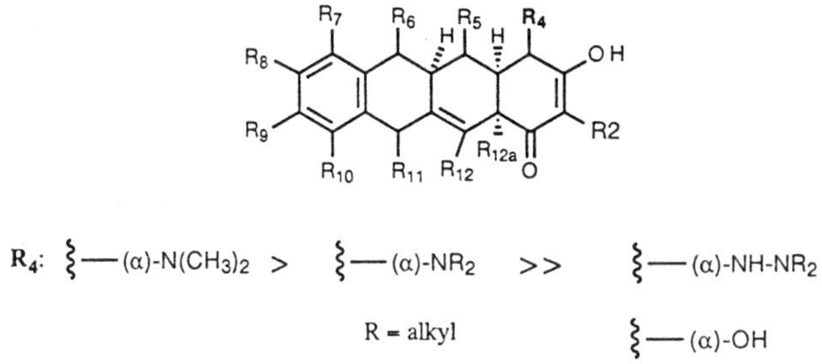

Figure 4.17 Tetracycline SAR: C-4 substituents.

source (*Nocardia* spp.) than the tetracyclines, and unlike tetracyclines is more potent against Gram negative bacteria than Gram positive organisms.

Quaternization of the C-4 dimethylamino moiety disposes the tetracycline to a variety of nucleophilic substitutions at this center. For example, various amines have been introduced at the C-4 position. The trend indicates that small C-4-amino moieties are preferred and the dimethylamino group remains optimal although other variations (e.g. ethylmethylamino and diethylamino substituents) are active. In addition to steric size, a key feature that appears to be operative with C-4 substituents is that of appropriate basicity. Replacement of the dialkylamine group with a hydrazone, oxime or hydroxy group leads to pronounced loss of activity. (Figure 4.17), probably due to the increase in heteroatom basicity. Tetracyclines with quaternary C-4 amines are also susceptible to intra-molecular ketal formulation or elimination (described later).

4.3.5 C-4a substituents

The (α)-hydrogen at C-4a position of the tetracyclines is necessary for useful antibacterial activity. The dactylocyclines (section 4.6) are structurally related natural products that have an α-OH group at the C-4a position; these compounds are relatively new additions to the tetracycline family and have yet to be developed.

4.3.6 C-5 substituents

Many naturally occurring antibacterial tetracyclines have an unsubstituted methylene moiety at the C-5 position. Howver oxytetracycline, the second tetracycline isolated, contains a C-5 α-hydroxyl group. Oxytetracycline is not only a potent derivative in its own right, but the functionality at the C-5 position has been chemically modified to prepare some semi-synthetic tetracyclines. Unfortunately, most C-5 variants offer no advantage over oxytetracycline or the C-5 unsubstituted tetracyclines (e.g. tetracycline itself) in terms of enhancing or expanding upon the antibacterial spectrum. Alkylation of the C-5 hydroxyl group results in loss of activity. Ester formation is only acceptable if the free oxytetracycline can be liberated *in vivo*: only small alkyl esters are useful. The C-5 formyl ester of oxytetracycline is noteworthy in that absorption is improved although activity is attributed to the parent C-5-hydroxytetracycline. Simple alkyl ester derivatives may retain some weak inherent activity but larger ester derivatives are inactive (Figure 4.18).

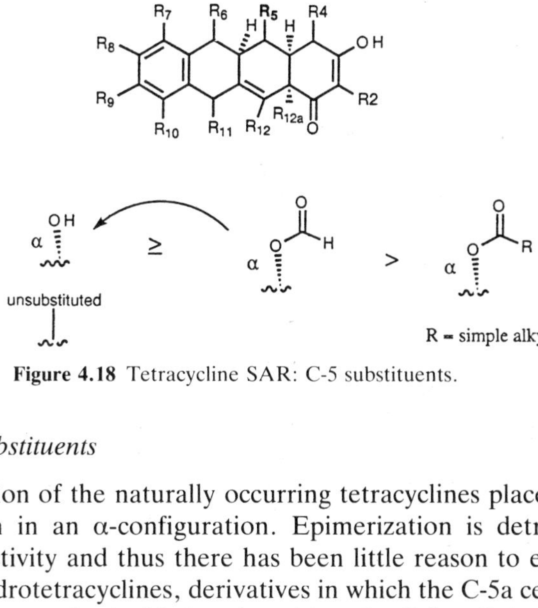

Figure 4.18 Tetracycline SAR: C-5 substituents.

4.3.7 C-5a substituents

The configuration of the naturally occurring tetracyclines places the C-5a hydrogen atom in an α-configuration. Epimerization is detrimental to antibacterial activity and thus there has been little reason to explore this position. Dehydrotetracyclines, derivatives in which the C-5a center is sp^2-hybridized and part of a double bond to either the C-5 or C-6 position, are inferior antibacterials.

4.3.8 C-6 substituents

The C-6 position is tolerant of a variety of substituents. The majority of tetracyclines have an (α)-methyl group and a (β)-hydroxyl group at this position. Demeclocycline **4** (Figure 4.4) is a naturally occurring C-6 demethylated chlortetracycline with an excellent activity. Thus the C-6 methyl group contributes little to the activity of the tetracycline family as a whole, although its presence does increase lipophilicity somewhat. Similarly, the C-6 hydroxyl group also appears to offer little in terms of antibacterial activity; removal of this group affords doxycycline **6** (Figure 4.6) which is a superb antibacterial. Tetracyclines that lack a C-6 hydroxyl group possess some advantages compared to their counterparts in that these compounds are generally more stable and more lipophilic in nature. Nevertheless, both demeclocycline and doxycycline are marketed agents. Interestingly, the dactylocyclines have both C-6 methyl and C-6 hydroxyl substituents present, but in reversed stereochemisty compared to the tetracycline series.

An important discovery pertaining to C-6 substituents was realized with the synthesis and subsequent investigation of the corresponding methylenotetracyclines. The prototype is methacycline **5** (Figure 4.5); the C-6 *exo*-methylene substituent is surprisingly stable under physiological conditions and this group is also amenable to chemical modifications. For the most part, addition reactions across the double bond have produced some interesting derivatives. For example, mercaptan addition has afforded some active C-6 alkylthiomethyl tetracyclines. Other groups have been added onto the *exo*-methylene moiety. In general, small alkyl moieties are preferred; larger groups (bulky alkyl or simple aryl moieties) are weak against Gram negative bacteria although significant Gram positive activity can be retained. As with C-4 and C-5 substituents, increased hydrocarbon size probably diminishes penetration into Gram negative organisms. Halogens have been introduced at the C-6 position

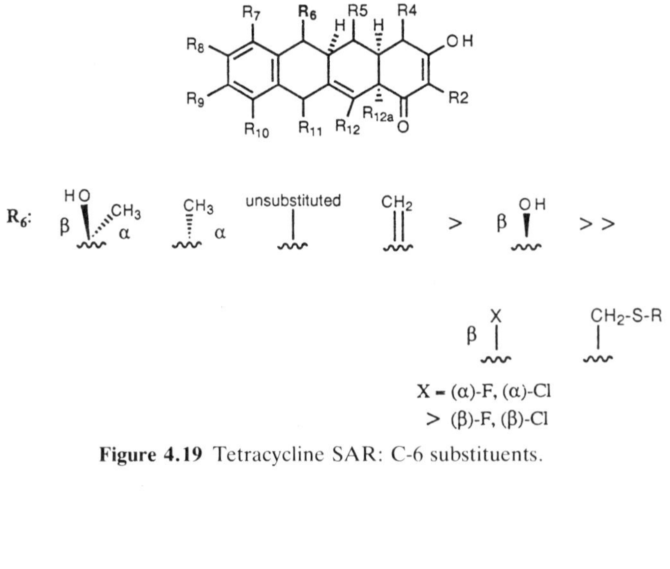

Figure 4.19 Tetracycline SAR: C-6 substituents.

and the α-chloro and α-fluoro analogs are superior to their epimeric counterparts. However these modifications offer no advantage in terms of potency and remain of academic interest only. A general summary of C-6 substituents is given (Figure 4.19).

4.3.9 C-7 substituents

The C-7 substituents have been explored with some outstanding success. Naturally occurring chlortetracycline **1** (Figure 4.1) has a C-7 chloro substituent and was the first member of the tetracycline family to gain widespread use in antibacterial chemotherapy. The first successful semi-synthetic C-7 derivative was tetracycline itself which was produced from chlortetracycline, although tetracycline is also a naturally occurring tetracycline. Other C-7 unsubstituted analogs such as oxytetracycline **2** (Figure 4.2) soon followed. The nature of the aromatic D ring of the tetracyclines predisposes the C-7 position to electrophilic substitution, and nitro and halogen groups have been introduced. Some C-7 nitro tetra-cyclines are among the most potent of all tetracyclines *in vitro*. Despite their promising potency, these agents have not been pursued as market candidates for a number of reasons (nitrated aromatic compounds are potentially toxic/carcinogenic). Halogenated derivatives such as the C-7 bromo-, fluoro-, and iodotetracyclines are less active. Unfortunately, a competing reaction plagues the introduction of most C-7 substituents; the C-9 position is similarly activated towards electrophilic substitution by the electron-donating properties of the C-10 phenolic moiety.

The C-7 acetoxy-, azido- and hydroxytetracyclines have been obtained through the intermediacy of a diazonium species; these derivatives are inferior in terms of antibacterial activity. The reduction of C-7 nitrotetra-cyclines to the corresponding C-7 aminotetracyclines is a key transformation in this approach. Although simple C-7 aminotetracyclines are extremely weak in terms of potency, reductive alkylation has afforded very potent C-7 dialkylamino congeners. This important breakthrough culminated in the development of minocycline **7** (Figure 4.7) which is the last tetracycline to be marketed. Minocycline remains a dominant tetracycline, in part due to its superior potency, as well as its activity against tetracycline-resistant organisms. Larger alkylamino groups at C-7 result in decreased anti-bacterial activity, reminiscent of the C-4 dialkylamino moiety. A summary of the tetracycline C-7 substituents is given below (Figure 4.20).

4.3.10 C-8 substituents

The tetracycline C-8 position is generally unreactive and total synthetic approaches have been impractical; accordingly this position has remained unexplored. The dactylocyclines have a methoxy group at this position.

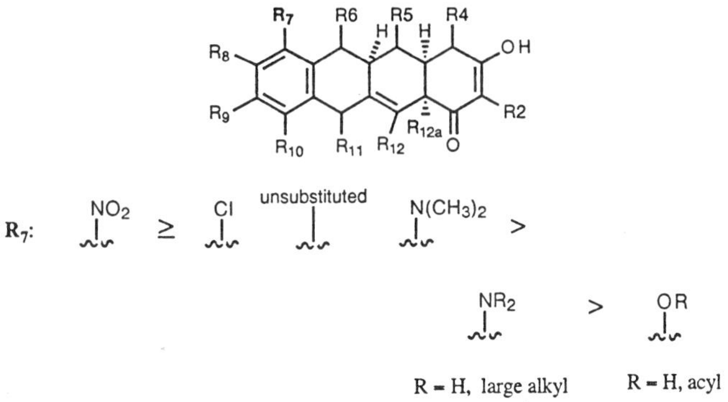

Figure 4.20 Tetracycline SAR: C-7 substituents.

4.3.11 C-9 substituents

The C-9 modifications have followed in parallel with those at C-7 in that aromatic electrophilic substitution has been utilized extensively. Consequently, the same substituents that have been incorporated at the C-7 position, have also been introduced at C-9 for the most part. However, the C-9 position is somewhat less reactive towards electrophiles as a result of electronic and steric factors caused by the presence of the C-10 phenol moiety. In many cases, both C-7 and C-9 isomers are produced under reaction conditions and must be painstakingly separated; oxytetracyclines are particularly problematic in electrophilic substitution reactions which introduce either a C-7 or C-9 substituent.

Although similar in chemical reactivity towards electrophiles, the C-7 and the C-9 positions are not identical with respect to antibacterial activity. A few noteworthy examples exist which highlight the impact of substituent placement upon the D ring. For example, C-9 amino (6-desmethyl-6-deoxy) tetracycline is an extremely potent analog while the corresponding C-7 isomer is a poor antibacterial. Conversely, the C-9 nitro (6-desmethyl-6-deoxy) tetracycline is inferior to the exquisitely potent C-7 nitrotetracycline analog. Despite the exceptional activity of minocycline, the corresponding C-9 dimethylaminotetracycline analog is essentially inactive. The most significant breakthrough has been the acylation of the C-9 amino group to afford the potent glycyltetracyclines (section 4.6). In summary, as with the C-7 substituents, the most potent C-9 analogs contain electron-withdrawing moieties, although there are exceptions depending upon other substituents.

4.3.12 C-10 substituents

The C-10 phenolic moiety is absolutely necessary for antibacterial activity. There have been few attempts to modify this position, both for this reason, and because of difficulties with the chemistry involved.

4.3.13 C-11 substituents

The C-11 carbonyl moiety is part of one of the conjugated keto–enol systems required for antibacterial activity.

4.3.14 C-11a substituents

In general, few modifications at the C-11a position of the tetracyclines have been tolerated. This is probably due to the detrimental effects exerted upon the keto–enol system, which is vital for magnesium cation binding and subsequent tetracycline uptake by the bacterial cell. Halogenation has been accomplished in the 6-demethyl-6-deoxy series. The bromo analog is more potent than the corresponding chloro derivative, but these compounds appear to be metabolically unstable, and the observed activity is at least partially due to the parent tetracycline. The fluoro derivative is metabolically stable and the least active of the three. No compound in this series can be considered as a potential therapeutic agent due to weak antibacterial activity, although halogenation can serve as a blocking agent for other chemical manipulations.

4.3.15 C-12 substituents

As with the C-11 position, the C-12 position is part of the keto–enol system vital for drug uptake, binding and observed antibacterial activity. There are no variations at this position.

4.3.16 C-12a substituents

The C-12a hydroxyl group is needed for antibacterial activity although this moiety can be esterified to provide tetracyclines with increased lipophilicity. Antibacterial properties are retained if the alkyl ester is small in size, and readily undergoes hydrolysis to liberate free tetracycline.

A summary of basic tetracycline structure–activity relationships is depicted (Figure 4.21).

4.4 Synthetic approaches to tetracycline antibacterial agents

The structural features of the tetracyclines that are necessary for antibacterial activity have been determined but many envisioned variations remain untested. This is due to the complexity of the molecules and inherent reactivity and/or instability towards reaction conditions required to introduce novel substituents at many positions on the tetracyclic nucleus. The handful of successful total syntheses have proven to be

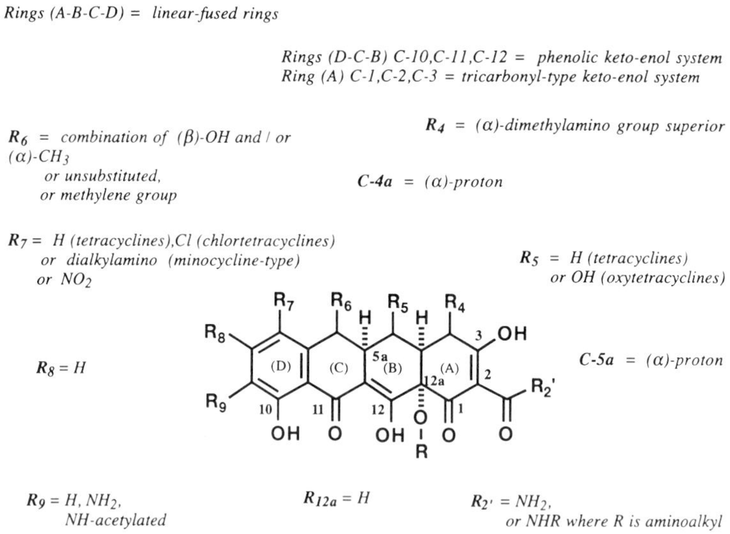

Figure 4.21 Summary of features which confer optimum antibacterial activity to the tetracycline nucleus.

inefficient, troublesome and complicated by the formation of unwanted products, undesired stereoisomers and degradation.

Most of the tetracyclines that are available for research study today are either obtained directly from fermentation or produced by semi-synthetic means. The mass production of marketed tetracyclines has been forced to rely upon fermentation processes. To date, there is no tetracycline on the market, nor in the clinic, that is obtained wholly via chemical synthesis. Despite this, there has been some elegant synthetic research that has proven to be a productive enterprise. The tetracyclines inspired a number of prolific academic and industrial laboratories to undertake projects aimed first at elucidating the structure of these compounds, and subsequently at developing synthetic methods to produce novel congeners. While all of the work in this area cannot be fully addressed here, several efforts are landmark and deserve consideration. Most notably, the reseach carried out by the Muxfeldt, Lederle, Woodward, Pfizer and Barton laboratories remain historic.

The early synthetic strategy of Muxfeldt is illustrated by synthesis of a biologically inactive tetracycline nucleus (F), however, later achievements include the first successful total synthesis of oxytetracycline. In the early work, ring annulations were accomplished via Stobbe chemistry, or by the addition of malonate anions onto activated centers alpha to the adjacent ring, followed by ring closure (Figure 4.22). Specifically, an aromatic D ring substrate (A) was elaborated into a fused bicyclic D-C ring

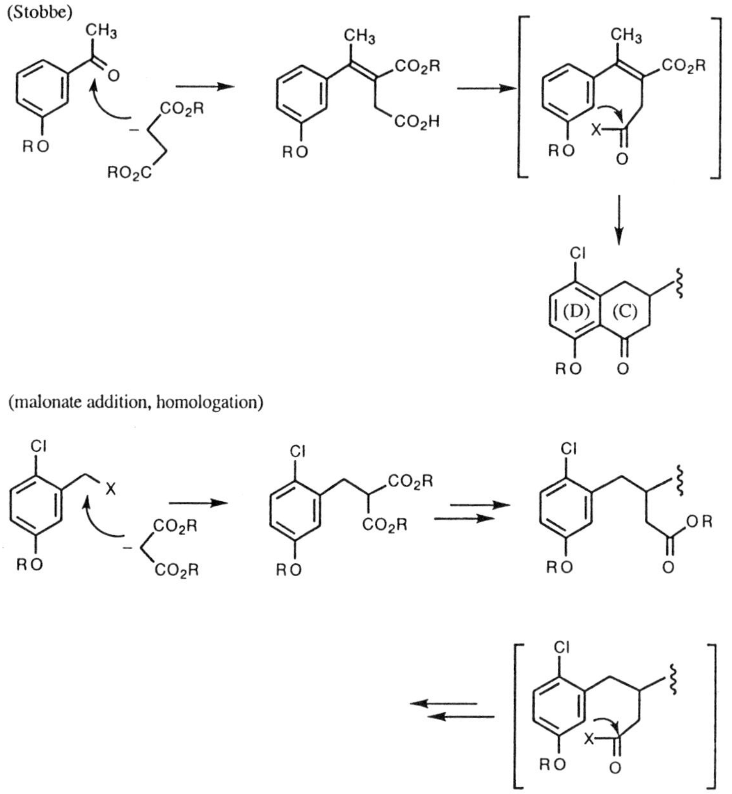

Figure 4.22 Cyclization method (shown for D-C ring).

intermediate (B) upon which the B ring (E) and A ring (F) were annulated in sequential fashion (Figure 4.23). The formation of the carbon–carbon bonds which closed each ring were accessed via Friedel–Crafts acylation, or via Aldol/Dieckman-type reactions. While this work is indeed elegant, it is limited in several important regards. The chemistry provides no means for introducing the C-4 dimethylamino substituent and carboxamide formation is extremely inefficient. As a result, the final anhydrotetracycline derivatives were inactive as antibacterials and obtainable only after an involved and complicated chemical synthesis.

A variation of this method was used to produce compounds in which the C-ring was formed with unsaturation and lacking other important moieties such as the desired C-1 carboxamide (Figure 4.24). However, one key feature in this approach was the masking of the C ring carbonyl moiety as a ketal until the A ring was introduced. Unlike the synthetic strategy outlined above, the addition of adjacent rings in sequential fashion was

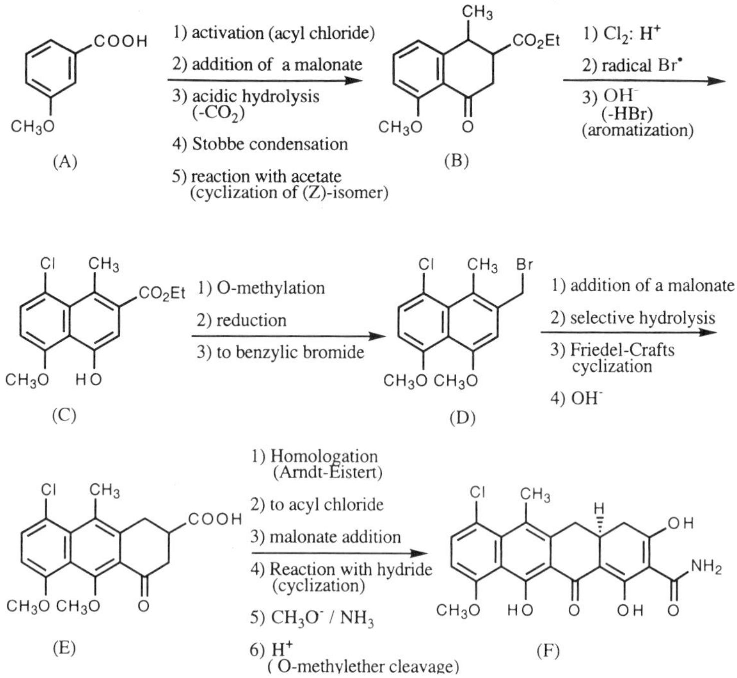

COOH
CH3O
(A)
1) activation (acyl chloride)
2) addition of a malonate
3) acidic hydrolysis (-CO2)
4) Stobbe condensation
5) reaction with acetate (cyclization of (Z)-isomer)
CH3
CO2Et
CH3O
O
(B)
1) Cl2: H+
2) radical Br•
3) OH- (-HBr) (aromatization)
Cl
CH3
CO2Et
CH3O
HO
(C)
1) O-methylation
2) reduction
3) to benzylic bromide
Cl
CH3
Br
CH3O CH3O
(D)
1) addition of a malonate
2) selective hydrolysis
3) Friedel-Crafts cyclization
4) OH-
Cl
CH3
COOH
CH3O CH3O
O
(E)
1) Homologation (Arndt-Eistert)
2) to acyl chloride
3) malonate addition
4) Reaction with hydride (cyclization)
5) CH3O- / NH3
6) H+ (O-methylether cleavage)
Cl
CH3
H
OH
NH2
CH3O
HO
O
OH
O
(F)

Figure 4.23 Muxfeldt approach to the total synthesis of a tetracycline.

Cl
CH3
CO2Et
CH3O
O
(G)
1) ketalization
2) ester reduction (to alcohol)
3) mesylation
4) displacement with cyanide
5) reduction of nitrile to aldehyde
Cl
CH3
CHO
CH3O
O
O
(H)
1) addition of a malonate
Cl
CH3
CO2R
CO2R
CH3O
O
O
(I)
1) reaction with acetoacetate ester (Michael addition - Dieckman cyclization)
2) ketal cleavage
3) hydrogenolysis
Cl
CH3
CO2R
O
RO2C
CH3O
O
O
(J)
1) Dieckman cyclization (base)
2) hydrolysis (ether and ester cleavage) (-CO2)
Cl
CH3
H
H
OH
HO
O
HO
O
(K)

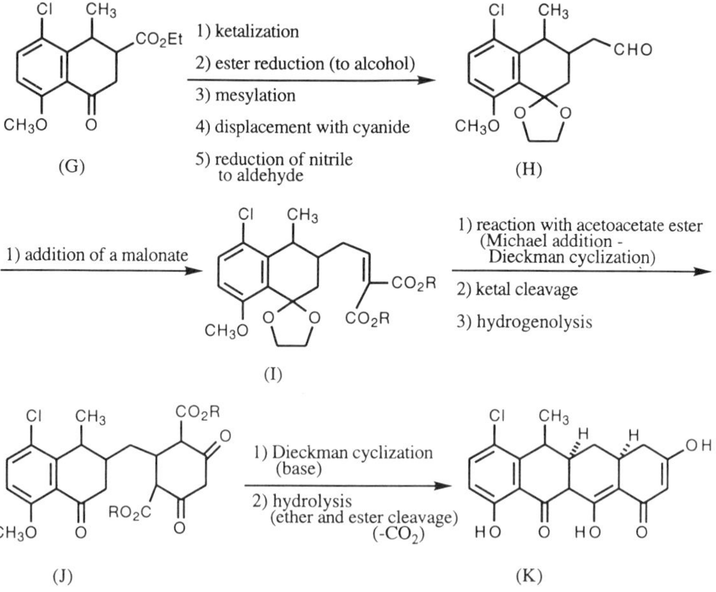

Figure 4.24 An alternative approach: B-ring formed last.

abandoned and C-B ring cyclization was performed after the synthesis of the A ring was completed.

The Lederle approach is reminiscent of the early Muxfeldt work in that the C ring was aromatized thus giving rise to the anhydrotetracycline product(s). Cyclization of the A-ring was unsuccessful if the C-ring was not aromatic. Ring annulations were accomplished in a sequential fashion with C ring construction (O) being followed by B ring formation (Q) and lastly, D ring synthesis (R). Another similarity is the use of malonate additions to electrophilic centers alpha to adjacent rings which are highlighted in each cyclization (Figure 4.25). However, in the case of the Lederle method,

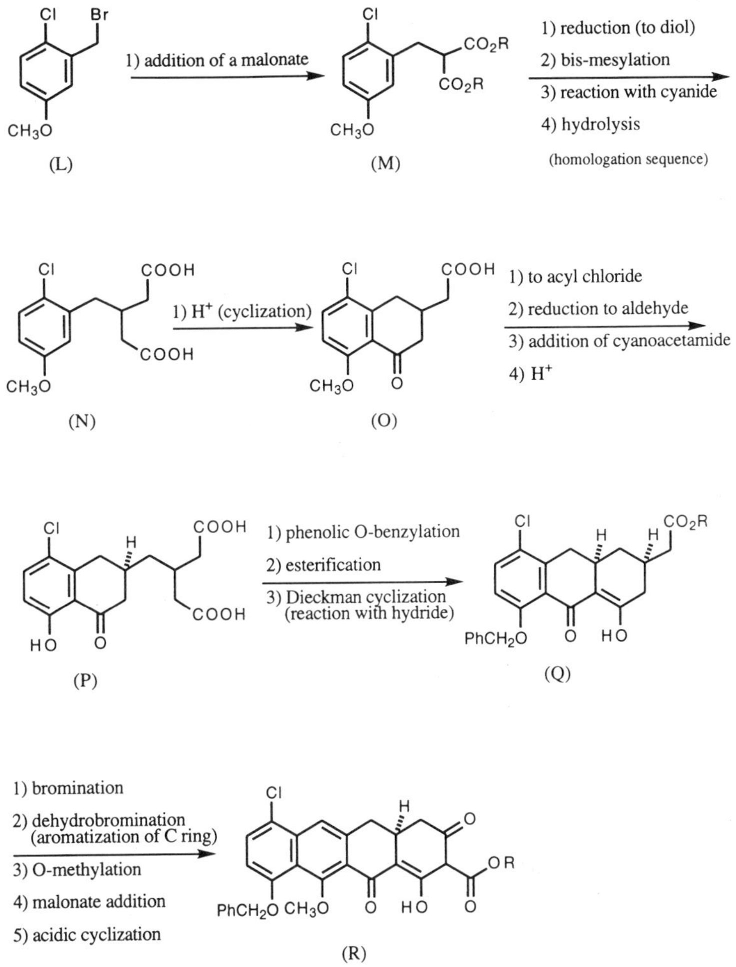

Figure 4.25 Lederle approach to the tetracyclines.

cyanoacetamide addition (and subsequent hydrolysis) was induced in order to extend the precursory C ring tether (P).

Barton adopted a clever strategy in which a nucleophilic D-C ring surrogate was condensed with an aromatic A ring. Barton recognized that a fused furan ring extending from the D-C ring substrate would not only serve as a means of protecting the D-ring phenolic oxygen moiety, but would also mask the neighboring C ring benzylic position until the desired C-11 ketone functionality was needed. The joining of the two intermediates occurred at the electrophilic benzylic center (S) and the neighboring isoxazole was later reduced to the aldehyde oxidation state and further activated (U) for an envisioned B-C ring closure. However all attempts to induce cyclization failed and so the approach was modified (Figure 4.26).

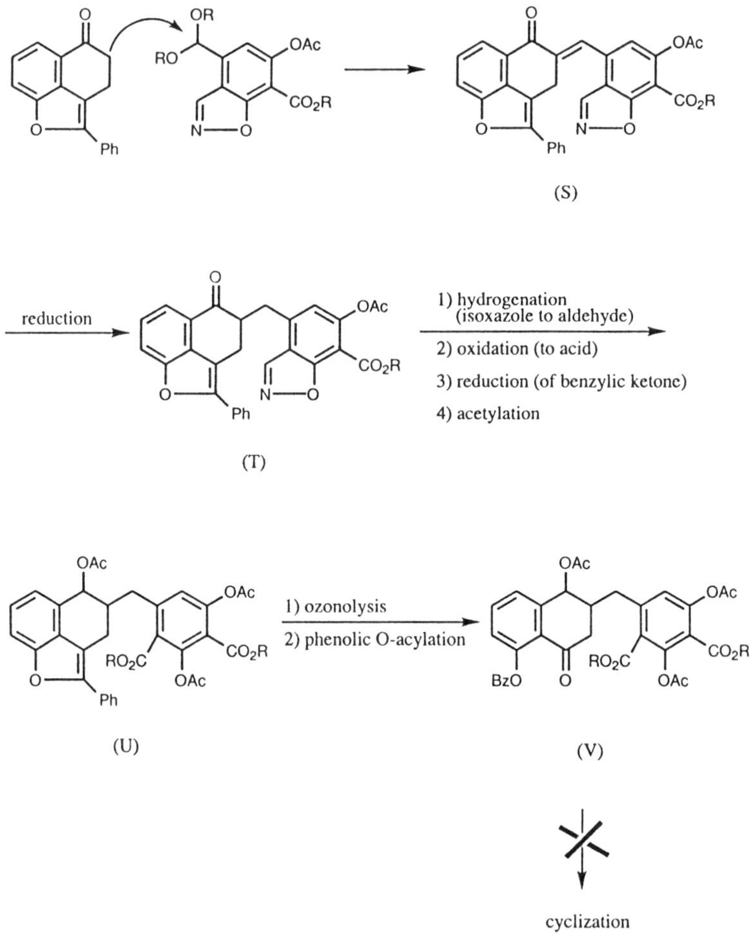

Figure 4.26 Barton method: D-C ring + A-ring.

As an alternative, the Barton group was able to adopt an umpolung strategy that complimented their earlier work. In this approach, the model system (W) was deprotonated giving rise to the corresponding cyanohydrin carbanion which underwent facile B ring closure via a Michael addition. While this chemistry was successful for the construction of saturated A ring tetracycline analogs (X) (Figure 4.27), the cyclization failed when an aromatic A ring precursor (Y) was used. Yet another attempt involved an elegant intramolecular [3+2] cycloaddition of a nitrone (Z) in order to close the B ring (AA). However, the aniline (BB) proved to be inert to

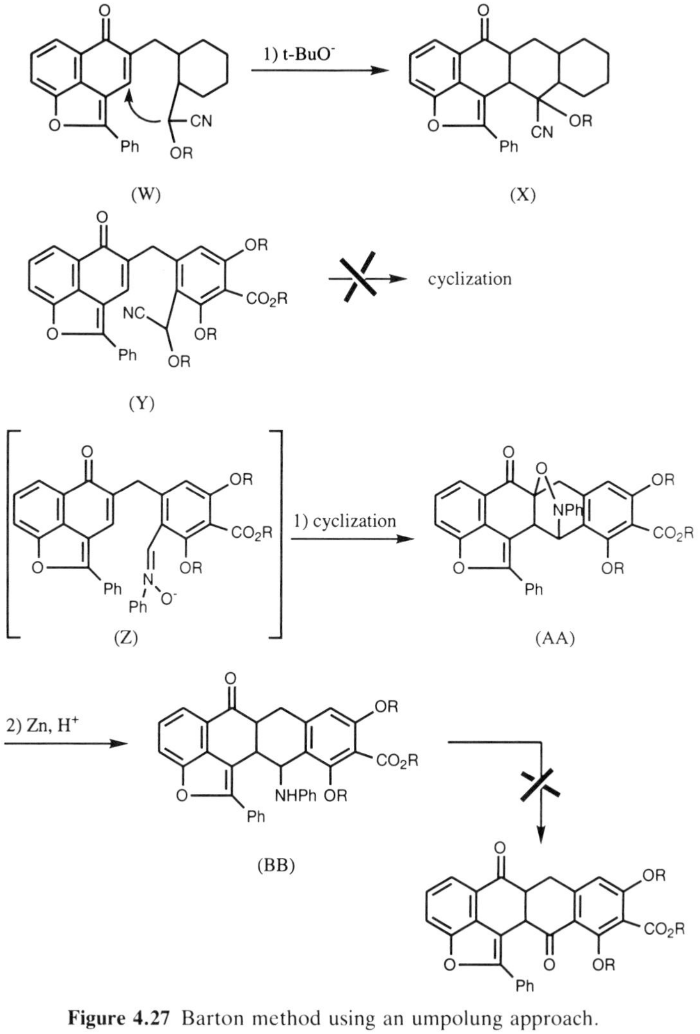

Figure 4.27 Barton method using an umpolung approach.

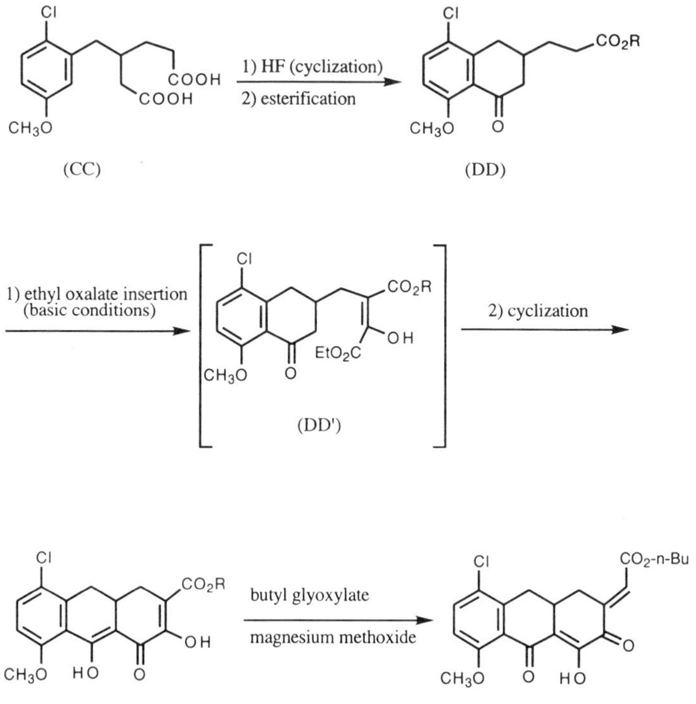

Figure 4.28 Woodward approach to the total synthesis of tetracyclines (part 1).

oxidation and the desired C-12 carbonyl functionality could not be introduced. As a result, active tetracyclines were not obtained via these approaches.

The first biologically active tetracycline antibacterial to be produced by total synthesis was carried by the Woodward group (Figures 4.28 and 4.29). Among the key reactions was an interesting insertion of ethyl oxalate (DD′) that allowed for B ring cyclization and established the necessary tricarbonyl keto–enol system (EE). The vinylogous ester analog (FF) was then ingeniously prepared which subsequently allowed for the introduction of a C-4 amine moiety via Michael addition. However, it was necessary to carry out a reduction in order to prevent the retro-Michael decomposition of this adduct, and in doing so the C-12a hydroxyl group was removed. Fortunately, this moiety could be successfully reintroduced at a later stage of the synthesis but required a tedious buffered cerrous ion-mediated oxidation. The A ring was formed by first activating the carboxylic acid (HH) as an anhydride and then reacting this electrophilic species with a mixed malonic ester–amide anion. Carbon–carbon bond

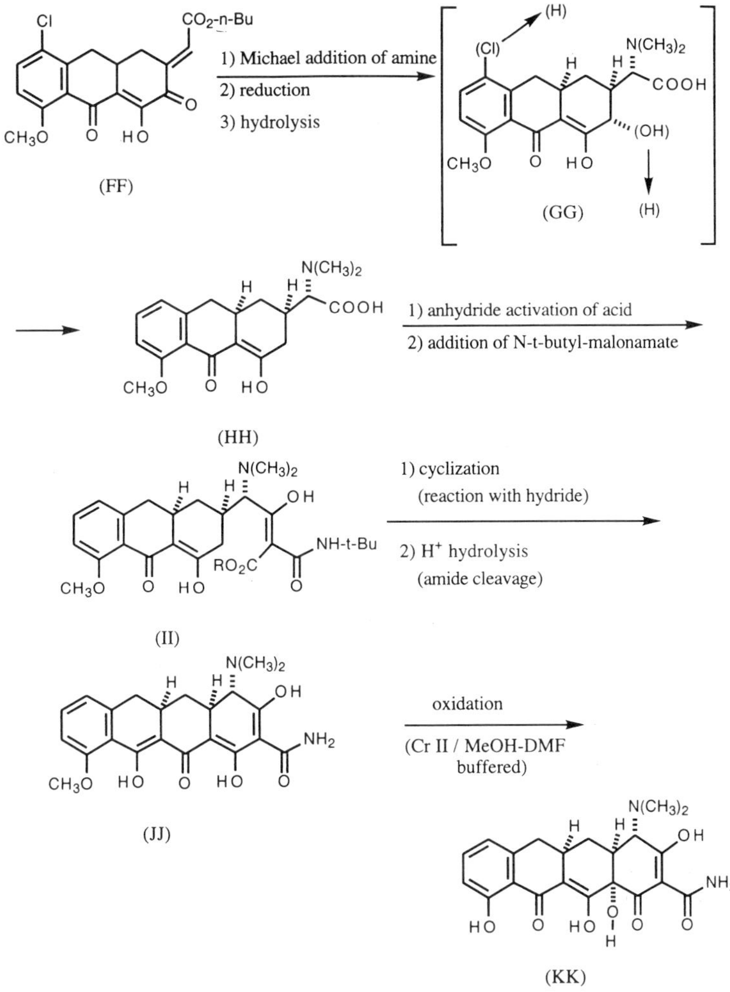

Figure 4.29 Woodward method (part 2).

formation occurred onto the ester center and the corresponding adduct was reached with hydride base to induce cyclization. A brilliant feature of this work is that the amide functionality derived from the malonamate was elaborated into the C-1 carboxamide moiety. In retrospect, the Woodward synthesis is successful where other approaches have failed for several reasons. Both the C-4 dialkylamino substituent and the C-1 carboxamide substituent are introduced as the A ring precursor is elaborated rather than after A ring formation. In addition, conditions capable of hydroxylating the C-12a position in the presence of a wide array of reactive moieties were discovered; this overcame the most obvious shortcoming of the entire synthesis.

Gurevich, Karpetjan and Kolosov devised a successful synthetic route to anhydrotetracyclines. Unlike the previously described syntheses that also produced the anhydrotetracycline nucleus, this approach allowed for the introduction of the C-4 dimethylamino moiety. As a result, the final products displayed activity, albeit weak in comparison to tetracyclines which do not have an aromatized C ring. The key feature of this work is the Michael addition of a nitroacetate ester onto the trycyclic enone (NN) to give an adduct (OO) which was reduced, N-protected (PP), condensed with a malonate amide and subsequently cyclized upon treatment with the

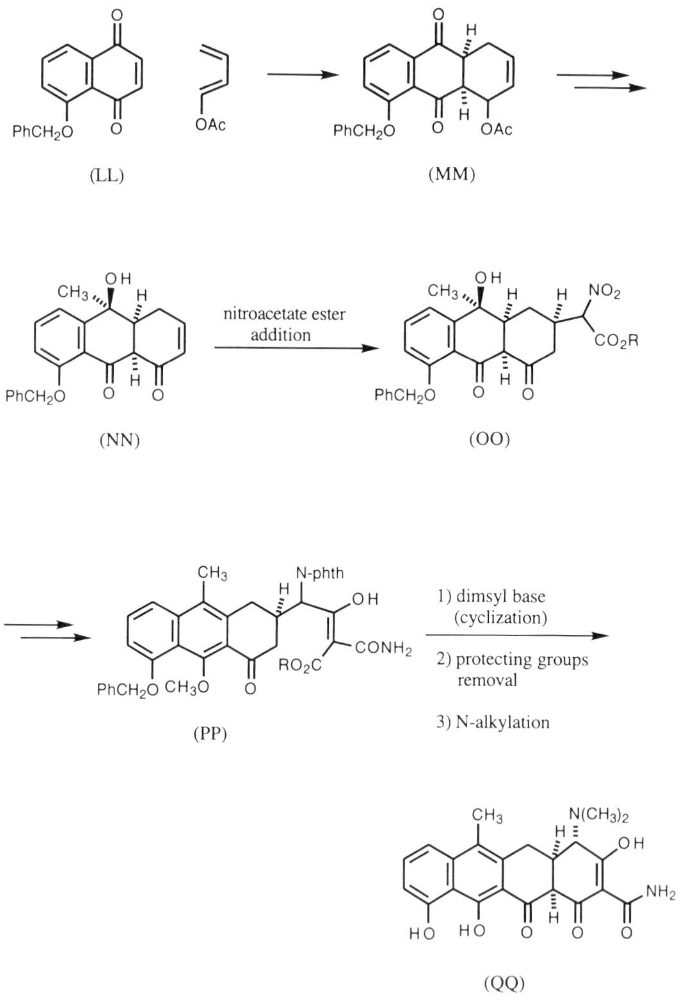

Figure 4.30 Gurevich method: D-C ring + B-ring + A-ring.

dimsyl base. Removal of the protecting groups and amine alkylation afforded the final product (QQ) (Figure 4.30).

The Muxfeldt research group was also able to come up with an improved route to anhydrotetracyclines and unlike the group's earlier work, this methodology did incorporate the C-4 amine functionality. An oxazolidinone cleverly served in this capacity as an A ring precursor. In a remarkable chemical transformation, the *t*-butyl amide of 3-ketoglutaric acid monoester deprotonated adjacent to the ester center, and subsequent Michael addition to the exocyclic methyleno-oxazolidinone (SS); the resultant anionic intermediate underwent A ring cyclization via cleavage of the oxazolidinone ring. A possible intermediate of this reaction, a diketo-stabilized carbanion, is shown in brackets in Figure 4.31 along with a

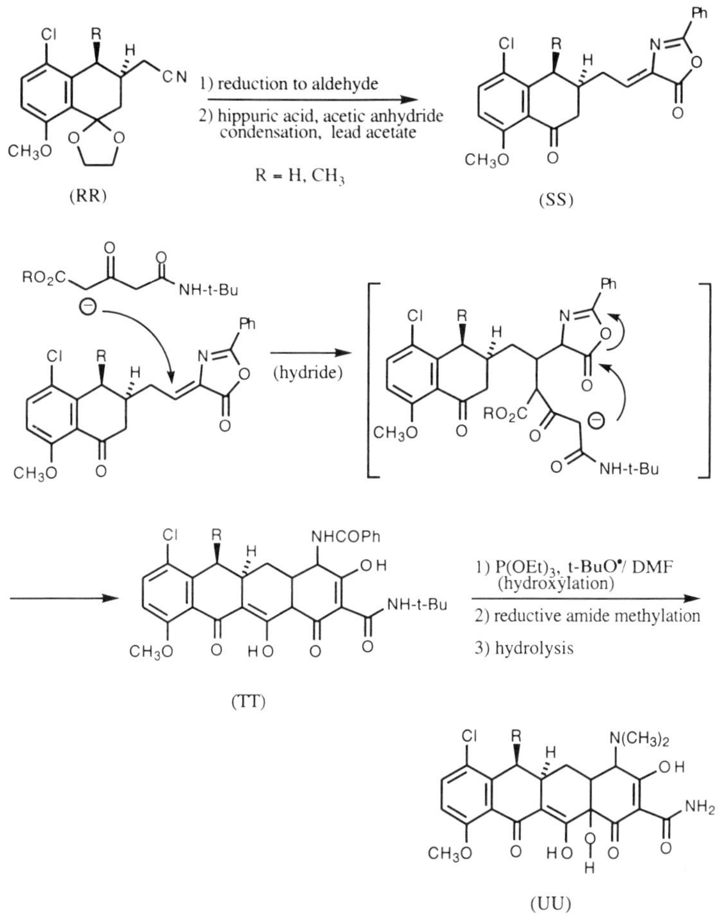

Figure 4.31 Muxfeldt method: D-C ring + oxazolidinone A-ring precursor.

postulated ensuing ring cyclization. Other crucial manipulations in this synthesis included the conversion of the benzamide into the desired C-4 dialkylamino group via a reductive alkylation; unfortunately the stereochemistry at this position was not set during B-A ring cyclization. The introduction of the C-12a hydroxyl group was accomplished using a phosphite-mediated oxygenation. The final product (UU) could be converted to the corresponding anhydrotetracycline via oxidation with an appropriate benzoquinone.

Perhaps the most acclaimed of all tetracycline syntheses is the landmark synthesis of oxytetracycline (Figure 4.32) by Muxfeldt, Vedejs and colleagues. This work showcases many of the advances in tetracycline synthesis that were developed in the Muxfeldt laboratories, but is striking for the different way in which the B ring was constructed in the presence of

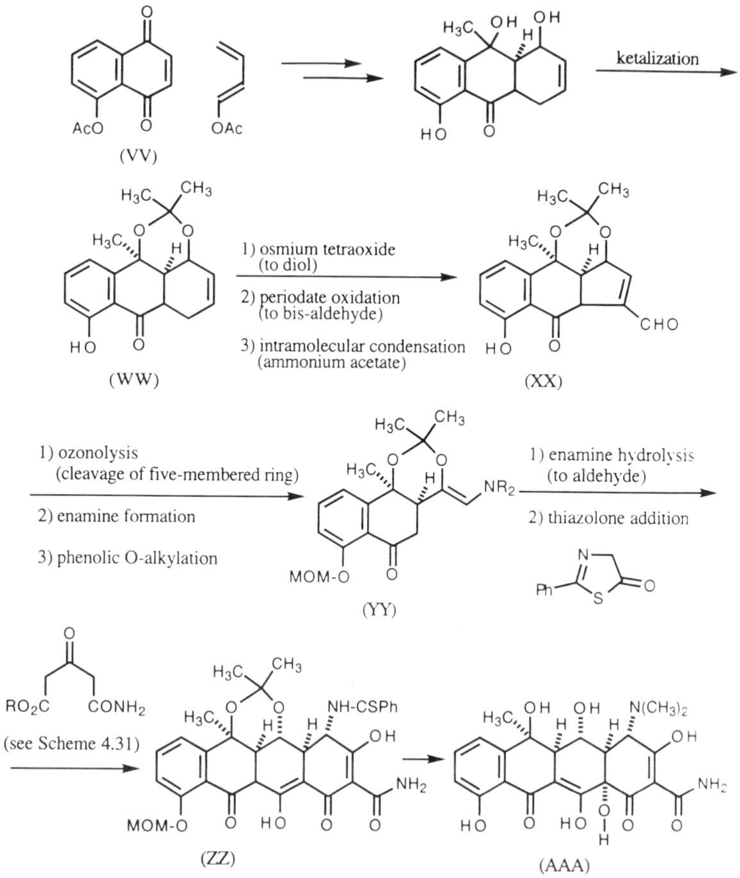

Figure 4.32 Muxfeldt total synthesis of oxytetracycline.

the C ring C-6 oxygen functionality. A Diels–Alder cycloaddition was used to construct the trycyclic system, and a crafty ketalization was then employed in order to protect both C-5 and C-6 oxygenated functionalities (WW). The B ring was then oxidatively cleaved to a bis-aldehyde species and reformed as a five-membered ring (XX). This allowed for a subsequent oxidative ring cleavage process and trapping of the intermediary aldehyde as an enamine (YY). A thiazolidinone was introduced as an A ring surrogate and left the position alpha to the C-11 carbonyl center unsubstituted and poised for B ring closure at a later stage of the synthesis. After the methylenothiazolidinone was accessed, the ketoglutaric carboxyester monoamide method described above (Figure 4.31) was applied, and the tetracycle (ZZ) was produced. The protecting groups were removed and reductive alkylation of the C-4 amino moiety completed the synthesis of the oxytetracycline (AAA).

Although all of the marketed tetracyclines originate from fermentation processes, several notable chemical transformations need to be addressed outside the context of total synthesis. These reactions typically result from an inherent instability of reactivity of the tetracycline nucleus and often complicate attempts to modify structurally a variety of tetracyclines. For example, the tetracyclines containing the normal oxygenation pattern at C-6 are inherently unstable in acidic media and undergo elimination to the C ring anhydrotetracyclines. At lower pH values, epimerization of the dimethylamino moiety at C-4 is observed, and both processes are detrimental in terms of antibacterial activity. In most cases (Figure 4.33), the elimination pathway is more facile and occurs at pH values at which C-4 epimerization does not occur. In basic media, tetracyclines undergo degradation (Figure 4.33) to the so-called isotetracyclines. This process occurs via the deprotonation of the C-6 hydroxyl group and subsequent intramolecular lactonization at C-11. Isotetracyclines are devoid of antibacterial properties.

Alkylation (or halogenation) of the C-4 amine functionality affords the corresponding ammonium species which activate the tetracycline C-4 position for intramolecular hemiketal formation via participation of the C-6 hydroxyl group. The C-4,C-6 bridged species is somewhat strained and inherently unstable under nucleophilic conditions. For example, dialkylated amines are able to undergo carbon–nitrogen bond formation with ketal ring opening, giving rise to series of C-4 alkylamino substituted tetracyclines (Figure 4.34). Strongly basic nucleophiles are able to react at other positions and so methodology of this type is limited, but useful nonetheless.

A positive halogen, preferably perchloryl fluoride in the presence of base, disrupts the normal enolic character of the C-11,C-12 functionality and adds halogen at the C-11a position. This activates the adjacent C-12 carbonyl moiety towards nucleophilic attack; the C-6 hydroxyl group can

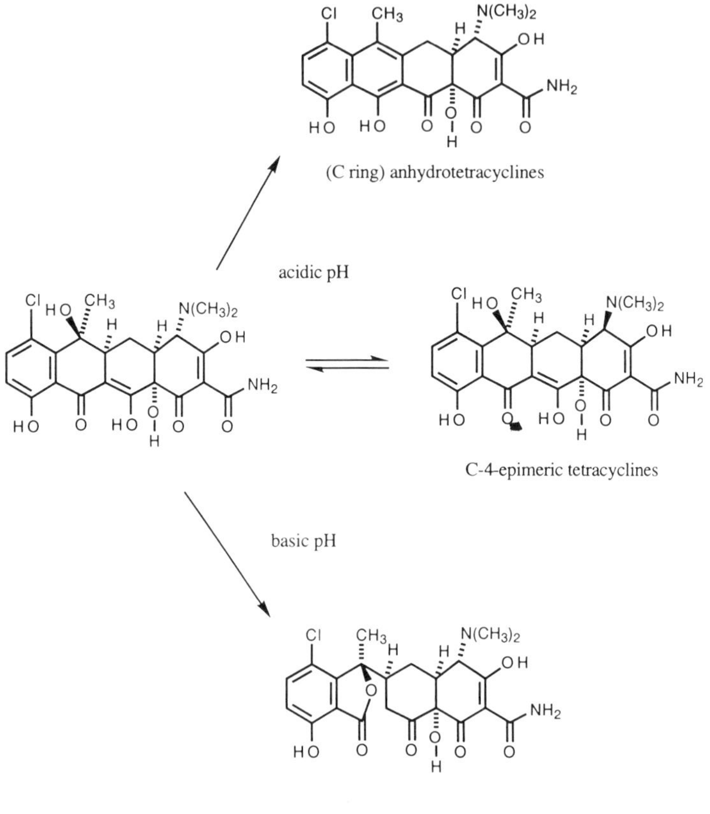

Figure 4.33 Reactive tendencies of tetracyclines dependent upon pH.

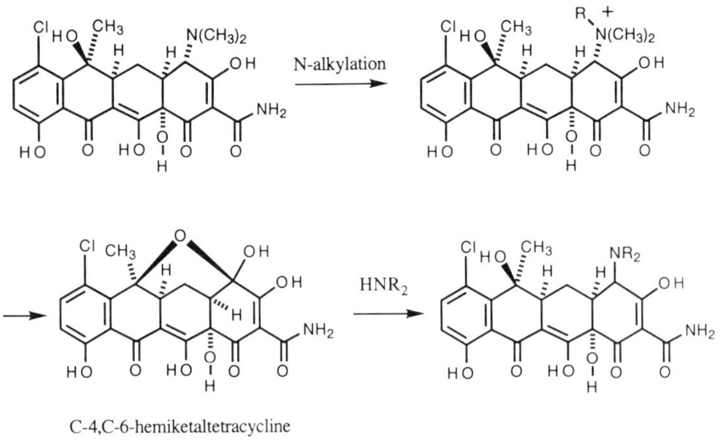

Figure 4.34 C-4,C-6-hemiketaltetracyclines.

add to produce C-6,C-12 hemiketals (Figure 4.35). In the case of the C-6-deoxytetracyclines, halogen addition occurs but hemiketal formation is impossible.

The C-7 and the C-9 substituted tetracyclines have been investigated and have produced some potent compounds *in vitro* (section 4.3). As mentioned earlier, both positions are activated towards electrophilic substitution. Therefore, it is desirable to block one of these positions in order selectively to introduce a substituent without forming isomeric products that are difficult to separate. In this regard, the C-9 position can be blocked by a *t*-butyl group and allow for C-7 functionalization as a nitro derivative. The most recognizable example illustrating this method is minocycline; the nitro group is reduced and alkylated to give the final

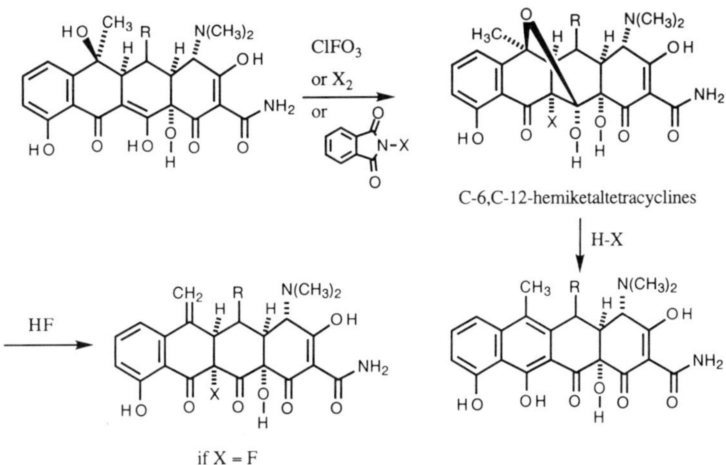

Figure 4.35 C-6,C-12-hemiketaltetracyclines.

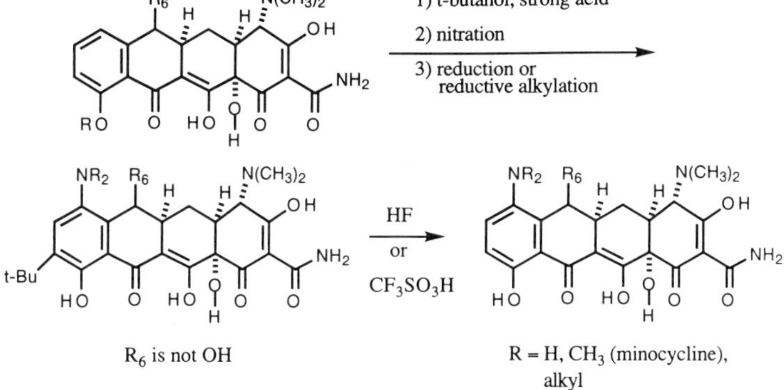

Figure 4.36 Protection of C-9 position, introduction of C-7 electrophile.

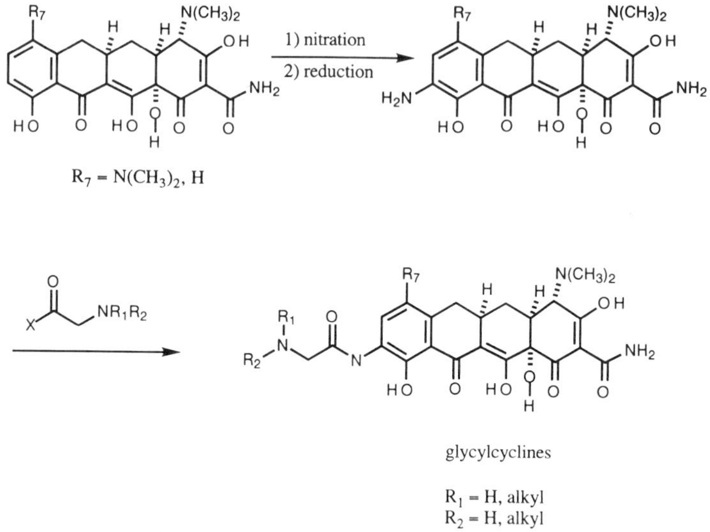

R₇ = N(CH₃)₂, H

glycylcyclines

R₁ = H, alkyl
R₂ = H, alkyl

Figure 4.37 Synthesis of glycylcyclines.

product (Figure 4.36). Anhydrous hydrofluoric acid removes the C-9 *t*-butyl group without competing side reactions.

Perhaps the most exciting area of tetracycline research today is that of the C-9 glycyl substituted tetracyclines (section 4.6). These agents are extremely potent and are able to accumulate within bacteria that normally pump out tetracyclines via efflux mechanisms (section 4.5). Accordingly, these compounds are effective against a number of tetracycline-resistant organisms. The synthesis of these compounds involves C-9 nitration, subsequent reduction to the corresponding amine derivatives and finally acylation with various glycine-derived acyl halides (Figure 4.37).

4.5   Uptake and bacterial resistance to tetracycline antibacterial agents

The success of the tetracycline antibacterials is currently being threatened by the emergence of resistant, often pathogenic bacteria. In some situations, tetracyline therapy can be regarded as a gamble, especially if the identity of the pathogen has not been firmly established, nor its susceptibility to tetracyclines. As a result, other antimicrobial drugs may be preferred and tetracyclines are increasingly becoming relegated to service as alternative agents. Although semi-synthetic tetracyclines such as minocycline have proven to be effective against some traditionally tetracycline-resistant bacteria, their introduction into the antibacterial market took place approximately two decades ago. During the intervening years, many bacteria have evolved mechanisms of resistance against this compound and related congeners. In addition, the tetracyclines are complex molecules that have been limited in terms of practical synthetic modifications and the number of derivatives which have progressed to the point of advanced clinical trials. Bacteria appear to have the 'upper hand' on the tetracyclines at this time; resistance is widespread and there are no compounds set to enter the market within the next few years. However, some recent advances offer promise; a glycyltetracycline is probably going to start clinical trials shortly. Despite this, bacterial resistance will probably remain the key issue that dictates the future use of currently marketed tetracyclines.

The mechanisms by which bacteria can exhibit resistance to the tetracyclines have been well established, although certainly not understood in their entirety. The most prevalent mode of tetracycline resistance can be considered in simple terms as the sum of two opposing phenomena, that of drug uptake and that of drug expulsion. Any agent that exerts its effects through some intracellular process must be taken in by the bacterial cell and achieve suitable concentrations if an antibacterial effect is to be produced. The tetracyclines are generally quite efficient at crossing the outer membrane of most Gram negative bacteria. Their molecular size of approximately 500 (amu) is below the apparent threshold limit (estimated approximately 700 molecular weight) at which porin channels exclude passage. For example, the porin protein Ia often provides a route of entry for the tetracyclines. This is not to say that the outer membrane offers no distinction between Gram negative and Gram positive bacteria in terms of tetracycline access to the cytoplasm; the more lipophilic tetracyclines are less effective against some Gram negative pathogens due, in part, to the outer membrane barrier.

Since the tetracyclines are not structurally modified (deactivated) by enzymes, they arrive intact at the cytoplasmic membrane. The cytoplasmic membrane then presents the major obstacle regarding intracellular uptake. Tetracycline uptake depends upon two distinct processes; the first is the initial binding of the drug to an appropriate membrane site. This process is energy independent and is likely to pose little hindrance for any given tetracycline. However, the ensuing transport of tetracycline across the cytoplasmic membrane is energy dependent and is biphasic in nature. A rapid uptake initially occurs followed by a period in which tetracycline accumulation continues in a slow manner. Several inherent properties of the tetracyclines are important in this process. Magnesium cation chelation of tetracyclines has been demonstrated to play a role in binding and transport of tetracyclines across the cytoplasmic membrane. In addition, tetracyclines, by virtue of their intricate conjugated systems and diverse functionality, are likely to exist in a variety of ionic or zwitterionic forms at physiological pH. The equilibrium between these species is probably crucial for overall cytoplasmic penetration. For example, minor structural changes to the tetracycline keto–enol systems and the C-4 dialkylamino substituent that alter pK_a values result in inactive congeners; these compounds are usually unable to accumulate within the bacterial cell and probably do not reach the cytoplasm.

Tetracycline movement across the cytoplasmic membrane is a true transport phenomenon in both Gram negative and Gram positive species. In Gram negative species there is a periplasmic protein binding transport that is energy driven by an accompanying phosphate (adenosine-5'-triphosphate (ADP)) hydrolysis mechanism. In Gram positive organisms, there are no periplasmic proteins (no periplasm) but there is an ATPase function that parallels that of the Gram negative species. After binding, tetracyclines undergo a membrane-bound transport across both Gram negative and Gram positive cytoplasmic membranes. An energy-dependent process is operative driven by the presence of an electrochemical gradient that transverses the membrane. This gradient has been well established as a primary uptake mechanism by which bacteria obtain intracellular supplies of various materials. The (electrochemical) gradient consists of two components, one being electronic in nature and resulting from transiently charged species, and the other being a protonmotive force, in which a pH gradient is operative. In the case of tetracycline transport across the cytoplasmic membrane, the protonmotive force appears to be intimately involved, which implicates the ability of the tetracycline nucleus to assume various ionic (zwitterionic) states in order to associate with and complement this gradient.

These observations suggest the presence of transport carriers and this phenomena can be viewed as the progressive 'passing on' of the drug, perhaps in various ionic states, from neighboring environments of the system to another until the interior of the membrane is reached. If this situation exists, then it is theoretically possible to saturate the carrier proteins with drug and produce a maximum intracellular concentration of drug. Indeed, tetracycline uptake appears to be saturable and so tetracycline–protein complexes are likely to be responsible for drug uptake into the cell interior. Tetracycline chelation of divalent magnesium cation is likely to play a role in association with the unidentified carrier species. However, high levels of magnesium actually decrease tetracycline accumulation and so an optimum ion concentration probably exists.

An energy source is needed for tetracyclines to accumulate to levels at which bacteriostatic effects are exerted; for example, the addition of a so-called energy inhibitor decreases intracellular drug levels. In fact, even tetracycline-susceptible bacteria only obtain meager intracellular tetracycline levels, comparable to those found in resistant organisms when the energy source is depleted. In general, energy independent or passive transport means are unable to provide sufficient uptake of tetracycline required to initiate and maintain inhibition of protein biosynthesis.

Tetracycline-resistant organisms can deviate from sensitive species in a number of ways. Some resistant bacteria diminish cellular permeability to the tetracyclines which results in a lessened uptake of drug compared to sensitive organisms. In some cases, both Gram negative and Gram positive resistant organisms are capable of blocking tetracycline transport through a partially inducible system; in the presence of subinhibitory concentrations of tetracycline, a mechanism of transport inhibition is operative. This phenomenon can be halted by the addition of a protein synthesis inhibitor, indicating that protein production is needed to establish this transport block.

The most common way in which bacteria exhibit resistance to the tetracyclines is through the phenomenon of drug efflux, a process by which the drug is removed from the cell. Resistant cells typically display a greater and more rapid loss of intracellular tetracycline than do sensitive cells. The use of everted membranes, vesicle preparations which expose the membrane interior to an extracellular supply of tetracycline (i.e. membranes are turned 'inside out') have been used to demonstrate this phenomenon in a different way. If everted vesicles transport drug, then the involvement of cytoplasmic proteins is inferred (perplasmic proteins are lost in the preparation of the vesicles). Everted vesicles derived from resistant bacteria take up tetracycline by an energy dependent process; these results indicate that in whole bacteria, the efflux of tetracycline is mediated by a cytoplasmic protein. In simple terms, resistant bacteria are able to establish a pronounced efflux system, mediated by cytoplasmic proteins, that eliminates tetracycline despite existing uptake processes (Figure 4.39). Consequently, even though tetracycline is taken in by the resistant cell, the level of tetracycline never reaches a level at which protein biosynthesis inhibition is operative. In contrast, tetracycline-susceptible bacteria readily accept drug into the cytoplasm but cannot remove the drug by efflux (Figure 4.38).

Tetracycline resistance determinants that cause efflux are carried on a variety of plasmids and have been found in many different species of bacteria. Both Gram positive and Gram negative bacteria can carry plasmids that produce tetracycline by efflux. Many of these determinants are readily passed onto other strains or species. The products of these plasmids are cytoplasmic proteins called Tet proteins (from *tet* genes). The Tet proteins are responsible for establishing the efflux mechanism by which bacteria remove tetracycline from the cytoplasm. There may be other factors at work but the Tet proteins are essentially the cause of tetracycline resistance of this type (Figures 4.38 and 4.39).

Not only do the Tet proteins remove tetracycline molecules from the cytoplasm, but these proteins have secondary roles. The Tet proteins have

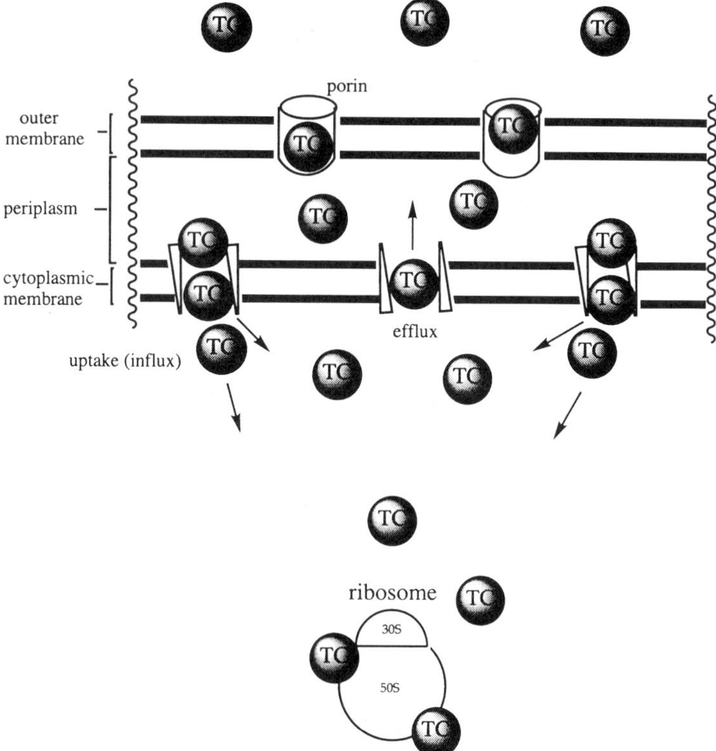

Figure 4.38 Tetracycline susceptibility: net drug uptake.

far-reaching implications in terms of the function and integrity of the cytoplasmic membrane; fatty acid and phospholipid metabolism appears to be altered and this may contribute to tetracycline resistance in some way. Tet proteins also apparently sequester tetracycline molecules away from the ribosome by direct association. In addition, the Tet proteins may exert other effects such as blocking or altering the energy dependent transport process. Even some energy independent transport processes may be affected by the Tet proteins. However, Tet proteins do not deactivate tetracyclines by chemical modification or degradation.

The determinants that encode for the Tet proteins are divided into groups (Table 4.1) based upon DNA cross-hybridization techniques (which will not discussed). Some classes show a high level of DNA homology and encode for Tet proteins that have similar characteristics. For example, the A, B, C and D classes produce efflux primarily in *Enterobacteriaceae*. Classes K and L are determinants responsible tetracycline efflux in resistant Gram positive pathogens. Sevral of the Tet proteins have been

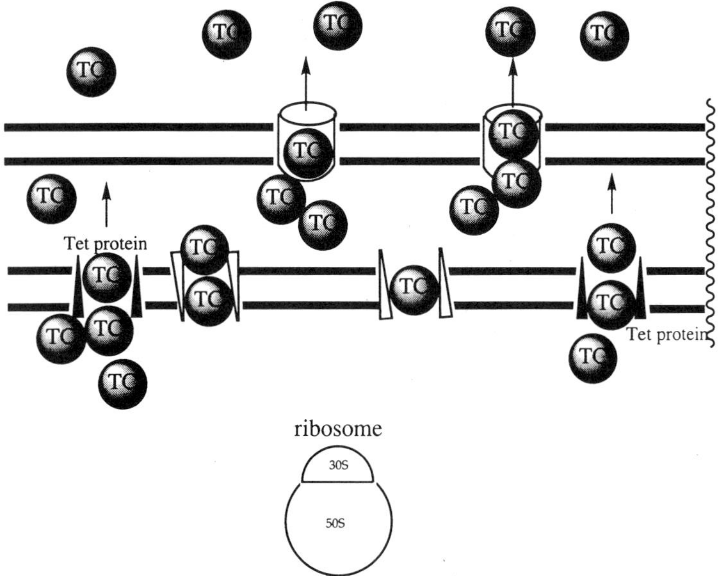

Figure 4.39 Tetracycline resistance: net drug efflux.

Table 4.1 Tet proteins

Determinant class	Mechanism of resistance
Tet A, Tet B, Tet C, Tet D	Efflux (primarily *Enterobacteriaceae*)
Tet E, Tet F	Efflux
Tet K, Tet L	Efflux (Gram-positive)
Tet M, Tet O	Ribosomal
Tet G, Tet N, Tet P	Unknown

isolated and studied in detail, particularly those of resistant Gram negative *E. coli* and the Gram positive pathogen *Staphylococcus aureus*. These proteins are moderate in size with molecular weights generally ranging from approximately 14 kDa to 50 kDa. However, genes of only slightly more than one thousand base pairs encode for some Tet proteins.

Efflux is not the only way a bacterium can acquire resistance to the tetracyclines; ribosomal protection has been recognized. Unfortunately, there is little known about the precise nature of the tetracycline–ribosome complex and so it has been difficult to discern the details of ribosomal protection. One model of tetracycline–ribosome binding invokes hydrogen-bond associations between the carbonyl, hydroxyl and amine groups of one side of the tetracycline nucleus with complementary amine and carbonyl moieties found in the r-RNA (ribosomal not t-RNA) polynucleotide.

Secondary hydrogen-bonding motifs may also involved the ribose ring itself and possibly the phosphoester linkage. (The bacterial ribosome contains RNA molecules of its own; see section 4.2). The fatty D ring of the tetracyclines is envisioned as occupying a relatively lipophilic pocket buried in the r-RNA molecule. Since this model remains speculative, it is even more difficult to surmise what ribosomal modifications could play a role in conferring resistance of this type. Nonetheless, two known determinants, Tet M and Tet O, produce tetracycline resistance by altering drug–ribosome interaction. In these cases, the drug reaches the bacterial ribosome but is unable to undergo binding; cell-free ribosomal preparations are likewise inert to tetracyclines. It is believed that either the r-RNA or a ribosomal protein is altered.

Lastly, there are some Tet determinant classes (N,P) whose functions which have yet to be elucidated.

4.6 Recent advances

4.6.1 New tetracyclines

Several recent discoveries have revitalized interest in tetracyclines and are ushering in an exciting new generation of compounds. The fermenation of decaying leaves housing a microbe (*Dactylosporangium*) afforded anti-bacterial substances identified as glycosidic tetracyclines. The dactylo-cyclines **11** (Figure 4.40) are the first known tetracycline glycosides and the first naturally occurring congeners that contain epimerically transposed C-6 methyl and hydroxyl moieties. The dactylocyclines are active *in vitro* against certain tetracycline-resistant organisms in addition to having good

Table 4.2 Tetracycline antibacterial chemotherapeutic agents

Generic name	Trade name	Administration
Chlortetracycline	Aureomycin	PO
Demeclocycline (HCl)	Declomycin	PO
Doxycycline	Monodox	PO
Doxycycline (hydrate)	Doryx, Vibramycin	PO
Doxycycline (calcium)	Vibramycin Calcium	PO
Methacycline	Rondomycin	PO
Minocycline (HCl)	Minocin	PO, IV
Oxytetracycline (HCl) and lidocaine	Terramycin, Uri-Tet	IM
Oxytetracycline (HCl) and sulfamethizole and phenazopyridine (HCl)	Urobiotic	PO
Tetracycline (HCl)	Achromycin, Sumycin, Tetracyn Kesso-tetra and others	PO

potency against typical Gram positive tetracycline-sensitive bacteria. While the dactylocyclines are essentially inactive against Gram negative bacteria, the corresponding aglycone, dactylocyclinone **12** (Figure 4.40) does possess significant Gram negative activity. The dactylocyclines as a class of tetracyclines are extremely acid sensitive, which would probably hinder their use as an oral agent, and they do not display broad spectrum potency. However, these compounds are extremely interesting and may serve as leads for future tetracyclines.

The C-9-glycinyltetracyclines, the so-called glycylcyclines, are the most promising new generation of tetracyclines. The placement of a glycylamino substituent at the C-9 position produces compounds with potent broad spectrum antibacterial activity. Apparently, the glycyl moiety enhances activity against organisms that are typically resistant due to tetracycline efflux. This finding is very significant and will probably revolutionize tetracycline research. A series of these compounds have been synthesized and tested for antibacterial activity. Of the congeners examined to date, the *N,N*-dialkylglycinyl derivatives **13** and **14** (Figure 4.41), are noteworthy; one is likely to enter clinical trials soon.

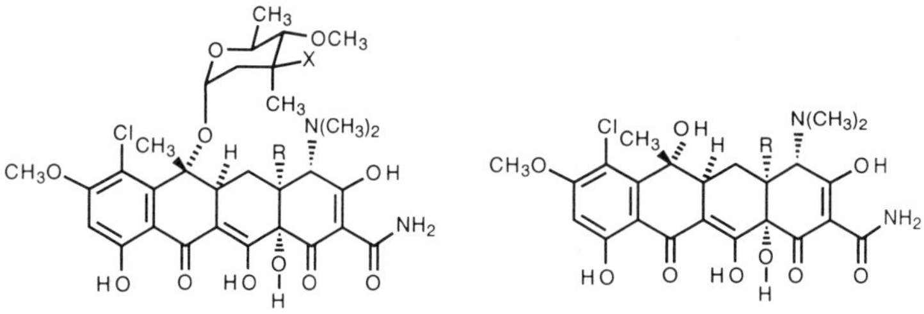

Figure 4.40 Dactylocyclines and the aglycone, dactylocyclinone.

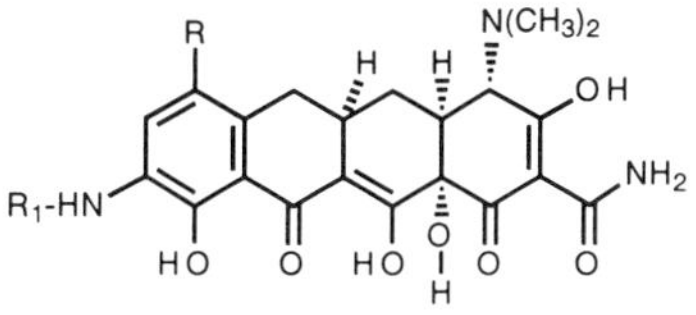

13: R = H; R_1 = COCH$_2$N(CH$_3$)$_2$

14: R = N(CH$_3$)$_2$; R_1 = COCH$_2$N(CH$_3$)$_2$

Figure 4.41 Promising glycylcyclines.

4.7 Summary

The tetracycline antibacterial agents have occupied a prominent position in infectious disease chemotherapy.

1. Tetracyclines are potent, broad spectrum antibacterial agents effective against a host of Gram positive and Gram negative aerobic and anaerobic bacteria. As a result, the tetracyclines are drugs of choice, or well-accepted alternatives for a variety of infectious diseases. Among these, their role in the treatment of sexually transmitted and gonococcal diseases, urinary tract infections, bronchitis and sinusitis remains prominent.

2. The tetracyclines are generally administered orally and are well tolerated. These properties make the tetracyclines among the safest and most easily used of all antibacterial agents. However, most tetracyclines need to be given several (four or three) times per day. In serious indications, some tetracyclines can be administered in an intramuscular or intravenous fashion.

3. The majority of the marketed tetracyclines (tetracycline, chlortetracycline, oxytetracycline and demeclocycline) are naturally occurring compounds obtained from fermentation of *Streptomyces* spp. broths. The generation of semi-synthetic tetracyclines (methacycline, doxycycline and minocycline) were introduced approximately two decades ago; they can offer the advantage of longer duration of antibacterial action. However, for the most part, all of the marketed tetracyclines exhibit a similar profile in terms of antibacterial potency. In general, this activity encompasses many strains of Gram negative *E. coli*, *Proteus*, *Klebsiella*, *Enterobacter*, *Neisseria* and *Serratia* species as well as Gram positive staphylococci and streptococci. Of particular interest is the potency of tetracyclines against *Haemophilus*, *Legionella*, *Chlamydia* and *Mycoplasma*.

4. The tetracyclines exhibit bacteriostatic effects on growing bacteria via the inhibition of protein biosynthesis. Their action occurs at the ribosomal level, where drug-binding to the 30S ribosomal subunit takes place on the ribosome–m-RNA complex. This phenomenon stops the attachment of aminoacylated t-RNA molecules and prevents peptide chain growth.

5. Tetracycline resistance is often caused by plasmid-mediated expression of cytoplasmic proteins that cause efflux of the drug and prevents intracellular tetracycline accumulation. In addition, bacteria can display resistance through some form of ribosomal modification.

Table 4.2 shows tetracycline antibacterials.

Further reading

J.J. Hlavka and J.H. Boothe (Eds) (1985) *The Tetracyclines*, Springer-Verlag, Berlin.

J.J. Hlavka and J.H. Boothe (1973) 'The tetracyclines', in *Progress in Drug Research*, Vol. 17, E. Jucker (Ed.), Birkhauser Verlag, Basel, pp. 210–240.

R.K. Blackwood and A.R. English (1977) 'Structure-activity relationships in the tetracycline series', in *Structure-Activity Relationships among the Semisynthetic Antibiotics*, D. Perlman, (Ed.), Academic Press, New York, pp. 397–426.

A.I. Laskin (1967) 'Tetracyclines', in *Antibiotics I, Mechanism of Action*, D. Gottlieb and P.D. Shaw (Eds), Springer-Verlag, New York, pp. 331–359.

A.I. Laskin and J.A. Last (1971) 'Tetracyclines', *Antibiotics Chemoth.*, **17**, 1.

I. Chopra, R.M. Hawkey and M. Hinton (1992) 'Tetracyclines, molecular and clinical aspects', *J. Antimicrob. Chemother.*, **29**, 245.

T.R. Tritton (1977) 'Ribosome-tetracycline interactions'. *Biochemistry*, **16**, 4133.

G. Hogenauer and F. Turnowsky (1972) 'The Effects of Streptomycin and Tetracycline on Codon-Anticodon Interactions', *FEBS Lett.*, **26**, 185.

J.J. Hlavka, G.A. Ellestad and I. Chopra (1992) 'Tetracyclines' (Antibiotics), in *Kirk-Othmer Encyclopedia of Chemical Technology*, 4th edn., John Wiley, New York, pp. 331–346.

L.A. Mitscher (1978) *The Chemistry of the Tetracycline Antibiotics*, Medicinal Research Series, Vol. 9, Marcel Dekker, New York.

D.L.J. Clive (1968) 'Chemistry of tetracyclines', *Quart Rev. (Chem. Soc.)*, **23**, 435.

F. Johnson (1973) 'The Total Synthesis of Antibiotics', in *The Total Synthesis of Natural Products*, Vol. 1, J. ApSimon (Ed.), John Wiley, New York, pp. 348–364.

G.C. Barrett (1963) 'Synthesis of tetracycline analogs', *J. Pharm. Sci.*, 309.

H. Muxfeldt, E. Vedejs, G. Haas, G. Hardman, F. Kathawala and J.B. Mooberry (1979) 'Tetracyclines. 9. Total synthesis of *dl*-terramycin', *J. Am. Chem. Soc.*, **101**, 689–701.

S.B. Levy (1984) 'Resistance to the Tetracyclines' in *Antimicrobial Drug Resistance*, L.E. Bryan (Ed.), Academic Press, Orlando, FL, pp. 191–240.

S.B. Levy (1992) 'Active efflux mechanisms for antimicrobial resistance', *Antimicrobial Agents Chemother.*, **36**, 695.

I. Chopra and T.G.B. Howe (1978) 'Bacterial resistance to the tetracyclines', *Microbial. Rev.*, **42**, 707.

M.L. Nelson, B.H. Park, J.S. Andrews, V.A. Georgian, R.C. Thomas and S.B. Levy (1993) 'Inhibition of the tetracycline efflux antiport protein by 13-thio-substituted 5-hydroxy-6-deoxytetracyclines', *J. Med. Chem.*, **36**, 370.

A.A. Tymiak, L.A. Mitscher, C. Aklonis, M.S. Bolgar, A.D. Kahle, D.R. Kirsch, J. O'Sullivan, M.A. Porubcan, P. Principe, W.H. Trejo, H.A. Ax, J.S. Wells, N.H. Andersen, P.V. Devasthale, H. Telikepalli, D.V. Velde and J.-Y. Zou (1993) 'Dactylocyclines: Novel tetracycline glycosides active against tetracycline-resistant bacteria', *J. Org. Chem.*, **58**, 535.

P.-E. Sum, V.J. Lee, R.T. Testa, J.J. Hlavka, G.A. Ellestad, J.D. Bloom, Y. Gluzman and F.P. Tally (1994) 'Glycylcyclines. 1. A new generation of potent antibacterial agents through modification of 9-aminotetracyclines', *J. Med. Chem.*, **37**, 184.

R.T. Testa, J.J. Hlavka, J.D. Bloom, G.A. Ellestad, Y. Gluzman, N.V. Jacobus, V.J. Lee, P.J. Petersen, P.-E. Sum, F.P. Tally and W.J. Weiss (1993) '*Glycylcyclines . . .*', *33rd Interscience Conference on Antimicrobial Agents and Chemotherapy* (sponsored by the American Society for Microbiology), New Orleans, LA, October, 17–20, poster abstracts 431, 432, 433, 442, 443.

5 Aminoglycoside antibiotics

5.1 History and overview

Like many other antibacterials, the origin of the aminoglycoside antibiotics can be traced to the practice of screening culture broths of soil microorganisms in search of antibacterial substances. In 1943, Waksman and colleagues isolated a potent antibacterial compound from a *Strepto-myces* species (*Streptomyces griseus*). This substance, streptomycin **1** (Figure 5.1), was structurally unique in containing an aminocyclitol and an aminosugar joined to a ribose unit. Streptomycin captured the attention of the scientific community due to its potency and spectrum of activity; eventually it became a successful drug for the treatment of mycobacterial infections (e.g. tuberculosis). Because of this property, researchers invested efforts to look for other structurally related compounds that were elaborated by other *Streptomyces* sources. A family of potent antibacterial aminoglycosides followed.

In 1949, the neomycins **2** (e.g. neomycins B and C, Figure 5.2) were isolated from *S. fradiae* and exhibited excellent antibacterial properties, as did the structurally related paromomycins **2** (Figure 5.2). The neomycins and paromomycins, although being aminoglycosides, are structurally different from streptomycin. These compounds contain an additional

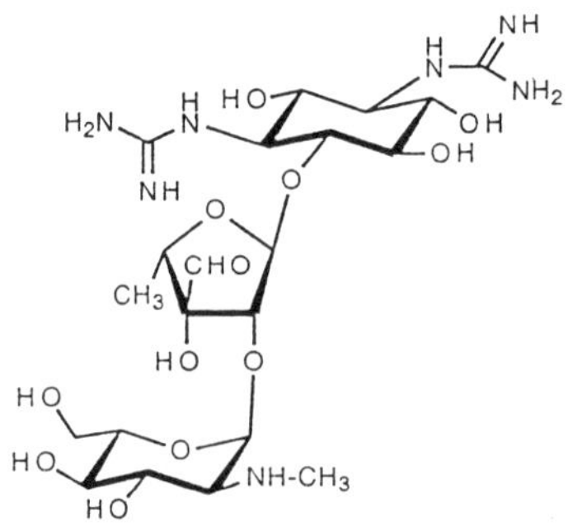

1

Figure 5.1 Streptomycin.

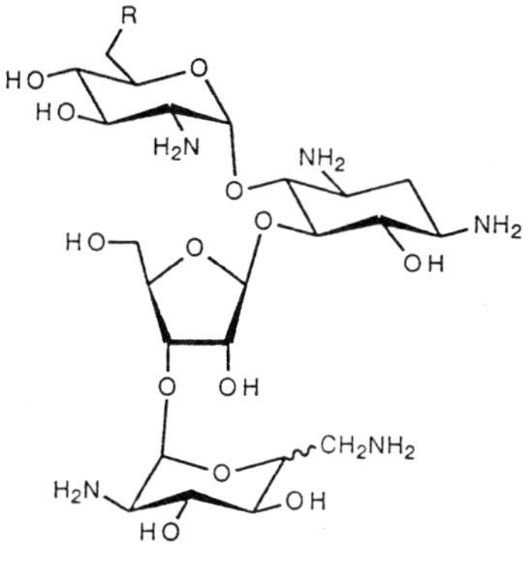

2

R = NH$_2$: neomycins
R = OH : paromomycins

Figure 5.2 The neomycins (neomycins B and C) and paramomycins.

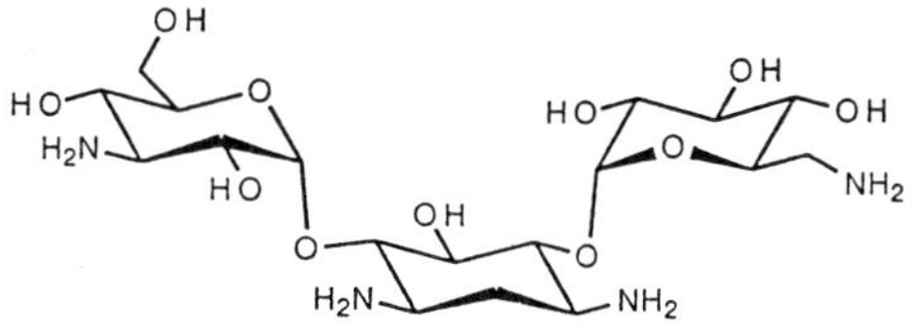

3

Figure 5.3 Kanamycins (kanamycin A shown).

glycoside unit and also have a different substitution pattern on the ribose unit. In 1957, Umezawa discovered the first of the kanamycins **3** (kanamycin A, Figure 5.3) and these substances presented yet another structural motif with respect to antibacterial aminoglycosides. The kanamycins have two glycosides joined to an aminocyclitol ring, but a ribose unit is absent. Regardless of the scaffolding and the individual composition of the sugar units, a common structural feature had emerged; the 'true' aminoglycoside antibacterials contain a characteristic amino-cyclitol nucleus (i.e. 2-deoxystreptamine, streptidine or streptamine itself) (section 5.2).

Other aminoglycosides were subsequently isolated and it was found that streptomyces microorganisms were not unique in producing structurally related compounds that paralleled the antibacterial action of the kana-mycins. *Micromonospora* species afforded the gentamicins **4** (gentamicins

(C_1, C_2, C_{1a}, C_{2a}, C_{2b}), Figure 5.4) and sisomicin **5** (Figure 5.5). In order to distinguish their origin, these derivatives were given a different suffix (-micin). Netilmicin **6** (Figure 5.6) is a semi-synthetic derivative of the naturally occurring aminoglycoside sisomicin.

In general, the research and development of the aminoglycosides has followed a somewhat modest course in that new agents have entered the market infrequently throughout the decades. However, tobramycin **7** (nebramycin factor 6, Figure 5.7) is an important aminoglycoside obtained from a *Streptomyces* (*S. tenebrarius*) and most closely resembles the kanamycins in structure.

Other additions to the aminoglycoside family of antibacterials have included semi-synthetic congeners belonging to the kanamycin subset. Amikacin **8** (Figure 5.8) is derived from kanamycin A and dibekacin **9**

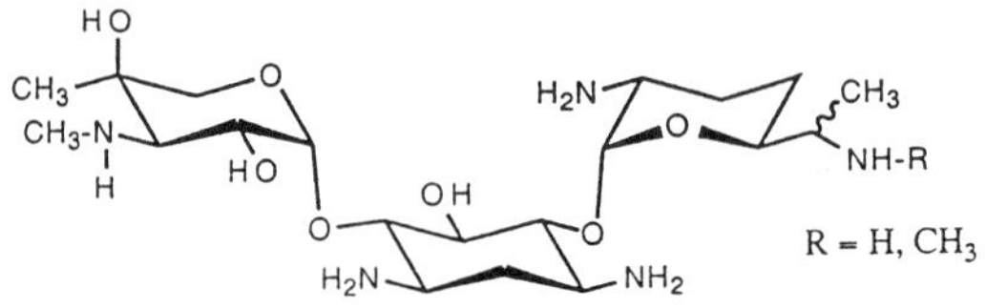

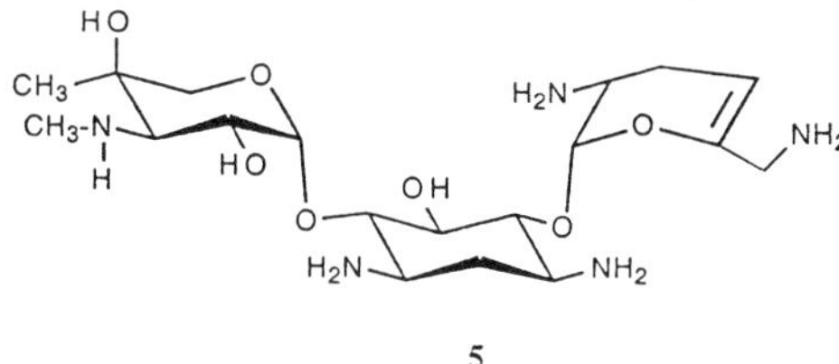

4

Figure 5.4 The gentamicins (C_1, C_2, C_{1a}, C_{2a}, C_{2b}).

5

Figure 5.5 Sisomicin.

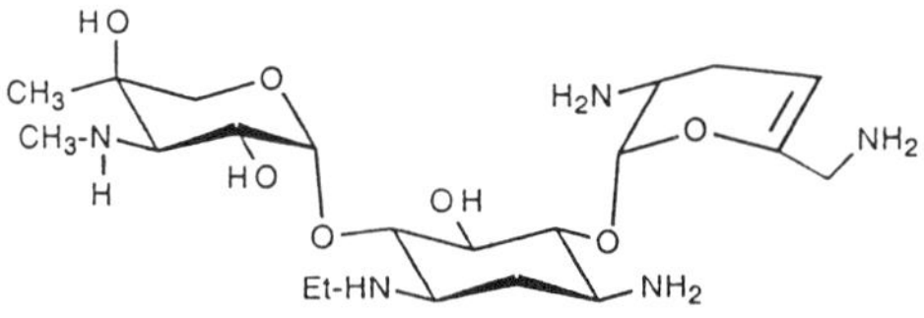

6

Figure 5.6 Netilmicin.

(Figure 5.9) is a congener of kanamycin B. These compounds contain replacements of key functional groups that are susceptible to enzymatic deactivation by bacterial enzymes (section 5.5). Accordingly, amikacin and dibekacin often offer activity against organisms resistant to older, conventional aminoglycosides (e.g. kanamycin).

A notable anomaly within the antibacterial aminoglycoside family is spectinomycin **10** (Figure 5.10) which contains an aminocyclitol moiety but no aminoglycoside unit(s) and so by definition, this compound is not a true aminoglycoside. However, spectinomycin displays antibacterial activity predominantly associated with the aminoglycoside family. Spectinomycin was first isolated from a fermentation broth of *Streptomyces* species (*S. spectabilis*) in 1961. It has been used therapeutically for decades, although

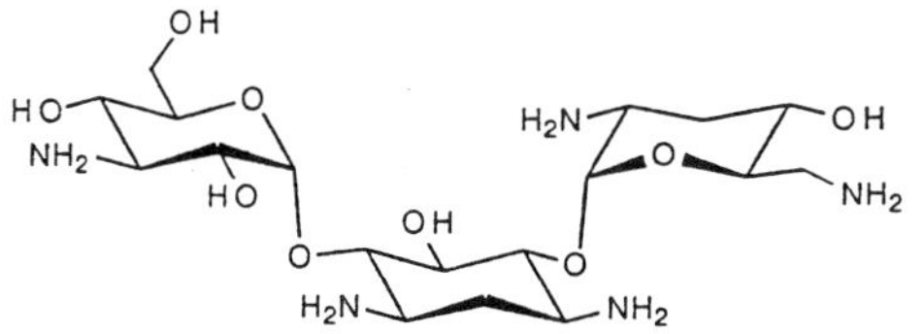

7

Figure 5.7 Tobramycin.

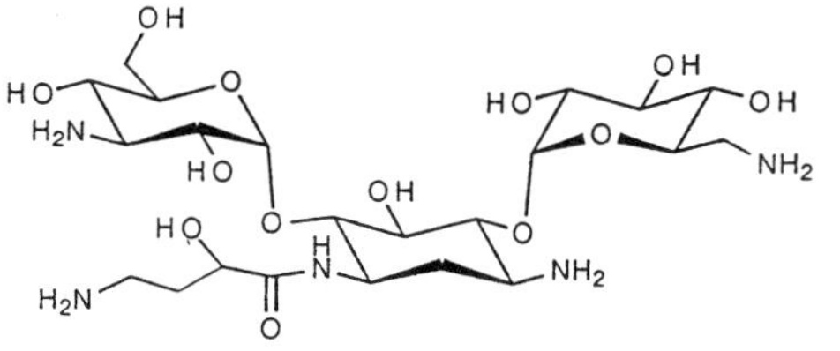

8

Figure 5.8 Amikacin.

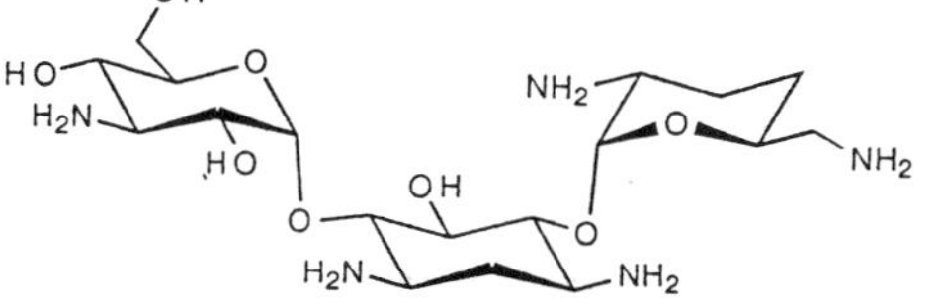

9

Figure 5.9 Dibekacin.

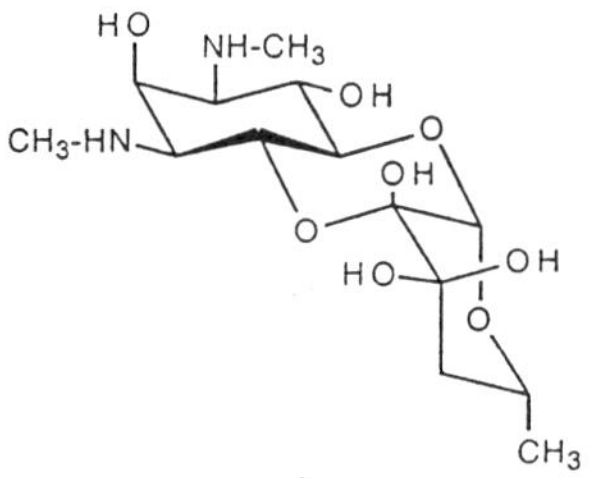

Figure 5.10 Spectinomycin.

today it is usually relegated to an alternative therapy in most indications.

Looking back, these prototypic compounds played a significant role in establishing the aminoglycoside family of antibacterials as valuable, marketed chemotherapeutic agents. Initially streptomycin paved the way with its success as an antituberculosis agent although its predominant reign was short lived. The neomycins and later the gentamicins and kanamycins, provided a needed therapy for certain clinical indications not covered by many other classes of antibacterial agents. Even before the first aminoglycoside was introduced, resistance was emerging in some serious infections traditionally treated with penicillins. At the time, the sulfa drugs were already regarded as inferior agents against many Gram negative pathogens. Not only did the aminoglycosides offer an alternative regimen, their potency against Gram negative pathogens was essentially unrivaled until the more modern β-lactam agents were developed. To this day, several aminoglycoside antibacterial agents are still highly regarded in serious infections involving Gram negative bacilli (particularly in immuno-compromised subjects), and their use in such indications remains prominent.

5.2 Mode of action: bacterial protein synthesis inhibition

The aminoglycosides exert profound effects on susceptible bacteria and several cellular components and processes are targeted. Exposing bacteria to aminoglycoside is usually not limited to inhibition of cellular growth and division (a bacteriostatic event); with most species, bacterial cell death occurs and therefore the aminoglycosides are bactericidal agents. The notable exception is spectinomycin which is not a true aminoglycoside and behaves primarily as a bacteriostatic agent. Nonetheless, despite decades of investigation, the precise antibacterial mechanism of the aminoglycosides remains unclear in many aspects and cannot be readily surmised by a lone action. Primarily, the aminoglycosides act as inhibitors of bacterial protein biosythesis but also produce effects on the bacterial cell membrane. Today, both phenomena are recognized as being important in producing the overall antibacterial action that is characteristic of these agents.

Most early research aimed at elucidating the mechanism of action of the aminoglycoside antibacterials was gathered from studying the effects of streptomycin on bacteria. Other aminoglycosides were subsequently investigated and in general many common features have emerged. However, it should be noted that there are some subtle differences between specific congeners; some particular features are not addressed here (interested readers will find useful references listed at the end of this chapter).

It is certain that the aminoglycosides must enter into the bacterial cytoplasm in order to exhibit antibacterial activity; these compounds do not produce any overwhelming effects on the bacterial cell surface alone that can account for their action. However, bacterial cell surfaces are disposed to weak interactions with aminoglycosides and this phenomenon can essentially constitute the first step in drug uptake. The aminoglycosides, being cationic under physiological conditions, are accepted by anionic environments on the cell surface. In Gram negative bacteria, anionic binding sites include the polar heads of phospholipids and lipopolysaccharides while in Gram positive species, teichoic acids and phospholipids fulfill this role. Association with aminoglycoside is predominantly electrostatic in nature and involves charge–charge interactions. Ionic binding of this type is non-specific, energy independent and reversible; nevertheless intimate drug–cell contact is achieved. As a result, some aminolgycoside molecules are subsequently able to pass through the cell wall with porin protein channels providing the entry in most susceptible Gram negative species. In addition, there may be defects in the cellular barrier which result from the initial association with aminoglycoside on the cell surface and these flaws, while perhaps being transient, can lead to the passage of aminoglycoside to the cell membrane. Upon arriving at the cell membrane, the aminoglycoside is poised for entry into the cell cytoplasm on its way to its destination, the bacterial ribosome.

An aminoglycoside transport system is responsible for uptake into the bacterial cell and must be operative in order for a bactericidal effect to be produced. Bacterial cells that do not allow for the transport of aminoglycoside into the cytoplasm are inherently resistant. Under normal situations, aminoglycosides are transported into the sensitive bacterial cell and intracellular levels achieve a greater concentration than that of the surrounding external media. Transport of aminoglycoside across the cytoplasmic membrane involves energy-dependent processes that rely upon an electrochemical gradient that is generated during respiration or adenosine-5′-triphosphate (ATP) hydrolysis. This gradient (section 4.5) consists of an electrical component and a chemical component, the latter of which represents proton concentration across the membrane. Aminoglycoside transport is primarily associated with the electrical potential.

There are two distinct phases of aminoglycoside uptake into the bacterial cytoplasm. The first, termed EDPI (energy-dependent phase I), is a slow rate of energized uptake that is dependent on the amount of aminoglycoside available to the cell. This phase relies upon electron transport and oxidative phosphorylation processes; for example, inhibition of electron transport blocks EDPI uptake. The EDPI probably represents the transport of aminoglycoside across the cytoplasmic membrane in response to a membrane potential. Another component may include slow uptake of drug through non-specific membrane channels or imperfections. The EDPI precedes inhibition of protein biosynthesis and hence is an event prior to ribosomal binding. In contrast, the EDPII (energy-dependent phase II) is an accelerated uptake of aminoglycoside across the cytoplasmic membrane. This process uses energy from electron transport and can be blocked by uncouplers of oxidative phosphorylation. Interestingly, some inhibitors of protein synthesis can abolish EDPII; for example chloramphenicol (chapter 8) inhibits EDPII. Protein synthesis is required for both the initiation and continuation of EDPII and this implies that aminoglycoside-sensitive ribosomes engaged in protein biosynthesis are similarly required.

Once the aminoglycoside reaches the cytoplasm, it is relatively unimpeded in its quest for the bacterial ribosome (assuming the microorganism is not resistant). Details of aminoglycoside–ribosome interactions remain poorly defined at the molecular level and individual congeners can vary with respect to the nature of binding. It is likely that no single universal mechanism of aminoglycoside–ribosome binding exists although there are obvious similarities among members of the aminoglycoside family. Streptomycin has been extensively studied in terms of ribosomal interactions and behaves somewhat differently to most other aminoglycosides; the non-aminoglycosidic aminocyclitol spectinomycin is also unique.

The principal target of the aminoglycosides is the bacterial 30S ribosomal subunit. Upon binding of aminoglycoside, the joining of the 50S subunit is impeded and so a functional ribosomal complex is not formed and bacterial protein biosynthesis is inhibited. Furthermore, the normal ribosome–polyribosome balance is disrupted and the accumulation of aberrant 30S subunits is likely to be toxic to the cell. Interestingly, aminoglycoside binding does not affect the ability of the 30S subunit to accommodate the m-RNA molecule and even a ribosomally bound formylmethionyl-t-RNA substrate can be accepted in the P site. Frequently, the aminoglycosides can also bind to the 50S subunit while the streptomycins exhibit an affinity for the intact 70S bacterial ribosome. For example, neomycin, gentamicin, kanamycin and tobramycin are able to bind to both the 30S and 50S subunits and yet do not compete with streptomycin-binding to the 30S subunit nor do they bind to the intact 70S complex. Binding to the 50S subunit or the 70S complex also arrests ribosomal dynamics; polysomal ribosome units are decreased while monoribosomes that contain m-RNA and aminoacylated t-RNA molecules are increased.

If the bacterial cell encounters aminoglycoside prior to protein synthesis or shortly thereafter, most active ribosomes are involved in either protein synthesis initiation or peptide chain elongation, and these processes are inhibited. If the intracellular concentrations of aminoglycoside become great enough, ribosomal binding is indiscriminate and nearly all ribosomes are affected; this halts virtually every aspect of protein biosynthesis. Furthermore, at least some aminoglycosides appear to be able to act upon more than one ribosomal entity and this produces a sort of catalytic antibacterial effect. For example, aminoglycoside molecules bound to protein-initiating ribosomal complexes are, in some instances, able to dissociate away thereby causing degradation of the complex with the release of free aminoglycoside. The liberated aminoglycoside, being fully intact and operational, can subsequently undergo binding to another ribosome.

Certain ribosomal proteins are affected by the presence of amino-glycosides and these phenomena are associated with the sensitivity of the mircoorganism. For example, the *E. coli* S12 protein of the 30S subunit is important in binding streptomycin; mutations of this protein often lead to resistance. In a similar fashion, the S5 protein plays a role in spectinomycin sensitivity/resistance, as do the S17 and S12 proteins in the case of neomycin and kanamycin, respectively. The L6 protein of the 50S ribosomal subunit is pivotal in condon–anticondon recognition; ribosomal binding of aminoglycoside can disrupt the fidelity of this process and cause codon misreading. Streptomycin and the aminoglycosides that contain a 2-deoxystreptamine nucleus such as kanamycin and gentamicin are usually able to induce codon–anticodon misreading, although there are subtle differences from agent to agent. For example, streptomycin induces a single base error, while neomycin can misread up to two bases per codon. Streptomycin affects pyrimidine bases more often than other amino-glycosides.

Translational misreading leads to the elaboration of faulty and non-functional proteins which can produce disastrous effects. For example, the incorporation of misread proteins into the cytoplasmic membrane can cause the membrane to lose integrity and become 'leaky'. The bacterial cell loses amino acids and nucleotide substrates while taking in increasing amounts of aminoglycoside. The culmination of these events is a so-called autocatalytic influx process: as intracellular levels of aminoglycoside increase, protein inhibition is more extensively inhibited and additional

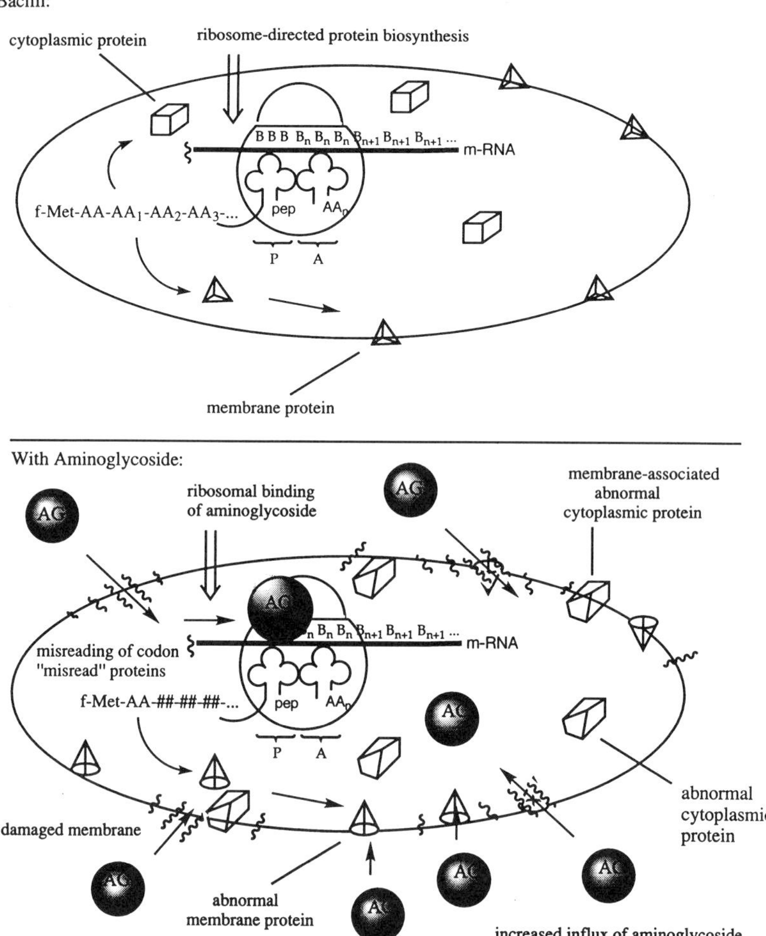

Figure 5.11 Effects of aminoglycosides on bacteria.

faulty proteins are formed. Even proteins normally bound for the cell cytoplasm or periplasm can accumulate in the membrane, causing further damage. This cycle is set in motion until intracellular aminoglycoside concentration reaches a level at which virtually all protein biosynthesis is halted. Bacterial cell death is usually irreversible at this point. A hypothetical model (Figure 5.11) has been advanced that accounts for the major effects of the aminoglycosides as described above.

Some aminoglycosides exert additional effects but these are poorly understood in terms of the overall response of a bacterial organism. For example, some aminoglycosides have been shown to be able to alter normal DNA and RNA synthesis and also to disrupt normal c-AMP (cyclic adenosine monophosphate) levels within the bacterial cell. (Cyclic adenosine monophosphate is an ubiquitous messenger molecule for a host for cellular activities.) These effects are probably yet another consequence of faulty protein production.

5.3   Structural features and structure–activity relationships of the aminoglycoside antibacterials

The aminoglycosides vary greatly in structure despite generally possessing similar antibacterial properties. The distinguishing structural component of all aminoglycosides is an aminocyclitol unit. Aminocyclitols are carbocycles that have amine functionality and hydroxyl groups emanating from the ring. By far the most commonly encountered aminocyclitols found among the antibacterial aminoglycosides are those derived from 1,3-dideoxy- and 1,2,3-trideoxy-(cyclic)inositols. Most aminoglycosides contain a 2-deoxy-streptamine (2-DOS) aminocyclitol nucleus. However, spectinomycin, in essence, contains a streptamine core and streptomycin has the bis-amidino congener streptidine as its aminocyclitol component (Figure 5.12). Aminosugars are nearly always present in the aminoglycoside antibacterials; the exception is the pseudosaccharide spectinomycin. Remarkably, simple monosaccharides and disaccharides that do not contain an aminocyclitol system can display notable, albeit weak, antibacterial properties. However, the aminoglycoside antibacterials are potent compounds that necessarily contain an aminocyclitol along with at least one saccharide unit; most clinically useful congeners contain two or more saccharide units.

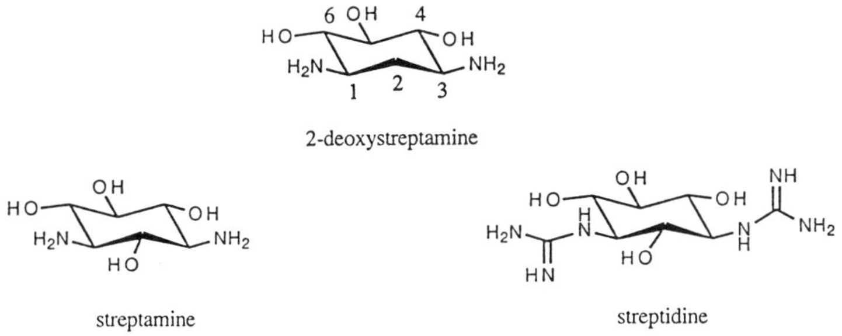

Figure 5.12 Aminocyclitols found in the aminoglycoside antibacterials.

5.3.1 2-Deoxystreptamine-derived aminoglycosides

The majority of the aminoglycosides that are clinically important possess a disubstituted 2-deoxystreptamine aminocyclitol (Figure 5.13). In each compound, six-membered cyclic sugar units are joined onto the C-4 and C-6 hydroxyl groups of the aminocyclitol through glycosidic linkages. The kanamycins **3** (Figure 5.3), gentamicins **4** (Figure 5.4), sisomicin **5** (Figure 5.5) and tobramycin **7** (Figure 5.7) are naturally occurring compounds belonging to this classification, and netilmicin **6** (Figure 5.6), amikacin **8** (Figure 5.8) and dibekacin **9** (Figure 5.9) are semi-synthetic congeners.

An extremely useful modification of the aminocyclitol unit is acylation or alkylation of the C-1 amine group as exemplified by some semi-synthetic aminoglycosides that have been marketed. Amikacin is the C-1 N-α-hydroxybutyryl derivative of kanamycin A and netilmicin is the N-1 ethyl derivative of sisomicin. Other C-1 acyl and alkyl derivatives have been prepared and many have demonstrated good activity. In general, the C-1 position is tolerant of modification if the acyl or alkyl group is rather small in size; typically C-1 substitutents containing fewer than five carbon atoms offer good antibacterial activity.

The C-3 amino substituent is necessary for antibacterial activity and all important aminoglycosides have this moiety. In addition, the 2-DOS-derived aminoglycosides necessarily have the C-2 position unsubstituted. The C-4, C-5 and C-6 hydroxyl moieties of the aminocyclitol are likewise crucial for activity and the stereochemistry at these positions is probably also important. However, few variations at the C-4, C-5 and C-6 positions have been investigated.

The sugar units that extend from the central aminocyclitol nucleus contain substituents that are also needed for antibacterial activity. Certain amine and hydroxyl groups that emanate from these sugars are important for ribosomal binding. Resistant bacteria can produce aminoglycoside-modifying enzymes (section 5.5) that chemically alter these key amine and hydroxyl moieties; the end result is that the modified aminoglycoside is unable to undergo ribosomal binding and thus antibacterial activity is lost. In addition, several analogs that lack these features have been chemically produced and have been shown to be inferior antibacterials.

Among a multitude of functionality, the C-2″ and C-4″ hydroxyl groups and the C-3″ amine moiety stand out among the C-6 sugar substituents of the 2-DOS-derived aminoglycosides. As in the case of the aminocyclitol C-3 amino group, the C-2″ hydroxyl group is also prone to structural modification by certain bacterial enzymes which results in loss of antibacterial activity. On the other hand, a C-4″ hydroxyl group is seemingly ubiquitous to deoxystreptamine-based aminoglycosides, however these are a number of variations on this theme. Most amino-glycosides contain a simple C-4″ carbinol although sisomicin and netilmicin have a tertiary alcohol at this position. Similarly, a C-3″ amine moiety is desirable, although again certain variations are permitted. The *Streptomyces* elaborate compounds with a simple amine group at C-3″ (e.g. kanamycins, tobramycin) whereas the amine group is methylated in compounds produced by *Micromonospora* (e.g. gentamicins and sisomicin).

The sugar or pseudosugar joined to the 2-deoxystreptamine C-4 hydroxy position can also vary considerably although certain key amine and hydroxyl groups appear to contribute to optimal ribosomal binding. The C-6′ position is usually a simple amine, but this group can be methylated as in the case of some of the gentamicins **4** (Figure 5.4). The C-3′ and the C-4′ centers are apparently tolerant of some modification. These positions are often both hydroxylated with appropriate stereochemistry such as that found with the kanamycins **3** (Figure 5.3). However, the C-3′ can be unsubstituted and C-4′ position hydroxylated, as evidenced by tobramycin **7** (Figure 5.7). Alternatively, both the C-3′ and C-4′ positions can be unsubstituted as illustrated by dibekacin **9** (Figure 5.9). The C-4′ center can be unsaturated providing the tetrahydropyranyl derivatives sisomicin **5** (Figure 5.5) and netilmicin **6** (Figure 5.6). Lastly, the C-2′ position appears to play a role in ribosomal binding, and contains either an amine or a hydroxyl group.

The common structural features of the 4,6-disubstituted 2-deoxystreptamine-derived aminoglycosides are summarized below (Figure 5.13).

Some aminoglycosides embody a 4,5-disubstituted aminocyclitol nucleus; the neomycins **2** (Figure 5.2) and the paromomycins **2** (Figure 5.2) are most significant in this regard. These compounds contain a different glycosidic scaffolding from that of the 4,6-disubstituted aminoglycosides; a ribose sugar emanates from the C-5 position of the 2-deoxystreptamine

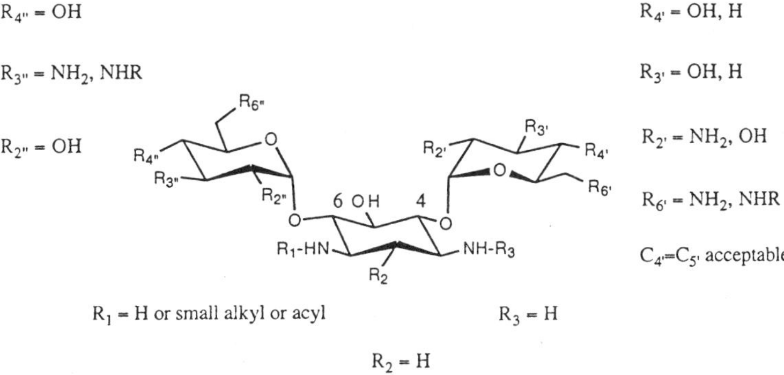

Figure 5.13 Common structual features of the 4,6-disubstituted-2-deoxystreptamine-derived aminoglycosides.

nucleus and usually contains a sugar substituent joined by way of a glycosidic linkage. In addition, the C-6 hydroxyl center of the 2-DOS core is unsubstituted. As with other aminoglycosides, there exists key amine and hydroxyl groups that are crucial for antibacterial activity.

As with the 4,6-disubstituted congeners, the 2-deoxystreptamine nucleus can be acylated at the C-1 amino functionality to provide *N*-α-hydroxybutyryl derivatives that retain good antibacterial activity. The butirosins (not shown) are illustrative of this feature; these congeners contain a ribose unit that lacks an additional sugar at C-3″. Other reported modifications to the 2-DOS system are rather scarce. Both the neomycins and the paromomycins contain C-3′ and C-4′ hydroxyl groups, the former position is targeted by bacterial enzymes that destroy activity. However, deoxygenation or dideoxygenation at this susceptible position has yet to procure active congeners; this is in contrast to sisomicin and netilmicin which represent useful modifications of the 4,6-disubstituted 2-DOS aminoglycosides. A C-2′ amine group is present in both the neomycins and the paromomycins and appears to be needed for useful activity. Lastly, the neomycins contain a C-6′ amine group while paromomycins posses a C-6′ hydroxyl moiety; the former position is susceptible to deactivation by bacterial enzymes (section 5.5). Thus when considered collectively, it has been the sugar moiety joined to the C-4 hydroxyl position of the 2-DOS nucleus of the 4,5-disubstituted aminoglycosides that has warranted the greatest attention. Modifications to the ribose-linked aminosugar joined at the C-5 streptamine hydroxyl position have remained essentially unexplored. In summary, only rather vague assessments of structure–activity relationships can be made for the 4,5-disubstituted 2-DOS-derived aminoglycosides (Figures 5.14).

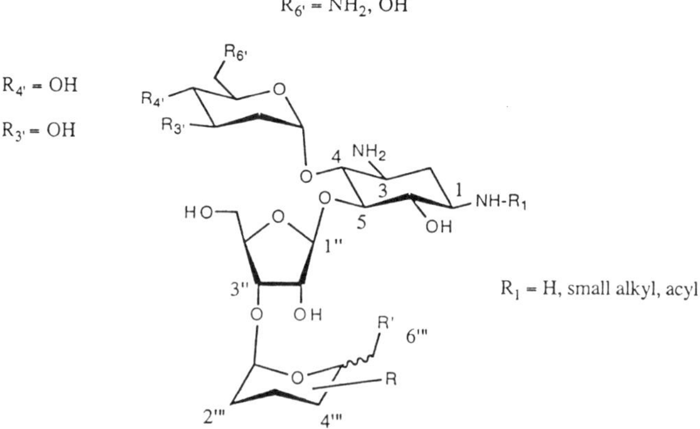

Figure 5.14 Common structural features of the 4,5-disubstituted-2-deoxysteptamine-derived aminoglycosides.

5.3.2 The streptomycins and spectinomycins

The streptomycins (e.g. streptomycin **1** (Figure 5.1)) have been limited in terms of faithful structural modifications. Unlike the 2-deoxystreptamine-derived aminoglycosides, the streptomycin aminocyclitol nucleus has yet to have been altered in any productive manner. Removal of the C-1 and C-3 amidino moieties from the streptidine unit, for example, affords the inactive analogs. Elaboration of the guanidino moieties by alkylation or acetylation has also been detrimental in terms of activity. The ribose-based aldehyde functionality is apparently important for activity, but allowing for some structural modification. For example, the reduction of the aldehyde to the corresponding alcohol affords dihydrostreptomycin **11** (Figure 5.15) which is an excellent congener. However, reductive amination of the ribose aldehyde results in loss of activity. Other modifications to the ribose ring have been few and uneventful. Similarly, the C-2 ribose aminosugar substituent addition has yet to be synthetically elaborated to an appreciable extent, although demethylation of the C-1′ methylamino group retains partial antibacterial activity.

The remaining agent that is considered as a member of the vast family of aminoglycoside agents is spectinomycin **10** (Figure 5.10). The spectino-mycins are pseudoaminoglycosides since they do not contain any glycosidic linkage but do possess a ring-fused aminocyclitol. Spectinomycin contains a ketone moiety that exists in equilibrium with the hydrated form (Figure 5.16). Spectinomycin does not induce codon misreading, presumably because it does not contain a 2-deoxystreptamine-derived nucleus. Its action is bacteriostatic which also differentiates it from most other aminoglycosides.

There have been few useful structural modifications of spectinomycin and in fact most alterations have produced derivatives which lack antibacterial properties. For example, deoxygenation of either the C-7 or

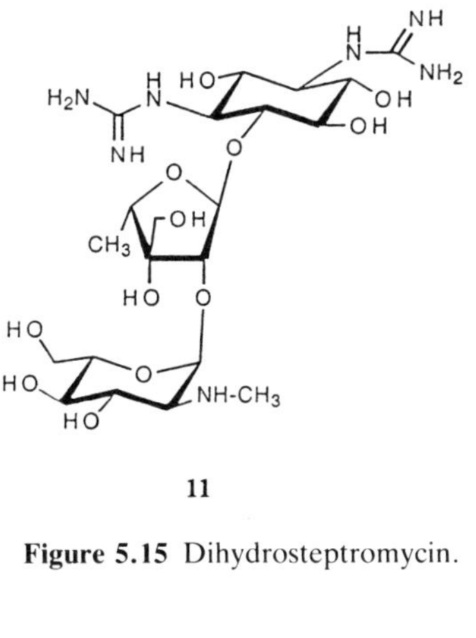

11

Figure 5.15 Dihydrosteptromycin.

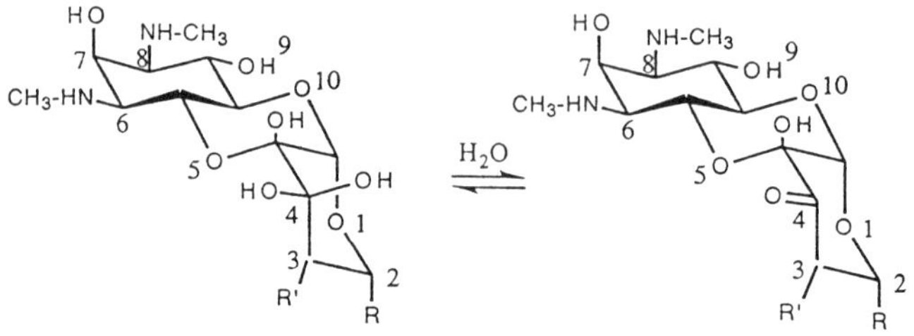

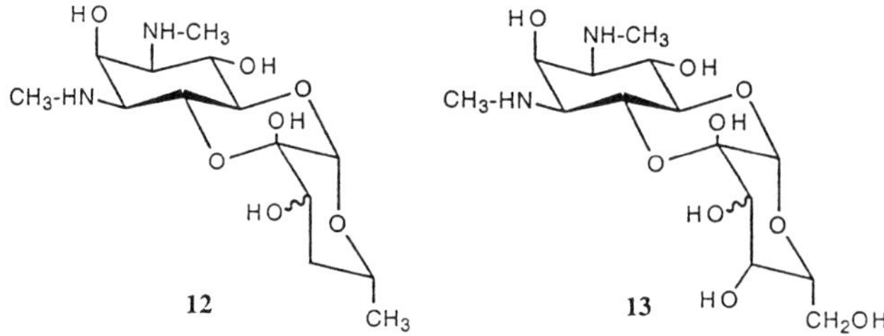

Figure 5.16 Spectinomycin in its hydrated and ketone form.

Figure 5.17 Dihydrospectinomycin and dihydroxyspectinomycin.

the C-9 hydroxyl groups of the aminocyclitol ring destroys activity. The relative stereochemistry at the ring junctions is necessary for activity, as is the absolute stereochemistry of the molecule as a whole. Enantiomeric spectinomycin is completely devoid of antibacterial activity. Epimerizations of several of the stereocenters have likewise yielded inactive compounds. Lastly, cleavage of any ring abolishes antibacterial properties, and as a result the tricyclic system is considered necessary for antibacterial activity.

There have been a few noteworthy modifications to the non-amino-cyclitol rings. Reduction of the ketone moiety at the C-4 position produces the corresponding dihydrospectinomycins **12** (Figure 5.17) which exhibit inferior antibacterial activity compared to the parent compound. The ring can also be bis-hydroxlated at the adjacent C-2′ and C-3′ positions to yield the corresponding dihydroxyspectinomycin **13** (Figure 5.17) which is only weakly active. The most recent promising advance has come from the realization that the C-2 methyl group can be replaced with larger alkyl substituents to produce derivatives that retain potency and spectrum of activity similar to those of the parent spectinomycin (section 5.6).

5.4 Synthetic approaches to aminoglycoside antibacterial agents

The marketed aminoglycoside antibacterial agents are either obtained directly from fermentation, or are structurally modified by chemical protocols. The neomycins, kanamycins, gentamicins and tobramycin, as well as streptomycin and spectinomycin are harvested from fermentation. In some cases, a mixture of factors are present and so commercial formulations can actually contain several structurally related compounds. For example, neomycin is marketed as a mixture of neomycin B and neomycin C (in roughly equal proportions), and gentamicin actually contains gentamicin C_1, C_{1a} and C_2. Semi-synthetic congeners typically contain modification of a single functional group; usually the C-1 amine moiety of the 2-deoxystreptamine nucleus is acylated or alkylated. As described above, examples include amikacin, dibekacin and netilmicin. The technology used to synthesize the non-natural aminoglycosides usually involves rather straightforward chemical transformations although some details are likely to remain secrets of the trade. For these reasons, specific reaction conditions used to access the semi-synthetic aminoglycosides will not be considered here.

In contrast, the aminoglycoside antibacterials are rather challenging in terms of total synthesis, as well as in terms of some structural modifications. As a family, these compounds are quite complex due to the presence of numerous reactive functional groups. The host of amine and hydroxyl moieties inherently pose problems in terms of chemical selectivity. Indeed, synthetic approaches using unmodified sugars and aminocyclitols are generally unrewarding due to the lack of chemoselective, regioselective or stereoselective control. A common strategy that has emerged avoids these problems; masking superfluous amine and hydroxyl functionality until late in the synthetic scheme properly directs the coupling of saccharide units to each other and to aminocyclitol. The formation of glycosidic linkages is generally carried out using protocols common in modern carbohydrate chemistry. Usually a saccharide is activated at its anomeric center in order to facilitate attack by the nucleophilic hydroxyl group of an aminocyclitol. In this context, the anomeric hydroxyl group is transformed to a leaving group; conversion to a halide or a thioether is common. A suitably protected aminocyclitol is then reacted with the saccharide substrate often in the presence of metal salts that increase the electrophilic character of the anomeric center. For example, silver salts possess an affinity for chloride leaving groups as well as thioethers; these properties (halophilic and thiophilic) allow cationic silver salts to assist in glycosidic bond formation. The Koenigs–Knorr reaction, and variations on this technology, have been successfully used to construct a variety of aminoglycosides (Figure 5.18).

The directing effects of a (C-2) nitroso group have been cleverly used to circumvent the problem of stereoselectivity inherent to the Koenigs–Knorr method. In this approach, a dimeric C-2-diazo(bis)N-oxide-C-1-halosugar is converted *in situ* to the electrophilic C-2-nitrososugar which readily undergoes condensation with a protected aminocyclitol to afford the corresponding ketoxime adduct (Figure 5.19). Acidic conditions cleave the oxime and generate the ketone which can further be reduced to the alcohol with high stereoselectivity. The corresponding aminosugar can also be prepared from the oxime adduct.

This strategy has been used to construct a number of aminoglycosides including spectinomycin. Spectinomycin, with its unique tricyclic aminocyclitol ring system, is well suited for this methodology since the resultant ketoxime (A) is poised for subsequent cyclization to form the central ring. This was accomplished by deoximation followed by hemiketal formation which, under appropriate conditions, afforded the desired spectinomycin precursor (B). Acetate hydrolysis and reduction completed the synthesis of spectinomycin (Figure 5.20).

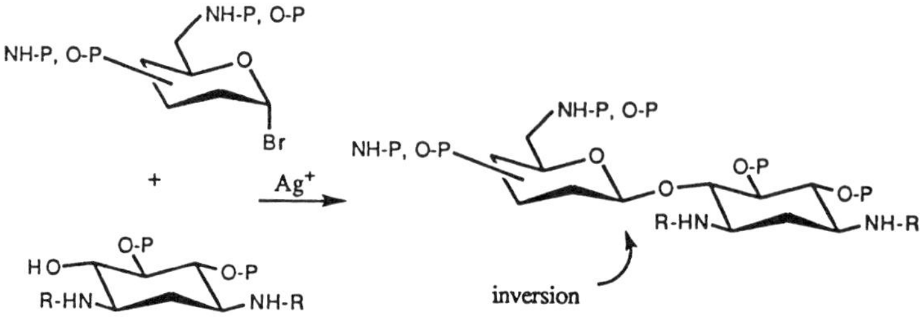

P = protecting group
R = H, P

Figure 5.18 Koenigs–Knorr reaction.

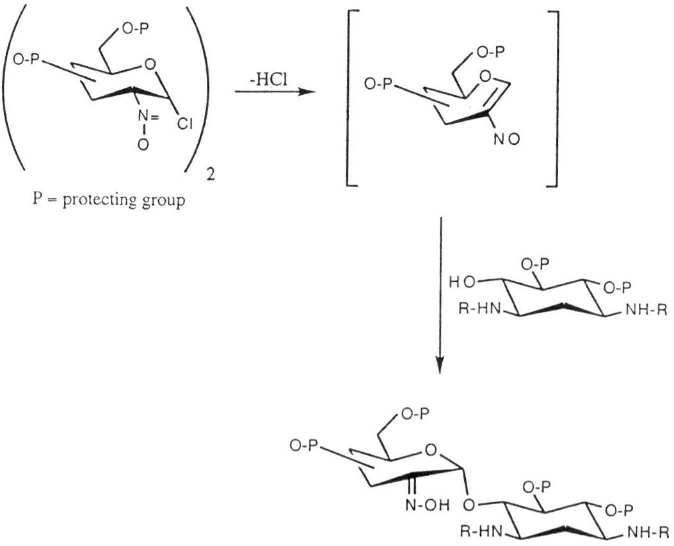

Figure 5.19 Nitroso group directed coupling to (protected) 2-deoxystreptamine.

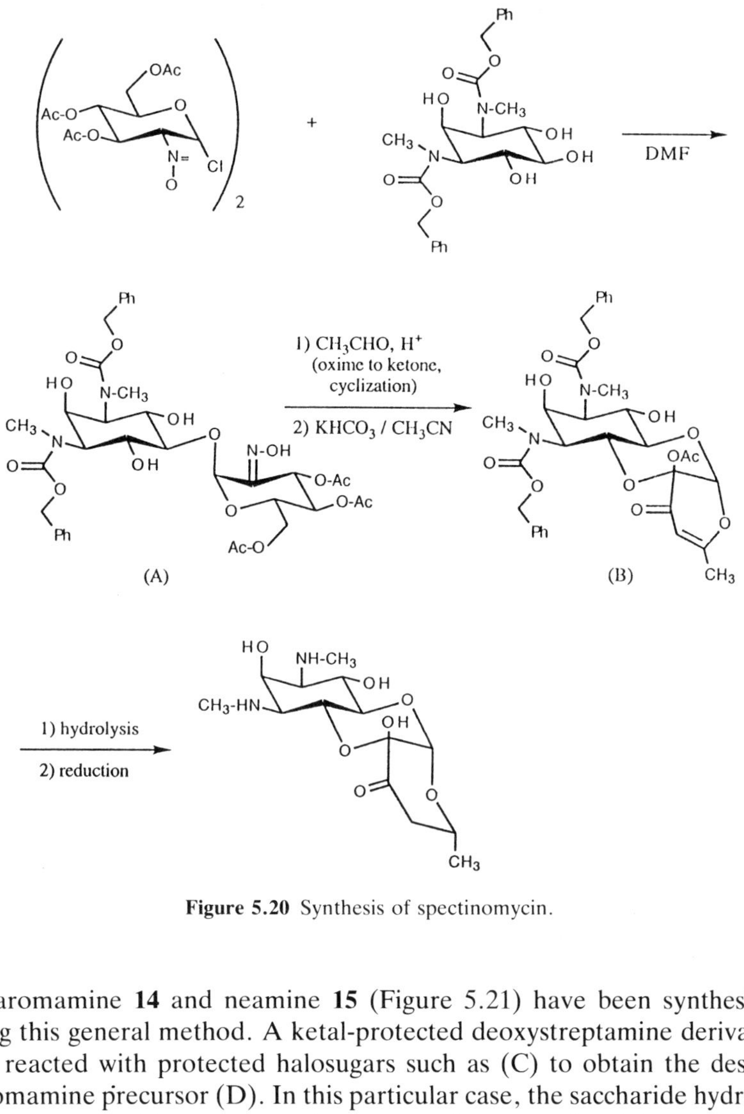

Figure 5.20 Synthesis of spectinomycin.

Paromamine **14** and neamine **15** (Figure 5.21) have been synthesized using this general method. A ketal-protected deoxystreptamine derivative was reacted with protected halosugars such as (C) to obtain the desired paromamine precursor (D). In this particular case, the saccharide hydroxyl groups were judiciously selected so as to differentiate the C-6′ position for further elaboration to neoamine. Azide displacement of the C-6′ tosylate provided for conversion to the C-6′ amine and a synthesis of neoamine.

Many aminoglycosides contain two (or more) saccharide components and a reiterative process has been used sequentially to attach each individual saccharide unit. The syntheses of the kanamycins illustrates this method (Figure 5.22). Kanamycin B has been synthesized from neamine and kanamycin C has been prepared from paromamine, both using variations of this approach. Tobramycin (C-3′-deoxykanamycin B) has

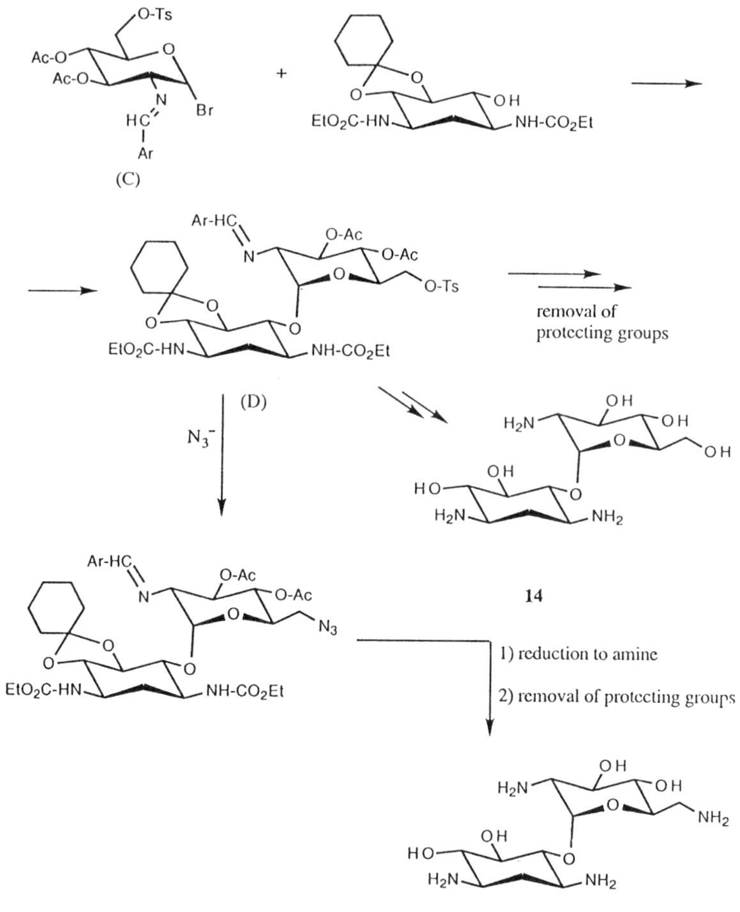

Figure 5.21 Synthesis of paromamine and neamine.

been synthesized in a similar fashion; the corresponding C-3′-tosylated derivative (E) allows for displacement with iodide and subsequent reduction to the deoxy product (F) (Figure 5.23).

The syntheses of the neomycins (e.g. neomycin C) have followed a similar protocol despite the presence of the ribose nucleus. At a first glance, the ribose sugar is amenable to most of the methods developed for the six-membered sugar congeners, although protecting group strategies must be addressed in a slightly different manner. For example, ketalization preferentially occurs at the ribose C-2 and C-3 hydroxyl groups which are not suitable for synthesizing neomycin. However the ribose C-3 hydroxyl

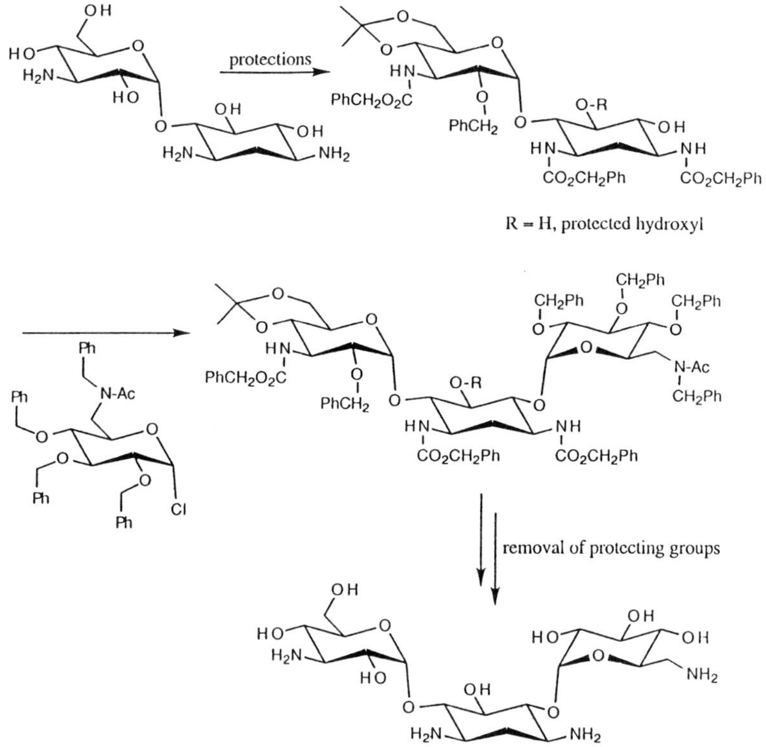

Figure 5.22 Synthetic approach to the kanamycins.

group is more reactive towards glycoside formation providing that the C-5 hydroxyl group is protected (G) (Figure 5.24).

The enzymatic deactivation of the aminoglycosides by resistant bacteria (section 5.5) has impacted significantly upon synthetic strategies. There have been several schools of thought aimed at preventing this phenomena. It has been envisioned that the replacement of complete saccharide units with chemically inert organic moieties could circumvent this problem. However, to date, no non-saccharide scaffolds have been discovered that can suitably mimic the naturally occurring aminoglycosides in keeping with potent antibacterial activity.

Therefore, there have been many efforts to attach substituents directly onto key amine and hydroxyl group functionality in the hope of suppressing enzymatic deactivation. The semi-synthetic aminoglycosides have been essentially limited to chemical elaboration of sites susceptible to the action of these bacterial enzymes (aminoglycoside-modifying enzymes, section 5.5). Usually, simple alkylation of hydroxyl groups and amine

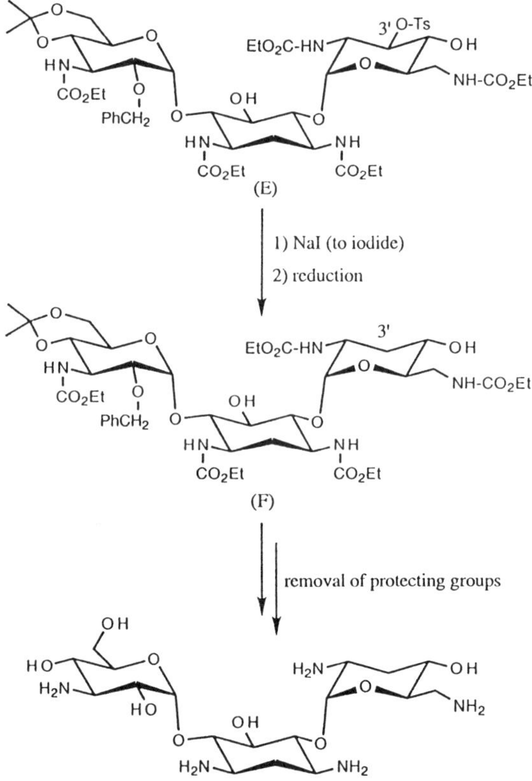

Figure 5.23 Synthetic approach to tobramycin(s).

functionality is unrewarding; one notable exception is the C-1 amine functionality of the aminocyclitol unit which does tolerate acylation and alkylation, as mentioned before. Netilmicin is illustrative; this compound can be prepared by treating sisomicin (H) with acetaldehyde under reductive conditions (Figure 5.25). Amikacin and other congeners can be similarly synthesized directly from naturally occurring aminoglycosides by acylating the 2-deoxystreptamine C-1 amine group.

The actual replacement of the susceptible hydroxyl and amine functionality with inert substituents has been essentially unrealized, except for the removal of certain groups. Tobramycin falls into this general category; as described above, one synthesis (Figure 5.23) relies upon the chemical removal of the C-3' hydroxyl group. The deoxykanamycins have also been obtained by coupling the deoxysugar directly to a suitable saccharide–aminocyclitol adduct (Figure 5.26). Despite this, tobramycin is a naturally occurring aminoglycoside and its total synthesis remains mostly of academic importance.

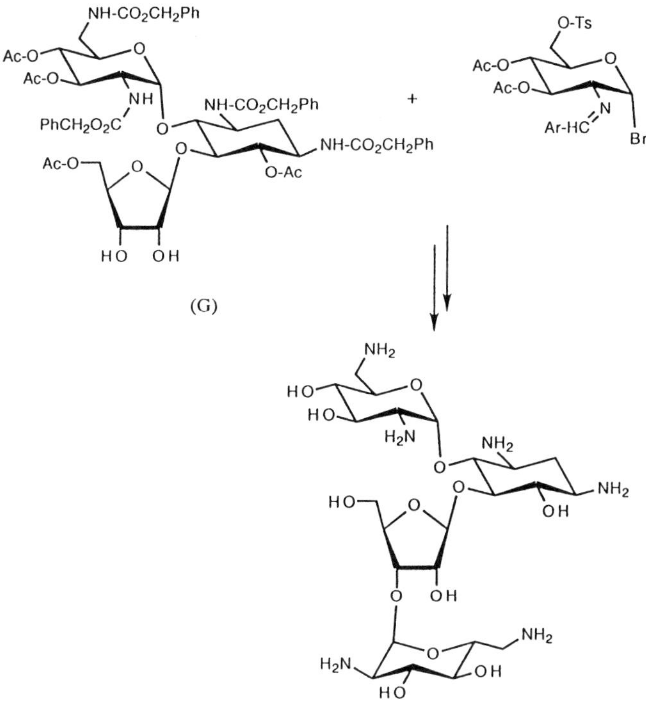

Figure 5.24 Synthetic approach to neomycins.

Figure 5.25 Preparation of netilmicin from sisomicin.

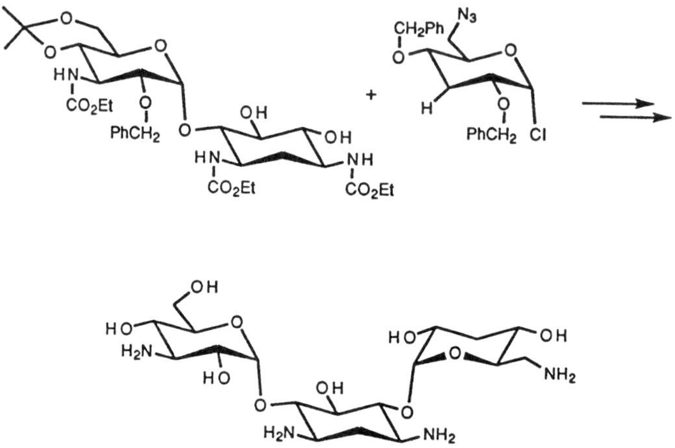

Figure 5.26 Synthetic approach to C-3′-deoxykanamycin.

Of the aminoglycosides, the streptomycins have been particularly challenging from a synthetic perspective; the α-hydroxyaldehyde functionality is extremely sensitive. A landmark synthesis encompassed protection of the streptose α-hydroxyaldehyde system as a cyclic ketal in conjunction with masking the ribose C-5 alcohol as a benzyl ether (I). This disposed the ribose C-3 hydroxyl moiety for coupling to the saccharide unit, thereby affording the desired adduct (J). After a series of protecting group manipulations, the streptose–saccharide adduct was activated (K) as the corresponding halide and coupled to a protected streptidine unit. Removal of the various protecting groups followed by oxidation to regenerate the necessary aldehyde completed the synthesis (Figure 5.27).

As can be gathered from the discussion presented above, synthetic approaches to the aminoglycosides have been hampered by protocols that are demanding in terms of efficiency and practicality. The total chemical synthesis of any aminoglycoside from its respective aminocyclitol and saccharide components remains primarily of academic interest. It has been more practical to harvest aminoglycosides from fermentation sources; however the semi-synthetic aminoglycosides such as amikacin, netilmicin and dibekacin are obtained from chemical procedures. The search for novel semi-synthetic analogs has been mostly abandoned at this time; however, a few significant congeners have been noteworthy (section 5.6).

Lastly, the technique of 'precursoring' has been successfully used, to some extent, to obtain novel aminoglycosides (recall that a similar technology was used to produce new early penicillins (chapter 3, section 3.1)). In such an approach, extraneous saccharides are fed into a fermentation culture in an attempt to induce the microorganism (*Strepto-myces*, *Micromonospora*) to utilize these additives in a biosynthetic fashion

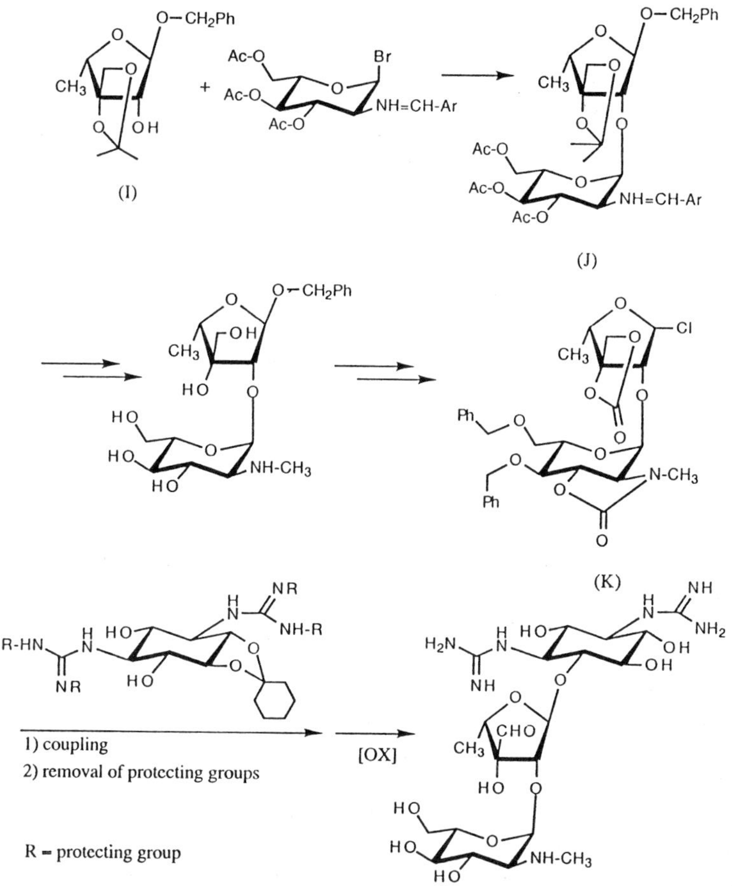

Figure 5.27 Total synthesis of streptomycin.

so as to produce novel aminoglycoside derivatives. Although there has been some noteworthy work in this area, this strategy has not challenged the existing technologies (fermentation and semi-synthetic); nevertheless, it has produced novel compounds, although less active than the established aminoglycosidic antibacterial agents.

5.5 Bacterial resistance to aminoglycoside antibacterials

There are several mechanisms by which bacteria can block the antibacterial effects of the aminoglycosides. In general terms, the major modes of resistance are modification of the bacterial ribosome, deactivation of the aminoglycoside by bacterial enzymes or ineffective uptake into the prokaryotic cell. Normally the aminoglycosides are not excluded by cellular barriers or membranes; these compounds are small polar molecules that can readily cross peptidoglycan and usually, lipopolysaccharide barriers. In some rather rare cases, resistant bacteria possess structural barriers that do not allow for penetration of the aminoglycoside. More often, a bacterial species is resistant due to an ineffectual aminoglycoside transport mechanism (see section 5.2 for a discussion of aminoglycoside uptake). For example, streptococci are inherently resistant (usually low level) to the aminoglycosides due to an incomplete electron transport chain that is unable to 'power' energy-dependent uptake. However, when administered in conjunction with an appropriate β-lactam antibiotic, cell penetrability is increased and under these conditions, aminoglycosides can exert bactericidal properties. However, the strictly anaerobic species such as *Chlostridium perfringens* and *Bacteroides fragilis* are highly resistant to most aminoglycosides since uptake is essentially nil.

Diminished aminoglycoside transport confers resistance to some Gram negative species. A number of resistant species (e.g. *E. coli*) have been selected and investigated; various small molecules and proteins involved in electron transport are either defective or absent compared to wild-type counterparts. For example, some quinones (ubiquinone, menaquinone) are absent in mutants that display resistance. Ubiquinone appears to play an integral part in the movement of aminoglycosides across the cytoplasmic membrane and depletion is detrimental to uptake. Alternatively, proteins involved in ATPase function, which fuels normal energy-dependent aminoglycoside uptake, may also be impaired.

Another mode of resistance involves modification of the molecular target of the aminoglycosides, the bacterial ribosome. An alteration of the ribosome or certain ribosomal proteins can diminish binding of amino-glycoside; consequently functional bacterial protein biosynthesis is retained and the organism is resistant. Resistance to streptomycin can be traced to an abnormal ribosomal protein (S12) resulting from a single step mutation. Some gentamicin resistant *E. coli* species contain an abnormal L6 ribosomal protein. Ribosomal modification can affect nearly the entire family of aminoglycosides; the most important mutation occurs in the 50S ribosomal subunit which can confer resistance to gentamicin and many of the 2-deoxystreptamine-derived aminoglycosides.

By far, the major contributor to aminoglycoside resistance is plasmid-mediated enzymatic deactivation of the drug. Bacterial enzymes that modify the aminoglycoside antibacterial agents are generally referred to as aminoglycoside-modifying enzymes; this family of proteins covalently attaches small molecular groups to the aminoglycoside. Conjugation can occur at several sites on the molecule in a variety of ways, depending upon the specific enzyme and substrate involved. Modified aminoglycosides are unable to bind to the bacterial ribosome and since regeneration of the parent drug via hydrolysis is negligible, ribosomal binding is lost.

The aminoglycoside-modifying enzymes are typically small proteins ranging from approximately 20 kDA to 70 kDA, and existing in multimeric states. They are encoded by small genes carried on (resistance) plasmids or transposable elements such as transposons. Some enzymes are particularly troublesome since they can recognize and deactivate several members of the 2-deoxystreptamine-derived aminoglycoside subset. The aminoglycoside-modifying enzymes catalyse one of three general chemical transformations. Accordingly, these proteins are categorized by the type of structural change they produce as well as by the substrate they act upon. The acetyltransferases catalyse acetylation of an amine group of an aminoglycoside substate; acetyl CoA enzyme serves as the acetate donor. The phosphotransferases convert a hydroxyl moiety of an aminoglycoside to the corresponding phosphate; adenosine triphosphate (ATP) provides

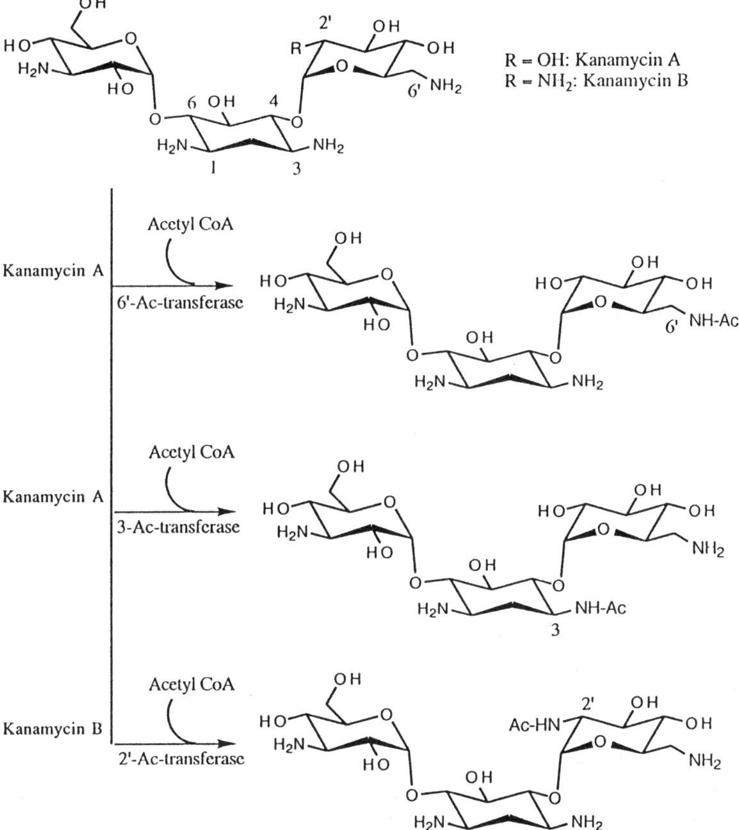

Figure 5.28 Action of aminoglycoside-modifying acetyltransferases.

the source of phosphate. Lastly, the nucleotidyltransferases and the more common adenylyltransferases attach a nucleotide group, specifically an adenylyl moiety in the latter case, onto a hydroxyl group of an aminoglycoside. The nucleotidyltransferases obtain nucleotide from an appropriate nucleoside triphosphate and adenylyltransferases rely upon ATP.

The aminoglycoside acetyltranferases target either the C-3, C-2′ or C-6′ amine groups depending upon the specific enzyme. Acetylation at the C-6′ position, mediated by aminoglycoside 6′-acetyltransferases, primarily occurs with the kanamycins and gentamicins (Figure 5.28). Less frequently, the semi-synthetic agents such as amikacin and dibekacin are deactivated in this manner. The kanamycins, gentamicins, neomycins, paromomycins and tobramycin can all serve as substrates for aminoglycoside 3-acetyltransferases (Figure 5.28). Lastly, aminoglycoside 2′-acetyltransferases act upon some of the kanamycins and gentamicins as well as neomycin and dibekacin (Figure 5.28).

Like the acetyltransferases, the phosphotransferases, as a family, can alter aminoglycosides at several positions. The 2-deoxystreptamine-derived congeners are prone to phosphorylation at C-3′ and C-2″ positions. The aminoglycoside 3′-phosphotransferases target the kanamycins,

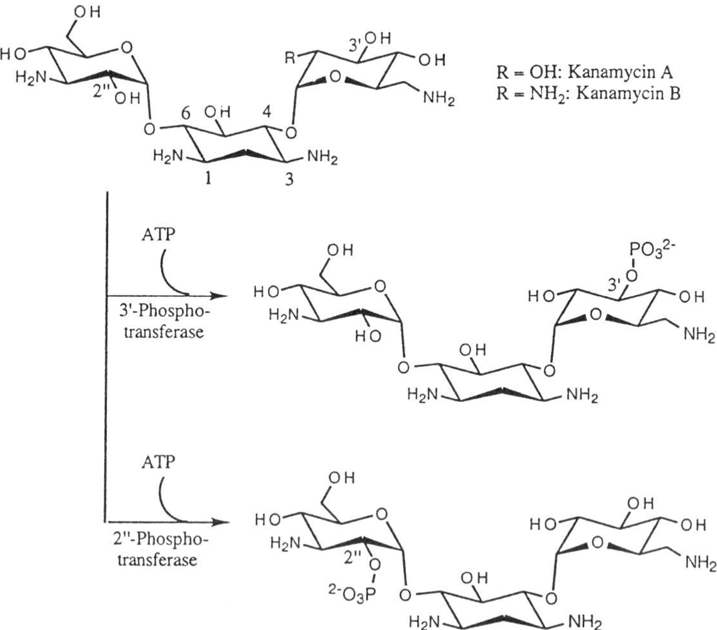

Figure 5.29 Action of aminoglycoside-modifying phosphotransferases.

neomycins and paromomycins, producing the corresponding inactive phosphates (Figure 5.29). The aminoglycoside 2″-phosphotransferases deactivate kanamycins, gentamicin (C) and dibekacin (Figure 5.29).

Several phosphotransferase enzymes are responsible for deactivating the streptomycins. Aminoglycoside 3″-phosphotransferase introduces the phosphate group at the C-3″ hydroxyl position whereas aminoglycoside 6-phosphotransferase phosphorylates the corresponding streptamine hydroxyl group (Figure 5.30).

Nucleotidylation is another common way by which the streptomycins are

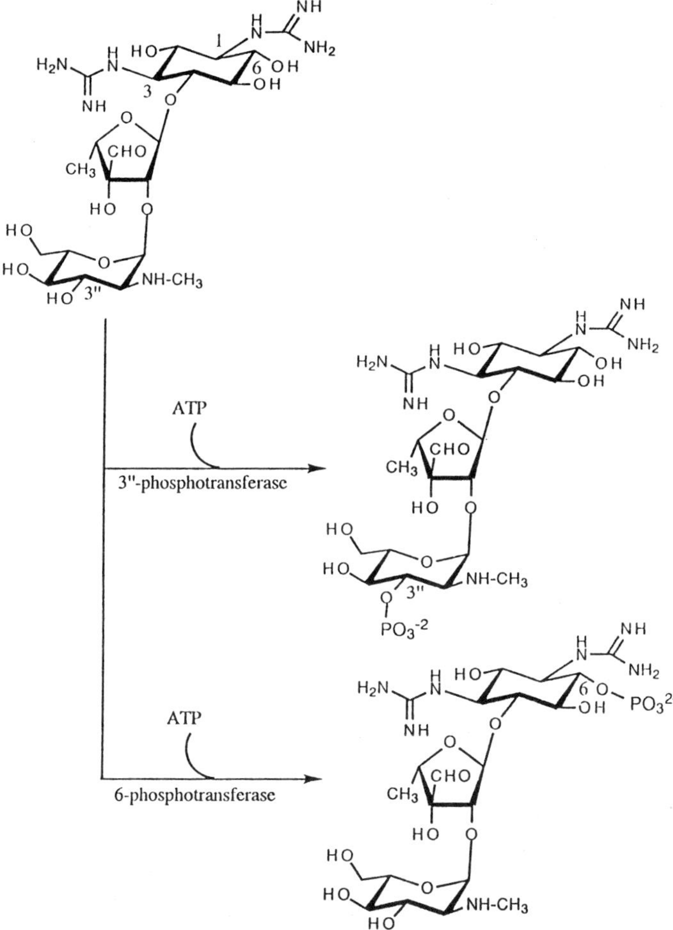

Figure 5.30 Action of phosphotransferases on streptomycin.

deactivated. As with the streptomycin phosphotransferases, these amino-glycoside adenylyltransferases can act at either the C-6 or the C-3″ hydroxyl moieties (Figure 5.31).

Other aminoglycoside adenylyltransferases, accept 4,6-disubstituted-2-deoxystreptamine-derived aminoglycosides as substrates. The kanamycins, gentamicins and tobramycin can undergo adenylation at the C-2″ or C-4′ hydroxyl moiety depending upon the particular enzyme involved (Figure 5.32). Under certain conditions, other nucleotides can also be transferred to the C-2″ position through the action of nucleotidyltransferases.

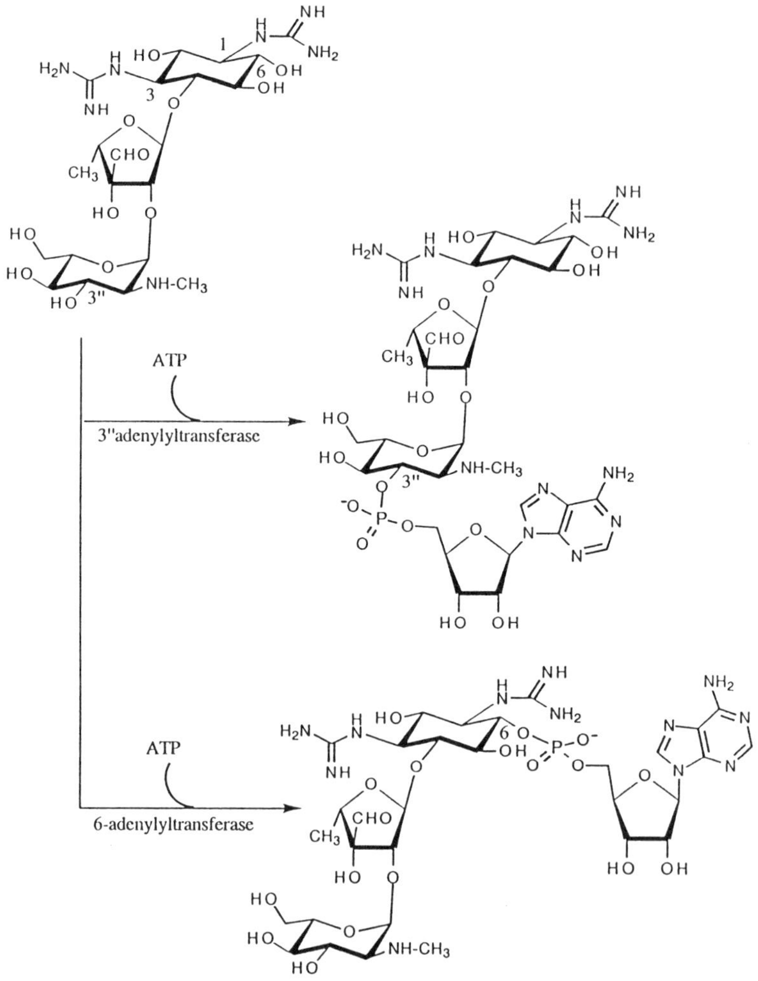

Figure 5.31 Action of aminoglycoside-modifying adenylyltransferases.

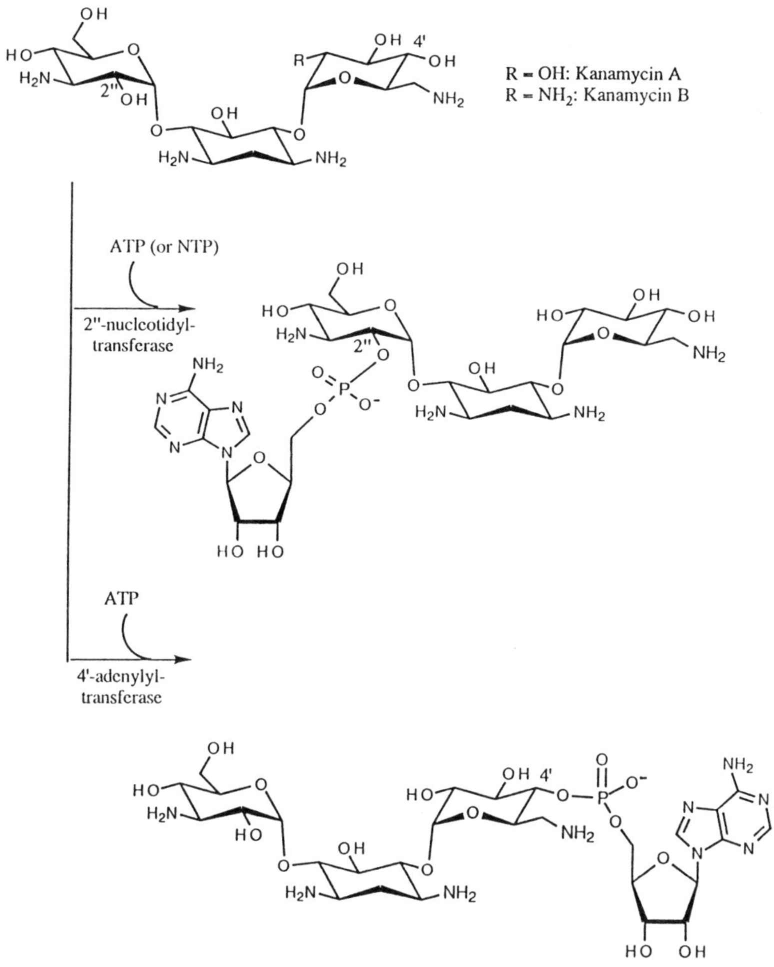

Figure 5.32 Action of aminoglycoside-modifying nucleotidyltransferases.

Lastly, the pseudoaminoglycoside spectinomycin is targeted by specific aminoglycoside adenylyltransferase enzymes (Figure 5.33).

Aminoglycoside-modifying enzymes pose a very serious threat to the success of many aminoglycoside therapeutic regimens. Efforts to overcome resistance of this type have been primarily focused upon the preparation of potent analogs that do not contain susceptible amine and hydroxyl functionality. To date, there has been no significant progress in developing inhibitors of aminoglycoside-modifying enzymes.

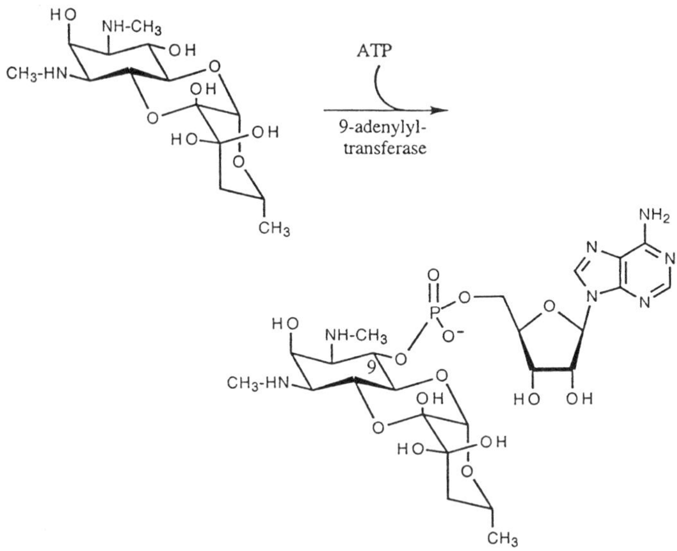

Figure 5.33 'Deactivation' of spectinomycin by adenylyltransferase.

5.6 Recent advances

There has been little research activity on the aminoglycosides in recent years for several reasons. First of all, as a family, the aminoglycosides can produce serious toxicities (ototoxicity and nephrotoxicity) which limit their use. Although some congeners (e.g. tobramycin) are regarded as being safer to use than earlier agents, the frequency and severity of toxic side effects remains troubling. As a result, other antibacterials are often chosen for use and the aminoglycosides are seldom first-line agents. Secondly, other classes of antibacterials offer a broader spectrum of activity, greater potency and a better therapeutic ratio than the aminoglycosides; for these reasons, there has been little reason to revisit the aminoglycosides from a research perspective.

Nevertheless, some more modern aminoglycosides have advanced to the clinic in recent years. Probably no new congener will challenge the existing agents, but some compounds are noteworthy in their own regard. For example, resistance caused by aminoglycoside-modifying enzymes has been addressed in a rather interesting way. Rather than masking or removing susceptible functionality (sections 5.2 and 5.5), it was found that some epimeric aminoglycosides do not act as substrates of the aminoglycoside-modifying enzymes and yet retain acceptable antibacterial activity. One such compound that was under investigation is 5-episisomicin

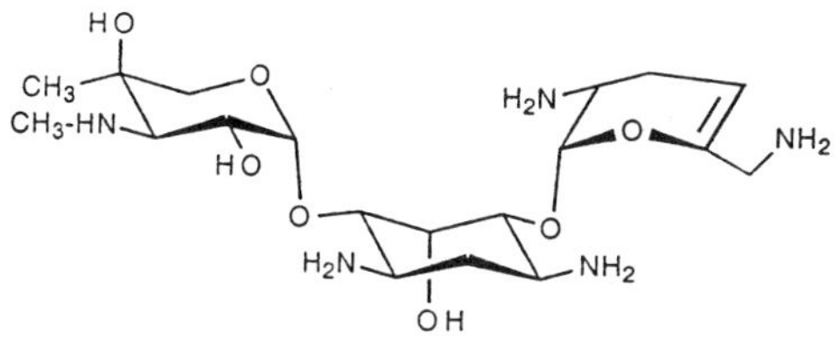

Figure 5.34 5-Episisomicin.

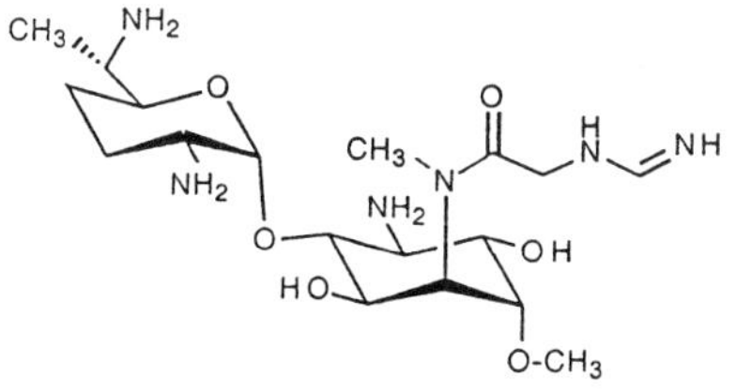

Figure 5.35 Dactimicin.

16 (Figure 5.34) which is less prone to 3-acetyltransferases and 2″-adenynyltransferases than the naturally occurring sisomicin.

A new aminoglycoside-like antibacterial has been obtained from screening naturally occurring microbial sources. Dactimicin **17** (Figure 5.35) exhibits good potency against some Gram negative bacteria. Interestingly, dactimicin lacks hydroxyl functionality normally associated with phosphotransferase and nucleotidyltransferase susceptibility. In addition, since the aminocyclitol amine is acylated, susceptibility to acetyltransferase should be avoided. Preliminary results show that dactimicin is active against a variety of amikacin- (kanamycin-) and gentamicin-resistant bacteria, but is less active against the important pathogen, *Pseudomonas aeruginosa*. Dactimicin is in clinical trials in Japan.

Certainly, the major focus in the design of improved aminoglycoside antibacterials has continued to entail the attachment of various acyl substituents onto the N-1 position of the 2-deoxystreptamine nucleus. Isepamicin **18** (Figure 5.36) and arbekacin **19** (Figure 5.37) are among the latest such derivatives; arbekacin has been launched in Japan. Isepamicin is a semi-synthetic derivative of gentamicin B and arbekacin is derived from tobramycin.

Finally, trospectomycin **20** (Figure 5.38), the C-6′-*n*-propyl analog of spectinomycin, has been recently advanced onto the market for the treatment of gonorrhea. Trospectomycin possesses a similar, albeit more potent, spectrum of antibacterial activity; it is more active (*in vitro*) than

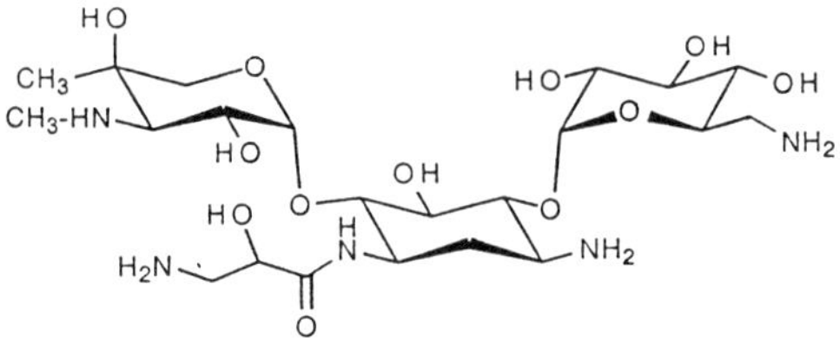

18

Figure 5.36 Isepamicin.

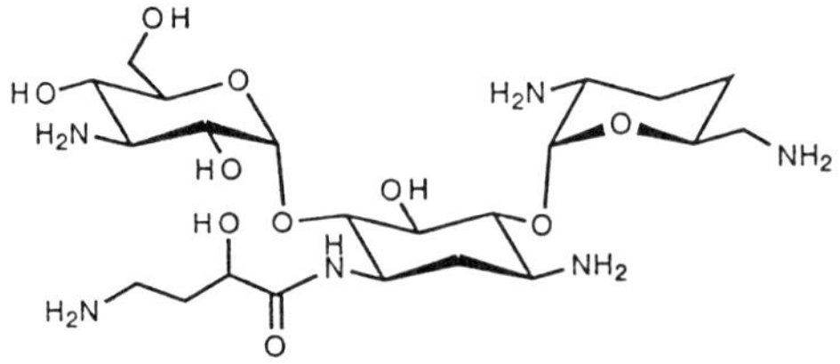

19

Figure 5.37 Arbekacin.

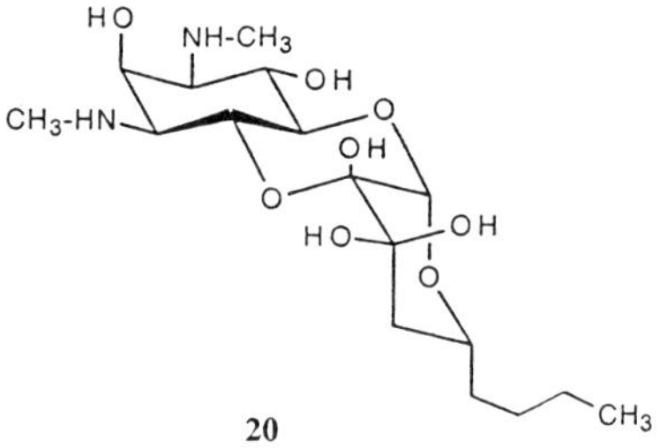

20

Figure 5.38 Trospectomycin.

spectinomycin against *Neisseria gonorrhoeae* as well as other common pathogens (*N. meningitidis*, staphylococci, streptococci and enterococci). This agent is also effective against many spectinomycin-resistant micro-organisms.

5.7 Summary

The aminoglycoside antibacterial agents occupy a restricted, but useful role in infectious disease chemotherapy.

1. The aminoglycosides are potent antibacterial agents being effective primarily against aerobic Gram negative bacilli. The spectrum of

activity of most (2-DOS-derived) aminoglycosides includes the troublesome *Pseudomonas aeruginosa* pathogen. Streptomycin is exquisitely active against *Mycobacterium tuberculosis*. The spectinomycins are effective agents against gonococcal diseases caused by *Neiserria gonorrhoeae*.

2. The aminoglycosides can produce severe adverse side effects which include nephrotoxicity, ototoxicity and neuromuscular effects. These properties have limited the use of aminoglycoside chemotherapy to serious systemic indications. Some aminoglycosides can be administered for ophthalmic and topical purposes.

3. The marketed aminoglycosides include naturally occurring compounds that are obtained from *Streptomyces* or *Micromonospora* species, and a generation of semi-synthetic aminoglycosides. The most frequently used agents today include gentamicin and tobramycin, (natural products) and amikacin, netilmicin and dibekacin (semi-synthetic derivatives).

4. The aminoglycosides exhibit bactericidal effects as a result of several phemonena. Ribosomal binding and translational misreading disrupt normal protein biosynthesis; cell membrane damage also plays an integral part in the ensuing bacterial cell death.

5. Aminoglycoside resistance is problematic and is most often caused by plasmid-mediated aminoglycoside-modifying enzymes. These proteins deactivate the aminoglycosides by attaching small molecular conjugates onto either an amine or hydroxyl functionality. Bacteria can display aminoglycoside resistance through ribosomal modification or by decreased uptake of aminoglycoside into the bacterial cell.

Major aminoglycoside antibacterials are shown in Table 5.1.

Table 5.1 Major aminoglycoside antibacterial chemotherapeutic agents

Generic name	Common trade names	Administration
Amikacin	Amikacin	IM, IV
Arbekacin	Harbekacin	IM
Dibekacin	Dibekacin, Panimycin	IM
Gentamicin (C_1, C_{1a}, C_2)	Garamycin, Gentacin	IM, IV
Kanamycin A	Kantrex	IV
Netilmicin (sulfate)	Netromycin	IM, IV
Neomycin	Neocin, Neomin	IM
Spectinomycin (HCl)	Trobicin	IM
Streptomycin (sulfate)	Streptsulfat, Streptomycin Sulfate	IM
Tobramycin (sulfate)	Nebcin	IM, IV
Trospectomycin (sulfate)	Spexil	IM

Further reading

H. Umezawa and I.R. Hooper (Eds) (1982) 'Aminoglycoside antibiotics', *Handbook of Experimental Pharmacology*, Vol. 62, Springer-Verlag, Berlin.

A. Whelton and H.C. Neu (Eds) (1982) *The Aminoglycosides*, Marcel Dekker, New York.

K.L. Rinehart, Jr, and T. Suami (Eds) (1980) *Aminocyclitol Antibiotics*, ACS Symposium Series 125, American Chemical Society, Washington, DC.

J. Reden and W. Durckheimer (1979) 'Aminoglycoside antibiotics: chemistry, biochemistry, structure-activity relationships', in *Topics in Current Chemistry*, Vol. 83, M.J.S. Dewar, K. Hafner, E. Heilbronner, S. Ito, J.-M. Lehn, K. Niedenzu, C.W. Rees, K. Schafer, G. Wittig and F.L. Boschke (Eds), Springer-Verlag, Berlin, pp. 106–170.

K.E. Price, J.C. Godfrey and H. Kawaguchi (1977) 'Effects of structural modifications on the biological properties of aminoglycoside antibiotics containing 2-deoxystreptamine', in *Structure-Activity Relationships among the Semisynthetic Antibiotics*, D. Perlman (Ed), Academic Press, New York, pp. 357–396.

S. Umezawa (1974) 'Structures and syntheses of aminoglycoside antibiotics', in *Advances in Carbohydrate Chemistry and Biochemistry*, Vol. 30, R.S. Tipson and D. Horton (Eds), Academic Press, New York, pp. 111–182.

K.L. Rinehart, Jr. (1969) 'Comparative chemistry of the aminoglycoside and aminocyclitol antibiotics', *J. Infectious Diseases*, **119**, 345.

J.D. Dutcher (1963) 'Chemistry of the amino sugars derived from antibiotic substances', in *Advances in Carbohydrate Chemistry*, Vol. 18, M.L. Wolfrom and R.S. Tipson (Eds), Academic Press, New York, pp. 259–308.

B.D. Davis (1988) 'The lethal action of aminoglycosides', *J. Antimicrob. Chemother.*, **22**, 1.

B.D. Davis (1987) 'Mechanism of bactericidal action of aminoglycosides', *Microbiol. Rev.*, **51**, 341.

K.L. Rinehart, Jr. (1977) 'Metasynthesis of new antibiotics', *Pure Appl. Chem.*, **49**, 1361.

K.L. Rinehart, Jr. and R.M. Stroshane (1976) 'Biosynthesis of aminocyclitol antibiotics', *J. Antibiotics*, **XXIX**, 319.

D.J. Cooper (1977) 'Comparative chemistry of some aminoglycoside antibiotics', *Pure Appl. Chem.*, **28**, 455.

H.W. Taber, J.P. Mueller, P.F. Miller and A.S. Arrow (1987) 'Bacterial Uptake of Aminoglycoside Antibiotics', *Microbiol. Rev.*, **51**, 439.

R.E.W. Hancock (1981) 'Aminoglycoside uptake and mode of action – with special reference to streptomycin and gentamicin. I. Antagonists and mutants', *J. Antimicrob. Chemother.*, **8**, 249.

R.E.W. Hancock (1981) 'Aminoglycoside uptake and mode of action – with special reference to streptomycin and gentamicin. II. Effects of aminoglycosides on cells', *J. Antimicrob. Chemother.*, **8**, 429.

N. Tanaka (1975) 'Aminoglycoside antibiotics', in *Antibiotics III*, J.W. Corcoran and F.E. Hahn (Eds), Springer-Verlag, Berlin, pp. 340–364.

D. Schlessinger and G. Medoff (1975) 'Streptomycin, dihydrostreptomycin and the gentamicins', in *Antibiotics III*, J.W. Corcoran and F.E. Hahn (Eds), Springer-Verlag, Berlin, pp. 535–550.

F.E. Hahn (1971) 'Streptomycin', *Antibiotics Chemother.*, **17**, 29.

F. Johnson (1973) 'The total synthesis of antibiotics' in *The Total Synthesis of Natural Products*, Vol. 1, J. ApSimon (Ed.), John Wiley, New York, pp. 364–390.

J.E. Davis (1983) 'Resistance to aminoglycosides; mechanisms and frequency', *Rev. Infectious Diseases*, **5**, S261.

H. Kawaguchi (1977) 'Recent progress in aminoglycoside antibiotics', *Jpn. J. Antibiotics*, **XXX**, S–190

D. McGregor (1992) 'Aminoglycosides' (Antibiotics), in *Kirk-Othmer Encyclopedia of Chemical Technology*, 4th edn, John Wiley, New York, pp. 904–925.

H.C. Neu (1976) 'Tobramycin: an overview', *J. Infectious Diseases*, **134** (Supplement), S3.

6 The non-peptidic macrocyclic antibacterials

6.1 History and overview

A large and structurally diverse group of macrocyclic substances exhibit useful anti-infective properties. These compounds are non-peptidic in composition and can be differentiated by general structure and anti-microbial profile. Some substances are principally active against bacteria while others mainly target fungi. The most important antibacterial members contain a characteristic macrocyclic lactone nucleus and are referred to as 'macrolides'. The macrolides are usually 14-membered or 16-membered lactones adorned with small alkyl groups and oxygen-based functionality (hydroxyl, alkyloxy and ketone groups); all antibacterially active congeners have sugar unit(s) extending from the macrocycle. The macrolides are constructed by soil microorganisms and many of these natural products are valuable chemotherapeutic agents. In addition, a number of semi-synthetic derivatives have also reached the market. The antifungal agents are inherently different from the macrolides and other macrocyclic antibacterials, and will only be mentioned for structural comparison with the antibacterial classes.

Methymycin **1** (Figure 6.1) and pikromycin **2** (Figure 6.2) are important from an historical perspective since these compounds first drew attention to the macrolide systems. Both compounds are produced by *Streptomyces* sources. Methymycin and pikromycin have remained as academic curiosities and have not had an impact upon antibacterial chemotherapy. However, since the original finding of pikromycin, over one hundred macrolides have been discovered and evaluated for antibacterial activity.

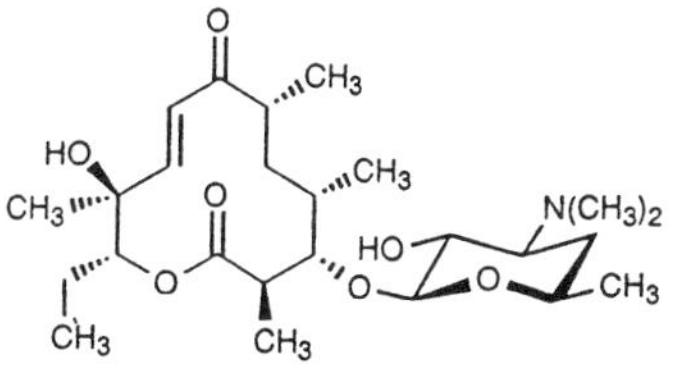

1

Figure 6.1 Methymycin.

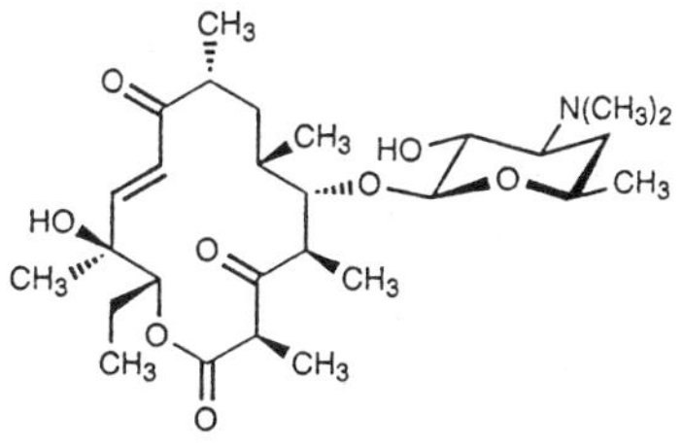

2

Figure 6.2 Pikromycin.

The most recognized macrolides are the erythromycins which were first discovered in 1952 from *Streptomyces erythraeus* (reclassified as *Saccharopolyspora erythraea*). Erythromycin A **3** (Figure 6.3) is the most important member and remains somewhat of a standard by which other antibacterial macrolides are judged. As with most macrolides, erythromycin A contains an aminosugar and a neutral sugar, *β*-D-desosamine and α-L-cladinose, respectively. Erythromycin B is a deoxy derivative, erythromycin C has a different neutral sugar (mycrose (3-*O*-demethylcladinose)) and erythromycin D combines both features in a single molecule. When introduced into practice, the erythromycins offered a welcomed alternative therapy to the penicillins which were being increasingly plagued by resistance. In general, the erythromycins are commonly used to treat respiratory, skin and soft tissue and genital tract infections. Erythromycin A exhibits excellent potency against many troublesome aerobic Gram positive bacteria including streptococci, staphylococci and *Corynebacterium diphtheriae* and good activity against Gram negative *Neisseria* (*N. meningitidis*, *N. gonorrhoeae*). Erythromycin has superb activity against *Mycoplasma* and recently, has become a valuable agent in combating increasingly prevalent *Legionella* and *Chlamydia* infections. The most obvious shortcoming in the spectrum of activity of most macrolides, including the erythromycins, is that many common Gram negative bacteria such as *Escherichia coli* and *Klebsiella*, *Salmonella* and *Pseudomonas* species are not susceptible.

Oleandomycin **4** (Figure 6.4), also produced by a *Streptomyces* (*S. antibioticus*), is similar in potency and spectrum of activity to the erythromycins. Oleandomycin contains an interesting structural variation in having a spiro-epoxide group extending from the macrocycle; this functionality has been a focus of some structural modifications. Another difference between the erythromycins and oleandomycin are the sugar components; oleandomycin has oleandose as its neutral sugar. This trend is recurring in the macrolide family; many members can be distinguished by the composition of the sugar groups and their positioning on the macrocyclic framework (section 6.2). Oleandomycin itself is not a

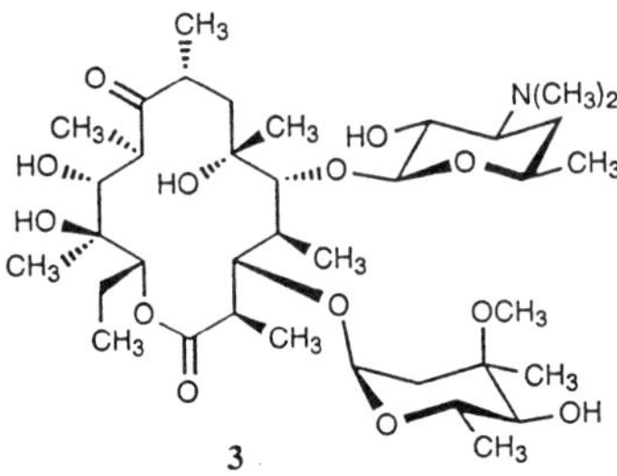

Figure 6.3 Erythromycin A.   **Figure 6.4** Oleandomycin.

significant macrolide; however, a triacetylated prodrug, troleandomycin, is a marketed product.

One drawback of the erythromycins is inherent instability in acidic media; upon reaching the stomach, extensive degradation can occur and antibacterial activity can be lost. As a result, ways in which to avoid this problem have been developed. Formulation of erythromycin A in enteric capsules has been productive since the drug is not released until the capsule reaches the basic environment of the intestine. Another approach has been to deliver erythromycin in a prodrug form that is less water soluble and more resistant to acidic degradation. A number of erythromycin ester derivatives have been particularly successful in this regard. Acetate, propionate and ethyl succinate ester derivatives are marketed erythromycin products. Alternatively, erythromycin esters can be converted to corresponding acid addition salts, and there have been two variations on this theme. Lipophilic acid addition salts tend to enhance absorption whereas hydrophilic acid addition salts are quite water soluble. These different properties can influence the route of administration. Erythromycin estolate (a propionate ester, lauryl sulfate salt) is widely used due to its good oral absorption whereas an erythromycin gluceptate salt is suitable for intravenous administration. Several other derivatives have reached the market or are under development (section 6.6). Being true prodrugs, the antibacterial activity comes from the parent erythromycin (A).

A number of other synthetic strategies have emerged in an attempt to improve the acid stability of erythromycin. In general, the key functional groups responsible for decomposition (section 6.3) have been structurally modified. One of the earliest derivatives in this regard is erythromycylamine **5** (Figure 6.5) which has an amine replacing the ketone functionality. Despite increased resistance to acidic conditions, erythromycylamine lacks good oral bioavailability and has not been marketed. However, the amine group has been derivatized in some interesting ways. The most significant variation are oxazine derivatives which act as prodrugs. Dirithromycin **6** (Figure 6.6) is well absorbed and generates erythromycylamine by *in vivo*

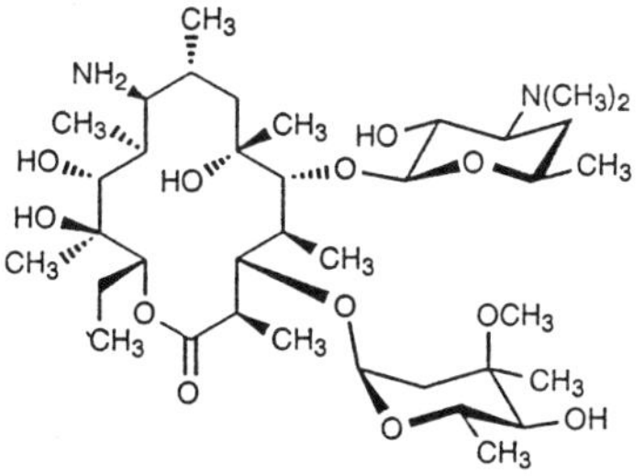

5

Figure 6.5 Erythromycylamine.

cleavage of the labile oxazine ring. Dirithromycin is the latest macrolide to be marketed and was first launched in 1994 in some countries (e.g. Spain).

Other modifications include conversion of the ketone functionality to various oxime derivatives. Roxithromycin **7** (Figure 6.7) is the most important compound of this type and has been marketed. Roxithromycin has similar potency to erythromycin, but is more acid stable and often exhibits good tissue penetration.

Other second generation semi-synthetic derivatives have recently been marketed. Clarithromycin **8** (Figure 6.8) is a methyl ether derivative of erythromycin that has quickly been established as a valuable macrolide. It

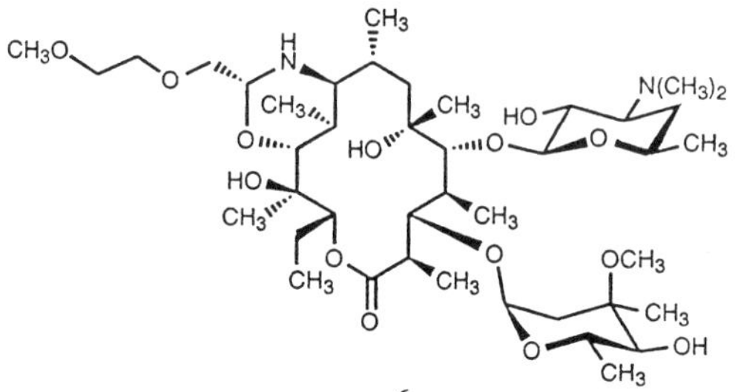

6

Figure 6.6 Dirithromycin.

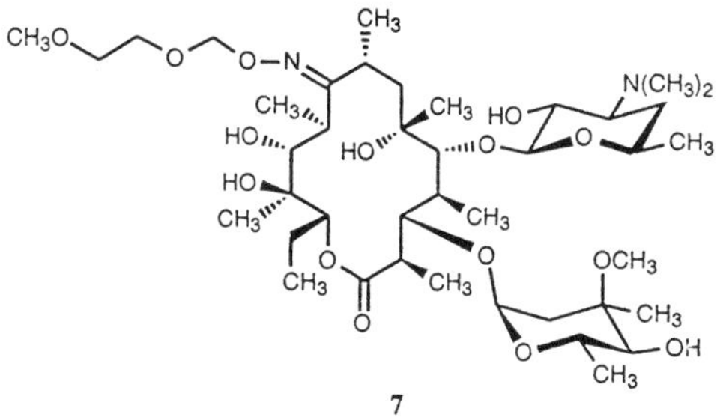

7

Figure 6.7 Roxithromycin.

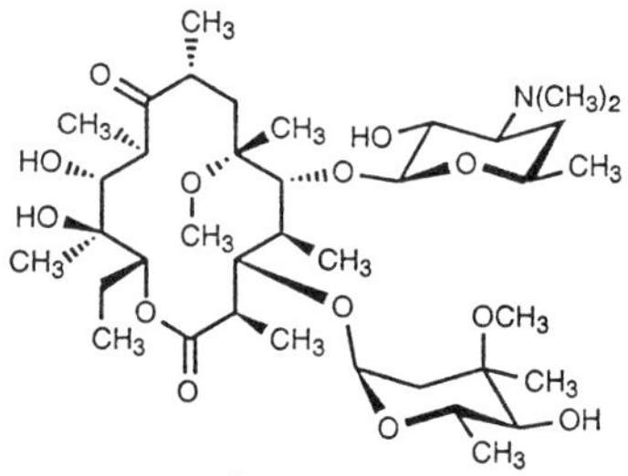

8

Figure 6.8 Clarithromycin.

is prescribed mainly for upper and lower respiratory infections caused by streptococci, *Haemophilus influenzae* and *Mycoplasma pneumoniae* and skin and tissue infections of streptococci and staphylococci.

One of the most significant advances in modern macrolide chemotherapy has been the introduction of the so-called azalide agents. The azalides are structurally unique among the macrolides since a 15-membered aza-lactone framework is present. The prototype, azithromycin **9** (Figure 6.9), has an extended spectrum of action compared to conventional macrolides such as erythromycin. Azithromycin has substantial *in vitro* activity against Gram negative bacteria, but this property appears to be rather academic at this time since superior, non-macrolide agents are available. Nevertheless, the azalides are intriguing and may represent a future generation of truly broad spectrum macrolides. Azithromycin is an important macrolide today. It possesses excellent stability and good oral Gram positive potency, and its pharmacokinetic properties allow for once-a-day dosing (after the first day dose).

A number of 16-membered macrolides are important antibacterial agents in markets outside the United States. In general, these congeners exhibit a similar antibacterial profile to erythromycin; Gram positive cocci and intracellular pathogens such as *Legionella* and *Chlamydia* are particularly sensitive. The 16-membered macrolides are usually divided into two subsets based upon structure. The leucomycin family is most important with regard to human chemotherapy and includes the midecamycins, carbomycins and spiramycins (see below). As with the 14-membered macrolides, the leucomycin-related compounds feature characteristic aglycone cores that bear alkyl and oxygen-based functionality and sugar substituents. However, the 16-membered macrolides often contain a disaccharide unit. The leucomycin complex (kitasamycin) was first isolated from *Streptomyces kitasatoensis* (reclassified as *Streptoverticillium*) as a mixture of ten factors; individual compounds vary in the number and type of acyl substituents. Leucomycin A_1 **10** (Figure 6.10) is representative; it has a disaccharide substituent and a single acyl chain that

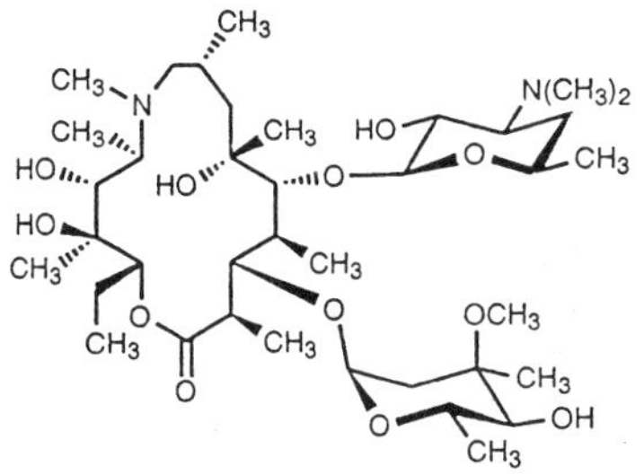

9

Figure 6.9 Azithromycin.

extends from the terminal sugar. Like most 16-membered macrolides, the macrocycle is partially unsaturated and a distinctive aldehyde group is present. Josamycin (leucomycin A$_3$) **11** (Figure 6.11) is among the most important of the 16-membered macrolides. Josamycin is an acetylated version of leucomycin A$_1$ and is well absorbed orally; it has been marketed since 1970.

The carbomycins embody some interesting structural alterations of the macrolide system, as exemplified by carbomycin A **12** (Figure 6.12). These congeners contain a macrocyclic ketone functionality and have an epoxide

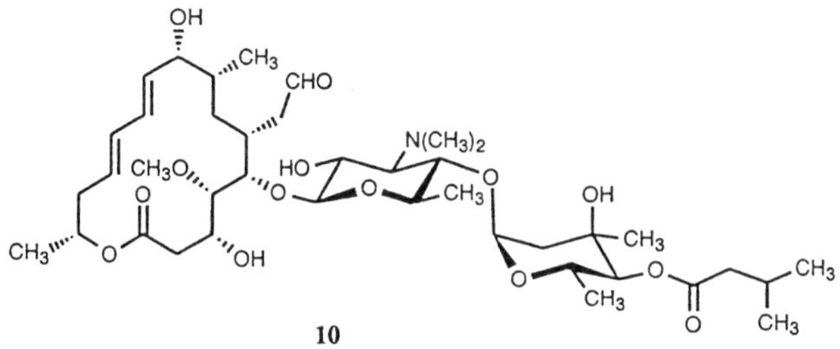

10

Figure 6.10 Leucomycin A$_1$.

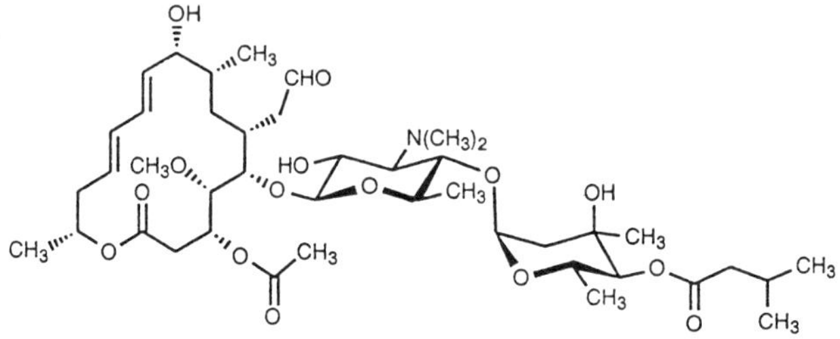

11

Figure 6.11 Josamycin.

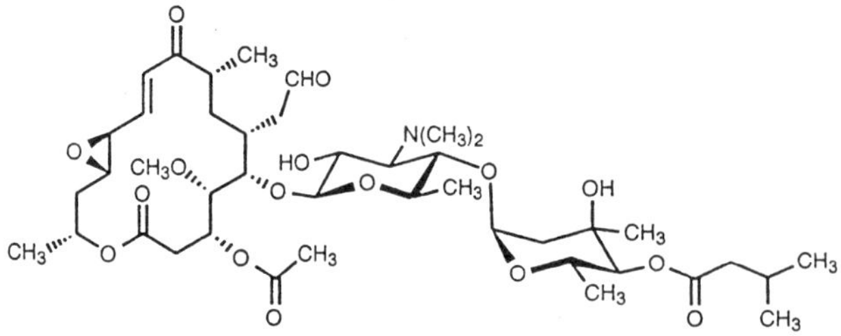

12

Figure 6.12 Carbomycin A.

group instead of a carbon–carbon double bond. The carbomycin complex is produced by *S. halstedii*. Although the carbomycins exhibit good antibacterial properties, their toxicological profile is inferior to some other 16-membered macrolides. The spiramycins, initially isolated as a complex from *S. ambofaciens*, also contain a unique feature in having an unusual sugar unit (*ß*-D-forosamine). Spiramycin I **13** (Figure 6.13), a prototype, has been used as an animal-feed supplement. More importantly, several acylated spiramycins, possessing enhanced *in vivo* activity, have undergone clinical evaluation for potential use in humans (section 6.6).

The midecamycins (from *S. mycarofaciens*) are also members of the leucomycin family. Midecamycin A_1 **14** (Figure 6.14) differs from josamycin and the leucomycins only in specific acyl substituents. A diacetylated semi-synthetic derivative, miokamycin (or miocamycin) **15** (Figure 6.15), has been launched in some markets. Miokamycin essentially parallels erythromycin in activity and tolerability but in some areas (e.g. Japan), it is preferred due to erythromycin resistance. A number of other acylated midecamycin derivatives are promising and are under investigation.

Rokitamycin **16** (Figure 6.16) is the latest 16-membered macrolide to be

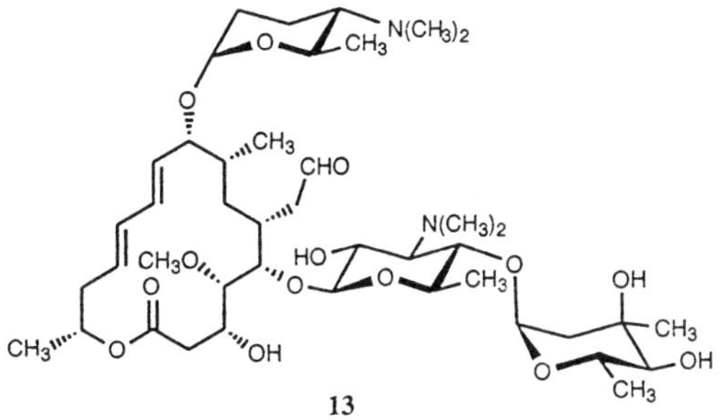

13

Figure 6.13 Spiramycin I.

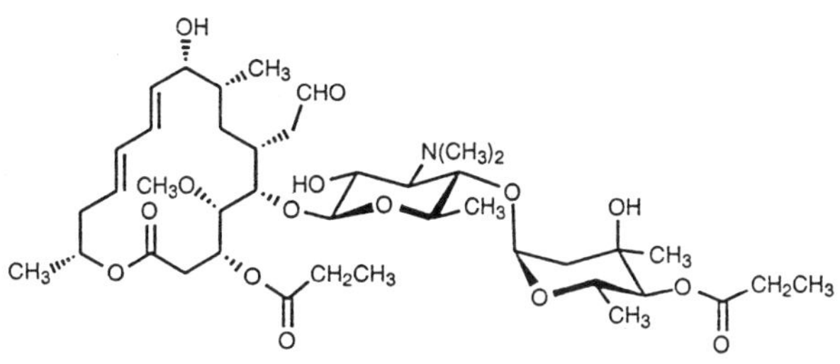

14

Figure 6.14 Midecamycin A_1.

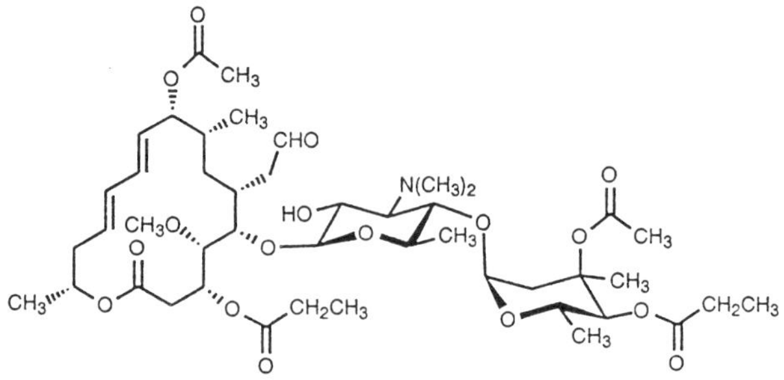

15

Figure 6.15 Miokamycin (miocamycin).

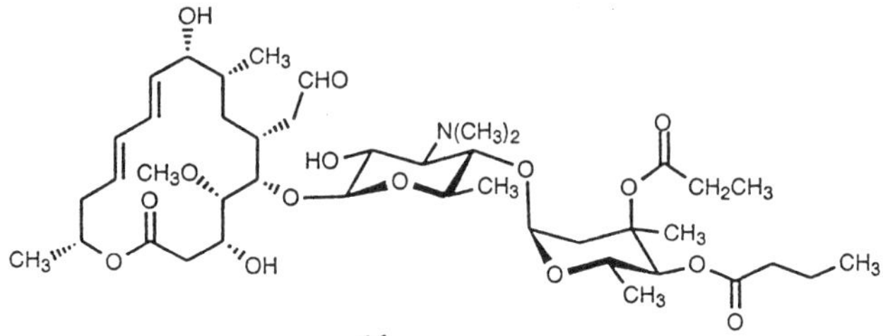

16

Figure 6.16 Rokitamycin.

commercialized (recently launched in Japan and Italy). It is a semi-synthetic derivative of a leucomycin factor (A_5) containing a propionyl ester. Rokitamycin exhibits excellent *in vitro* activity and typically gives high blood levels. Various other 16-membered macrolide prodrug derivatives, primarily esters, have undergone at least preliminary evaluation (section 6.6). In each case, antibacterial activity is attributed to the parent compound which is produced upon cleavage of the ester moiety *in vivo*.

The tylosins are a distinct family of the 16-membered macrolides by virtue of a glycosylated hydroxymethyl substituent that extends from the macrocycle. The naturally occurring tylosins are produced by a *Streptomyces* (*S. fradiae*). Tylosin **17** (Figure 6.17) is a widely used veterinary antibacterial agent. An acylated derivative, tilmicosin, is now a commercial product for use in animals, and several other congeners are under investigation. Tylosin-derived agents have not found use in human chemotherapy to date.

Although the macrolides comprise the majority of the non-peptidic macrocyclic agents, a family of unrelated substances are certainly valuable in antibacterial chemotherapy. The ansamycins bear no resemblance to the macrolides; these compounds possess an aromatic moiety incorporated into a macrocyclic framework. The rifamycins are the most significant of the ansamycins. Rifamycin B **18** (Figure 6.18) was first isolated in 1957 from a *Streptomyces* source (*S. mediterranei*). Interestingly, rifamycin B possesses no antibacterial activity, but it is readily oxidized to its antibacterial congener rifamycin S **19** (Figure 6.19). From this finding, as well as the structural novelty of the rifamycins, over a hundred semi-synthetic derivatives have been prepared.

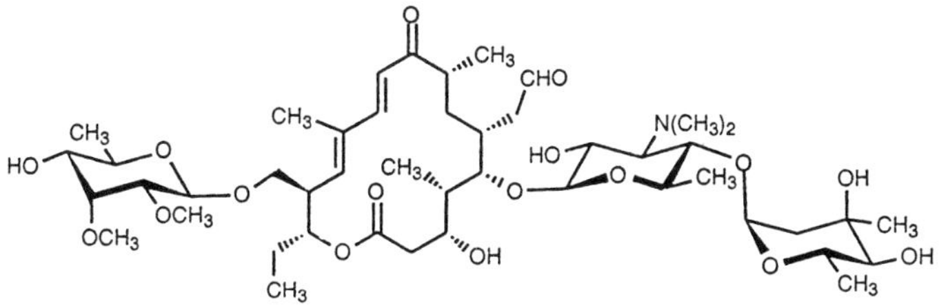

17

Figure 6.17 Tylosin.

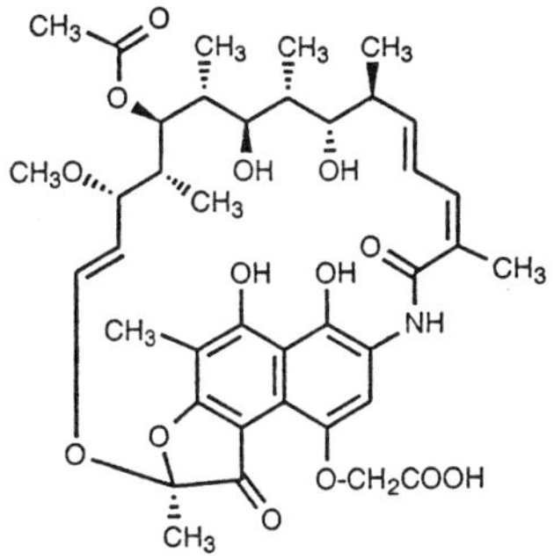

18

Figure 6.18 Rifamycin B.

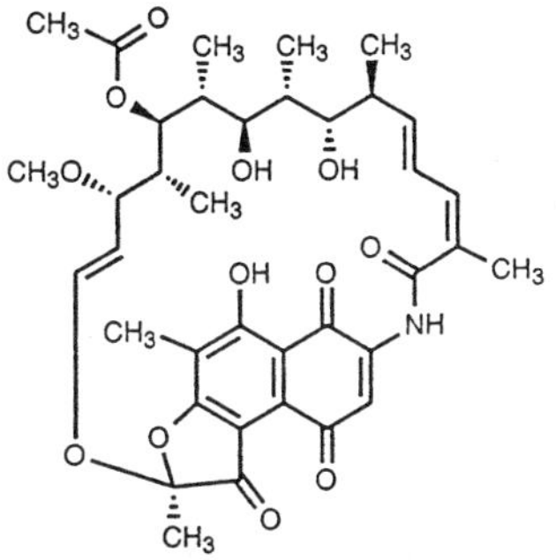

19

Figure 6.19 Rifamycin S.

Rifampin **20** (Figure 6.20), a semi-synthetic derivative of rifamycin B, remains an important oral and IV (intravenous) agent for the treatment of tuberculosis (*Mycobacterium tuberculosis*). Multi-drug resistant mycobacteria have become problematic and so rifampin is often administered with tuberculostatic agents such as isoniazid (chapter 9). The rifamycins also exert antibacterial effects against staphylococci and *Legionella* and *Haemophilis* species. In some cases rifampin can be used for the treatment of Legionnaire's disease, endocarditis and the prevention of meningitis. Several newer rifamycins have undergone clinical evaluation (section 6.6). Rifabutin **21** (Figure 6.21) is the latest rifamycin to be introduced to the marketplace. Its main attribute at this time is its potency against *Mycobacterium avian* complex in AIDS patients.

Lastly, a separate group of non-peptidic macrocyclic compounds,

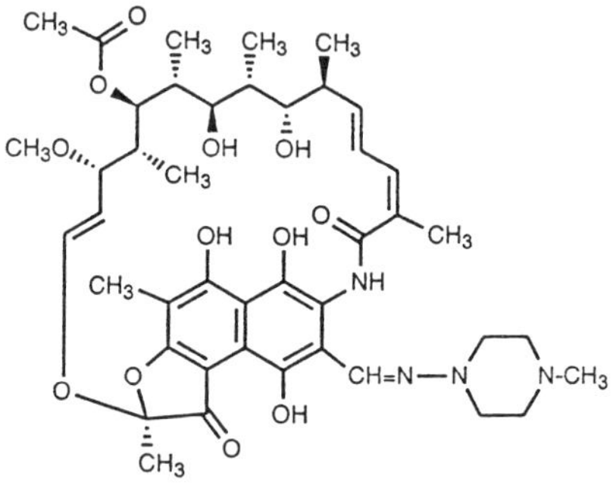

20

Figure 6.20 Rifampin.

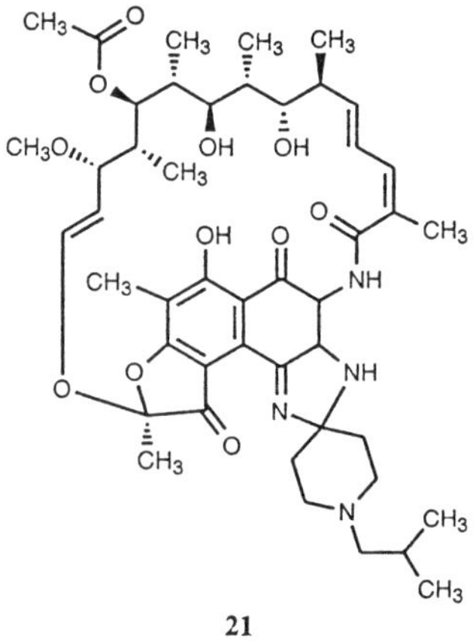

21

Figure 6.21 Rifabutin.

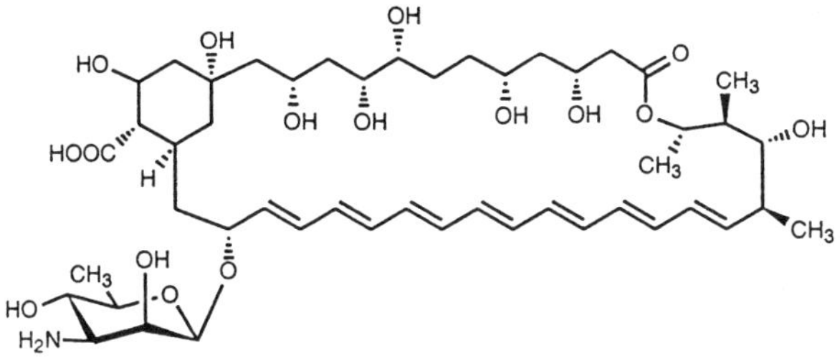

22

Figure 6.22 Amphotericin B.

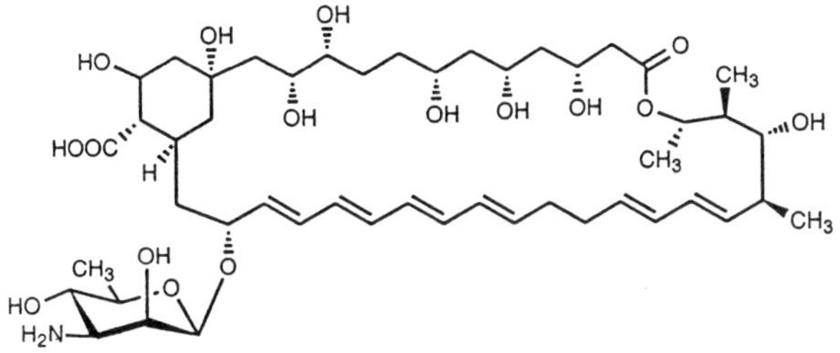

23

Figure 6.23 Nystatin.

unrelated in origin and composition to the macrolides and rifamycins, are important antimicrobials. The polyene family is exquisitely active against fungi. Amphotericin B **22** (Figure 6.22) and nystatin A_1 **23** (Figure 6.23) are landmark agents. In these compounds, the macrocyclic system is compromised of a lipophilic polyene portion complemented by an opposing hydrophilic region.

6.2 Mode of action of the non-peptidic antibacterials

6.2.1 The macrolide antibacterials: inhibition of bacterial protein synthesis

The antibacterial mode of action of the macrolides has been elucidated in considerable detail. The erythromycins have been the most extensively investigated of all macrolides to date. Although there are subtle differences between members, the macrolides essentially exert identical gross effects upon susceptible bacteria. For example, the macrolides are predominantly bacteriostatic although some important exceptions exist. Erythromycin is bactericidal towards some streptococci and *Cornyebacterium diphtheriae*.

In culture, some macrolides become bactericidal at concentrations five- to ten times greater than the bacteriostatic threshold, depending upon the particular species. Usually, drug concentrations of this magnitude are not normally obtainable *in vivo* and so the macrolides can be regarded as being primarily bacteriostatic in nature. In addition, the spectrum of activity among individual members is fairly uniform and in some cases, similar *in vitro* potency is observed. The macrolides are best described as possessing a 'medium' spectrum of activity which mostly encompasses certain important Gram positive bacteria.

The macrolides target the bacterial ribosome and cause a pronounced inhibition of bacterial protein biosynthesis. This phenomenon is supported by a number of observations. Growing bacteria normally experience a noticeable increase in cell mass as high molecular weight proteins are synthesized and incorporated into cellular components. However, in the presence of erythromycin and other macrolides, a decrease in culture cell mass, commonly associated with inhibition of protein biosynthesis, is apparent. The leucomycins, carbomycins, spiramycins, tylosins and other macrolides share this feature. Inhibition of bacterial protein biosynthesis has also been demonstrated in cell-free ribosomal preparations in which polynucleotide substrates are presented to the ribosome and the ensuing construction of peptidic substrates is monitored. In the presence of macrolides, fully elaborated peptides are not formed although small peptidic fragments can be produced (see below). (See chapter 4, section 4.2 for a discussion on bacterial protein biosynthesis.)

The nature of macrolide interaction with the bacterial ribosome is speculative although it is certain that macrolides must intimately bind to the ribosome in order for inhibition of protein biosynthesis to take place. Many macrolide molecules can associate with a bacterial ribosome but most interactions are weak, being reversible and easily removable upon dilution of drug or by cell-washing techniques. Upon saturation, bacterial ribosomes optimally accept only one molecule of macrolide each; under physiological conditions, less than one molecule of macrolide per bacterial ribosome is sufficient to inhibit protein biosynthesis.

A number of studies (e.g. with radiolabelled compounds) have unequivocally determined that the target of the macrolide antibacterials is the 50S ribosomal subunit, Figure 6.24). Covalent bond formation does not occur; release from the ribosome affords intact and structurally unaltered macrolide as well as full ribosomal function. Drug–ribosome binding probably encompasses a variety of electrostatic forces such as hydrogen bond interactions as well as lipophilic associations. Part of this hypothesis can be surmised from structural modifications of specific macrolides that produce a loss or decrease in binding affinity with the ribosome (section 6.3). In addition, macrolides (i.e. erythromycin) interfere with the binding of other antibacterial compounds such as the lincosamides (chapter 9) which are known to occupy an overlapping binding site on the bacterial ribosome.

The macrolides arrest protein biosynthesis at relatively early stages of the process, but do not necessarily cause an immediate shutdown of peptide formation. Depending upon the specific agent and bacterial species, the construction of small, ribosomally bound peptidic fragments may proceed but elaboration to completed protein precursors is not carried out. Individual macrolides differ in their tolerance towards peptide formation. Some macrolides actually enhance formation of ribosomally bound peptides that consist of only a few amino acid residues, but other congeners inhibit virtually any peptide bond formation. For example, the erythromycins facilitate the construction of some dipeptides and tripeptide chains; pronounced protein inhibition does not occur until the chain incorporates four or five amino acid residues. At the other end of this spectrum, the carbomycins and leucomycins essentially inhibit any elongation of the peptide, regardless of length. The spiramycins and some other macrolides fall in the middle of this spectrum; dipeptide formation is substrate dependent; some dipeptides are readily produced while others are not. A pronounced inhibition is operative when peptide formation has advanced to incorporate a third or fourth amino acid residue.

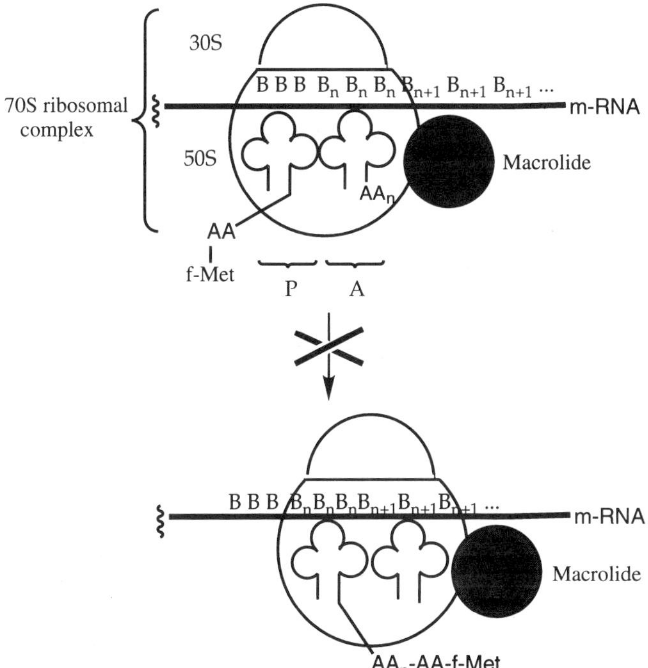

Figure 6.24 Macrolide inhibition of chain elongation.

As alluded to above, inhibition of bacterial protein biosynthesis is partially dependent upon the nucleotide template. This has been readily demonstrated in polynucleotide-dependent cell-free ribosomal systems. For example, the erythromycins preferentially inhibit processing of polycytosine and polyadenine substrates (preventing polyproline and polylysine peptide formation, respectively) but polyphenylalanine formation from polyuridine substrates is unaffected. Under similar conditions, the carbomycins and spiramycins halt polyphenylalanine production from polyuridine substrates.

Ribosomally bound macrolide disrupts chain elongation by inhibiting the formation of new peptide bonds (Figure 6.25). Specifically, the macrolides target peptidyl transferase, the ribosomal component responsible for joining each amino acid unit on to the peptide chain. Recall (chapter 4, section 4.2) that peptidyl transferase action is provided by the 50S ribosomal subunit, and that catalytic activity requires potassium and ammonium cations. The macrolides have been shown to require these same cations for optimal activity and a direct involvement of macrolides upon peptidyl transferase activity has been inferred. Furthermore, the macrolides bind to the 50S ribosomal subunit in a region that is proximal to

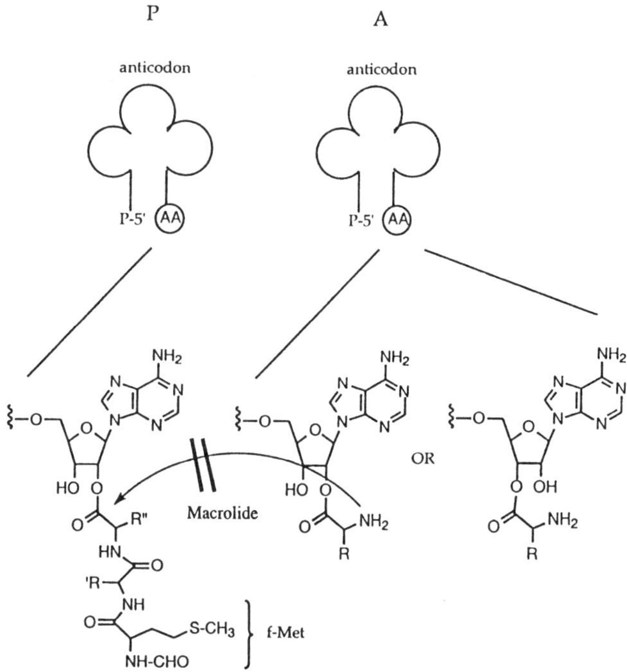

Figure 6.25 Macrolide inhibition of peptidyl transferase activity.

the peptidyl tranferase center. The loss of peptidyl transferase activity does not affect binding of aminoacyl-t-RNA molecules to the ribosomal sites; rather, the AA-t-RNA in the acceptor site is unable to react with the ribosomally bound peptide chain (P site).

In addition to inhibiting peptidyl transferase activity, some macrolides are able to affect the movement of the ribosome along the m-RNA stand. If a peptide is to be fully constructed, the ribosome must be able to move along the m-RNA template in order properly to read each and every codon (chapter 4, section 4.2). This operation, ribosomal translocation, is intimately associated with the release of the 'spent' (deacylated) t-RNA molecule from the ribosome. Some macrolides appear to be able to hinder the release of t-RNA from the ribosome and in doing so, translocation is inhibited. Consequently, the ribosome is unable to continue protein biosynthesis since the acceptor site is not vacant and cannot bind another AA-t-RNA. Since translocation occurs after peptide bond formation (peptidyl transferase), it is inherently more difficult to evaluate the significance of this phenomenon; inhibition of translocation is likely to be an important secondary effect that is probably not operative with all macrolides.

Both inhibition of peptidyl transferase activity and ribosomal translocation are attributed to binding to the 50S subunit. The macrolides do not target the 30S ribosomal subunit whether complexed to the 50S subunit (thereby as the 70S complex), nor when free from the 50S entity. Recall (chapter 4, section 4.2) that it is the 30S ribosomal subunit that binds to the m-RNA molecule region containing the start codon. The macrolides do not disrupt this association and so initiation of protein biosynthesis is able to proceed; the attachment of formylmethionyl-t-RNA to the 30S m-RNA complex at the P site is unencumbered. Formation of the 70S complex is also not affected by the macrolides and the acceptor site is accommodating to its first aminoacylated t-RNA substrate. Since small peptides are elaborated by the ribosome in the presence of macrolides, codon–anticodon recognition is not perturbed. Furthermore, the macrolides do not disrupt cellular supplies of amino acids nor do they inhibit the enzymes responsible for the formation of the aminoacylated t-RNA molecules. Lastly, the actions of various initiation, elongation and termination factors are also not directly affected by the macrolides.

6.2.2   *The rifamycins: inhibition of bacterial RNA polymerase*

The rifamycins are the most important antibacterial ansamycins at this time. Their site of action is upstream from protein biosynthesis and therefore the rifamycins are unrelated to the macrolides both in structure and mode of action. The rifamycins are somewhat unique among bacterial agents since they do not target the bacterial ribosome, cell wall or metabolic enzymatic pathways. Rather the rifamycins act at the level of genetic transcription although their mechanism of action is primarily confined to inhibition of a key enzyme in this process. For this reason, a brief overview on transcription follows before more details are presented.

6.2.2.1 Transcription. The transfer of genetic information from DNA to RNA is a central doctrine in fundamental molecular biology. Transcription, an early stage in gene expression, is the process in which genetic data from a region (gene) of a DNA molecule is transcribed, or transferred, to a m-RNA molecule. Once transcription is completed, the genetic information contained within the m-RNA template can be further processed by the ribosome into functional protein. Since protein is not directly synthesized from DNA, transcription is vital for cell survival.

The enzymes responsible for transcription are the DNA-dependent RNA polymerases. These proteins catalyse the formation of RNA from a DNA template and a pool of phosphorylated ribonucleotides (NTPs). The first step is the binding of the RNA polymerase enzyme to the DNA molecule. After the initial DNA–RNA polymerase complex is formed, the first ribonucleotide triphosphate is able to associate with the complex and initiate the synthesis of the RNA chain in a 5′ to 3′ direction. The chain is subsequently elongated by the sequential addition of single ribonucleotide triphosphates as encoded for by the DNA template (Figure 6.26). Upon completion of the polyribonucleotide, chain termination occurs and the completed RNA chain is released from the DNA.

RNA polymerases are rather ubiquitous enzymes that are present in both bacterial and mammalian cells. In general, the prokaryotic enzymes differ significantly from the eukaryotic proteins in homology, composition, size and structure. The rifamycins selectively target the bacterial RNA polymerases (Figure 6.27) and at low concentrations eukaryotic cells are usually unaffected. This allows the rifamycins often to display favorable therapeutic profiles; inhibition of mammalian RNA polymerase activity would impose grave consequences on the host.

The rifamycins interact with DNA-dependent RNA polymerase; one molecule of rifamycin directly binds with high affinity of the β-subunit of the protein. This results in loss of enzyme function and inhibition of RNA biosynthesis. Interestingly, the rifamycins are unable to stop polymerases actively involved in chain elongation. However, they readily thwart chain initiation that starts the construction of the RNA chain. Complete RNA biosynthesis involves several RNA polymerases acting on a single DNA template, and so initiation of nucleotide polymerization, as well as nucleotide chain elongation, must be repeated several times in order to produce a fully elaborated RNA product. Since the rifamycins are potent inhibitors of the initiation of ribonucleotide chain formation, these compounds are able to halt RNA synthesis and induce bactericidal effects.

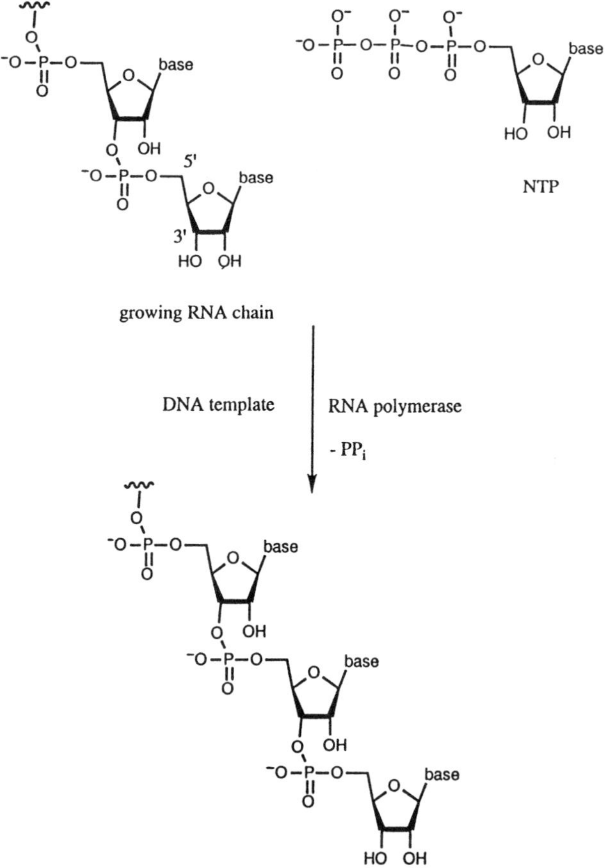

Figure 6.26 The action of (DNA-dependent) RNA polymerase.

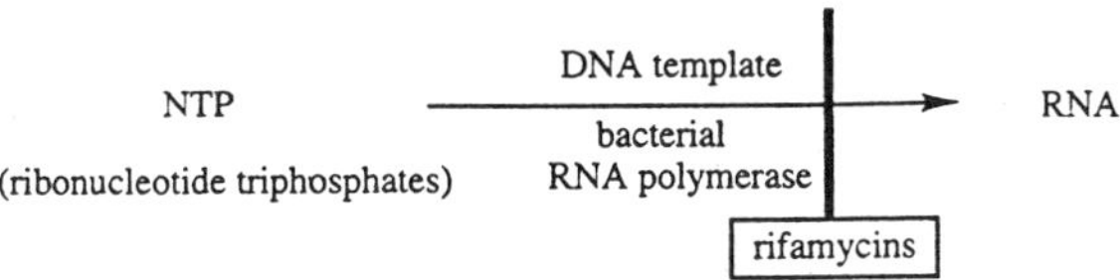

Figure 6.27 Action of the rifamycins.

6.3 Structural features and structure–activity relationships of the macrocyclic antibacterials

6.3.1 The macrocycles and sugar substituents

The macrolides and rifamycins both have non-peptidic macrocyclic frameworks that are vital for antibacterial activity. In each family, cleavage of the macrocycle results in loss of activity; however, the macrocycle alone does not impart antibacterial properties. A number of key substituents that extend from the macrocyclic core are also needed.

The macrolide antibacterials are characterized by the presence of a large lactone ring. The naturally occurring members, as well as many semi-synthetic analogs, are 12-membered (methymycin), 14-membered (erythromycins and derivatives) or 16-membered (leucomycin-related, tylosin-related) lactones. The azalides (e.g. azithromycin) are semi-synthetic 15-membered congeners in which a nitrogen atom has been introduced to expand a 14-membered precursor. Since ring size is variable among the antibacterial macrolides, the macrocycle probably serves as scaffolding for key functionality that is needed for ribosomal recognition and binding.

In addition to the cyclic system, nearly all macrolides contain a neutral sugar and a basic aminosugar. The aminosugar is absolutely essential for antibacterial activity while the neutral sugar can be removed and appreciable activity retained. The 14-membered macrolides and the 15-membered azalides have the neutral sugar attached at the aglycone C-3 position, and the basic aminosugar is joined to the C-5 position. The 16-membered macrolides usually have a disaccharide group extending from the aglycone C-5 position. The neutral sugar is frequently acylated and joined to the aminosugar through a glycosidic linkage; the aminosugar is in turn connected to the aglycone. Some 16-membered macrolides have an additional sugar substituent. For example, the spiramycins have an aminosugar connected to the C-9 hydroxyl group and the tylosins have a glycosylated C-14 hydroxymethyl substituent. Smaller macrolides such as the 12-membered methymycins, usually contain only the basic aminosugar residue at the C-3 position (Figure 6.28).

All macrolides contain alkyl and hydroxyl groups that arise from biogenesis and often a ketone functionality is present. The 16-membered macrolides have a distinguishing aldehyde group extending from the C-6' position of the aglycone. Most of these groups cannot be chemically introduced, nor replaced, unless extensive synthetic approaches are undertaken. However, the total synthesis (section 6.4) of antibacterial macrolides is not a practical alternative at this time, and all marketed macrolide agents originate from microbial sources. For this reason, the naturally occurring macrolide systems are cautiously regarded as being vital features, but are essentially unexplored.

6.3.2 Modifications of hydroxyl group substituents

Despite their structural complexity, the naturally occurring macrolides can be chemically modified and some of these approaches have afforded valuable compounds. The esterification or acylation of hydroxyl groups is illustrative; many semi-synthetic and naturally occurring congeners are

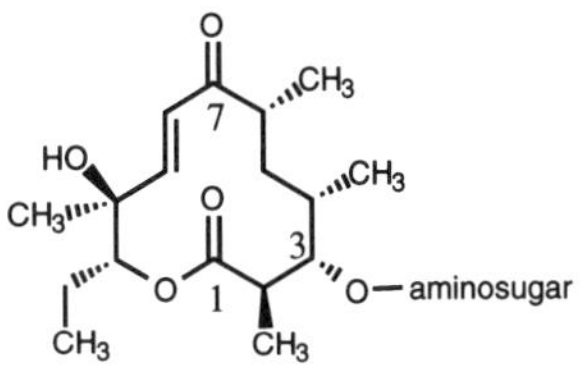

12-membered lactone

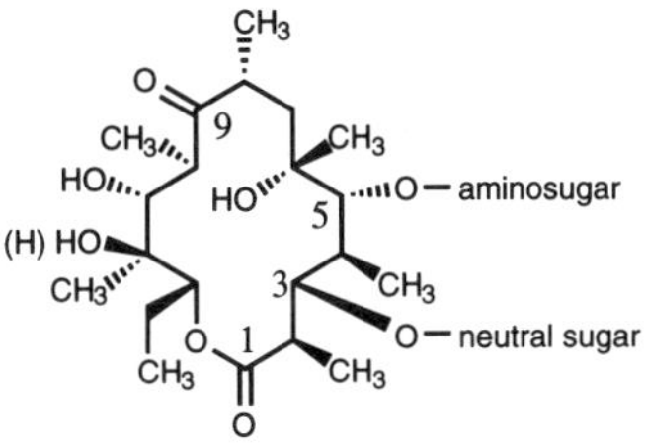

14-membered lactone

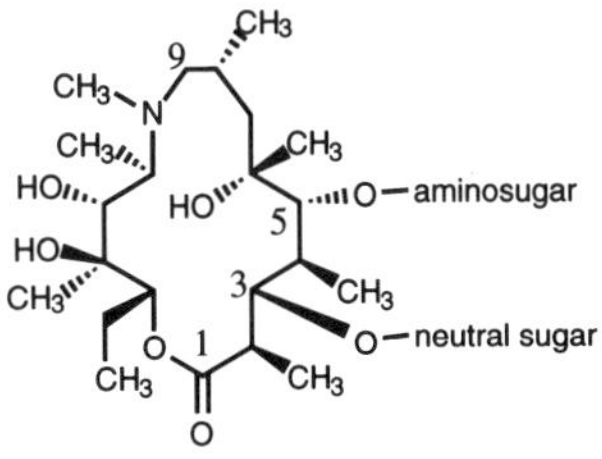

15-membered lactone

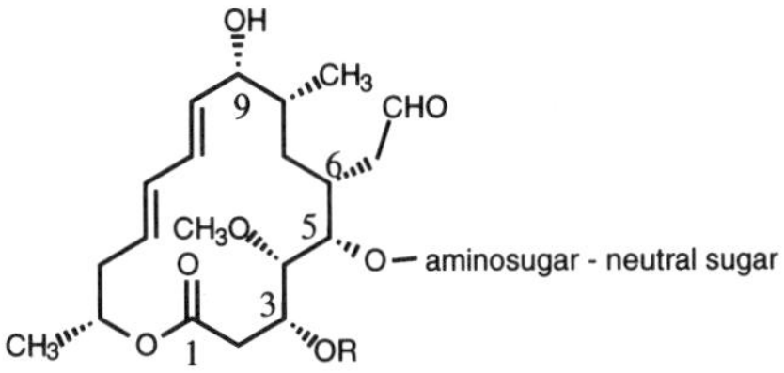

16-membered lactone

Figure 6.28 Macrolides: basic structural features.

acyl or ester derivatives of other macrolides. The erythromycins have been synthetically esterified at the C-11 hydroxyl position, the C-4″ cladinose hydroxyl group and the C-2′ desosamine moiety. In general, small alkyl esters are preferred as prodrugs; long chain alkyl esters or the bulkier aromatic esters are generally less favorable. Esterification of the C-2′ hydroxyl group is selective under appropriate reaction conditions; the neighboring dimethylamino group directs esterification in the absence of base. Many erythromycin C-2′ esters are important clinical (section 6.1) or preclinical agents. Multiple ester groups can be introduced and if labile, can act as macrolide prodrugs. Triacetyloleandomycin **24** (Figure 6.29) is an oleandomycin derivative that is acetylated at the macrolide C-11 hydroxy position as well as at both hydroxyl moieties on each sugar group. It has achieved clinical success despite being metabolized to a variety of deacylated derivatives.

This trend carries over to the 16-membered macrolides. Many leucomycin-related macrolides vary merely by the length of the terminal acyl group at the C-4″ hydroxyl functionality (neutral sugar) or by whether or not the aglycone C-3 hydroxyl group and the aminosugar C-2′ hydroxyl group is acylated. There are many examples, encompassing both naturally occurring and semi-synthetic products. Josamycin is a C-3 acylated leucomycin, and other members of the leucomycin complex are distinguished by having either a butanoate or isovaleroate side chain at C-4″.

Alternatively, those 16-membered macrolides that contain free hydroxyl functionality can be chemically acylated. Acetylspiramycin and diacetylated midecamycin are among the most promising semi-synthetic derivatives (section 6.6).

Other changes can be detrimental, particularly if a hydroxyl moiety that plays an intimate role in ribosome binding is modified in such a way that the parent can not be regenerated (*in vivo*). For example, alkylation (e.g. methylation) of the erythromycin C-11 hydroxyl group results in decreased antibacterial activity. Similarly, alkylation of sugar hydroxyl functionality is not tolerated. Interestingly, methylation of the erythromycin C-6 hydroxyl group is acceptable (e.g. clarithromycin), apparently since such an alteration has little effect on ribosomal binding.

One model of the erythromycin–ribosome complex implicates a binding motif in which both hydrophilic and hydrophobic portions of the macrolide are involved. The hypothesis is that the hydrophilic groups of the macrolide, namely the lactone carbonyl, the ketone carbonyl and the C-11 and C-12 hydroxyl groups of the macrolide, orient to form a hydrophilic

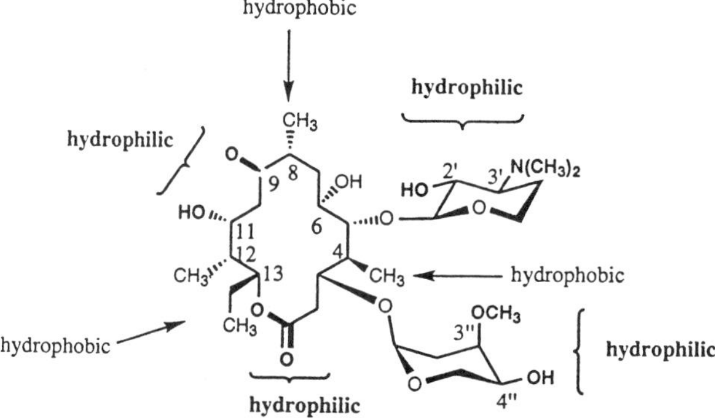

Figure 6.29 Oleandomycin triacetate.

Figure 6.30 Macrolide substituents implicated in ribosomal binding.

domain that establishes important hydrogen bonding interactions. The C-2′ hydroxyl group and the C-3′ dimethylamino substituents of the desosamine unit, along with the C-3″ methoxy group of the cladinose sugar, are believed to form hydrogen bonds with specific nitrogenous bases of the ribosomal (RNA) nucleotides. A consequence of this binding configuration is that the hydrophobic macrolide substituents, namely the C-4, C-8 and C-12 methyl groups and the C-13 ethyl substituent are all aligned away in a lipophilic pocket probably stabilized by the ribosome. These key groups are summarized below (Figure 6.30). Although first postulated for the erythromycins, there is evidence to suggest that similar interactions are operative in binding other macrolides to the bacterial ribosome.

6.3.3 Modification of the erythromycin C-9 ketone functionality

Some essential structural features of the macrolides have also been inferred from the behavior of macrolides under varying pH environments. The instability of the erythromycins in acidic media has been thoroughly studied; the main pathway is shown below (Figure 6.31). The C-6 hydroxyl group reversibly attacks the C-9 ketone giving rise to a hemiketal intermediate. Dehydration prevents regeneration of the parent erythromycin and the C-12 hydroxy group can subsequently add to produce the spiroketal species. The cladinose sugar group is cleaved from the macrocycle and more harsh conditions lead to the release of desosamine. Useful antibacterial activity is lost upon dehydration of the hemiketal and the spiroketal is also only very weakly active.

As alluded to earlier (section 6.1), a number of strategies have been utilized to improve the acid sensitivity of the erythromycins (i.e. erythromycin A). The past decade has been teeming with attempts to modify the portions of the erythromycin molecule that are responsible for degradation. Since this area remains such an important one in macrolide research, some additional comments are warranted. The (erythromycin) C-9 ketone moiety has been extensively modified. This group is attractive from a chemical standpoint since ketones are versatile centers that can be elaborated to a variety of derivatives, often under conditions inert to hydroxyl groups and other functionality present in the erythromycins.

Since the C-9 center can ultimately undergo bond formation with the neighboring C-5 and C-12 hydroxyl groups (Figure 6.31), attempts to reduce or eliminate the electrophilicity of the ketone have been explored. In this regard, the C-9 ketone has been subjected to four general types of modification.

The first approach has been reduction of the C-9 ketone which eliminates the electrophilicity of this center; however, the corresponding 9-dihydroerythromycins are only weakly active. The second strategy has

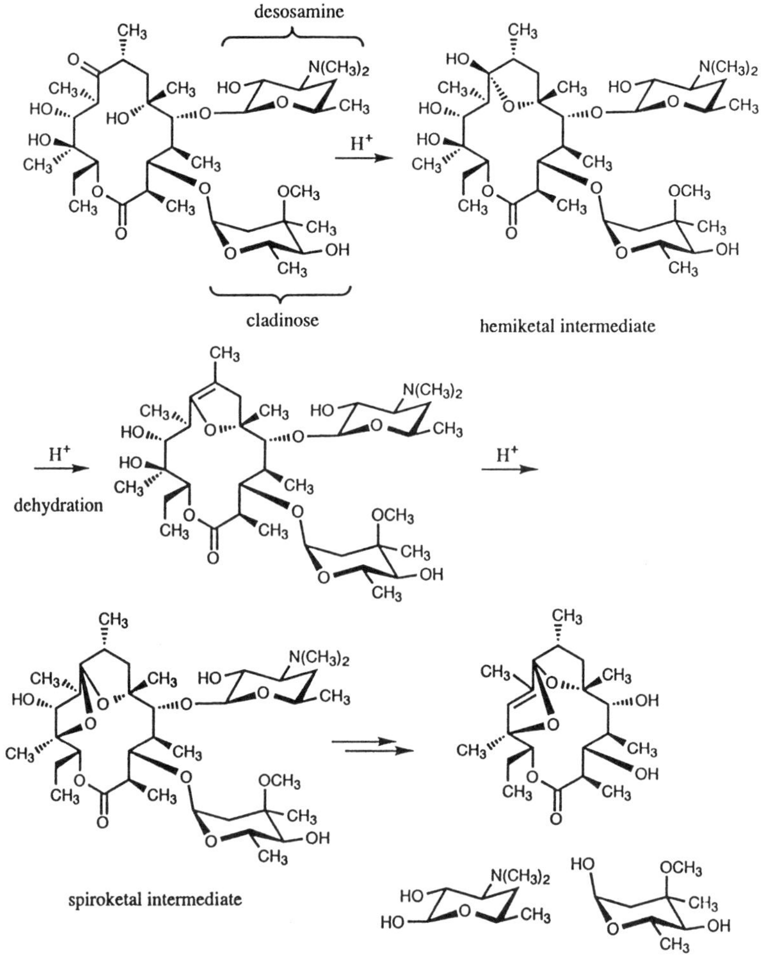

Figure 6.31 Acidic degradation of erythromycin.

involved the addition of an amine to the ketone in order to introduce novel substituents joined at a C-9 sp^2-hybridized carbon center. In this regard, the simple erythromycyl-C-9 oxime (product of hydroxylamine addition) had generated interest, but is now best regarded as a synthetic intermediate. However, a number of other oxime derivatives have been more promising. For example, roxithromycin **7** (Figure 6.7) is a prototypic oxime derivative, being both potent and acid stable; C-9 oximination continues to be explored.

The third approach has been the replacement of the C-9 ketone with a substituent that is joined to an sp^3-hybridized C-9 carbon center. The introduction of an amine functionality at the C-9 position has been particularly rewarding. For example, the C-9 aminoerythromycin derivative, 9(*S*)-erythromycylamine **5** (Figure 6.5), is nearly as active as erythromycin A *in vitro*, and is also a valuable semi-synthetic intermediate. The C-9 amine group readily undergoes condensation with aliphatic and aromatic aldehydes to give a host of corresponding Schiff bases. These derivatives are generally inferior in terms of *in vitro* potency compared to the parent erythromycin (A), but can display good *in vivo* activities. The erythromycyl amines can also be acylated or alkylated, or condensed with alkylhaloformates to give carbamates. Interestingly, some 9(*R*)-dialkylaminoerythromycin derivatives have been reported to exhibit antibacterial properties comparable with erythromycin A, but to date none of these derivatives have been extensively developed. An extremely promising variation on this theme has been the erythromycin-derived C-9,C-11 oxazines. These congeners incorporate an oxazine bridge that joins the C-11 hydroxyl functionality to a C-9 amine group through hemiaminal formation. Dirithromycin **6** (Figure 6.6) is the prototype but other oxazine derivatives appear to be promising.

The fourth approach has been aimed at replacing acidic hydrogen atoms alpha to the ketone with inert substituents. The rationale is that the removal of acidic hydrogen atom(s) will prevent the ensuing dehydration step (Figure 6.31), the stage at which regeneration of the active erythromycin is no longer possible (energetically disfavored). Halogenation alpha to the ketone group has been particularly fruitful; if the halogen substituent is properly disposed, the loss of water is not possible. Flurithomycin, 8-fluoroerythromycin, is the most promising congener (section 6.6) and is undergoing clinical evaluation.

Alternatively, alterations of the C-6 hydroxyl group, the nucleophilic functionality which initiates erythromycin degradation, have been productive. Modifications that remove the nucleophilic nature of this hydroxyl group can retain excellent antibacterial properties if the size of the group is kept small so as to not affect ribosomal binding. As mentioned above, clarithromycin **8** (Figure 6.8) is the prototype.

6.3.4 *Modifications of the sugars*

The hydrogen bonding motifs of the macrolide sugar units appear to be rather exquisite requirements for optimal ribosomal binding. Most of the research in this area has focused upon the desosamine and cladinose sugars of erythromycin, although many trends have carried over to the 16-membered macrolides. In this regard, the aminosugar unit is important in all macrolides; the C-3'-dimethylamino moiety plays an important role in ribosomal binding. For example, in the erythromycin series, oxidation of the desosamine dimethylamino moiety to the corresponding *N*-oxide destroys appreciable activity. Dialkylation of the amine (via Hoffmann elimination and subsequent reduction) to the didemethyldesosamine derivative also reduces activity. Monoalkylated amine substituents as well as acetylated derivatives also afford ineffectual antibacterial compounds. These results clearly demonstrate the significance of the dimethylamino group; basicity is also a substantial factor since the simple amino and monoalkylated or monoacetylated amino derivatives are unfavorable.

Other groups that extend from the saccharide are preferred for optimal activity in the erythromycin series. For example, the C-3″ methoxy group of the cladinose sugar moiety is significantly more active than the corresponding free hydroxy derivative. The conversion of the C-4″ hydroxyl group to various derivatives such as the corresponding amino analog as well as the O-sulfonyl or oxime derivatives, affords active compounds, but none which rival the parent erythromycin.

6.3.5   *Other modifications of 16-membered macrolides*

The structural features needed for optimal antibacterial activity of the 16-membered macrolides, in many ways parallels those issues addressed above for the 14-membered relatives (pertinent examples have been mentioned above). One of the most obvious structural components that distinguishes the 16-membered macrolides is the presence of an aldehyde moiety extending from the C-6′ position of the macrocycle. This feature has been tolerant of few changes to date. For example, reductive amination, reduction to the corresponding alcohol or oxidation to the carboxylic acid, all lead to inferior agents. A host of Schiff bases derived from the aldehyde have been prepared and investigated but are also unfavorable in terms of potency. In some instances, antibacterial activity comparable to the parent macrolide has been observed, but this has been attributed to the generation of the free aldehyde.

The most notable finding has been the discovery that the C-6 center can be extended to incorporate dialkylamino groups that retain good antibacterial activity. Tilmicosin **25** (Figure 6.32) is a new veterinary macrolide that contains a piperidinylethyl C-6 substituent.

The 16-membered macrolides also contain a region of carbon–carbon conjugated double bonds which is in contrast to the smaller macrolides. Reduction of the double bond(s) affords compounds with decreased ribosomal binding and poor or negligible antibacterial activity. It appears that these double bonds are important in restricting the conformation of the macrolide to provide for an optimum fit into the ribosomal pocket.

6.3.6   *Structural features and issues of structure–activity relationships of the rifamycin ansamycins*

The rifamycins, like the macrolide antibacterials, are complex in structure (Figure 6.33) and as a result have been limited in terms of structural

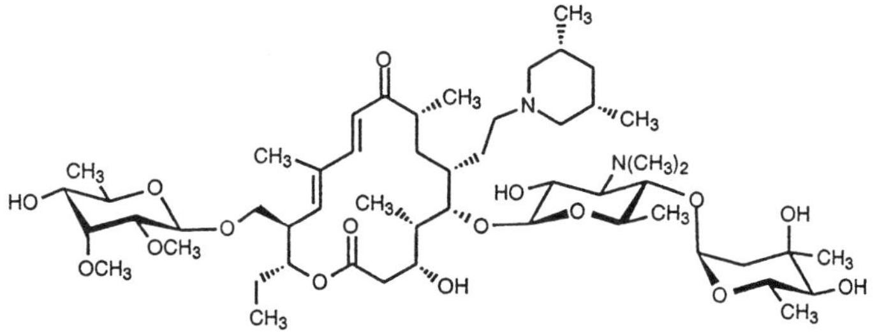

25

Figure 6.32 Tilmicosin.

modifications. Since the rifamycins are obtained as naturally occurring products or are elaborated in semi-synthetic fashion, many centers and positions of the macrocycle framework have not been extensively modified. The central structural feature is a naphthalene-derived nucleus (rings A and B) and a long tether (C-16 to C-28) that spans the aromatic nucleus. The aromatic system can be oxidized to a naphthoquinone such as that found in S rifamycin(s). Other important rifamycins contain non-natural substituents that extend from the naphthalene ring system; an additional ring can be present. The streptovaricins (Figure 6.33) represent yet another important member of the ansamycin family. In these compounds the B ring exists in an oxidized state as a *para*-quinone methide species. Some streptovaricins have undergone extensive clinical evaluation in Europe.

The hydrocarbon tethers of the rifamycins and related ansamycins contain a locus of repeating alkyl and hydroxy-based functionality that is attached to the aromatic nucleus at distal positions. Interestingly, the tether length varies among the rifamycins, the streptovaricins (and geldamycin) and other ansamycins. In general, there are three distinct regions that make up the tether (Figure 6.33). The first is a hydrocarbon region consisting of two conjugated double bonds that extend from the lone amide center (C-15). This section of the tether appears to be important in properly constraining the macrocycle so that a tight fit into the RNA polymerase enzyme is possible. The second region is a short segment in which hydroxyl groups are embedded within a stretch of alternating methyl substituents. The hydroxyl groups probably contribute to enzyme binding through hydrogen bonding forces, and the methyl substituents may form an opposing lipophilic side of the tether that is stabilized by a complementary portion of the enzyme. The third region is a short segment which includes a carbon–carbon double bond that extends the tether to the aromatic nucleus.

The most promising new rifamycins have resulted from modifications at

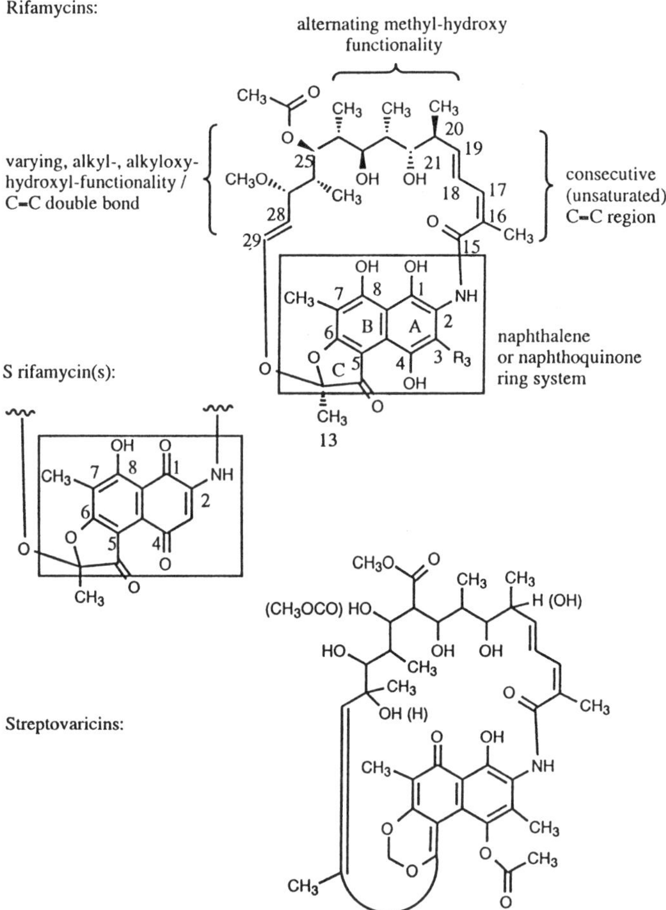

Figure 6.33 Rifamycins and streptovaricins: basic structural features.

the C-3 and the C-4 positions of the aromatic naphthalene or naphthoquinone nucleus. Such compounds are accessible by semi-synthetic manipulation of naturally occurring prototypes. There have been a host of C-3 and/or C-4 variants that have generated interest in recent years. The C-3 position, in particular, appears to be tolerant of various alterations while maintaining potent activity against bacterial RNA polymerase. Unfortunately, antibacterial activity can be adversely affected in some cases primarily due to cell impermeability. Several compounds are undergoing clinical evaluation (section 6.6).

In contrast, the reduction of the double bonds to saturated analogs has produced inferior antibacterial compounds; in simple terms, the greater the degree of saturation (i.e. conversion of carbon–carbon double bonds to carbon–carbon single bonds), the greater the loss of RNA polymerase inhibition. These observations indicate that the C-16–C17, C-18–C-19 and perhaps the C-28–C-29 carbon–carbon double bonds collectively play a role in binding to the RNA polymerase enzyme. The hydroxyl functionality at the C-21 and C-23 centers of the rifamycins (and perhaps the C-20 position of the streptovaricins) also appears to be crucial. For example, oxidation of the C-21 hydroxyl group to the ketone moiety abolishes activity. The (other) hydroxyl groups have also been implicated in hydrogen bonding motifs that contribute to enzyme binding.

6.4   Synthetic approaches to the non-peptidic macrocyclic antibacterials

All antibacterial macrolides in use today, or under clinical investigation, are derived from naturally occurring products. Many macrolides are marketed without chemical modification; this is particularly true with the 16-membered macrolides, although recently a number of superb semi-synthetic analogs have been synthesized. The semi-synthetic macrolides typically embody a single chemical transformation such as esterification or acylation (erythromycin C-2′ esters, leucomycin-related derivatives), alkylation (clarithromycin), modification of the erythromycin ketone system (roxithromycin, erythromycylamine, dirithromycin, flurithromycin) or expansion of the macrocycle (the azalides). This chemistry can often be carried out in a relatively straightforward manner using modern protocols; some pertinent issues have already been discussed but more detail will be presented in this section.

In contrast, the total synthesis of the macrolides remains mostly of academic interest; nevertheless, some approaches are certainly worth considering since sophisticated organic chemistry is showcased, and some work has influenced the development of new congeners to this day.

6.4.1   Classical strategies to macrolides

The total synthesis of the antibacterial macrolides has proven to be one of the most challenging tasks in organic chemistry. The macrocyclic framework contains an imposing array of stereocenters that embrace a variety of functionality. In addition, there are sugar substituents that extend from the macrocycle through glycosidic linkages and which possess additional stereocenters and functionality. For example, the prototypic erythromycin macrolides have ten asymmetric carbon centers embodied in the macro-cycle, along with two distinct sugar substituents that contain six more asymmetric centers and two glycosidic linkages.

Synthetic strategies must address three key issues. The first challenge has been to develop a methodology to construct the macrocycle in a way that allows for the stereocenters to be set and embellished with the appropriate alkyl and hydroxyl group substituents. In this regard, synthetic approaches have been cumbersome compared to the biogenesis of the macrolides; microbial organisms utilize simple acetate, propionate and butyrate materials. Traditionally, it has been easier to piece together fragments that contain a collection of preset stereocenters rather than attempt to set each position sequentially. Historically, optically active fragments were derived from suitable chiral molecules such as the Prelog–Djerassi lactone (Figure 6.34). Contemporary chemistry includes a

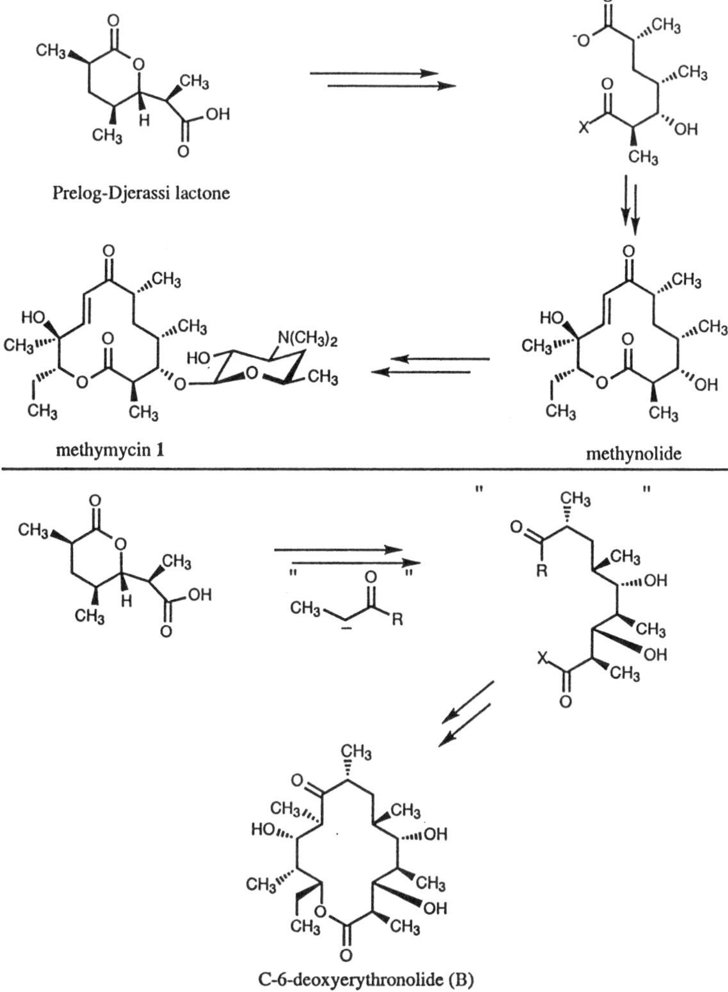

Figure 6.34 Prelog–Djerassi lactone as precursor to macrolides.

number of methods that create carbon centers with a high degree of stereospecificity; these advances have been more recently applied to the synthesis of macrolides. Such techniques have typically included (boron) enolate additions to optically active aldehydes (Figure 6.35). Other approaches have constructed acyclic macrolide precursors containing stereocenters derived from fused six-membered (Figure 6.36) or seven-membered (Figure 6.42) systems via ring-opening processes.

The second challenge facing total synthetic approaches to the macrolides is closure of a (usually) highly functionalized precursor to a macrolactone. Cyclizations to form large rings are frequently plagued by competing intermolecular reactions. Fortunately, a number of ways have been developed that overcome the reluctant nature of (ω)-hydroxycarboxylic acids towards intramolecular cyclization. Some of these macrolactonization

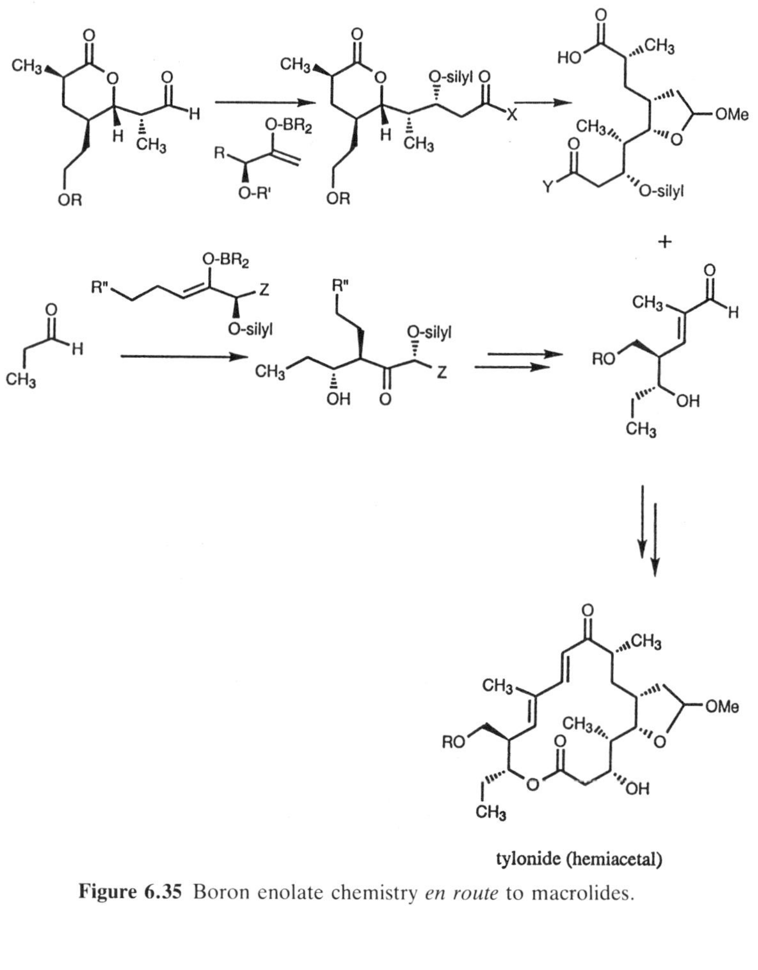

tylonide (hemiacetal)

Figure 6.35 Boron enolate chemistry *en route* to macrolides.

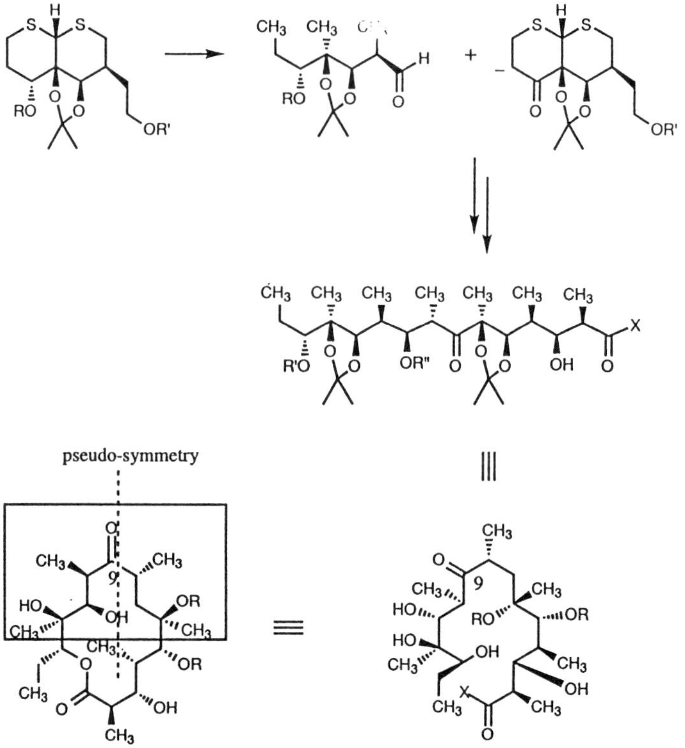

Figure 6.36 Macrolide precursors from six-membered templates.

protocols are summarized below (Figure 6.37). Typically, high-dilution techniques are used to maintain low concentrations of the acyclic precursor so as to minimize competing intermolecular reactions. The carboxylic acid is usually activated to facilitate the attack of the hydroxyl functionality; in this regard, conversion to a thioester, pyridinium ester or to an anhydride, is noteworthy.

There are several macrolactonization protocols that are particularly attractive in the context of a total synthesis of a macrolide. One procedure involves formation of an anhydride; those best suited for cyclization to macrolide systems contain electron-withdrawing substituents that increase the tendency towards ring closure (Figure 6.37). For example, the derivatization of ω-hydroxycarboxylic acids as the corresponding 2,4,6-trichlorobenzyl anhydride species has been extensively used. An alternative protocol involves activation as a thioester; in the presence of thiophiles such as mercury(II), silver(I) or trivalent phosphorous species, the electrophilicity of the carbonyl center is increased and macrocyclization is usually facile (Figure 6.37). Thioesters are reluctant towards cyclization in the absence of thiophilic additives. The thioester precursors are usually obtained by the reaction of a thallium thiolate with an acyl halide (usually chloride). Alternatively, the carboxylic acid can be converted to an acyl phosphonate and subsequently reacted with a thallium thiolate. Thallium thiolates are particularly amenable to macrolide synthesis since basicity is decreased compared to conventional sodium thiolates; therefore sensitive functionality is more likely to be tolerated.

One of the most successful macrocyclization protocols is the Mukaiyama reaction which is widely regarded as a premier technology for the construction of large ring organic substrates. Reaction of a hydroxy-carboxylic acid with a chloropyridinium salt affords the corresponding

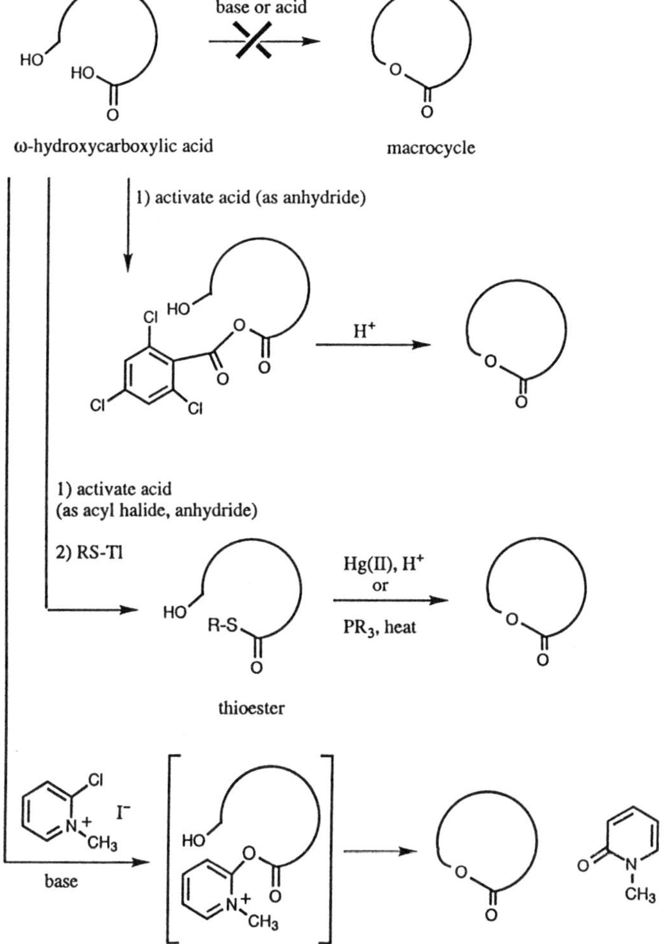

Figure 6.37 Macrolactonization protocols.

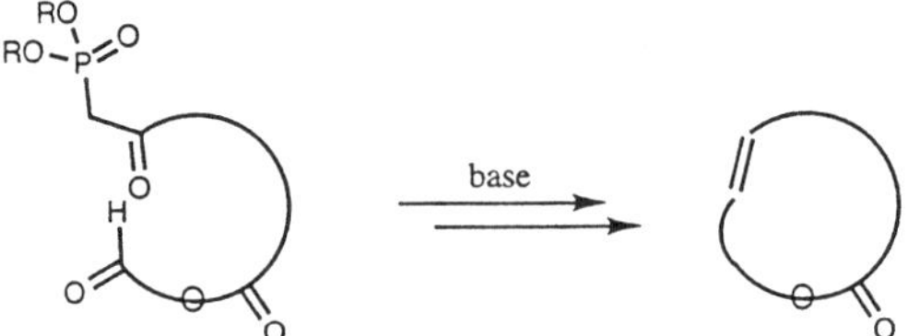

Figure 6.38 Horner–Emmons type macrocyclization.

pyridinium ester. The carbonyl center is activated for attack by the terminal hydroxyl group due to the adjacent positive charge and the propensity to release a non-charged and highly stabilized pyridone species. Macrocyclizations of this type occur under mild conditions and are particularly well suited for macrolide syntheses (Figure 6.37).

Lastly, some macrolides have also been obtained using other intra-molecular processes such as Horner–Emmons type macrocyclizations (Figure 6.38). This approach has been particularly useful for the construction of 16-membered macrolides since endocyclic double bond(s) are formed.

The final synthetic issue that has complicated the total synthesis of the antibacterial macrolides has been the attachment of the sugar moieties to the aglycone ring. Since glycosidic bond formation is required, the sugar units are usually incorporated at the latter stages of the approach. In general, reactive functionality is first masked and the sugar subsequently activated towards glycosidic bond formation. A number of strategies pertaining to sugar chemistry and glycoside bond formation were outlined in chapter 5 (section 5.4, the aminoglycosides); some of this chemistry is revisited briefly in presenting total synthesis of macrolides.

The 12-membered macrolides serve as an appropriate starting point; the prototype, methymycin **1** (Figure 6.1), is among the simpler of the macrolides in terms of structure and functionality. The Masamune synthesis of methymycin, as outlined below (Figure 6.39), is classic and illustrates many of the issues that were previously introduced. Specifically, the Prelog–Djerassi lactone (A) ultimately provided the C-1–C-7 segment of the macrocyclic framework. The remaining portion of the macrocycle was derived from an optically active epoxy aldehyde and was introduced via a Wittig reaction which afforded the α,β-epoxyenone (C). Epoxide opening yielded the ω-hydroxythioester which was cyclized (D) via a mercury(II)-assisted lactonization with concomitant loss of thiolate. Lastly, the desosamine sugar was attached onto the C-3 hydroxyl group via reaction of the aglycone with the corresponding *O*-acetylated bromosugar.

The use of protecting group chemistry in this synthesis is technically simple but the rationale is not necessarily straightforward. In the case of

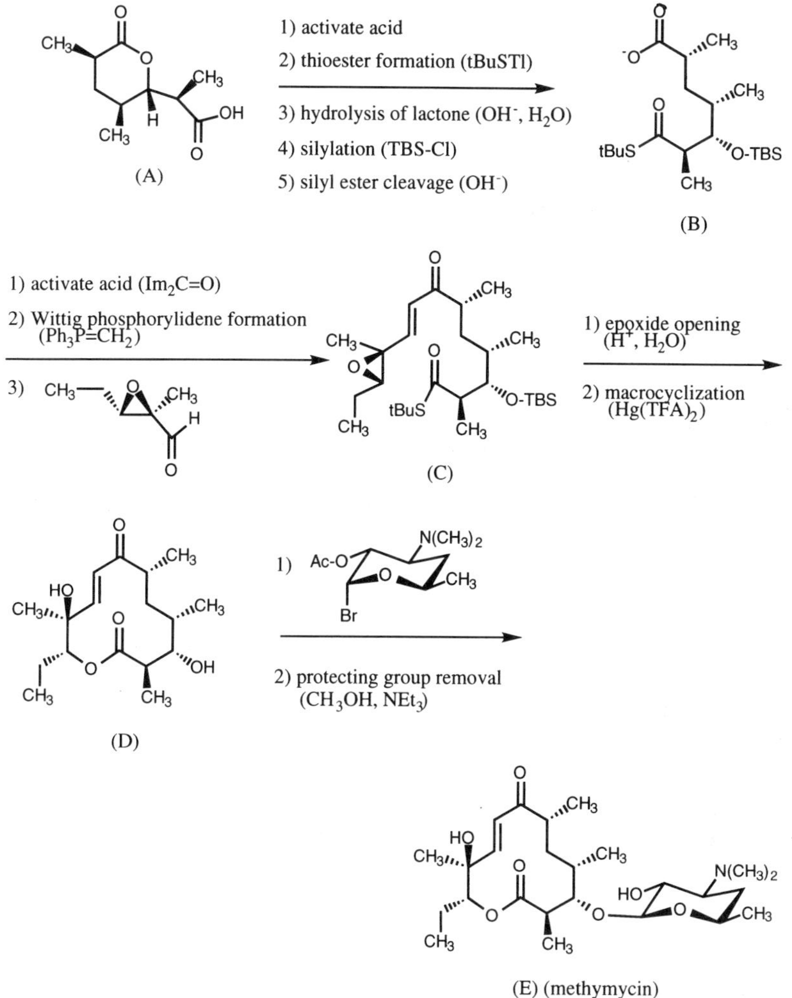

Figure 6.39 Masamune synthesis of methymycin.

methymycin, the reactive C-3 hydroxy functionality is masked as a silyl ether so as to isolate the carboxylic acid for activation and subsequent conversion to the stabilized Wittig phosphoranylidene. In addition, this protecting group also prevents undesired reactions such as β-elimination during other operations including macrocyclization. The C-10 hydroxyl group is less problematic and does not require protection; it is less reactive than the C-3 hydroxyl group since it is tertiary and not adjacent to acidic hydrogen atoms so β-elimination is not possible.

An interesting synthetic variation (Figure 6.40) used an alkyne fragment

(G) that was added to the C-7 center in order to introduce the C-8–C-12 segment of the macrolide. In this example, the Prelog–Djerassi lactone (A) was opened to the corresponding carboxylic acid and then activated by conversion to the corresponding acyl bromide (F). The presence of the carbon–carbon triple bond did not hamper subsequent macrolactonization (I) which was accomplished through the intermediacy of the trichlorobenzyl anhydride. Oxidation furnished the ynone that was reduced to the *trans*-double bond system thus completing the synthesis of the aglycone of methymycin, methynolide (D).

Similar methodology has been used to synthesize the 14-membered macrolides from the Prelog–Djerassi lactone; additional steps are required since only the C-1–C-7 fragment can be derived from the lactone starting material. The Masamune synthesis (Figure 6.41) of 6-deoxyerythronolide B utilizes highly stereoselective aldol chemistry to introduce the C-8–C-9 segment of the macrocycle. Boron enolate addition to the aldehyde

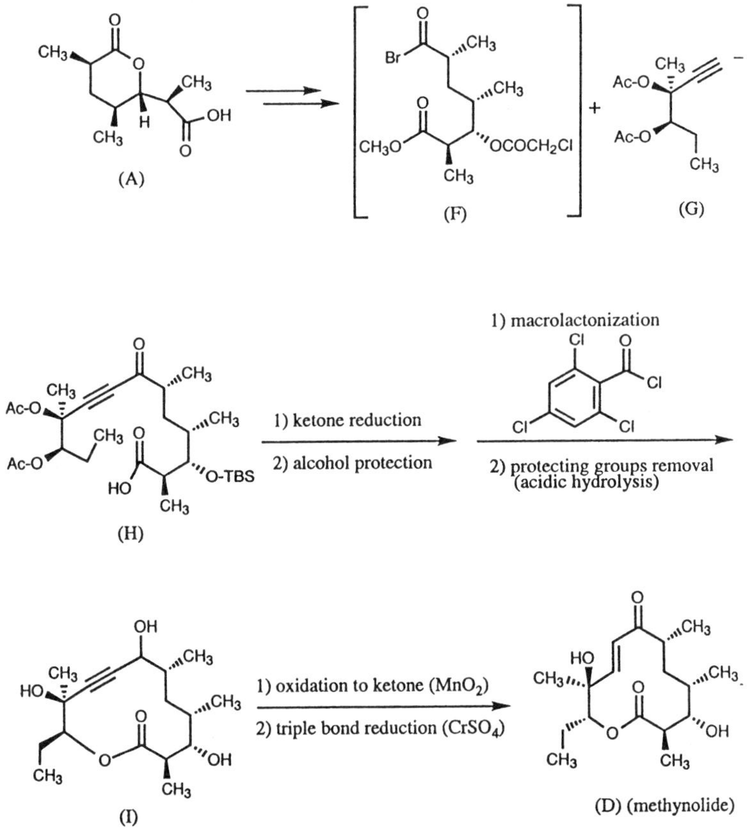

Figure 6.40 Synthesis of methynolide: C-8–C-12 fragment added as an alkyne.

derived from the Prelog–Djerassi lactone afforded the hydroxyacid (J) after oxidative cleavage to remove the (silyl)oxy moiety that directed enolate addition. The carboxylic acid was activated by conversion to the acyl halide and quenched with a thallium thiolate to produce the thioester. Subsequent lactone opening and protecting group chemistry afford the C-1–C-9 fragment (K). The C-10 center was introduced by activation of the C-9 center as the corresponding acyl chloride, followed by reaction with ethyl cuprate which yielded the ethyl ketone (L). Deprotonation with lithium hexamethyldisilazide occurred at the C-10 position and

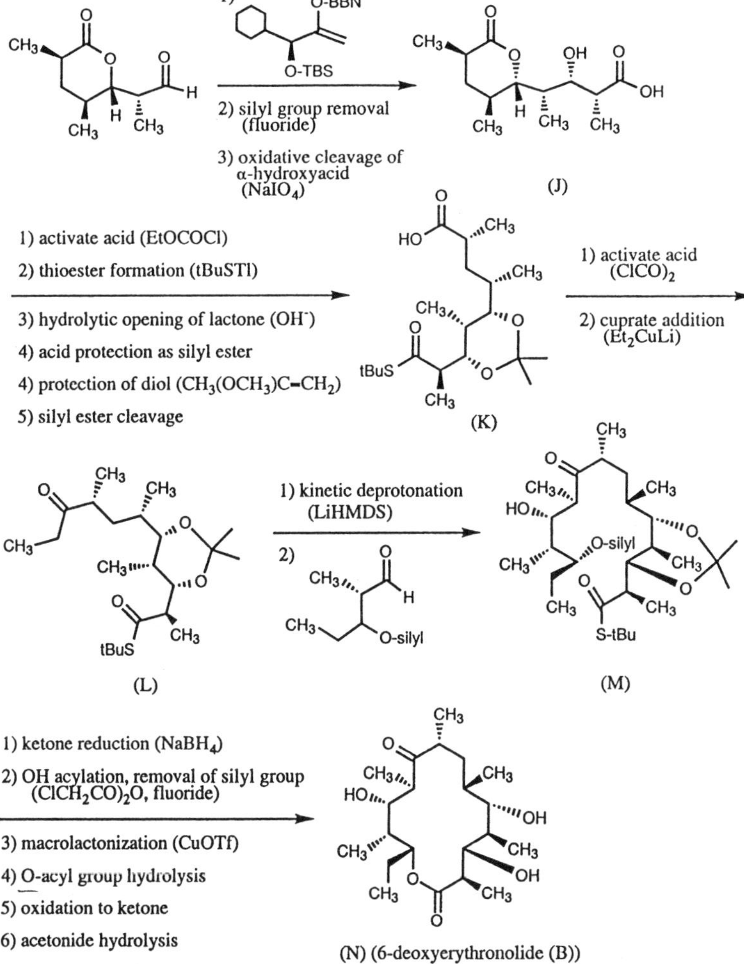

Figure 6.41 Masmune synthesis of 6-deoxyerhthromycin B. BBN-9-borabicyclo(3.3.1)nonane.

quenching with the optically active aldehyde completed construction of the 14-membered macrolide tether (M). A series of protecting group manipulations and modification of oxidation states, along with copper(I) mediated lactonization of the thioester, completed the synthesis of 6-deoxyerythronolide (N).

The syntheses outlined so far have used the Prelog–Djerassi lactone (A) to form the C-1–C-7 fragment of the macrocycle. Although this strategy has been quite successful, it is ironically dependent upon the availability of macrolides; the Prelog–Djerassi lactone is most readily obtained by degradation of certain macrolides. However, there are a number of synthesis of macrolides that do not rely upon the Prelog–Djerassi lactone. The Corey synthesis (Figure 6.42) makes use of a six-membered dienone (O) which was elaborated to bicyclic substrate and resolved to afford an optically active intermediate (P). Unfortunately, a series of subsequent reactions gave diastereomeric mixtures but undesired products could either be recycled or separated, and so the lactone ring opened compound (Q) was obtained. Oxidation of this highly substituted cyclohexanol afforded the corresponding ketone which underwent Baeyer–Villager oxidation to the ring-expanded seven-membered lactone. The thioester derivative (R) added one equivalent of a vinyl Grignard reagent (S) and yielded the enone–lactone (T) that contained all 14 carbon stereocenters required for the construction of erythronolide B. Reduction of the enone led to ring-opening (U) to a ten-membered macrocycle which, upon hydrolysis produced the acyclic erythronolide precursor (V). Macrolactonization was induced via closure onto an imidazoylthioester, and reduction of the carbon–carbon double bond and protecting group manipulations furnished erythronolide (B).

The Corey synthesis is remarkably skilleful in that all the stereocenters are set prior to the ring expansion cascade: a six-membered cyclohexanol was converted to a seven-membered lactone which was elaborated to a ten-membered system, and finally hydrolysed to the 14-membered acyclic tether. This synthesis is also remarkable in that the original scaffolding was provided by an achiral, (C_2) symmetrical starting material. This methodology was also used to synthesize the oxygenated congener, erythronolide (A).

The Woodward synthesis (Figures 6.43–6.46) is also a landmark for the same basic reason; all stereocenters are either set within the constraints of a dithiadecalin system or by resolution of diastereomeric mixtures early in the synthesis. The synthesis begins with elaboration of the mercaptothiane (W) to the diastereomeric bicycle (Y) using a stereoselective Robinson annulation. Introduction of the hydroxyl functionality (Z) at the ring conjunction was accomplished by dehydration to the enone by mesylation and elimination, followed by ketone reduction, and finally (bis)hydroxylation of the carbon–carbon double bond. This dithiadecalin intermediate (Z) ingeniously served in two contexts, ultimately providing

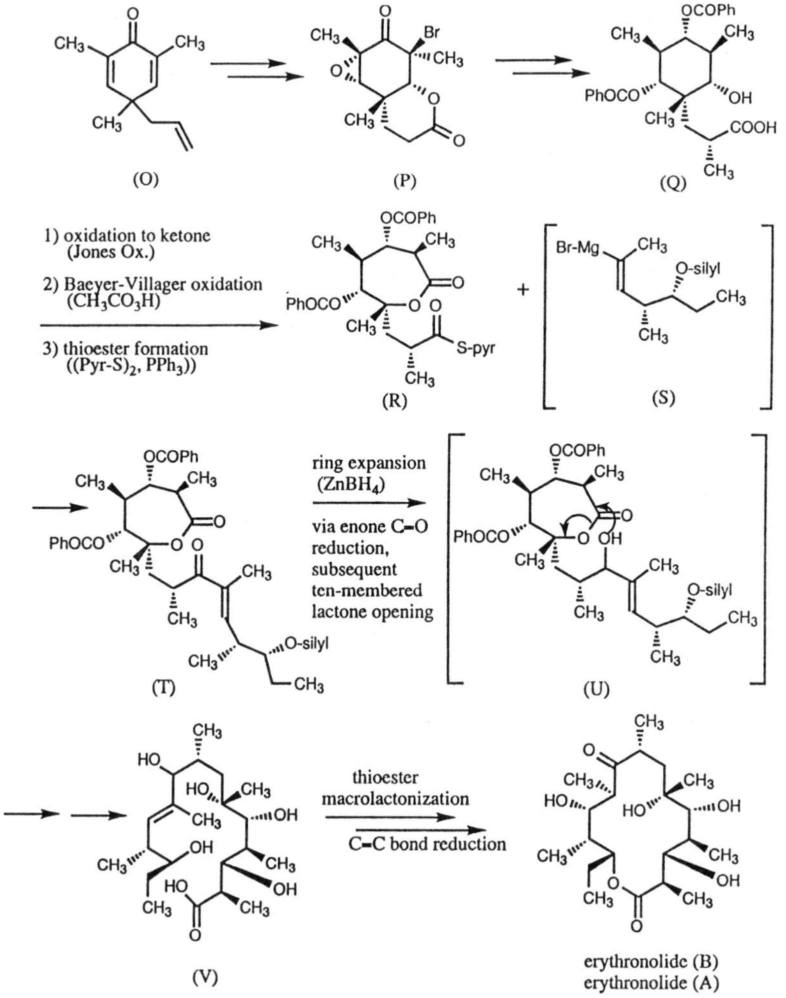

1) oxidation to ketone
(Jones Ox.)

2) Baeyer-Villager oxidation
(CH$_3$CO$_3$H)

3) thioester formation
((Pyr-S)$_2$, PPh$_3$)

ring expansion
(ZnBH$_4$)

via enone C=O
reduction,
subsequent
ten-membered
lactone opening

thioester
macrolactonization

C=C bond reduction

erythronolide (B)
erythronolide (A)

Figure 6.42 Corey synthesis of erythronolides.

for both the C-9–C-14 and the C-3–C-8 portions of the macrocycle. The C-9–C-13 fragment was derived from the dithiadecalin via desulfurization followed by a protocol in which the exocyclic alcohol substituent was oxidatively cleaved to afford the functionalized heptanal (BB). The C-3–C-8 portion was obtained from the dithiadecalin (Z) by deprotection of the alcohol and oxidization to the corresponding ketone. Deprotonation with an aryllithium base furnished enolate (AA) and quenching with the aldehyde (BB) gave adduct (CC). This dithiadecaline was elaborated by a rather complex sequence which resulted in ketone reduction and the

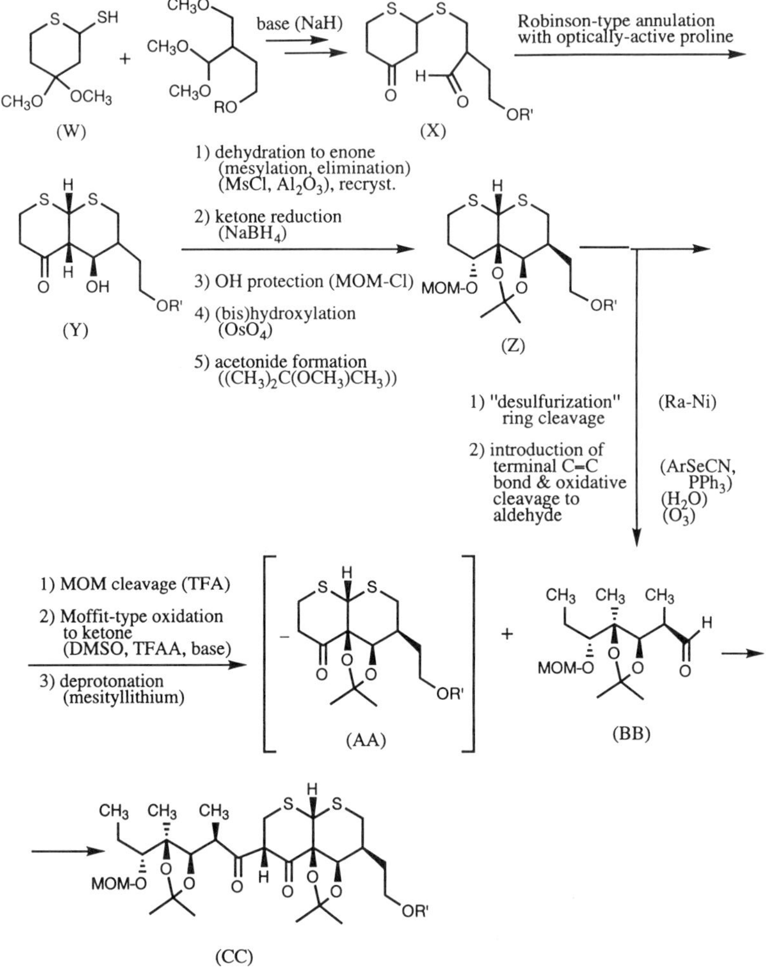

Figure 6.43 Woodward synthesis of erythromycin A (part 1). TFA–trifluoroacetic acid. TFAA–trifluoroacetic anhydride.

subsequent introduction of a benzylthioether at this position (DD). The purpose of placing a sulfur substituent onto the bicyclic ring system cleverly allowed for reductive elimination at this position which accompanied desulfurization of the dithiadecalin construct (DD). The same oxidative cleavage process described earlier afforded the corresponding aldehyde (EE) which contained 12 of the required 14 centers. The remaining carbon centers were introduced by thioester enolate addition onto the aldehyde (EE). The resultant thioester (FF) was then subjected to a lengthy sequence that ultimately placed an amino substituent at the C-9 position.

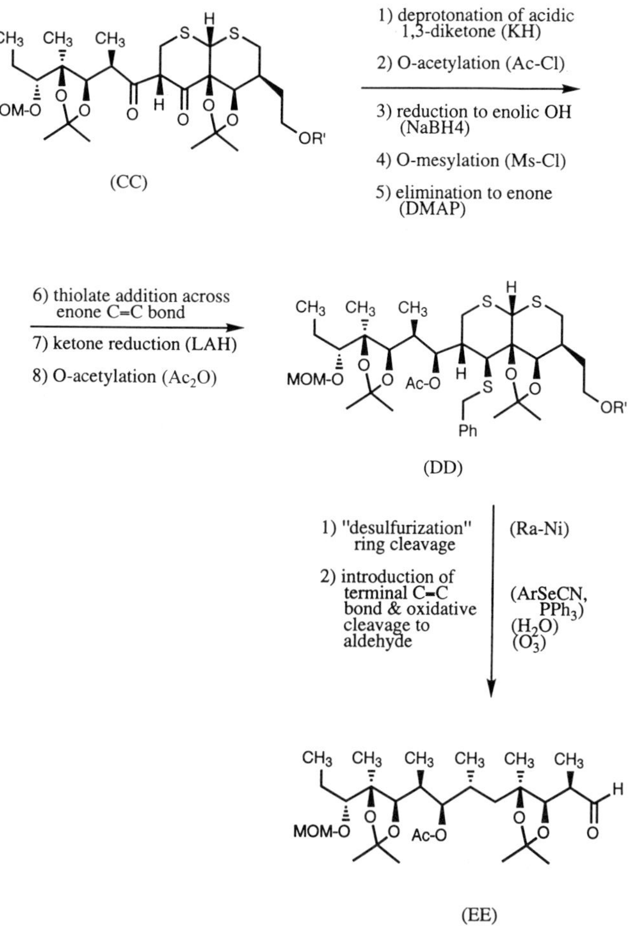

Figure 6.44 Woodward synthesis of erythromycin A (part 2).

This was accomplished by cleavage of the C-9 acetate to the alcohol followed by mesylation and displacement with azide. The azide was reduced to the amine which was converted to carbamate (GG) en route to the cyclic oxazolidinone (HH); macrocyclization of the corresponding (pyridyl)thioester precursor afforded the 14-membered lactone. The cyclic oxazolidinone avoided hydroxyl group protection but necessitated the regeneration of the C-9 ketone group at a later step of the synthesis. In this context, the oxazolidinone was cleaved and the C-9 amine protected as a biphenyl amide (II) prior to the attachment of each sugar moiety. After protecting group manipulations, a hydroxyl-protected desosamine derivative, activated as a pyrimidinylthioether, underwent silver(I)-assisted coupling to the C-3 hydroxyl group of the erythronolide surrogate, and was

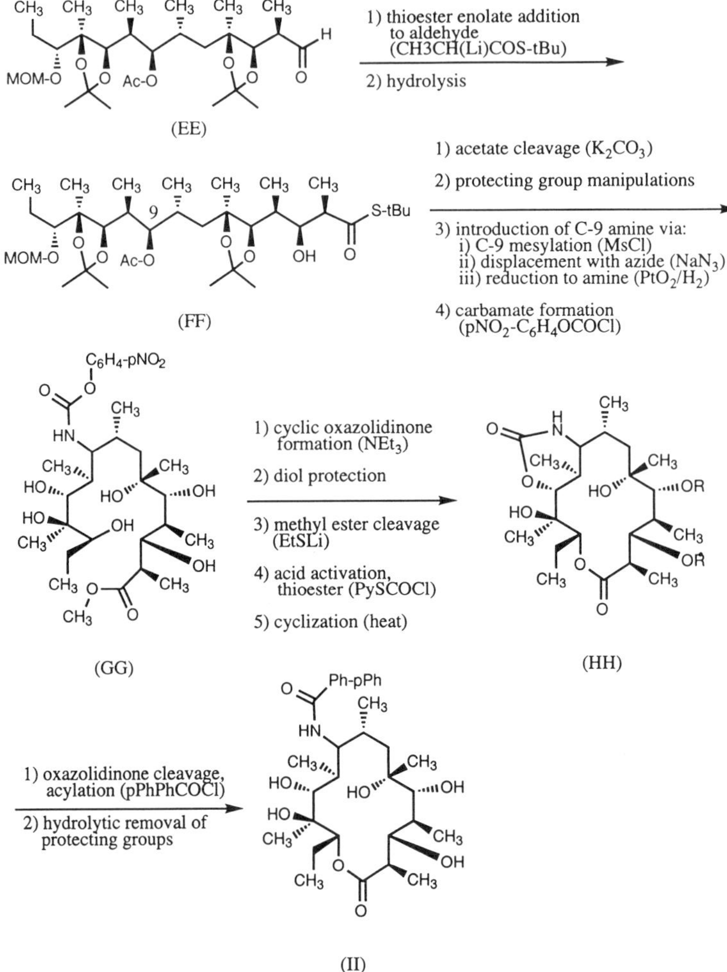

Figure 6.45 Woodward synthesis of erythromycin A (part 3).

properly deprotected. The C-5 sugar was incorporated by utilizing a similar protocol; a protected cladinose-derived thiopyridine was coupled to the macrocycle. The last major transformations entailed reductive cleavage of the biphenylamide, followed by oxidation of the free amine to the imine and subsequent hydrolysis to the C-9 ketone, thereby completing the synthesis of erythromycin A (KK).

The Woodward synthesis of erythromycin A must be regarded as one of the most elegantly conceived organic syntheses of all time. Despite requiring approximately five dozen individual chemical transformations, the synthesis showcases the art of dissecting complex organic substrates

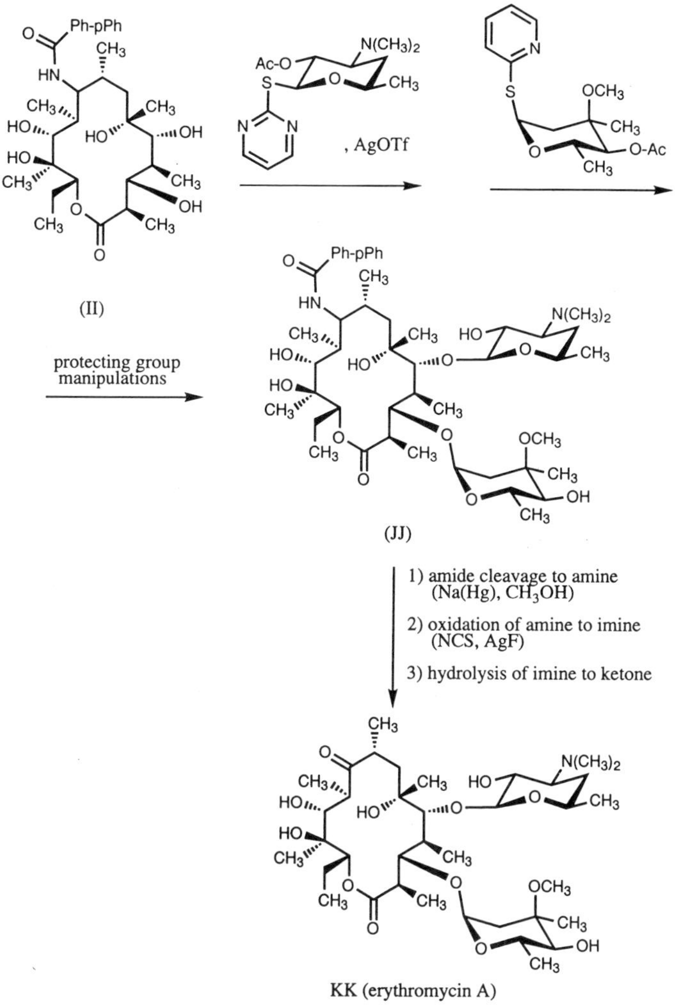

Figure 6.46 Woodward synthesis of erythromycin A (part 4).

into smaller molecular targets that embody common stereochemical features. Woodward's recognition of a relatively simple dithiadecalin system as a means to access nearly all of ten carbon stereocenters of the erythromyin macrolide is brilliant.

Synthetic approaches to the 16-membered macrolides bear some resemblance to the synthesis of the 12- and 14-membered congeners. Most approaches have likewise focused upon the construction of the C-1–C-7 macrolide portion early in the synthesis. Also, the macrolactonization protocols and the attachment of the sugar moieties to the macrocycle have been key features relegated to the latter stages of the synthesis. However, unlike the methymycin and erythromycin syntheses described above, synthetic approaches to the 16-membered macrolides must address formation of the C-6′ aldehyde functionality, as well as the double bond systems at C-10–C-11 and C-12–C-13 in many cases. For these reasons, a C-5–C-6 hemiacetal functionality is prevalent in order to introduce the sensitive aldehyde functionality as the last step in the synthesis. Due to the carbon–carbon double bond systems, Wittig-type chemistry is another common theme.

A particularly illustrative example is that of the Tatsuta synthesis of carbomycin B (Figure 6.47). After construction of the thioacetal (LL) a Wittig reaction was used to extend the chain giving the methyl ketone (MM) which ultimately served as the C-1–C-10 macrolide fragment. Aldol

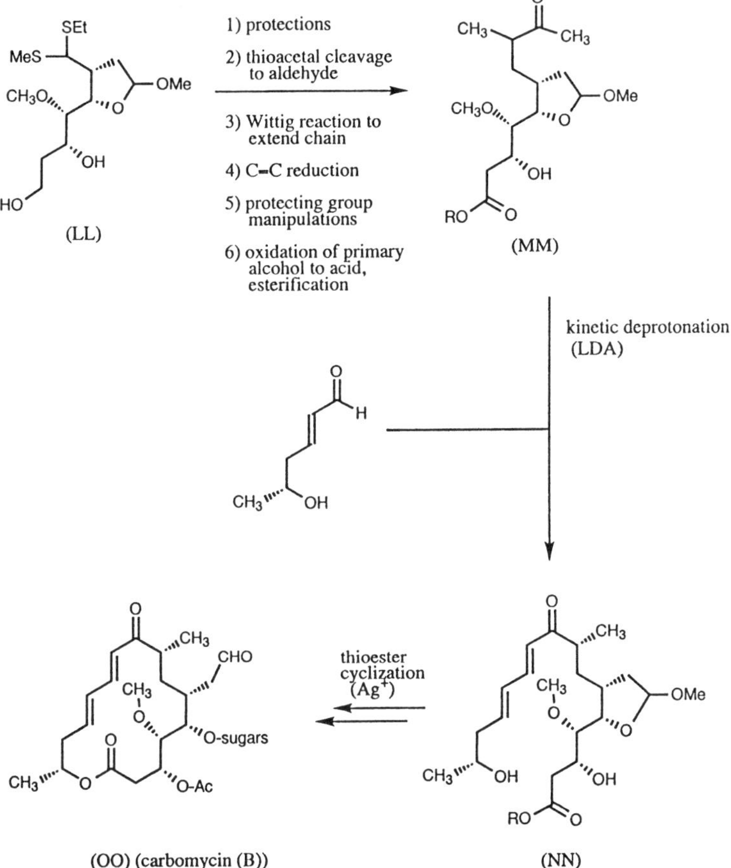

Figure 6.47 Tatsuta synthesis of carbomycin B. LDA–lithium diisopropylamide.

condensation with an α,β-unsaturated aldehyde completed the formation of the 16-membered acyclic macrolide precursor (NN). After ketone reduction, this substrate was taken on to a thioester and subsequently cyclized (silver(I)-assisted). A series of reactions first unmasked the C-5 hydroxyl functionality and then allowed for the attachment of the glycosides at the macrolide C-5 position. Lastly, the C-6′ aldehyde was generated thus completing the synthesis of carbomycin (B) (OO).

6.4.2 Semi-synthetic macrolides

Although elegant, any total synthesis of the antibacterial macrolides is not a practical means to make new analogs at this time (as evident by the lengthy approaches described above). For these reasons, the naturally occurring macrolides normally serve as starting materials in semi-synthetic endeavors. Some of the more general transformations include modifications to the erythromycin ketone system and the formation of the azalides. (Esterification (acylation) and alkylation have already been considered). The erythromycin ketone group is versatile and can be converted to oximes, amines and imines (Figure 6.48). Erythromycylamine **5** (Figure 6.5), an important synthetic intermediate, is obtained through a formal reductive amination, and C-9 oxime derivatives are produced by if the reduction is omitted. Roxithromycin **7** (Figure 6.7) is produced by the reaction of erythromycin with hydroxylamine (R = OH).

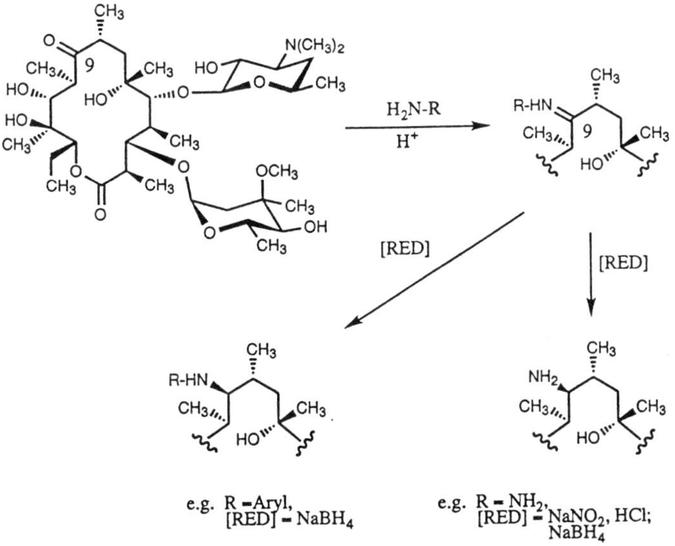

Figure 6.48 Amine additions to the erythromycin C-9 ketone.

Erythromycylamine can also be condensed with a variety of aldehydes to afford C-9–C-11 bridged-oxazine erythromycin derivatives. The synthesis of the dirithromycin **6** (Figure 6.6) consists of the reaction of methoxyethoxyaldehyde with erythromycylamine **5** (Figure 6.49).

The C-9 ketone system can be modified in other useful ways; in some cases, compounds arising from the acidic degradation of erythromycin are useful intermediates. For example, flurithromycin **26** (Figure 6.50) can be made from the reaction of 8,9-anhydroerythronolide-6,9-hemiketal (see hemiketal in Figure 6.31) with trifluoromethylhypofluorite.

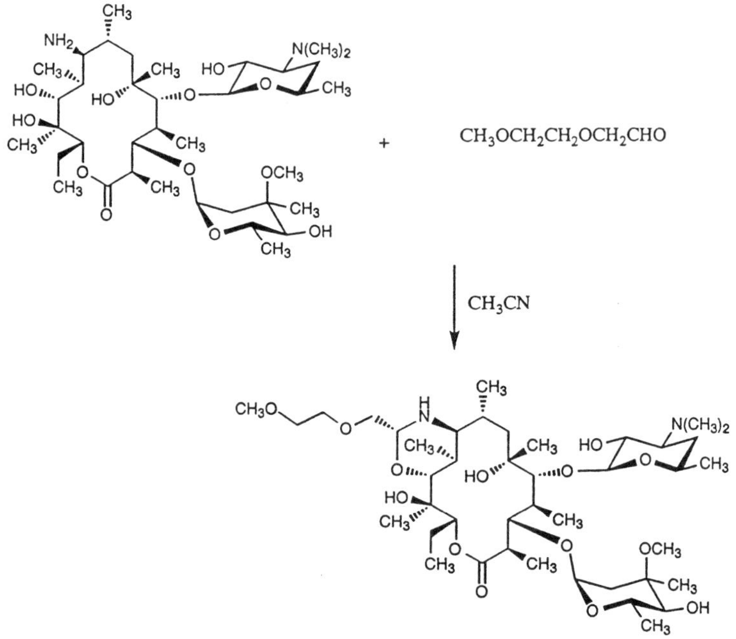

Figure 6.49 Synthesis of C-9, C-11-oxazine derivatives.

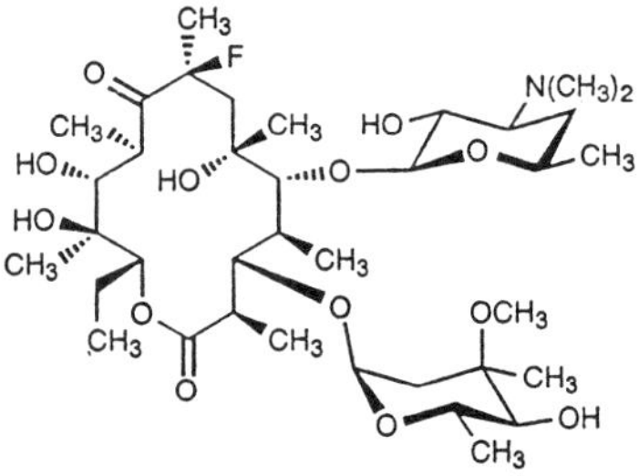

26

Figure 6.50 Flurithromycin.

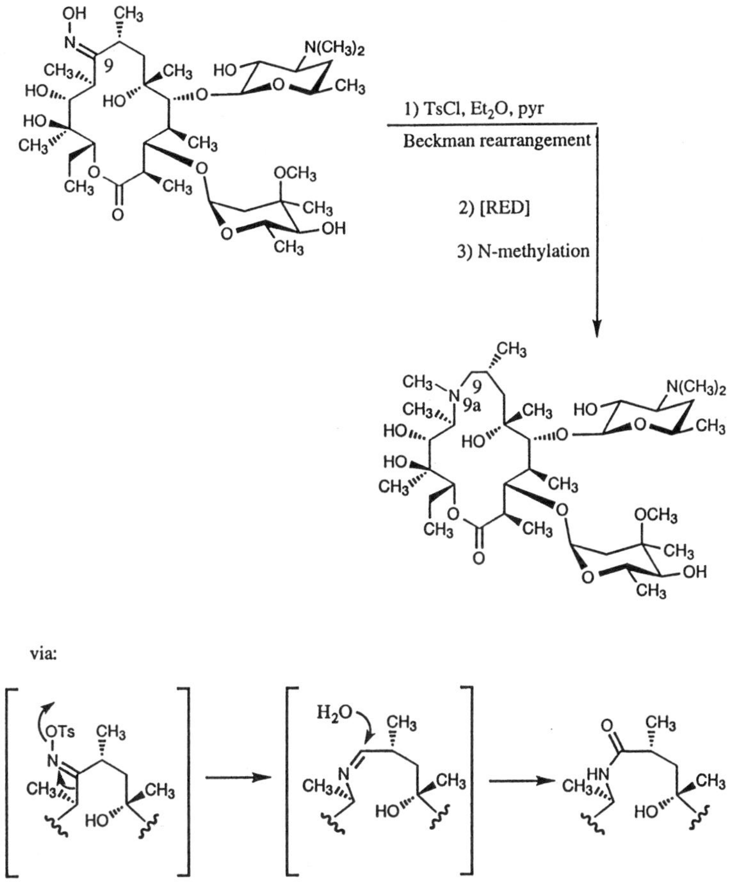

Figure 6.51 Synthesis of azalides.

The latest innovation in macrolide SAR comes from the discovery that certain 15-membered aza-lactones are potent antibacterial compounds. The azalides are constructed using a stereospecific Beckman rearrangement. For example, azithromycin **9** (Figure 6.9) is made from the rearrangement of the erythromycin C-9 oxime, subsequent imine reduction and nitrogen alkylation (Figure 6.51). A similar strategy has been developed that allows for the production of isomeric azalides in which the nitrogen atom is introduced adjacent to the C-8 center (section 6.6).

6.4.3 Semi-synthetic modifications of the rifamycins

The total synthesis of rifamycin S was first reported by Kishi in 1980; this and subsequent synthetic approaches are not considered here.

A number of rifamycins are closely related in structure, therefore some semi-synthetic congeners are readily acquired using straightforward oxidation or reduction protocols. For example, the naphthalene aromatic core can be oxidized and subsequently hydrolysed to the naphthoquinone unit thus providing entry to both the rifamycin B (naphthalene) and rifamycin S (naphthoquinone) series of compounds (Figure 6.52).

Several new promising rifamycins contain non-natural substituents at the C-3 and C-4 positions of the naphthalene/naphthoquinone ring system. The C-4 *O*-(carboxymethyl) rifamycin (rifamycin B) has been used to make a host of amides and hydrazides; these congeners are prepared by activating the carboxylic acid and subsequently coupling it to amines or hydrazines. Primary and secondary amines and mono-, di- and trisubstituted hydrazines are all amenable to this methodology. The *N,N*-diethylamide rifamide **27** (Figure 6.53) (section 6.6) is under development. Some

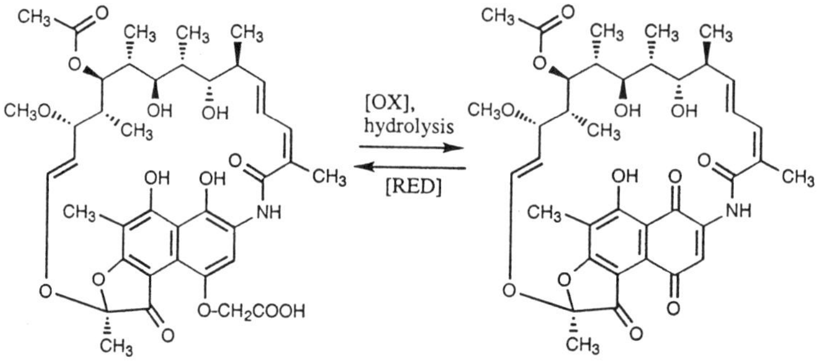

rifamycin B   rifamycin S

Figure 6.52 Rifamycin B and rifamycin S analogs can frequently be prepared from each other.

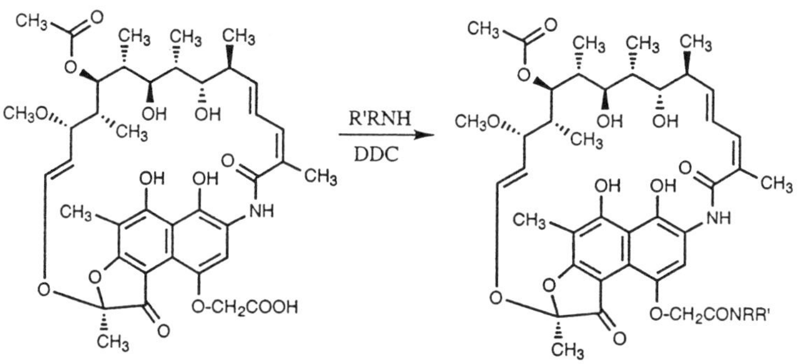

R = alkyl, aryl, H;
R' = alkyl, (disubstituted amine)
H, (monosubstituted amine)
amino (hydrazine)   27 (R = R' =Et)

Figure 6.53 Preparation of rifamide and other rifamycin(B) derivatives.

adducts, such as amidino rifamycins, have been oxidized to the corresponding *N*-oxides and retain good *in vitro* antibacterial activity.

The C-3 position in the rifamycin S series undergoes Mannich condensation giving rise to a host of disubstituted aminomethylated derivatives. Mercaptides also add to this center producing C-3 alkylthio rifamycins. In the latter case, thioester formation has proven to be a more general route to introduce a sulfur substituent (Figure 6.54). Simple aliphatic amines or aromatic monoamines have also been incorporated at the C-3 position; some are extremely potent agents that may undergo clinical evaluation in the near future.

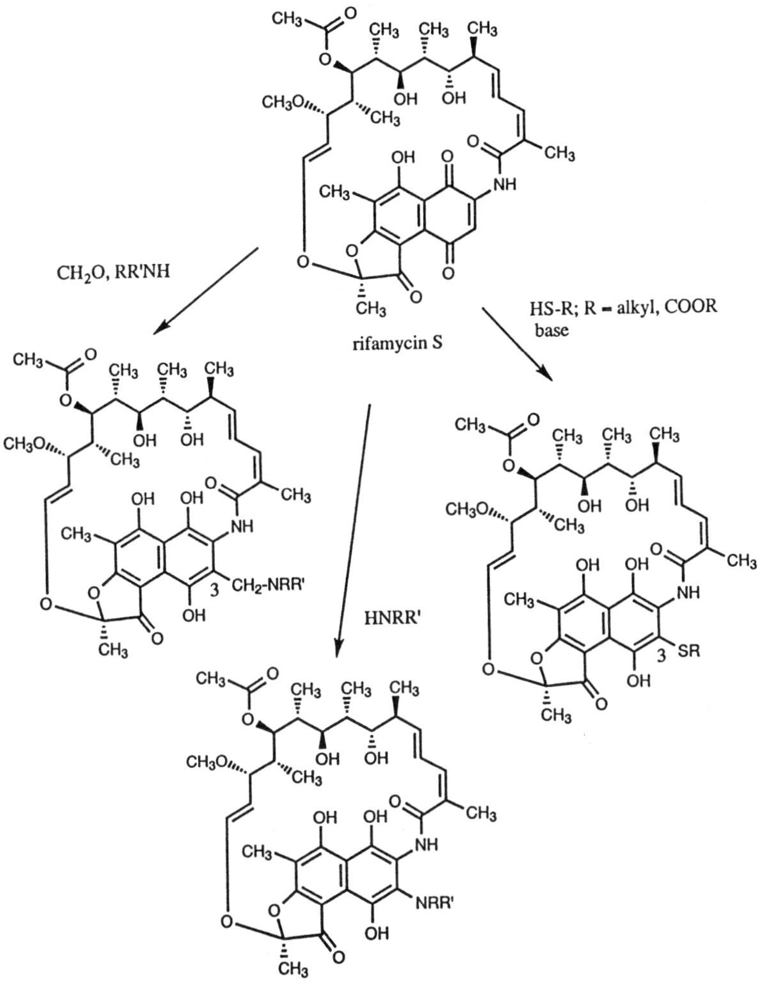

Figure 6.54 The introduction of C-3 substituents in the rifamycin S series.

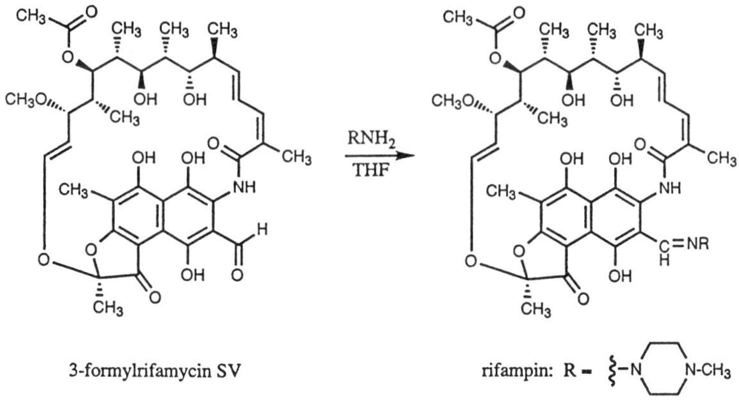

Figure 6.55 Synthesis of rifampin and other derivatives.

Rifampin and other related C-3 alkyliminomethylated rifamycins are potent antibacterial compounds. The synthesis of these compounds is accomplished by introduction of the formyl group onto the C-3 position of rifamycin (SV) followed by condensation with primary and secondary amines (Figure 6.55). Similarly, 3-formylated rifamycins have been condensed with heteroaryl amines and hydrazines to afford a vast series of intriguing analogs.

Lastly, the C-3 and C-4 substituents can also be joined together in a cyclic structure. In general, such compounds are produced by brominating the C-3 position of rifamycin S, followed by azide displacement and conversion to the C-3 amine. During this sequence, the corresponding iminoquinone is formed since the C-4 (quinone) position is also a reactive site. Subsequent reaction with a ketone leads to intramolecular cyclization at the C-3 and C-4 positions (Figure 6.56). The rifamycin can also be thiolated or oxygenated to incorporate sulfur or oxygen into the newly formed fused heterocycle ring. A variety of phenthiazinyl-, phenazino- and phenoxazino rifamycin derivatives have been obtained using this general methodology. Rifabutin **21** (Figure 6.21) is a phenazino prototype that contains a C-3–C-4 spirofused piperazinyl substituent. A number of other congeners are being considered for development at this time.
time.

6.5 Bacterial resistance to non-peptidic macrocyclic antibacterial agents

6.5.1 Resistance to the macrolides

Although the macrolide antibacterials have been successful agents for decades, the presence of resistant microorganisms remains problematic.

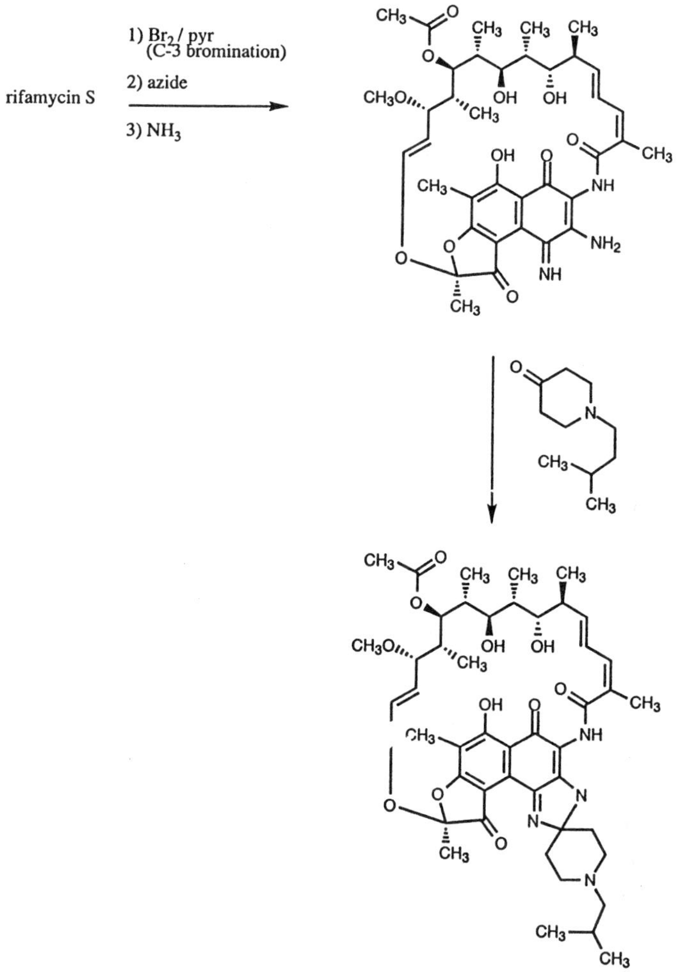

Figure 6.56 Synthesis of C-3–C-4 spirofused rifamycins.

Erythromycin continues to be one of the most widely used of all antibacterial agents despite the fact that resistance surfaced within years of its introduction to the market. In general, erythromycin has remained as an effective agent against many common strains of streptococci, staphylococci as well as *Legionella* and *Chlamydia* species and *Mycoplasma*. Erythromycin-resistant bacteria have been thoroughly studied and other macrolides are prone to the same kinds of resistance.

An antibacterial effect can only be produced if the macrolide can first enter the bacterial cell and then subsequently travel to the site of the bacterial ribosome. The outer membrane is sometimes able to exclude macrolide and thereby exhibit an inherent resistance. Many Gram negative bacteria possess outer membranes which prevent or limit uptake of macrolide. In such instances, the outer membrane can be experimentally removed or ruptured and the corresponding cell-free ribosomal preparations (e.g. from *E. coli*) then readily bind macrolide (e.g. erythromycin). In essence then, outer membrane impermeability merely prevents the macrolide from entering the cell and reaching its site of action; bacterial ribosomal function is not disrupted. Consequently, most macrolides are only weakly active, or inactive, against many Gram negative pathogens; in particular, many of the *Enterobacteriaceae* and *Pseudomonas* species are resistant to macrolides even at high concentrations.

Assuming that the macrolide is able to reach the bacterial ribosome, several events must occur in order for an antibacterial effect to be initiated. First of all, the macrolide must be able to establish weak interactions with the surface of the ribosome. The ensuing event is that of binding of one molecule or macrolide to the ribosome in a tight, non-covalent manner (section 6.2). In some resistant species, the ribosome simply does not accept the macrolide molecule and ribosomal binding does not occur. This mechanism of resistance is independent of other factors. Uptake into the bacterial cell may be sufficient and although the intact drug arrives in proximity to the ribosome and is able to establish weak associations with the ribosome surface, tight binding is never achieved. As a result, antibacterial effects are not induced (i.e. inhibition of protein synthesis). In such cases, resistance is conferred by modification of the bacterial ribosome.

There are several distinct features that are characteristic of this type of resistance. First of all, it has been shown that methylation of the 23S r-RNA molecule of the ribosome is responsible. This alteration apparently changes the shape of the ribosome in a way that prevents binding of macrolides. Resistance caused by ribosomal modification can be acquired by a variety of common pathogens including staphylococci, streptococci and *Corynebacterium diphtheriae*. This process frequently confers resistance to some other structurally unrelated antibacterial agents that also normally target the 50S ribosomal subunit. The lincosamides and streptogramins are often similarly affected; this phenomenen is referred to as MLS (macrolide-lincosamide-streptogramin) (phenotypic) resistance.

The details surrounding methylation of ribosomal 23S RNA molecule have been elucidated. The N-6 position of a specific adenine residue is targeted and this amine group is converted to the corresponding dimethylamino conjugate. In some *E. coli* carrying the MLS resistance (*erm*C), the methylation site has been pinpointed to the base position 2058. The genetic site responsible for methylation has been traced to a highly conserved sequence of genes that encode a methylase enzyme; expression can be inducible or constitutive. The ribosomal methylase delivers methyl group(s) to the specific adenine residue but does not affect the macrolide itself.

Genes that carry methylase enzymes are referred to as the *erm* (erythromycin resistance methylase) genes. To date there are at least eight such genes that have been shown to lead to MLS resistance (*erm*A, *erm*AM, *erm*A', *erm*C, *erm*D, *erm*E, *erm*F, *erm*G). These genes are either borne on plasmids, transposons or chromosomes. All *erm* genes probably have a common origin since their corresponding methylase proteins have a high degree of amino acid homology, although their genetic support can vary. For example, *Staphylococcus aureus* resistance can be traced to the *erm*A gene localized on a transposon (e.g. Tn554), or to the *erm*C gene localized on a plasmid (pE194). A gene that produces ribosomal methylation may also be contained within the bacterial chromosome, as exemplified by the *erm*D gene of *Bacillus lichenforms*. The *erm*A, *erm*AM and *erm*C genes are clinically common. The *erm*A and *erm*C genes are usually chromosomal or plasmid borne and are responsible for a variety of macrolide-resistant staphylococci that are particularly troublesome at this time. The *erm*AM gene, often present in resistant streptococci and enterococci, is often borne on transposons.

Small (approx. 2–4 kb), non-conjugative, multicopy plasmids are often responsible for MLS resistance in staphylococci and other Gram positive bacteria (e.g. streptococci). These plasmids usually confer resistance to the erythromycins and some other macrolide antibacterials. Larger conjugative plasmids have been identified and associated with multiple resistance including some aminoglycosides (kanamycin, gentamicin, tobramycin, spectinomycin) and penicillins. For example, a plasmid (pSJ24, 35.8 kb in size) confers resistance to erythromycin, gentamicin, kanamycin and penicillin in staphylococci. MLS resistance caused by these larger plasmids is less common at this time.

Transposable elements responsible for macrolide (MLS) resistance occur in a variety of Gram positive bacteria such as staphylococci, streptococci and enterococci. A number of transposons of varying size and construction have been identified on plasmids or within bacterial chromosomes. Some of these transposons contain genetic elements for resistance to other structurally unrelated antibacterials such as chloramphenicol, kanamycin and some tetracyclines. Larger transposons of this type are usually conjugative as exemplified by a transposon (Tn1545, 25.3 kb in size) that also confers tetracycline and kanamycin resistance.

Within the past decade, there have been reports of resistance that involves structural alteration of the macrolide itself. Some species are able to produce enzymes that can deactivate a 14-membered macrolide such as erythromycin, by either destroying the macrocyclic framework, or by attaching a molecular conjugate onto a vital moiety. Resistance of this type has been restricted to members of the *Enterobacteriaceae* genus. There appear to be at least two separate classes of enzymes capable of modifying the macrolides. The first type of macrolide-modifying enzymes are esterases which attack the lactone carbonyl and result in a ring-opened product that is inactive. Erythromycin esterases can be traced to *ere* (erythromycin resistance esterases) genes that encode for the esterase enzyme. Most information about erythromycin modifying esterases has come from the study of macrolide-resistant *E. coli*. At least two esterases have been identified; a 349 amino acid protein (designated as type I) that is a product of the *ere*A gene, and a 419 amino acid protein (type II) from expression of the *ere*B gene. Both *ere* genes are carried on plasmids.

Interestingly, at least one strain of macrolide-resistant *E. coli* has been shown to contain a phosphotransferase enzyme that catalyses the phosphorylation of the C-2′ hydroxyl group of the cladinose sugar substituent. The resultant phosphorylated erythromycin apparently possesses a greatly diminished affinity for the bacterial ribosome and thus is ineffective in initiating the inhibition of protein biosynthesis.

The remaining mechanism of acquired macrolide resistance involves the ability of a bacterial microorganism to remove macrolide so as to prevent intracellular concentrations from reaching harmful levels. Certain erythromycin-resistant *S. epidermidis* species have been shown to establish an efflux mechanism, attributed to the formation and placement of a membrane-bound protein. One such species has been identified as a 60 kDa protein that is encoded by the *erp*A (erythromycin resistance permeability) gene carried on a (26.5 kb) plasmid. The exact manner in which this protein escorts the macrolide from the cell interior is unknown at this time. It seems reasonable to hypothesize that the protein may be capable of binding to macrolide in some fashion, and/or that the protein may be involved in providing a channel across the membrane through which the macrolide is expelled. There has been a supporting example of this phenomenon in some bacteria resistant to streptogramin. In these instances, a protein expressed by the *msr*A (macrolide streptogramin resistance) gene is responsible; an ATP-driven protein pump has been hypothesized as being the operative.

Considering the many elegant ways in which bacteria have been able to assemble mechanisms which confer macrolide resistance, it might appear that these agents have fallen from chemotherapeutic use over the years. However, the macrolides continue to be valuable antibacterial agents for the treatment of respiratory, skin and soft tissue infections, as well as sexually transmitted diseases. Even more importantly perhaps, there has been a resurgence of interest in the antibacterial macrolides with the discovery of new agents that can be effective against erythromycin-resistant bacteria.

6.5.2 Resistance to the rifamycins

Although rifamycin-resistant bacteria have been known since virtually the discovery of the first rifamycins, the current concern is focused upon resistant mycobacteria. In particular, resistant tuberculosis (*Mycobacterium tuberculosis*) has been on the increase and even rampant within immuno-compromised sectors of the population (e.g. HIV-positive, drug abusers, etc.). Within the past years, episodes of fatal tuberculosis caused by highly resistant mycobacteria (*M. tuberculosis*) have alarmed the medical community. In some of these cases, the mycobacteria (*M. tuberculosis*) proved to be resistant to an extensive battery of antituberculosis agents including rifamycins and several other classes of antibacterial agents.

Rifamycin resistance is usually caused by the replacement of a single amino acid within the RNA polymerase protein. This modification is the result of a single step mutation that alters the rifamycin binding site without compromising enzymatic function. This resistance occurs at a frequency of approximately 10^{-7}–10^{-8} depending upon the particular bacterial species. The mutation has been traced to the β-subunit of the DNA-dependent RNA polymerase which is encoded by the *rpo*B gene. This gene has been cloned and sequenced; the amino acid replacements occur within a small region, approximately two dozen residues, of the polymerase protein. Two of the more common modifications, found in resistant *M. tuberculosis*, are the replacement of the amino acids His-526 or Ser-531 with a variety of amino acids (e.g. Tyr, Asp, Gln, Asn, Arg, Pro at His-526; Leu, Gln, Trp at Ser-531). Other residues are targeted; Leu-533 (replaced with Pro in both species), Gln-513 (replaced with Leu, in addition to Phe in *E. coli*) and Ser-522 (replaced with Leu in *M. tuberculosis* and with Phe in *E. coli*). In most cases, a single nucleotide of the corresponding codon is changed, resulting in the incorporation of the abhorrent amino acid.

6.6 Recent advances

6.6.1 New macrolides

There has been a notable resurgence of interest in the antibacterial macrolides. Many semi-synthetic agents have been introduced recently: roxithromycin, clarithromycin, azithromycin, dirithromycin are all marketed second-generation 14-membered macrolides and miokamycin and rokitamycin are 16-membered macrolides. Tilmicosin is a new 16-membered veterinary agent. Although a number of other congeners are undergoing some form of preclinical assessment, the C-2'-ethylsuccinate ester of flurithromycin has advanced into a clinical trial.

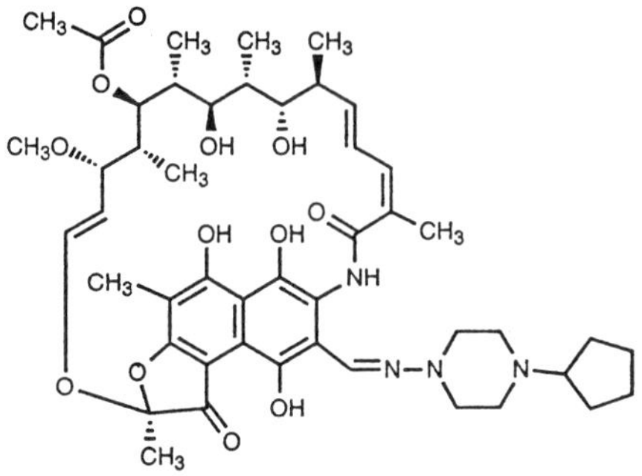

28

Figure 6.57 Rifapentine.

6.6.2 New rifamycins

Rifapentine **28** (Figure 6.57) has been chosen from a series of hydrazono rifamycins and is in advanced clinical trials as an antituberculosis agent.

6.7 Summary

1. The macrolide antibacterial agents are extremely useful chemotherapeutic agents for the treatment of a variety of infectious disorders and diseases caused by a host of Gram positive bacterial pathogens. These agents, as exemplified by erythromycin, are generally effective against staphylococci and streptococci (methicillin-susceptible), as well as *Chlamydia* and *Legionella* species and *Mycoplasma*. As a result, the macrolides are commonly administered for respiratory, skin and tissue and genitourinary infections caused by these pathogens.

2. Most of the macrolides that find use in current chemotherapy are 14-membered congeners; erythromycin is still the most commonly prescribed of all macrolides for human use. A number of 16-membered macrolides are being used or investigated outside the United States. The latest generation of macrolides are the 15-membered azalides which offer an extended spectrum in being relatively potent *in vitro* against some Gram negative pathogens; the clinical significance of this activity has yet to be determined.

3. The antibacterial macrolides are generally well tolerated, and are usually administered orally, although some parenteral preparations are available for severe indications. The macrolides are obtained

directly from a variety of microbial sources such as streptomyces, or may be produced in a semi-synthetic fashion from an appropriate naturally occurring progenitor.

4. The macrolides exert their antibacterial effects, which are usually bacteriostatic, via inhibition of bacterial protein biosynthesis. Specifically, the macrolides target the 50S ribosomal subunit; different members stop protein synthesis at varying stages of peptide chain elongation. The macrolides inhibit ribosomal peptidyl transferase activity. Some macrolides are also able to inhibit the translocation of the ribosome along the m-RNA template.

5. Bacterial resistance to the macrolide agents can occur through several different mechanisms. The most problematic involves ribosomal modification; a specific adenine base of the 23S RNA molecule is methylated which prevents macrolide binding to the ribosome. This type of resistance can also confer resistance to other antibacterial agents that normally target the 50S subunit (a.k.a., MLS resistance). Resistance can also occur via efflux of the drug from the bacterial cell, or from structural modification of the macrolide itself.

Major antibacterial macrolides are shown in Table 6.1.

6. The rifamycins exert bactericidal effects on susceptible bacteria such as mycobacteria, and as such are a mainstay in the treatment of

Table 6.1 Major antibacterial macrolide chemotherapeutic agents

Generic name	Common trade names	Administration
Azithromycin	Zithromax	PO
Clarithromycin	Biaxin	PO
Dirithromycin	Dynabac	PO
Erythromycin	E-mycin, Ethril, Erypar, ERYC, Ery-Tab and others	PO
Erythromycin (gluceptate)	Ilotycin Gluceptate	IV
Erythromycin (estolate)	Ilosone	PO
Erythromycin (ethyl succinate)	E.E.S., Pediamycin, EryPed	PO
Erythromycin (ethyl succinate) and sulfisoxazole acetyl	Pediazole	PO
Erythromycin (stearate)	Erythrocin and others	PO
Erythromycin-11,12-carbonate	Davercin	PO
Josamycin	Josamycin	PO
Miokamycin	Miocamycin	PO
Rokitamycin	Ricamycin	PO
Roxithromycin	Rulid	PO
Troleandomycin (triacetyloleandomycin)	TAO	PO
Veterinary		
Tylosin	Tylan	IM, PO
Tilmicosin	Micotil	IM, PO

Table 6.2 Antibacterial rifamycin chemotherapeutic agents

Generic name	Trade name	Common administration
Rifampicin	Rifadin, Rimactane	PO, IV
(with isoniazid)	Rifamate	PO
Rifabutin	Mycobutin, Ansamycin	PO

tuberculosis. In addition, rifampin can be effective for the prevention of (meningococcal) meningitis.

7. The rifamycins inhibit bacterial DNA-dependent RNA polymerase which results in rapid inhibition of bacteria growth and reproduction. Bacterial resistance to the rifamycins occurs by a single mutation to the gene encoding for the β-subunit of the enzyme. Resistance species contain modified RNA polymerase enzymes in which a single amino acid is altered within a distinct region of the protein.

8. The emergence of multidrug-resistant *Mycobacterium tuberculosis* has redirected an interest in the development of new rifamycins.

9. Antibacterial rifamycins are shown in Table 6.2.

Further reading

A.J. Bryskier, J.-P. Butzler, H.C. Neu and P.M. Tulkens (Eds) (1993) *Macrolides: chemistry, pharmacology and clinical uses*, Arnette Blackwell, Paris.

S. Omura, (Ed.) (1984) *Macrolide Antibiotics: Chemistry, Biology and Practice*, Academic Press, Orlando, FL.

H.A. Kirst, J.P. Leeds, J.W. Paschal and J. Martynow (1993) 'Stereospecific modifications of erythromycin derivatives', in *Recent Advances in the Chemistry of Anti-infective Agents*, P.H. Bentley and R. Ponsford (Eds), Royal Society of Chemistry, Cambridge, UK, pp. 67–78.

H. Kirst (1993) 'Semi-synthetic derivatives of erythromycin', in *Progress in Medicinal Chemistry*, Vol. 30, C.P. Ellis and D.K. Luscombe (Eds), Elsevier Science, pp. 57–88.

H. Kirst (1994) 'Semi-synthetic derivatives of 16-membered macrolide antibiotics', in *Progress in Medicinal Chemistry*, Vol. 31, C.P. Ellis and D.K. Luscombe (Eds), Elsevier Science, pp. 265–295.

H. Kirst and G.D. Sides (1989) 'New directions for macrolide antibiotics: structural modifications and *in vitro* activity', *Antimicrob. Agents Chemother.*, **33**, 1413.

H. Kirst and G.D. Sides (1989) 'New directions for macrolide antibiotics: pharmacokinetics and clinical efficacy', *Antimicrob. Agents Chemother.*, **33**, 1419.

R. Leclercq and P. Courvalin (1991) 'Bacterial resistance to macrolide, lincosamide and streptogramin antibiotics by target modification', *Antimicrob. Agents Chemother.*, **35**, 1267.

F.J. Antosz (1992) 'Macrolides' (Antibiotics), in *Kirk-Othmer Encyclopedia of Chemical Technology*, 4th edn, John Wiley, New York, pp. 926–961.

D. Vazquez (1975) 'The macrolide antibiotics', in *Antibiotics III, Mechanism of Action of Antimicrobial and Antitumor Agents*, J.W. Corcoran and F.E. Hahn (Eds), Springer-Verlag, New York, pp. 459–479.

D. Vazquez (1967) 'Macrolide antibiotics – spiramycin, carbomycin, angolamycin, methymycin and lancamycin', in *Antibiotics I, Mechanism of Action*, D. Gottlieb and P.D. Shaw (Eds), Springer-Verlag, New York, pp. 366–377.

F.E. Hahn (1967) 'Erythromycin and oleandomycin', in *Antibiotics I, Mechanism of Action*, D. Gottlieb and P.D. Shaw (Eds), Springer-Verlag, New York, pp. 378–386.

N.L. Oleinick (1967) 'The erythromycins', in *Antibiotics I, Mechanism of Action*, D. Gottlieb and P.D. Shaw (Eds), Springer-Verlag, New York, pp. 396–419.

E.E. Schmid (1971) 'The macrolide group of antibiotics', in *Antibiotics and Chemotherapy*, Vol. 17, Karger, Basel, pp. 52–66.

R.K. Boeckman, Jr. and S.W. Goldstein (1988) 'The total synthesis of macrocyclic lactones', in *The Total Synthesis of Natural Products*, Vol. 7, J. ApSimon (Ed.), John Wiley, New York, pp. 1–140.

S. Masamune (1978) 'Recent progress in macrolide synthesis', *Aldrichimica Acta*, **11**, 23.

S. Masamune, G.S. Bates and J.W. Corcoran (1977) 'Macrolides. Recent progress in chemistry and biochemistry', *Angew. Chem. Int. Ed. Engl.*, **16**, 585.

T.G. Back, (1977) 'The synthesis of macrocyclic lactones, approaches to complex macrolide antibiotics', *Tetrahedron* (Report 46), **33**, 3041.

K.C. Nicolaou (1977) 'Synthesis of macrolides', *Tetrahedron* (Report 27), **33**, 683.

R.R. Wilkening, R.W. Ratcliffe, G.A. Doss, K.F. Bartizal, A.C. Graham and C.M. Herbert (1993) 'The synthesis of novel 8a-Aza-8a-homoerythromycin derivatives via the Beckman rearrangement of (9Z)-erythromycin A oxime', *Bioorg. Med. Chem. Lett.*, **3**, 1287.

W.D. Celmer (1971) 'Stereochemical problems in macrolide antibiotics', in *Symposium on Antibiotics*, S. Rakhit (Ed.), Organic Chemistry Division Section on Medicinal Chemistry in conjunction with The Medicinal Chemistry Division of the Chemical Institute of Canada and the Canadian Society for Microbiologists, pp. 413–453.

Rifamycin antibacterial agents

P. Sensi (1983) 'History of the development of rifampin', *Rev. Infectious Diseases*, **5**, Suppl. 3, S402.

G. Lancini and W. Zanichelli (1977) 'Structure–activity relationships in rifamycins', in *Structure–Activity Relationships among the Semisynthetic Antibiotics*, D. Perlman (Ed.), Academic Press, New York, pp. 531–600.

W. Wehrli and M. Staehelin (1974) 'Rifamycins and other ansamycins', in *Antibiotics III, Mechanism of Action of Antimicrobial and Antitumor Agents*, J.W. Corcoran and F.E. Hahn (Eds), Springer-Verlag, Berlin, pp. 252–268.

W. Wehrli and M. Staehelin (1971) 'Actions of the rifamycins', *Bacteriol. Rev.*, **35**, 290.

W. Wehrli, F. Knusel, K. Schmid and M. Staehelin (1968) 'Interaction of rifamycin with bacterial RNA polymerase', *Proc. Nat. Acad. Sci. Biochem.*, **61**, 667.

7 Quinolone antibacterials

7.1 History and overview

One of the most promising and vigorously pursued areas of contemporary anti-infective chemotherapy is that of the quinolone antibacterials. Broad spectrum, potent activity is available from a relatively simple molecular nucleus which is amenable to many structural modifications. Unlike antibiotics which originate from a natural source such as a fungi or a mold, the quinolone antibacterials are purely synthetic in origin.

The birth of the quinolone antibacterial agents can be traced to the discovery of an antibacterial by-product that was formed during a synthesis of the antimalarial agent chloroquine. An isomer of a key intermediate, namely 7-chloro-1-ethyl-1,4-dihydro-4-oxo-3-quinolinecarboxylic acid **1** (Figure 7.1), was isolated, subsequently prepared in quanity and screened in a variety of biological assays. Although this compound possesses only modest *in vitro* antibacterial activity against some Gram negative bacteria, this finding stimulated efforts to design and synthesize new analogs. In 1962, Lesher described the 8-azaquinolone nalidixic acid **2** (Figure 7.1), which is the first quinolone to be marketed and is generally considered to be the progenitor of the family of quinolone antibacterial agents that followed. Nalidixic acid displays good activity against certain Gram negative pathogens and has been successfully used for some kinds of urinary tract infection. However, bacterial resistance is problematic and furthermore, blood levels of nalidixic acid are typically poor and so the drug is ineffective for the treatment of many systemic infections. For these reasons, it became apparent shortly after its introduction that nalidixic acid would serve only as a complement to existing chemotherapies (e.g. sulfonamide treatment of urinary tract infections) and with these shortcomings the research and development of new quinolones would be slow to come.

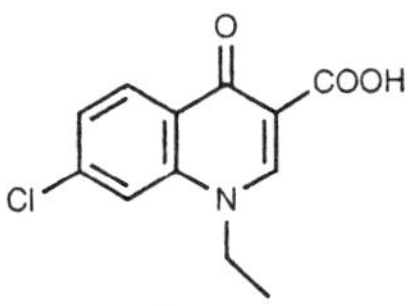

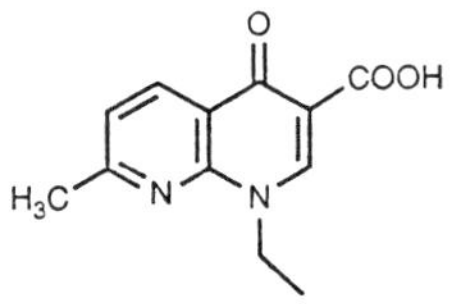

Figure 7.1 The first antibacterial quinolone and nalidixic acid.

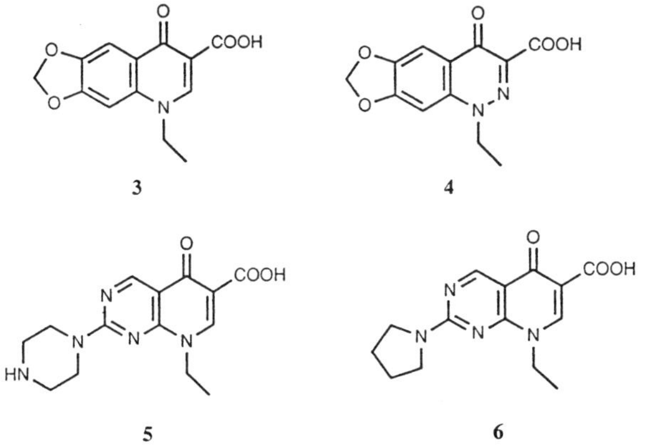

Figure 7.2 Oxolinic acid, cinoxacin, pipemidic acid and piromidic acid.

During the late 1960s and early 1970s, oxolinic acid **3**, cinoxacin **4**, pipemidic acid **5** and piromidic acid **6** (Figure 7.2) were developed. These compounds are somewhat more potent than nalidixic acid but otherwise constituted only marginal improvements with respect to clinical efficacy. The incorporation of additional nitrogen atoms within the bicyclic nucleus (e.g. pipemidic acid and piromidic acid) does not significantly improve activity although the introduction of nitrogen-containing heterocyclic substituents can be notable (see below). As with nalidixic acid, these agents were primarily limited to urinary tract infections, but research in the area became brisk during this time and other structural modifications of the quinolone nucleus soon followed. By 1977, over five thousand quinolone analogs had been prepared and many were investigated in terms of antibacterial activity.

Without doubt, the most significant modification has proven to be the introduction of a fluorine substituent onto the quinolone nucleus. Flumequine **7** (Figure 7.3) was the first fluoroquinolone and clearly demonstrated enhanced spectrum of activity and improved pharmacokinetic properties but it was not developed as a drug. However, the combination of a fluoro substituent and a basic nitrogen heterocycle in a single congener has markedly changed the course of history of the quinolone antibacterials to this day. The early work in this area focused upon the synthesis of piperazinyl-substituted fluoroquinolones. Norfloxacin **8** (Figure 7.3) was the first quinolone belonging to this class and served as a prototype for enoxacin **9**, pefloxacin **10** and ciprofloxacin **11**, all of which are successfully marketed quinolones with potent oral activity (Figure 7.4). These compounds possess a broader spectrum of antibacterial activity and more favorable pharmacokinetic properties compared to older, non-fluorinated derivatives.

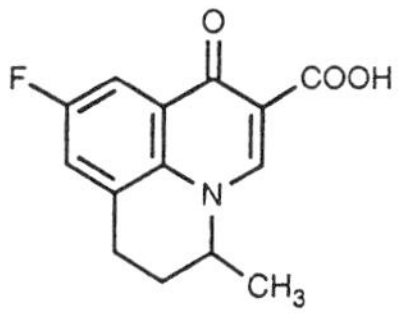

Figure 7.3 Flumequine and norfloxacin.

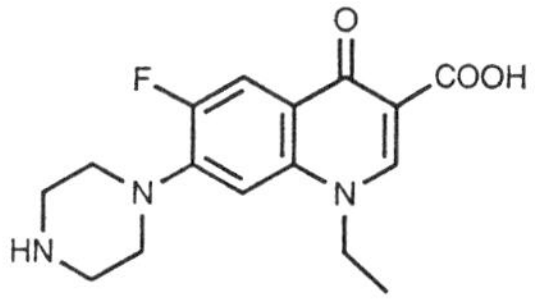

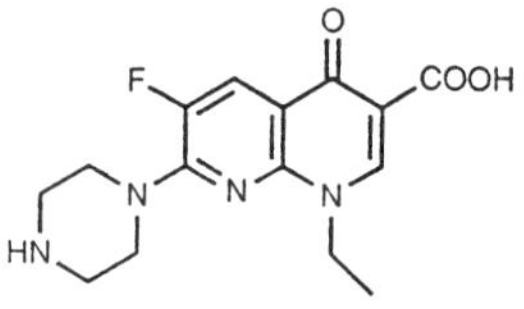

Figure 7.4 Enoxacin, pefloxacin and ciprofloxacin.

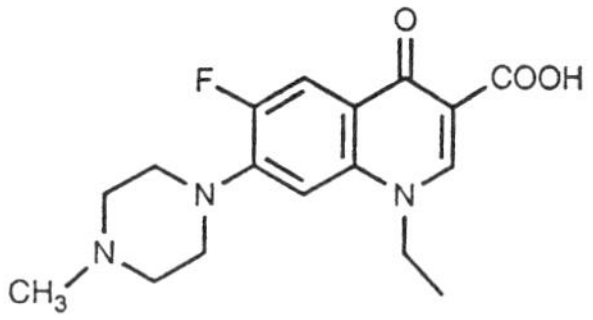

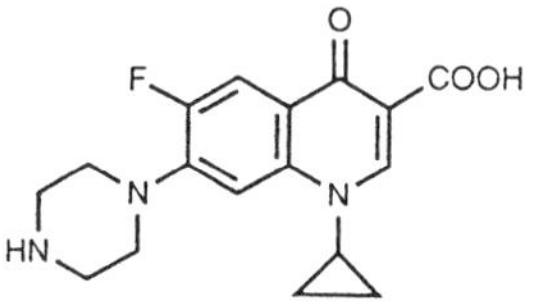

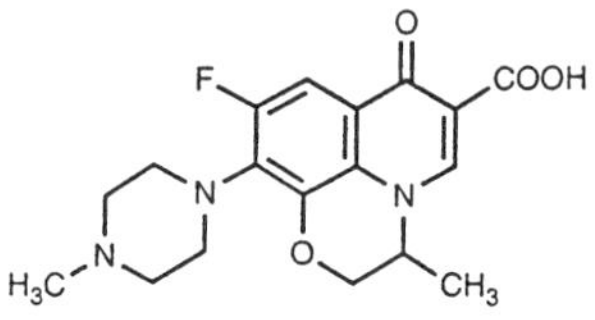

Figure 7.5 Ofloxacin.

Other structural variations were eventually investigated and many have proven to be beneficial in terms of pharmacokinetic properties or potency. The fluoroquinolone ofloxacin **12** (Figure 7.5) contains an additional fused ring and is a very successful agent due to its efficacy using a once-a-day dosing. Other quinolones containing annulated ring systems have been synthesized and some exhibit excellent properties as antibacterial agents.

The nitrogen substituent of the quinolone nucleus traditionally has been an ethyl group although the cyclopropyl group has since become most common. A number of other promising structural variations have been discovered recently. The incorporation of fluoroalkyl and fluoroaryl substituents has afforded fleroxacin **13** and difloxacin **14** (Figure 7.6), respectively, which are potent antibacterials. In addition, alkylamino and alkyloxy modifications are acceptable as evidenced by amifloxacin **15** and miloxacin **16** (Figure 7.6). Recently, the *t*-butyl and various alkynyl derivatives have come under scrutiny.

Research aimed at investigating the role of the heterocyclic appendage has been particularly intense and fruitful. The aminopyrrolidines have been shown to be excellent piperazine surrogates, and many of the newer quinolones fall into this general category. In addition, other heterocycles have been utilized; rosoxacin **17** has a pyridine substituent, irloxacin **18** has

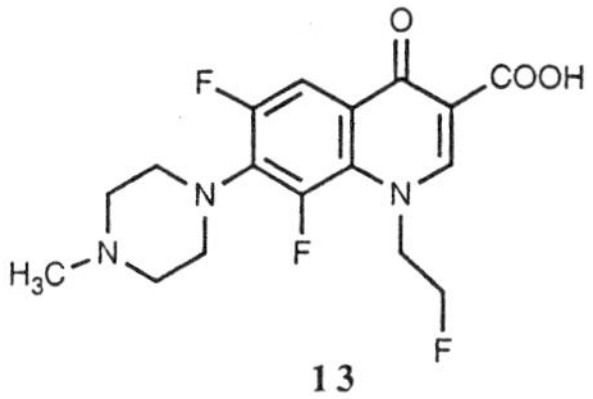

13

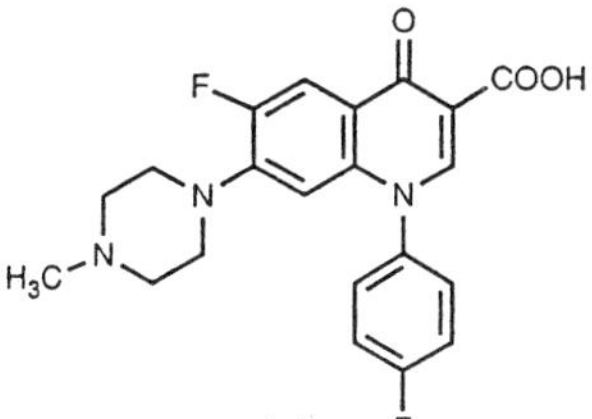

14

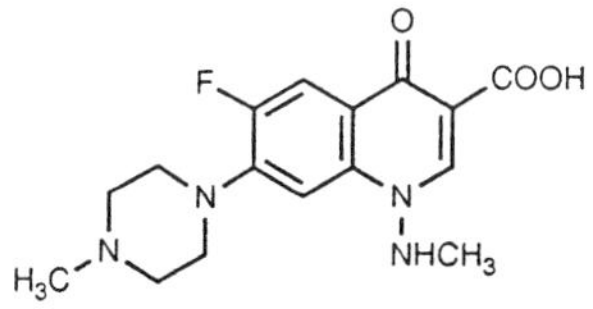

15

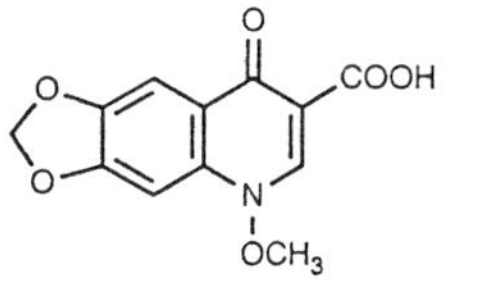

16

Figure 7.6 Fleroxacin, difloxacin, amifloxacin and miloxacin.

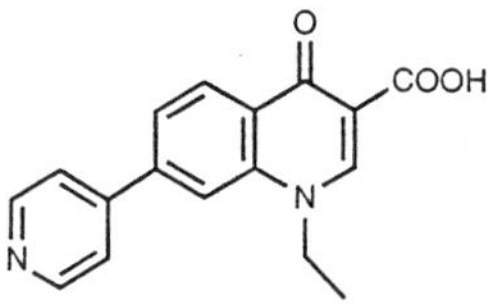

17

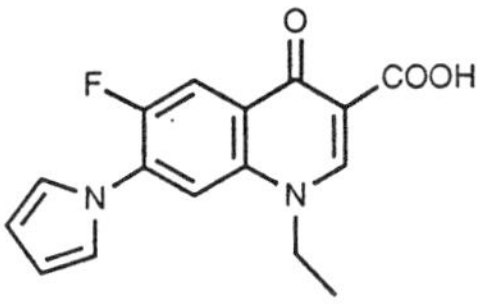

18

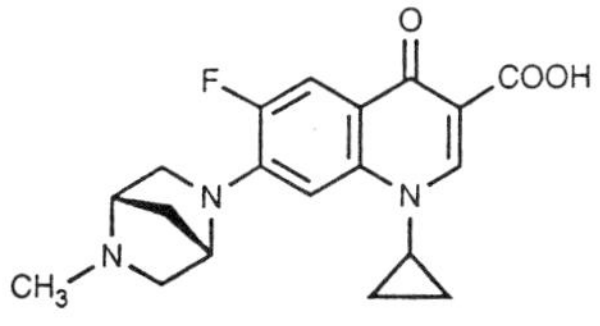

19

Figure 7.7 Rosoxacin, irloxacin and danofloxacin.

a pyrrole ring and danofloxacin **19** (Figure 7.7) has a bicyclic system. The stereocontrolled introduction of small alkyl, hydroxyl and amino groups onto a variety of these ring systems has produced additional variations that are of interest.

The recent generation of ultra potent quinolones such as sparfloxacin **20**, lomefloxacin **21**, tosufloxacin **22** and temafloxacin **23** (Figure 7.8) contain either piperazine or pyrrolidine heterocyclic substituents and all are fluorinated. Unfortunately, temafloxacin was removed from the market shortly after its introduction due to isolated cases of lethal toxicities. Despite the misfortunes of temafloxacin, succeeding quinolone congeners are likely to retain these structural features (the fluoro substituent and a cyclic amine substituent) since derivatives lacking these features usually have substantially reduced potency.

Other findings are reshaping the design of novel quinolone agents. First of all, the inspired use of carboxylic acid isosteres has redefined traditional quinolone structure–activity relationships. Quinolones containing a fused isothiazolone ring are extremely potent but disappointingly, are rather toxic. Secondly, it has become apparent that the choice of the aromatic substituents can be crucial in terms of potency, spectrum of activity and toxicological profile. For example, amino substitution or the incorporation of fluoro or methoxy groups, if properly placed, can afford quinolones with enhanced antibacterial activity compared to their unsubstituted counterparts.

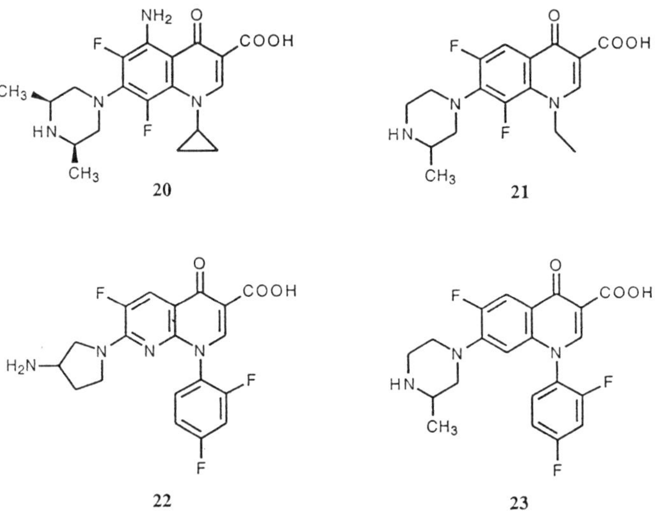

Figure 7.8 Sparfloxacin, lomefloxacin, tosufloxacin and temafloxacin.

7.2 Mode of action: bacterial DNA gyrase inhibition

The mechanism of action of quinolone antibacterial agents has been elucidated in considerable detail but is not understood in its entirety. There remain a number of issues that are poorly defined and that will need to be addressed in the future for an understanding that can lead to superior, rationally designed analogs. However, in order to proceed with a discussion of the mechanism of action of quinolone antibacterials, an overview of topoisomerases and DNA gyrase inhibition is helpful. A great amount of the current knowledge resulted from investigations of type-I and type-II topoisomerases from both bacterial and mammalian systems, and so these issues will be addressed.

It has been demonstrated that quinolones inhibit the action of bacterial DNA gyrase proteins. This heart-shaped enzyme is needed to supercoil and compact bacterial DNA molecules that can exceed 1000 μm in length into the bacterial cell, which is of much smaller dimensions (often approximately 1 μm by 2 μm). This remarkable feat is accomplished by modifying the topology of DNA via supercoiling and twisting of these macromolecules. In order for proper DNA functions (translation, repair, replication, etc.) to occur, the supercoiling must be reversible; specific enzymes control these processes. The discovery of enzymes that are capable of altering topological states of DNA followed from the observation that closed, circular DNA molecules are more compact than their nicked or linear counterparts. When examined by electron microscopy, these compacted molecules exist in a twisted state. Clearly, some mechanism exists that allows DNA molecules to untwist or relax in order for genetic material to be processed. Proteins responsible for regulating DNA topology were identified from both prokaryotic and eukaryotic sources, and were given descriptive names such as the DNA twisting enzymes, swivelases, untwisting enzymes and nick-closing enzymes. These proteins and others that regulate DNA supercoiling and relaxation belong to a family of enzymes now referred to as the topoisomerases.

Supercoiling is just one of the many conformational states that double-stranded DNA can achieve. Common duplex DNA consists of poly-nucleotide strands that wind around each other producing a right-handed helix designated as positive. In this conformation there is one helical turn about every ten base pairs. Left-handed helical DNA as well as other conformational states (e.g. Z-DNA) are well recognized today. The twisting of the DNA duplex helix itself, regardless of its 'handedness', into a superhelical state is called supercoiling. All supercoiled DNA exists in a high energy state relative to its relaxed or extended form. By convention, positive supercoiling has the same handedness as the duplex helix whereas negative supercoiling has a reverse handedness. In bacteria, supercoiled DNA is twisted in the oposite direction to that of the duplex helix.

Closed circular double-strand DNA, whether right- or left-handed helical in nature, is theoretically limited in the possible conformations it can adopt compared to a linear double strand. This constraint results from the elimination of the free ends of the polynucleotide strands and the loss of unlimited spacial freedom in molecules whose termini are joined together in a loop. A mathematical model of this DNA topology has been advanced and is described in terms of the writhe (W) and its relationship to the linking number and twist: $W = L - T$. Twist (T) is the number of helical turns in the DNA molecule in its native conformation. The linking number (L) is the number of times one strand of the double helix passes over the other if the molecule were constrained to lie on a plane. The algebraic sum of the number of supercoils and the number of double helical turns is a constant called the linking number. The writhe (W) represents the geometric contortion of the helix axis and qualitatively, supercoiling is a representation for W that arises when L does not equal T.

Topoisomerases change the linking number of DNA by modifying the degree of coiling. This is accomplished by a three stage process. In the first stage after binding, the enzyme introduces a 'nick' or break in one or two strands of the DNA. In the next stage, the enzyme is able to pass DNA strands through the transient nicks previously formed. This process is normally intramolecular but can be accomplished in an intermolecular fashion to give hybrid DNA molecules under special conditions. The final manipulation is the resealing of the broken DNA strand(s) to reform a duplex which is either 'relaxed' or 'supercoiled' relative to the native DNA substrate and the release of free DNA and enzyme.

A topoisomerase that operates via the introduction and resealing of a single DNA strand break is designated as a type-I topoisomerase. These enzymes are capable of promoting various topological interconversions of DNA that result in a change in the linking number. Prokaryotic (bacterial) topoisomerase-I was originally called the ω protein and the eukaryotic species was often referred to as the DNA twisting enzyme. Both eukaryotic and prokaryotic topoisomerases (type-I) are able to relax closed, circular negatively supercoiled double-stranded DNA. The eukaryotic enzymes act upon either positively or negatively supercoiled duplex DNA whereas the prokaryotic (*Escherichia coli*) topoisomerase-I only removes negative supercoils (unless under special conditions).

At least two mechanistic models describing this process of DNA relaxation have been advanced, both of which have implications for the antibacterial action of the quinolones. The first model is called the 'free rotation model' since it requires that one end of the broken duplex strand remain unattached to the enzyme throughout the relaxation process. A consequence of this assumption is unlimited rotation about the phospho-diester bond in the unbroken strand. The second model is termed the 'enzyme-bridging model' since the enzyme is joined onto both ends of the broken strand and one strand is subsequently passed through a sort of 'gate' formed in this enzyme–substrate complex. Unfortunately it has proven difficult to discern, at the molecular level, the exact sequence of events responsible for type-I topoisomerase-induced relaxation of super-coiled duplex DNA. Part of this issue is complicated by other actions of topoisomerase-I enzymes such as intertwining and renaturation of two single-strand, complementary DNA circles. Prokaryotic topoisomerase-I can also catalyse knotting and unknotting of single-strand DNA circles. In addition, topoisomerases can catenate duplex circular DNA in the presence of DNA aggregation.

Many of the experimental protocols used to study topoisomerase-I enzymes, have carried over to the studies of other topoisomerase species and deserve some comment here. In general, the relaxation process can be halted. This is accomplished by first allowing the enzyme to undergo binding and reaction with DNA and then adding a protein denaturant. The result is the formation of a cleaved species in which the enzyme is found to be covalently attached to one end of the DNA. Prokaryotic topoisomerase-I (*E. coli*) is attached to the 5'-end whereas eukaryotic topoisomerase-I is joined to the 3'-end of the cleaved DNA. In both cases, the newly formed covalent linkage is a consequence of transesterification of a tyrosine hydroxyl residue of the enzyme with the phosphodiester linkage of the DNA and concomitant strand cleavage.

These intermediates are often referred to as 'cleavage complexes' since the toposiomerase is stalled in its action and although successful in strand cleavage and relaxation, is unable to secure its release from the substrate. Studies have shown that specific tyrosine residues are crucial to the formation of the enzyme–substrate covalent complex. For example, topoisomerase-I from *E. coli* undergoes the transesterification reaction at Tyr-319. The replacement of this tyrosine residue with phenylalanine prevents transesterification from occurring and renders the enzyme inactive. Furthermore, even though transesterification is intuitively possible with the serine derivative (since a free hydroxyl group is present), the replacement of tyrosine with serine fails to produce an active enzyme. Lastly, on the basis of homology, the active transesterification site of human topoisomerase-I is Tyr-723; this amino acid is likewise needed for topoisomerase activity.

Type-I topoisomerases vary considerably in terms of substrate specificity and ability to relax supercoiled DNA. Prokaryotic topoisomerase-I (*E. coli*) appears to need a region of single-strand duplex DNA for binding. This requirement is likely to be fulfilled by regions of double helix destabilization that are inherent in highly negatively supercoiled DNA. The enzyme (*E. coli*) is able to relax negative supercoils but not completely, and positive supercoils are unaffected. These activities oppose those of the bacterial topoisomerase-II DNA gyrase (discussed below), but such antagonistic functions are probably essential for maintaining an equilibrium of topological states of DNA necessary for storage and replication. Prokaryotic topoisomerase-I enzymes, while sensitive to inhibition by single-strand DNA, are not targeted by any specific inhibitory drug to date. It is conceivable that this protein may provide a future means for therapeutic intervention in bacterial disease pathogenesis. Inhibition of eukaryotic topoisomerase-I, on the other hand, is the molecular action of the camptothecin class of antitumor agents.

Topoisomerase-II enzymes utilize a transient double-strand break in duplex DNA in order to alter DNA topology which confers a change in the DNA linking number by an integral of two. Because there are two sites of cleavage, more complex mechanistic issues need to be addressed than those posed for topoisomerase-I. In principle, the action of cleaving and regenerating the duplex could be asynchronous and/or could join two different DNA molecules together. However, the resealing of the two ends of the double-strand break appears to be essentially concerted and its integrity highly conserved. Illegitimate strand recombination is possible in certain cases such as in the presence of oxolinic acid and other quinolones.

Interaction of DNA with topoisomerase-II enzymes produces a staggered double-strand break. The 5'-end extends a distance of four nucleotides from the recessed 3'-end bearing free hydroxyl groups. The 5'-end is joined to the enzyme by way of a covalent linkage between the DNA phosphoryl group and a specific tyrosine residue of the protein. This acylation process is analogous to the transesterification described previously for the topoisomerase-I. For example, it is Tyr-122 of a subunit (A) of *E. coli* DNA gyrase protein that becomes covalently attached to the DNA substrate.

There appears to be a substrate length requirement in order for a DNA molecule to be processed by some (prokaryotic) topoisomerase-II enzymes. In some studies, DNA gyrase did not act upon either a 34 or a 77 base pair fragment. Experiments in which the bound enzyme–substrate was subjected to nuclease digestion afforded intact DNA segments of approximately 110–160 base pairs. This suggests that the topoisomerase is able to protect part of the DNA molecule from this enzymatic destruction. Non-specific DNA nuclease (DNAase) is able to cleave nucleotide fragments of approximately 10 base pairs from a 50 base pair region that flanks each side of the protected region described above. It is widely believed that the DNA substrate is wrapped once around on the surface of the topoisomerase. Interesingly, this binding does not protect the DNA molecule from solvent-mediated effects since, for example, DNA alkylation occurs readily in solution.

A unique topoisomerase-II enzyme called DNA gyrase is present in bacteria and is the only topoisomerase known that is able to introduce negative superhelical turns into duplex DNA. The introduction of supercoils into DNA is an energy-demanding process. The DNA gyrase enzyme is able to couple the energy released from the hydrolysis of ATP (adenosine 5′-biphosphate) to drive the formation of supercoils. Gyrase can also remove positive supercoils in the presence of ATP and relax negatively (but not positively) supercoiled DNA in the absence of ATP (Figure 7.9).

Bacterial DNA gyrase is a tetrameric (A_2B_2) 400 kD enzyme composed of two subunits. The 105 kD gyrase A protein is encoded by the *gyr*A gene (also known as *nal*A) and is responsible for the DNA strand breaking and resealing process. Upon denaturation, this protein subunit undergoes covalent bond formation via transesterification with the DNA substrate. The 95 kD gyrase B protein is a product of the *gyr*B gene (also known as

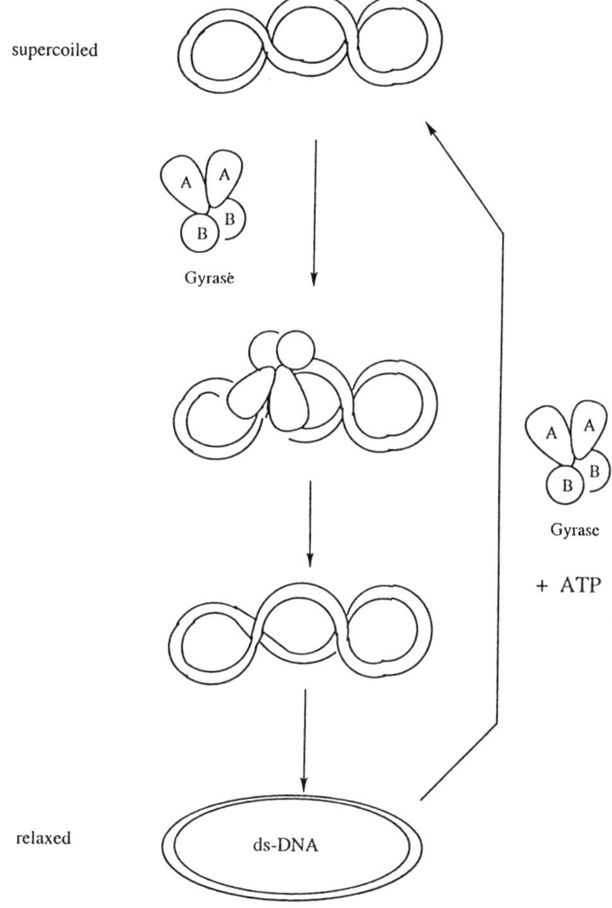

Figure 7.9 DNA gyrase function.

cou) and provides ATPase function (the hydrolysis of adenosine triphosphate with release of energy) which fuels DNA supercoiling. Both functions are necessary for enzymatic activity. Mixing of the two subunits under appropriate conditions reconstitutes activity. Subunit A is the more abundant of the two in cells, and it has been hypothesized that this protein may be incorporated in another topoisomerase since otherwise the subunit ratio (A:B) should be 1:1.

A crystal structure of the N-terminal fragment of the DNA gyrase B protein has recently been reported, and sheds some light on how gyrase works. A 43 kD peptide, obtained by overexpression, was complexed with an non-hydrolysable ATP surrogate, ADPNP (5′-adenylyl-β-γ-iminodiphosphate). Subsequent crystallization and resolution revealed that two distinct domains exist. Domain 1 consists of an array of eight stranded, β-sheet and five helical regions; domain 2 contains four stranded β-sheets that are neither antiparallel nor parallel. A similar motif is found in some common enzymes such as some carboxypeptidases and pancreatic lipase. This gyrase protein fragment forms a dimer and interestingly, the C-terminal domains form the sides of a hole approximately 10 Å across. Presumably this opening can accommodate the passage of a DNA helix, which incidentally is about 20 Å in diameter.

Quinolone antibacterials inhibit bacterial DNA gryase activity and the gyrase A subunit is targeted since resistance is predominantly associated with alterations in the *gyr*A gene. Resistance traced to the *gyr*B gene appears to be an artifact of subunit interaction. Quinolones do not undergo high affinity binding to native double-stranded DNA but rather tightly bind to the DNA substrate–DNA gyrase complex. A ternary complex results consisting of a quinolone–DNA–gyrase entity. A region of highly strained DNA is formed upon the binding of gyrase to DNA, a phenomenon that probably prepares the DNA stand(s) for cleavage. The pocket is essentially single stranded in nature and appears to be crucial for the establishment of a complex incorporating tightly bound quinolone. This supposition is supported by the observation that some topoisomerases exert their effects on single-stranded DNA.

While the details of this ternary complex remain somewhat speculative, at some stage of DNA strand separation, the quinolone agent is able to insert itself within the gyrase–DNA pocket and undergo binding through a number of attractive forces. Hydrogen bonding interactions between the quinolone and the nucleotide bases of the DNA strand appear to be a strong contributor and lipophilic and van der Waals interactions are likely to be established which further stabilize the complex. It is proposed that the C-7 substituent of the quinolone may be pivotal in direct quinolone-enzyme associations. Lastly, the N-1 substituents may contribute to a drug–drug self-association in which quinolone molecules assume a stacked configuration within the DNA strands. This model, as envisioned by Shen and Mitscher addresses many of the issues regarding the structure–activity relationships of quinolones and is regarded as a valuable tool for the design of future agents. Modified depictions of this elegant hypothesis are provided in Figures 7.10 and 7.11.

In this model, the quinolone antibacterial nucleus can be segregated into three different domains (Figure 7.11) that are important for enzyme- and DNA-binding and the formation of the DNA substrate–gyrase–quinolone complex. The dihydro-4-pyridone-3-carboxylic acid A ring provides a hydrogen bonding motif that allows for association with the DNA (nucleotide) bases. Alteration of this moiety (section 7.3) results in loss of antibacterial activity. This aromatic (B) quinolone ring along with the

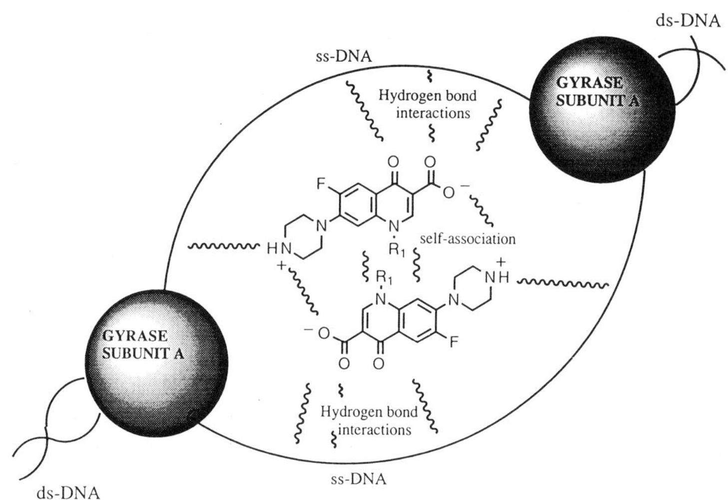

Figure 7.10 Proposed quinolone interaction at binding site of gyrase–DNA complex.

Figure 7.11 Proposed binding motifs of the quinolone antibacterials.

(C-7) heterocyclic substituent that extends from this ring, appear to be responsible for drug–enzyme interactions through lipophilic interactions and hydrogen bonding forces. The aromatic system is essential for activity and the various amine groups attached to this ring, such as a piperazine or aminopyrrolidine, produce greatly enhanced activity and binding. The quinolone nitrogen substituents are implicated in drug self-association within the complex; spacial orientations as well as orbital electronic effects may be operative in producing this effect (see section 7.3 for a supporting discussion of structure–activity relationships).

Not only are quinolones potent gyrase inhibitors but their mammalian counterpart, topoisomerase-II, is several magnitudes less sensitive to the effects of most conventional antibacterial congeners. In essence, bacterial topoisomerases and mammalian topoisomerases have evolved to provide somewhat different functions although both utilize the same mode of action. Since it is the ternary complex of DNA substrate, topoisomerase enzyme and quinolone that appears responsible for inhibition of enzymatic activity, the difference between bacterial DNA and mammalian DNA becomes an issue that needs to be addressed briefly.

The bacterial DNA substrate consists of a single chromosome that is not enclosed by a nuclear envelope. The DNA content accounts for approximately 70% of the chromosome; the remainder consists of 20% RNA and 10% protein, predominantly RNA polymerase. With these components, bacterial DNA gyrase has evolved the capability to supercoil DNA to accommodate it within the dimensions of the bacterial cell. Mammalian DNA, in particular human DNA, is also compacted in shape and in size, in order to fit within the eukaryotic cell. Human DNA is comprised of chromatin (and 46 chromosomes) containing relatively small histone proteins that 'package' human DNA through mechanisms involving aggregation and association. Hence, human topoisomerase neither needs to introduce supercoils into DNA nor appears able to supercoil DNA. Its main function may be that of decatenation of DNA, a process essential to separation of daughter chromosomes following replication.

With these differences of substrate and of enzyme, a pronounced selectively is operative in which quinolones preferentially exert their action upon bacterial DNA gyrase. Quinolones appear to trap the cleavable complex that is formed by the action of gyrase on DNA substrate. In this context, quinolones actually inhibit the catalytic turnover that is expressed in native bacterial cells. Upon denaturation, the gyrase enzyme can be found covalently bound to the 5'-end of a DNA strand through a phosphoryl tyrosine linkage, implicating an transesterification process as previously described. However, since denaturation is required to reveal such adducts, the intimate intracellular associations of enzyme to substrate and to quinolone are not clearly understood.

The result of bacterial DNA gyrase inhibition mediated by quinolones can be observed in terms of gross bacterial cell morphology as well as mechanisms of killing. Bacterial cells treated with quinolone agents often undergo an elongation process which results in the characteristic formation of filaments. In addition, biphasic killing curves are usually observed and there exists a single most bactericidal concentration of quinolone that is unique to each quinolone. Paradoxically, higher concentrations of quinolone can actually lessen bacterial killing. This is probably due, in part at least, to inhibition of RNA biosynthesis, an event that has been shown to be a progressively bacteriostatic effect. Indeed, in the presence of an inhibitor of RNA synthesis such as rifampicin, the bactericidal effects of certain quinolones is abolished. While nalidixic acid suffers this fate and displays only bacteriostatic effects, under these conditions ciprofloxacin and ofloxacin remain essentially unaffected and bactericidal in nature (at high concentrations of quinolone). This observation implies that more than one mechanism of bacterial killing can be operative. Quinolones appear able to act through a blend of these different mechanisms and individual pathogens may possess varying degrees of susceptibility to each mechanism of bacterial killing. The scenario can be quite complicated and many of the newer quinolone agents need to be investigated more thoroughly before any rational conclusions are possible. It is worth noting that the quinolones are not alone in exerting antibacterial properties due to interaction with the bacterial DNA gyrase enzyme. Novobiocin (a coumarmycin) is an antibacterial agent that targets the gyrase B subunit. Its effect is the inhibition of the ATP hydrolysis that drives the supercoiling machinery.

Structure–activity relationships of quinolones have been well established and will be discussed below. Interestingly, certain variations in the traditional quinolone nucleus have afforded compounds with greatly increased mammalian topoisomerase-II poisoning. This effect is detrimental to use in antibacterial chemotherapy but may provide new agents to treat cancer or possibly other hyperproliferative diseases. One of the most interesting advances in this area is the design and synthesis of dihydroxy-anthraquinone-quinolone hybrids **24** (Figure 7.12) that combine the features of the anticancer agent such as the aglycone moiety of daunorubicin with a typical quinolone nucleus.

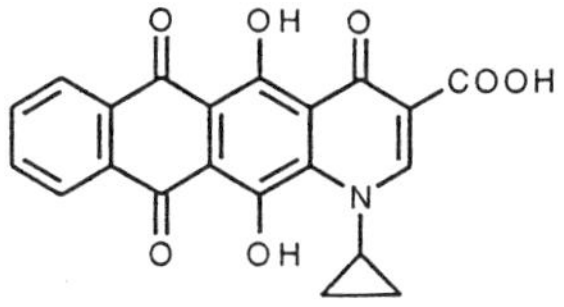

24

Figure 7.12 A dihydroxyanthraquinone–quinolone hybrid: enhanced mammalian topoisomerase II activity.

7.3 Structural features and structure–activity relationships of quinolone antibacterials

The structural requirements of the quinolone antibacterial agents that are necessary for biological activity are rather simple. The crucial feature is merely an *N*-substituted 4-oxo-1,4-dihydroquinoline which contains an acidic functionality in the C-3 position (Figure 7.13). Deviation from this simple heterocyclic assemblage results in loss of activity.

The replacement of the nitrogen atom at position 1 with oxygen, sulfur or carbon (methylene group) leads to inactive derivatives; hence the quinoline system is vital. An N-substituent other than hydrogen needs to be present; compounds containing a simple free amine (N-H) at this position are, at best, weakly active, probably due to their existence in a tautomeric form (4-hydroxyquinoline). Reduction of the aromatic (B) ring system results in loss of antibacterial activity, but interestingly, (B) ring benzene isosteres, such as a 2,3-fused thiophene, can yield compounds retaining some activity. The 4-oxo moiety is a requirement; any alteration to date has led to a loss of antibacterial activity. The presence of an acidic group, usually a carboxylic acid attached to the C-3 center, is needed for gyrase inhibition and antibacterial activity. Certain carboxylic acid surrogates have been favorably substituted for the acid group and have led to a new generation of quinolones that possess potent antibacterial activity.

Specific heteroatom substitutions within this quinolone nucleus are well tolerated in terms of antibacterial activity. In fact, the incorporation of nitrogen at positions 2, 6 or 8 can afford agents which display useful antibacterial activity (Figure 7.14). The replacement of carbon with nitrogen at position 2 affords the cinnoline heterocycle. Cinoxacin **4** (Figure 7.2) is the prototypic aza-quinolone containing this heterocyclic nucleus. The incorporation of nitrogen at both positions 6 and 8 gives rise to pyridopyrimidine heterocycles such as that contained in pipemidic acid **5** and piromidic acid **6** (Figure 7.2). The most favourable variation consists of the introduction of nitrogen at position 8 which affords the 1,8 naphthyridine heterocycle present in the potent agent enoxacin **9** (Figure 7.4). Isomeric 1,5-naphthyridines as well as 1,6-naphthyridines can afford

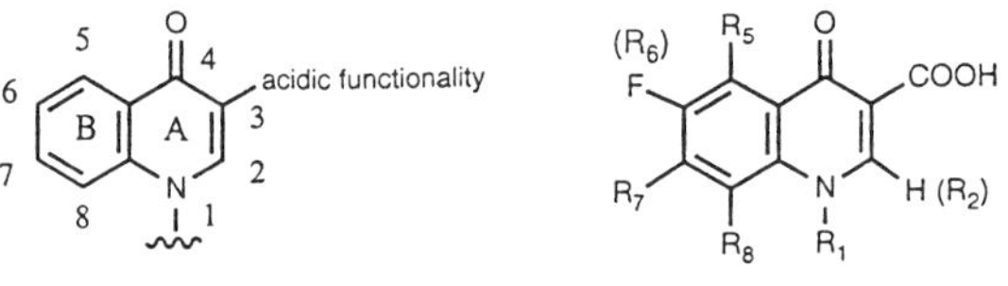

a 4-oxo-1,4-dihydroquinoline containing an acidic functionality at C-3

4-oxo-1,4-dihydroquinoline-3-carboxylic acid

Figure 7.13 Basic quinolone SAR.

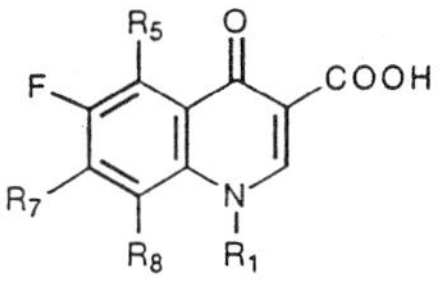

quinolone

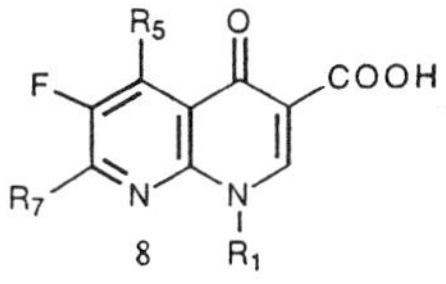

1,8-naphthyridine

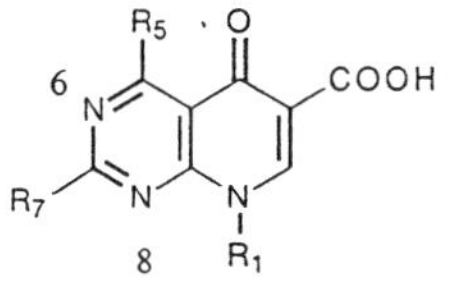

pyridopyrimidine

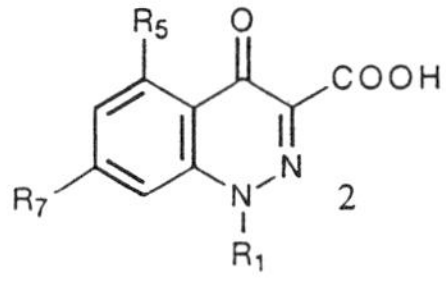

cinnoline

Figure 7.14 Variations of the quinolone nucleus which can retain antibacterial activity.

active compounds, whereas 1,7-naphthyridine analogs are essentially inactive. Incorporation of nitrogen at positions 5 and 8 affords the pyridopyrazine heterocyclic system and analogs of this type can possess notable antibacterial activity. Despite the fact that these variations formally belong to different heterocyclic systems than the 4-oxo-1,4-dihydroquinolines, these analogs are often regarded as part of the family of quinolone antibacterial agents if they possess antibacterial activity and a mode of action characteristic of quinolones.

7.3.1 Substituents at N-1

Until the mid 1980s, the choice of substituents at the N-1 position of the quinolone nucleus was primarily dictated by steric factors. Agents possessing the best activity in a given series incorporated an ethyl moiety or a group of similar size. Many established quinolones contain this characteristic, evident in norfloxacin **8**, enoxacin **9** and pefloxacin **10** (Figure 7.3 and 7.4). Fleroxacin **13** has a fluoroethyl substituent, amifloxacin **15** contains an aminomethyl group and miloxacin **16** has a methoxy group at N-1 (Figure 7.6). Vinyl, chloroethyl, difluoroethyl and trifluoroethyl groups, among others, have been shown also to be suitable N-1 moieties. A study of QSAR (qualitative structure–activity relationship) indicated that a STERIMOL length of $L = 4.2$ Å of the N-1 substituent was optimal and most of these groups fall within approximately 20% of this value (ethyl: $L = 4.1$ Å; methylamino: $L = 3.6$ Å).

In the mid 1980s, the cyclopropyl group was investigated as an N-1 substituent and found to afford compounds capable of exerting extremely potent inhibition of DNA gyrase. Superb *in vitro* antibacterial activity was also observed for both the quinolone and 1,8-naphthyridine series. To this day, the cyclopropyl group is among the first N-1 moieties to be considered in the design of new quinolone antibacterial agents. Ciprofloxacin **11** (Figure 7.4) possesses this group and is the leading quinolone agent on the market today. This is due to its outstanding potency and broad spectrum of activity, properties which set ciprofloxacin as the gold standard of antibacterial quinolones. Interestingly, a cyclopropyl group is predicted to be a worse N-1 group than the ethyl analog accordingly to the earlier QSAR report. The enhancement of activity mediated by the cyclopropyl group has been attributed to a combination of steric and spacial factors. In addition, through-space interactions caused by the likely skewing of this group away from the plane of the quinolone nucleus are probably important. Substituted cyclopropyl groups (e.g. fluoro) are promising variations of this feature. The choice of the C-7 substituent is important when directly comparing N-1 ethyl and N-1 cyclopropyl analogs since certain combinations have proven particularly attractive in each case.

Other factors are operative and not yet fully understood in the selection of an N-1 substituent. Certain bulky fluoro-substituted phenyl N-1 groups have been introduced and afford compounds with excellent potency, despite an obvious violation of the steric argument described above. In particular, 4-fluorophenyl and 2,4-difluorophenyl substituents are notably exceptional compared to other isomeric (2-fluoro or 3-fluorophenyl) analogs. Difloxacin **14** (Figure 7.6) is a prototypic agent belonging to this class of quinolones and naphthyridines. While many agents that have a fluorinated aryl group at N-1 are extremely potent, their future may be overshadowed by the withdrawal of temafloxacin from the market after reports that comparatively high doses of this agent may have contributed to the deaths of several patients with impaired renal function.

Other unusual N-1 substituents have proven to be promising as exemplified by the N-1 *t*-butylquinolone **25**, the alkynyl derivative **26**, as well as the bicyclopentyl **27** and oxacyclobutyl **28** congeners (Figure 7.15). In general, these novel compounds display good antibacterial activity, and have provided a number of companies with a protected position regarding patent rights. A compilation of active N-1 quinolone substituents is shown in Figure 7.16 with an emphasis on overall (and general) *in vitro* potency.

In summary, the N-1 cyclopropyl group probably confers the best broad spectrum activity (*in vitro*) although in some cases, the ethyl, vinyl, (di)fluorophenyl and fluoroethyl moieties can be comparable (or even better depending upon the rest of the substitution pattern). Other N-1 substituents discussed here can bestow notable activity against certain organisms. For example, the N-1 *t*-butyl group imparts enhanced activity against Gram positive bacteria and the fluorophenyl N-1 group generally improves activity against anaerobic pathogens.

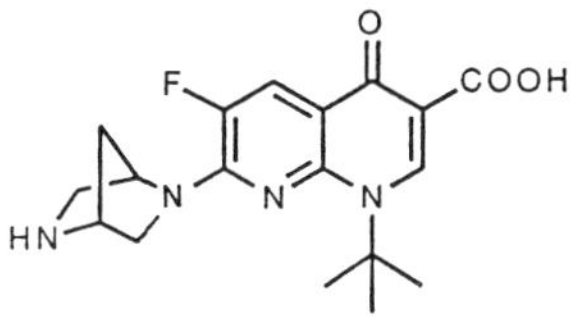

25

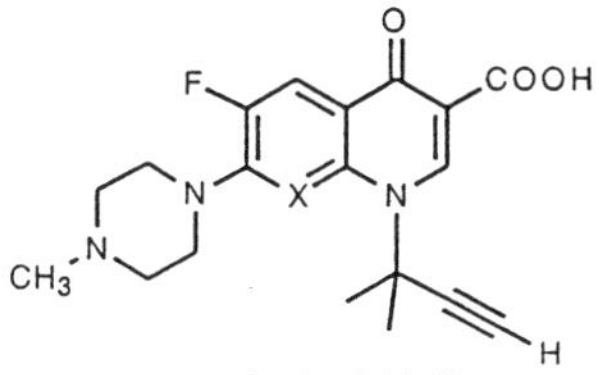

26: X = C-H, N

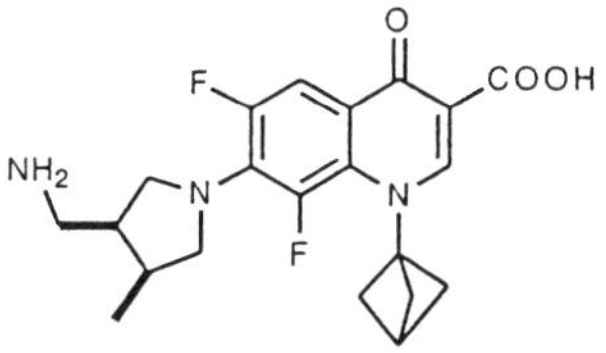

27

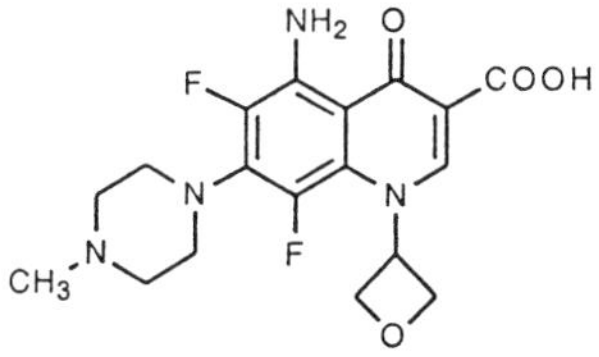

28

Figure 7.15 Quinolones with novel N-1 substituents.

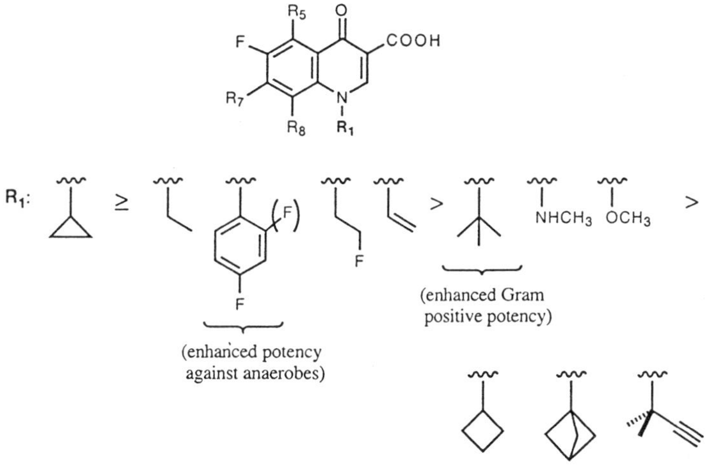

Figure 7.16 Quinolone SAR: N-1 substituents.

7.3.2 C-2 substituents

The simple replacement of C-2 hydrogen has generally proven to be disadvantageous. For example, C-2 methyl and C-2 hydroxyl quinolones are useless as antibacterial agents. However some derivatives containing a suitable C-1, C-2 ring have recently been shown to possess notable activity. Specifically, the incorporation of a sulfur atom directly onto the C-2 position has been especially attractive as evident in the sulfur-bridged quinolones **29** (Figure 7.17) and corresponding benzothiazolone derivatives

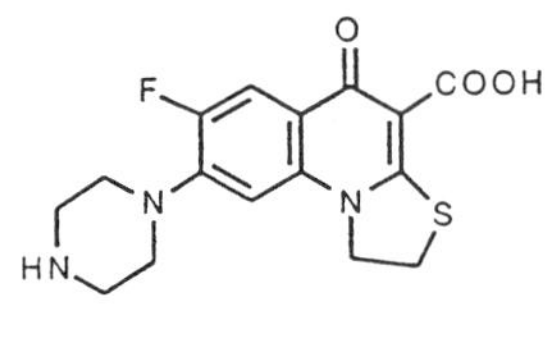

29

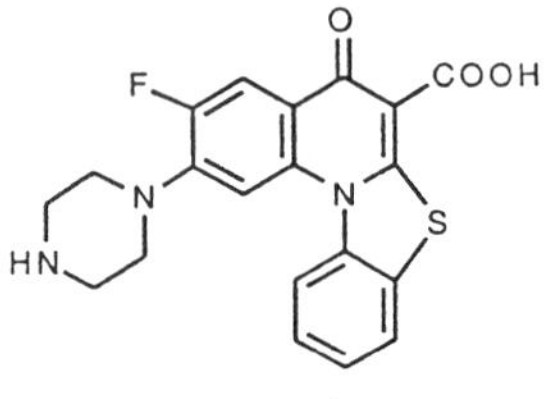

30

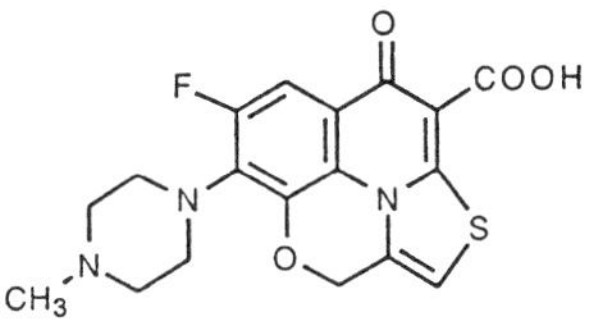

31

Figure 7.17 Quinolones with sulfur substituents at C-2.

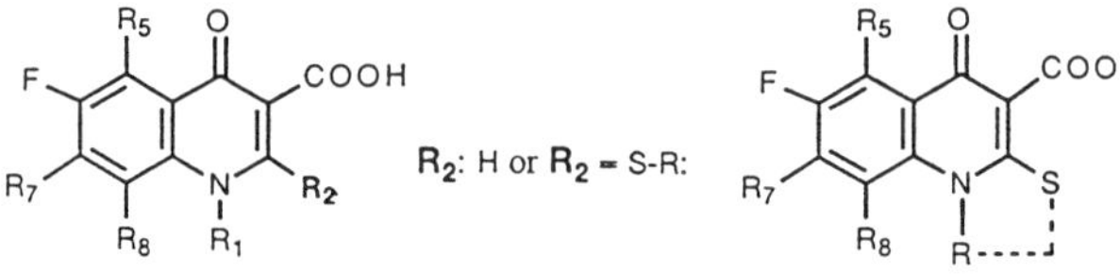

R_2: H or R_2 = S-R:

Figure 7.18 Quinolone SAR: C-2 substituents.

30 (Figure 7.17). Some encouraging analogs also have a C-1, C-8 bridge in place, such as the tetracyclic quinolone **31** (Figure 7.17). Appropriate C-2 substituents are summarized by the representation given in Figure 7.18. A hydrogen atom at the C-2 position is the best choice but a C-2 sulfur atom is obviously tolerated if part of an additional ring system.

7.3.3 C-3 substituents

Without doubt, the C-3 carboxylic acid moiety is most commonly encountered. Other acidic groups such as sulfonic acid, phosphonic acid, as well as derivatization as an ester results in loss of antibacterial activity. Even the well-recognized carboxylic acid surrogate, the 1H-tetrazol-5-yl group, is an unacceptable variation. The incorporation of an aldehyde (formyl) group at the C-3 position, as well as certain labile carboxylate esters, can afford derivatives which display antibacterial properties but activity is due to (*in vivo*) conversion to the corresponding carboxylic acids.

In the late 1980s, a successful modification was described that eliminated the need for a C-3 carboxylic acid. A fused isothiazolone ring was identified as serving as a carboxylic acid mimic. In fact, C-2, C-3-isothiazoloquinolones **32** (Figure 7.19) have been shown to be more potent

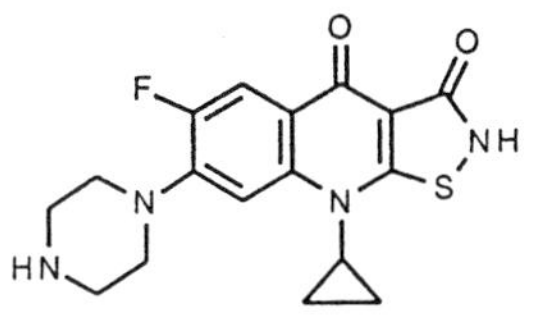

32 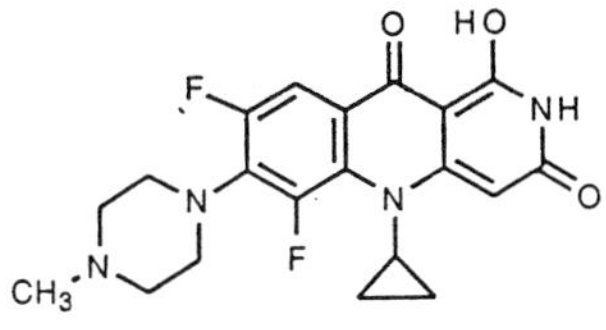   **33**

Figure 7.19 Quinolones with carboxylic acid surrogates.

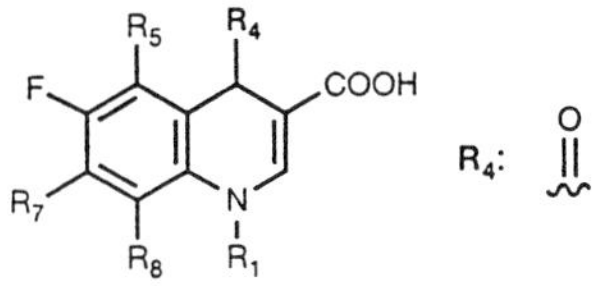

Figure 7.20 Quinolone SAR: C-4 position.

than their conventional carboxylic acid analogs, but appear to suffer from some undesirable properties such as insolubility and mammalian toxicity. The future of compounds containing this feature is unclear but certainly this modification should be regarded as a landmark in quinolone SAR. Other carboxylic acid surrogates (**33**, Figure 7.19) have since been synthesized but are also extremely insoluble.

To date, a free carboxylic acid at the C-3 position, or a C-2, C-3 fused isothiazolone ring, are acceptable with regard to antibacterial activity and there are few other options available, or likely to be contemplated.

7.3.4 C-4 substituents

The C-4-oxo group of the quinolone nucleus appears to be essential for antibacterial activity (Figure 7.20). Replacement with a 4-thioxo group or with a 4-sulfonyl group leads to loss of activity. A few other variations have been briefly investigated with similarly disappointing results. Consequently, research into other modifications has yet to be undertaken or reported with success. Although it may be possible that a suitable 4-oxo replacement exists, current research efforts are likely to be focused in other directions that hold more promise.

7.3.5 C-5 substituents

The choice of a quinolone C-5 substituent appears to be dictated mostly by steric regulations, and perhaps by other factors not yet fully appreciated. The incorporation of an amino group at the C-5 position has proven beneficial in terms of antibacterial activity in the case of some quinolone derivatives **34** (Figure 7.21). However a C-5 amino group is detrimental if introduced onto a pyridopyrimidine nucleus. Naphthyridine analogs are being investigated but there are no reports substantiated at this time. Alkylated or acylated C-5 amino quinolones are at best only weakly active, providing the alkyl or acyl groups are small in size. Larger groups at this position give compounds devoid of antibacterial activity. Furthermore, the presence of a C-5 amino group in conjunction with a somewhat labile C-8 substituent (chloro) is unfruitful. Other heteroatom substitution (oxygen, sulfur) at the C-5 center has yet to produce any derivatives with properties that surpass the variations described above.

Alkyl substitution at the C-5 position is beneficial only when the substituent is a methyl group or of similar size; larger groups result in loss of antibacterial activity. Several C-5 methylated quinolones are exciting in terms of potency and spectrum of activity (**35** and **36**, Figure 7.21). Lastly, fluorination at the C-5 position is tolerated in some cases but is more likely to be used as a synthetic handle upon which further elaboration (nucleophilic addition/fluoride elimination) can be performed. A simple depiction of these findings is given in Figure 7.22.

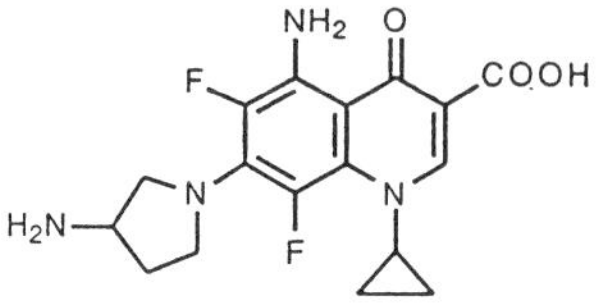

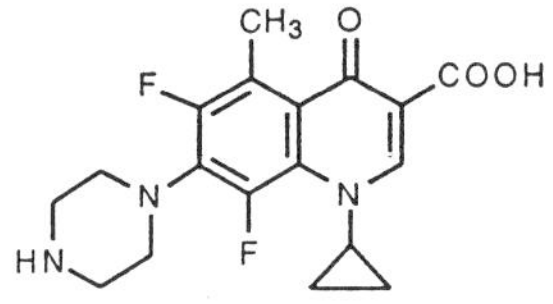

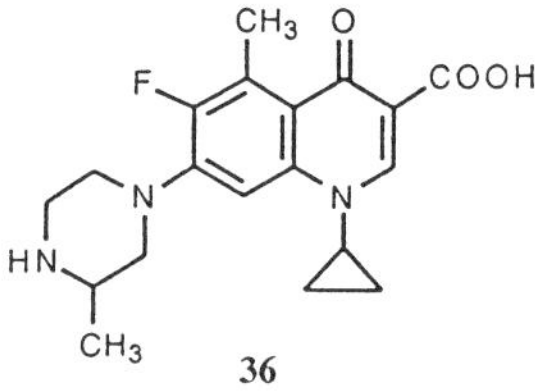

Figure 7.21 Quinolones with a C-5 amino or C-5 alkyl substituent.

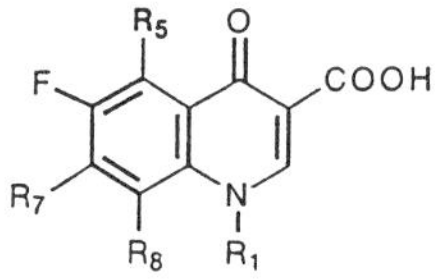

Figure 7.22 Quinolone SAR: C-5 substituents.

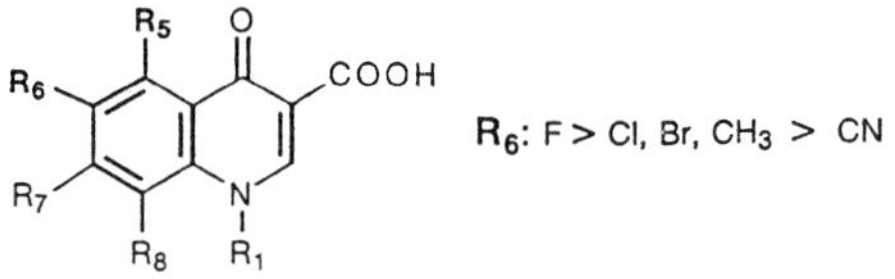

R_6: F > Cl, Br, CH$_3$ > CN

Figure 7.23 Quinolone SAR: C-6 substituents.

7.3.6 C-6 substituents

Without doubt, the incorporation of a fluorine atom at the C-6 position of a quinolone was monumental. Both gyrase inhibitory potency as well as *in vitro* antibacterial activity are enhanced in congeners containing a C-6 fluorine as compared to their unsubstituted (C-6 hydrogen) counterparts. Chloro and bromo substituents are also tolerated as are C-6 methyl or cyano groups, in terms of antibacterial activity but these variations are overshadowed by C-6 fluoroquinolones (Figure 7.23).

7.3.7 C-7 substituents

The choice of the C-7 substituent is a key issue which continues to guide the design of new agents. In general, small alkyl, heteroatomic (alkyloxy, alkylamino, alkylthio) or halogen substituents at C-7 are unfavorable. Despite nalidixic acid and flumequine, most small C-7 groups such as methyl, chloro, amino, alkylamino and others are undesirable. The introduction of a piperazine group at the C-7 position was a landmark development and most successful quinolones to date (**5** (Figure 7.2), **8** (Figure 7.3), **9, 10, 11** (Figure 7.4), **12** (Figure 7.5), **13, 14, 15** (Figure 7.6)) have this feature or structurally related modifications. Aminopyrrolidine (**22** (Figure 7.8), **34** (Figure 7.21)) derivatives have been shown to be among the most promising of the piperazine surrogates investigated so far, and several C-7 aminopyrrolidinyl quinolones are candidates for the market of the future. For example, quinolone **37** (Figure 7.24) is one of the most potent quinolones discovered to date. The utilization of piperazine and aminopyrrolidine groups containing various small alkyl substituents are under scrutiny. Aminoalkylpyrrolidines are of particular interest as exemplified by the *gem*-dimethyl C-7 pyrrolidinyl naphthyridine **38** and its desmethyl analog **39** (Figure 7.24).

Rather surprisingly, smaller cyclic C-7 groups can furnish compounds with worthwhile activity indicating that this position is quite tolerant. The C-7 azetidinylnaphthyridine **40** and the C-7 aminocyclopropylquinolone **41** are among the most promising of these agents (Figure 7.25). Bicyclicalkylamine (e.g. **19** Figure 7.7) and spirocycloalkylamine C-7 substituents are among the latest variations.

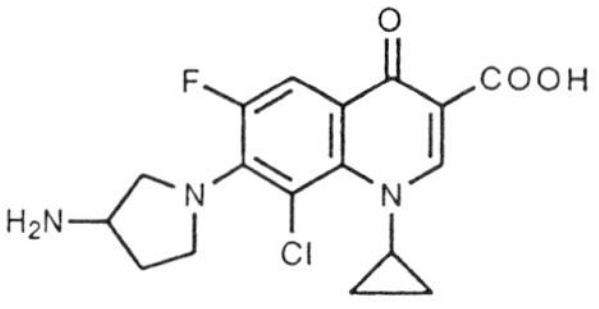

Figure 7.24 Quinolones with pyrrolidine C-7 substituents.

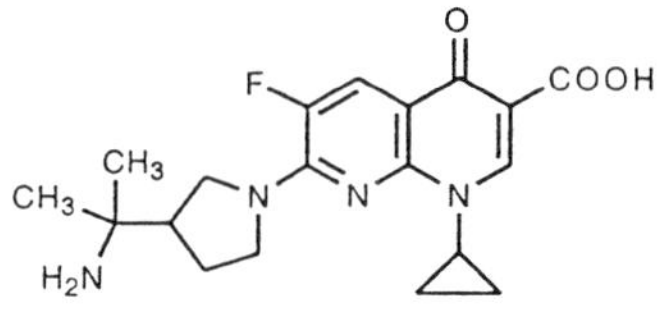

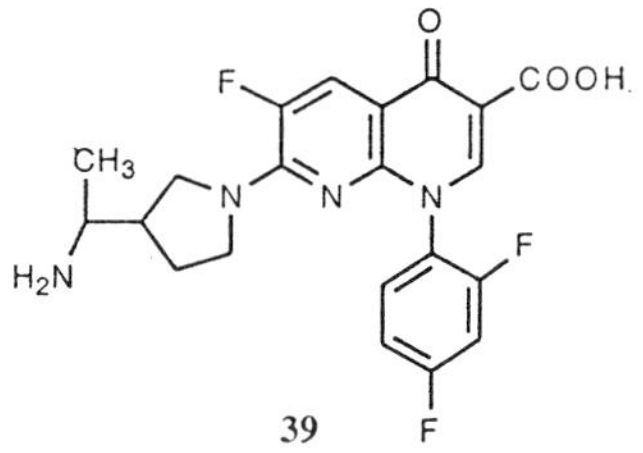

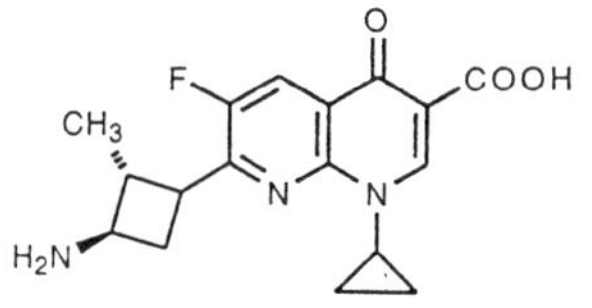

Figure 7.25 Quinolones with novel small ring C-7 substituents.

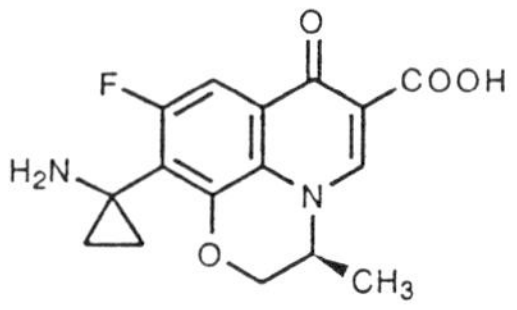

Other heterocyclic systems such as thiazole, imidazole, pyrrole and pyridine (**17** and **18**, Figure 7.7) can be suitable C-7 substituents in terms of antibacterial activity. More elaborate versions have recently emerged, the more potent being the C-7 dimethylpyridinyl moiety quinolone **42** and the C-7 isoindolinylquinolone, **43** (Figure 7.26). However these C-7 groups containing aromatic moieties often produce agents that display significant mammalian cytotoxicity. Quinolone **44** (Figure 7.26) possessing a *para*-hydroxyphenyl C-7 substituent is more active against mammalian topo-isomerase than bacterial DNA gyrase.

The C-7 substituent has also been elaborated in several ways in the hope of improving physical and pharmacokinetic properties. Amino acids have been joined to the free nitrogen atom of aminopyrrolidine C-7 groups to give quinolone prodrugs (e.g. **45**, Figure 7.27) with improved solubility. In another approach, the nitrogen atom attached to the C-7 position has been replaced with an sp^2 carbon atom giving active quinolones and naphthyridines (**46**, Figure 7.27). The most effective C-7 groups, consider-ing overall *in vitro* potency, are generalized below (Figure 7.28).

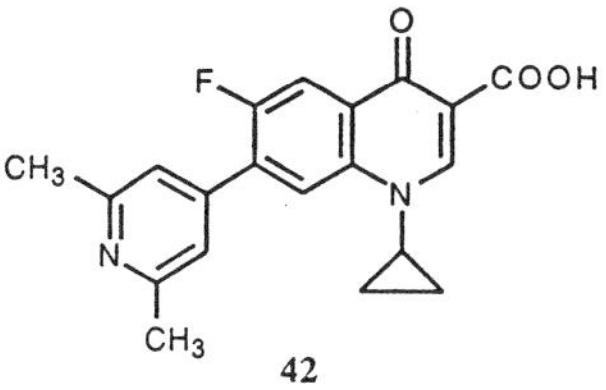

Figure 7.26 Quinolones with C-7 substituents that enhance activity against mammalian topoisomerase II.

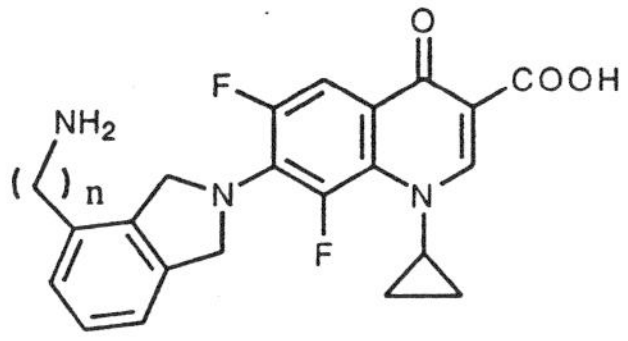

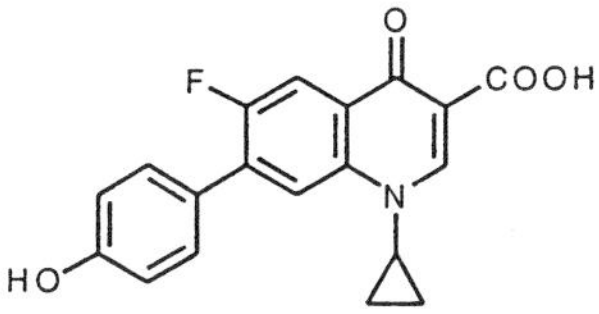

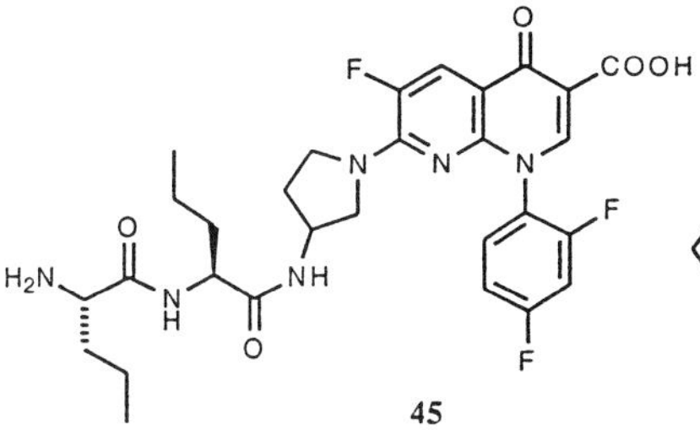

Figure 7.27 Other variations: an amino acid quinolone prodrug and an sp^2 carbon C-7 linkage.

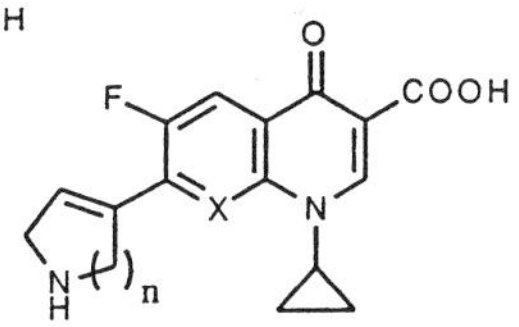

In general, C-7 aminopyrrolidine groups confer enhanced activity against Gram positive bacteria compared to piperazinyl analogs, but tend to suffer from insolubility. Alkylation of the distal piperazine amino functionality can often improve activity *in vivo* but at the expense of potency against some Gram negative pathogens.

7.3.8 C-8 substituents

A hydrogen atom at the C-8 position or a nitrogen atom (a naphthyridine) are the most common variations at this position of the quinolone nucleus. However, compact, lipophilic groups such as C-8 fluoro, trifluoromethyl and methoxy groups have recently gained recognition due to enhancement of activity (Figure 7.29). In general, a C-8 fluoro substituent offers superb potency against Gram negative pathogens while a C-8 methoxy moiety is

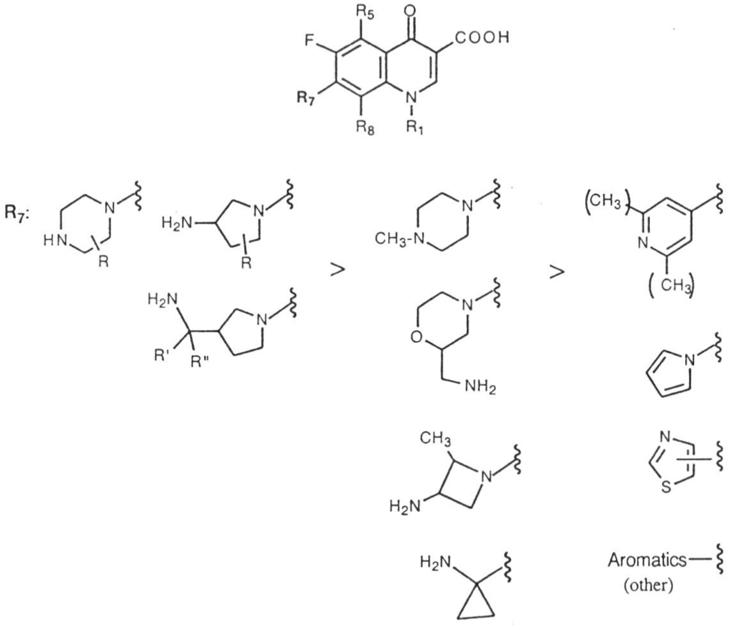

Figure 7.28 Quinolone SAR: C-7 substituents.

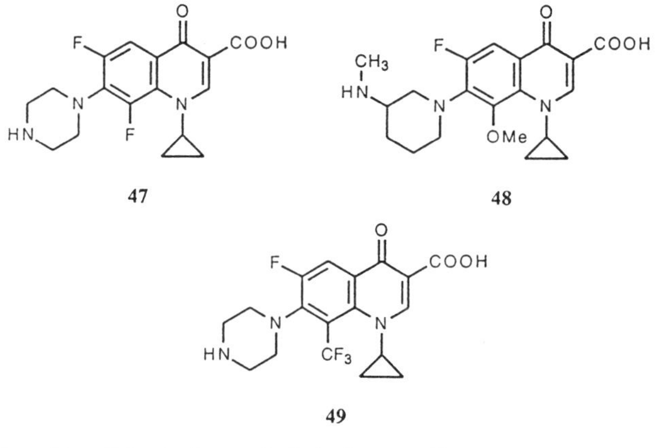

Figure 7.29 Promising C-8 substituents include fluoro, trifluoromethyl and methoxy groups.

exquisitely active against Gram positive bacteria. The 8-chloroquinolone **37** (Figure 7.24) is extremely potent, as is the 8-fluoro analog of ciprofloxacin **47** and the 8-methoxyquinolone **48** (Figure 7.29). Larger homologs are less desirable, illustrated by 8-ethoxyquinolones which show reduced activity against Gram negative pathogens (compared to 8-methoxy analogs). Trifluoromethyl groups have been introduced at this position (e.g. **49** (Figure 7.29)) and often demonstrate reduced phototoxicity which is problematic with the corresponding halogenated derivatives.

Alkylation at the C-8 position is tolerated especially if the substituent is lipophilic or small in size. In this context, vinylic, propargylic and methyl groups can furnish compounds possessing modest antibacterial properties but amino, hydroxyl or nitro groups at this position abolish activity (Figure 7.30).

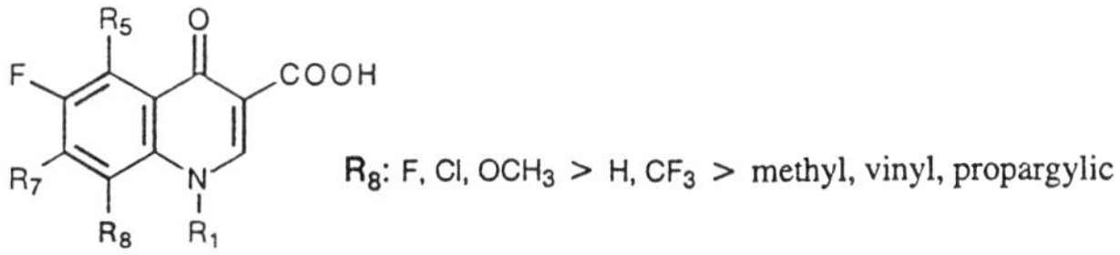

R_8: F, Cl, OCH$_3$ > H, CF$_3$ > methyl, vinyl, propargylic

Figure 7.30 Quinolone SAR: C-8 substituents.

7.3.9 C-1,C-8 bridged variations

The joining of the N-1 group to the C-8 position is often a favorable modification. Examples include the oxazine derivatives flumequine **7** (Figure 7.3) and ofloxacin **12** (Figure 7.5). An oxydiazine ring analog **50** (Figure 7.31) is in development for animal use, and rufloxacin **51** (Figure 7.31), a thiazine derivative has been marketed. Other modifications such as piperazinyl fused ring systems can also produce potent antibacterial agents. With most 1,8-bridged species of this type, methylation of the distal nitrogen of the C-7 piperazinyl substituent is necessary for acceptable (*in vitro*) activity.

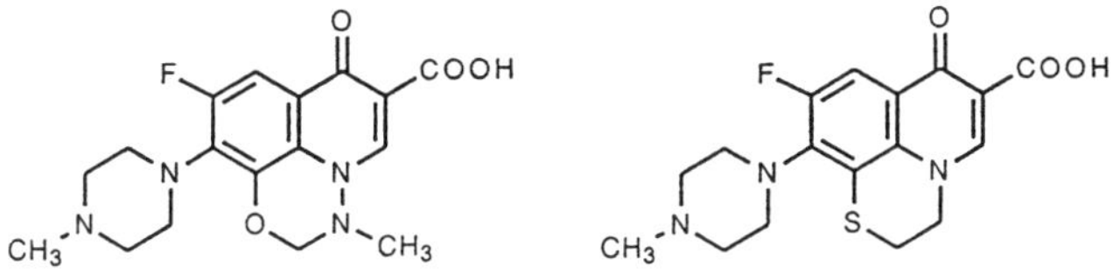

50

51

Figure 7.31 C-1,C-8 bridged quinolones.

7.3.10 Miscellaneous

Substituted C-7 groups as well as C-1,C-8 bridged groups can contain chiral centers which complicate the discussion of quinolone SAR. Some quinolones have yet to be studied in optically pure form since only the most potent racemates are targeted for purification or synthesis as an enantiomer (or diastereomer, in the case in which more than one chiral center is present). However, a few general findings can be briefly described here. The (S)-enantiomer of ofloxacin **52** (Figure 7.32) is considerably more potent *in vitro* (8–64 times with respect to most organisms) than the (R)-enantiomer, and this trend is similarly observed in analogous tricyclic quinolone nuclei. However, the effects exerted by a chiral center appear to diminish somewhat as the stereogenic center is moved away from the quinolone nucleus. For example, the (S)-enantiomer of simple C-7 aminopyrrolidinyl quinolones such as **53** (Figure 7.32) display the greatest potency but the effect is notably less in magnitude (2–8 times).

Novel quinolone agents are being designed and synthesized at a rapid pace and it is impossible to address all issues and future prospects within the scope of this chapter. Certainly it can be stated that quinolone SAR, despite extensive investigation, is still undergoing exciting changes. For example, the benzonaphthyridine **54** (Figure 7.33) deviates from conventional quinolone SAR at positions 1, 6 and 7 but unquestionably bears a striking resemblance and activity to traditional quinolone agents.

Another interesting class of quinolone-like antibacterial compound analogs has recently been reported. The traditional quinolone β-keto-carboxylic acid was structurally modified to incorporate additional hydrogen-bond donor–acceptor motifs which are hypothesized as playing a key role in binding to the nucleotides of the DNA strands within the DNA

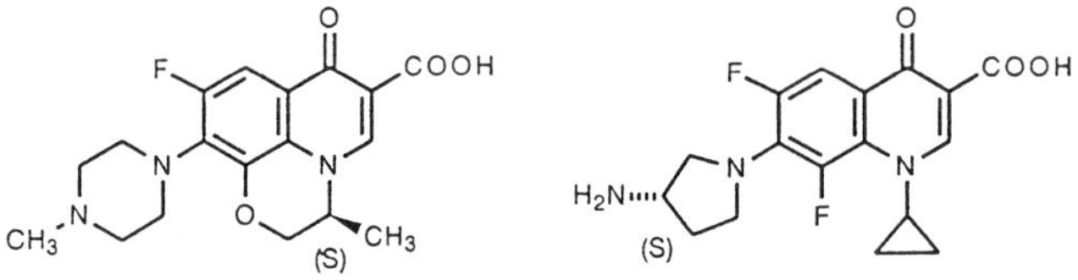

Figure 7.32 Most potent enantiomers of ofloxacin and C-7 aminopyrrolidinyl quinolones.

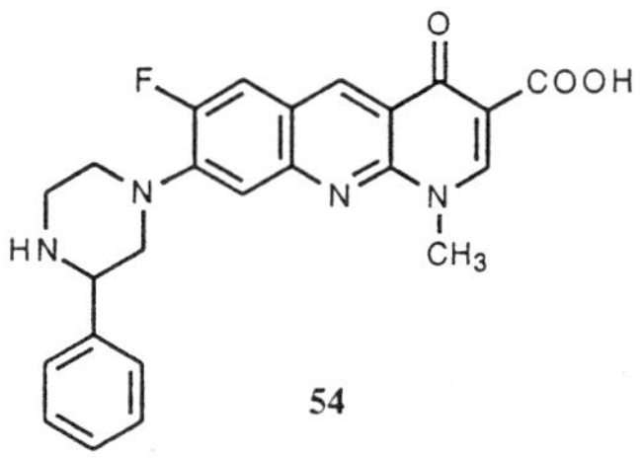

Figure 7.33 A benzonaphthyridine antibacterial.

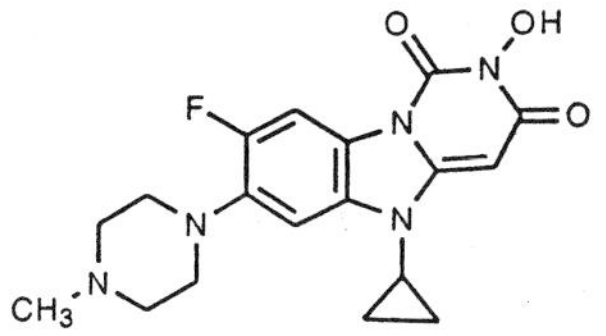

55

Figure 7.34 A pyrimidobenzimidazole antibacterial.

substrate–gyrase–drug complex. Pyrimidobenzimidazoles **55** (Figure 7.34) have been synthesized and in general are broad spectrum antibacterial compounds. This elegant work further refines quinolone SAR and will probably stimulate new research and potentially superior chemotherapeutic agents.

In summary, a comprehensive overall of quinolone SAR is given below (Figure 7.35).

7.4 Synthetic approaches to quinolone antibacterial agents

Quinolone antibacterial agents are wholly of synthetic origin. An incredible amount of elegant chemical synthesis has been accomplished to date that has allowed for the manufacture of thousands of compounds. While it is not possible to consider every effort, landmark achievements will be discussed here so as to provide an appreciation and insight into the methods used for the construction of quinolone antibacterial agents.

There are two major routes to synthesize the quinolone nucleus and from these procedures a wide variety of quinolone antibacterials has been obtained. The first approach is accomplished through the use of a Gould–Jacobs cyclization (Figure 7.36). An appropriate unsubstituted aniline (A) is reacted with dialkylalkoxymethylene malonate to produce the resultant anilinomethylene malonate (B). A subsequent thermal process induces Gould–Jacobs cyclization to afford the corresponding 4-hydroxyquinoline-3-carboxylate ester (C). The following operation is the alkylation of the quinoline amine which is usually accomplished by reaction with a suitable alkyl halide to produce the quinolone-3-carboxylate ester (D). A limitation of this method arises when bulky alkyl groups, such as *t*-butyl or even cyclopropyl are to be introduced as the N-1 substituent. The quinoline nitrogen atom, being rather weakly nucleophilic, is frequently inefficient in its attack on such crowded electrophiles. Another limitation of this approach is that N-1 aryl congeners cannot be obtained.

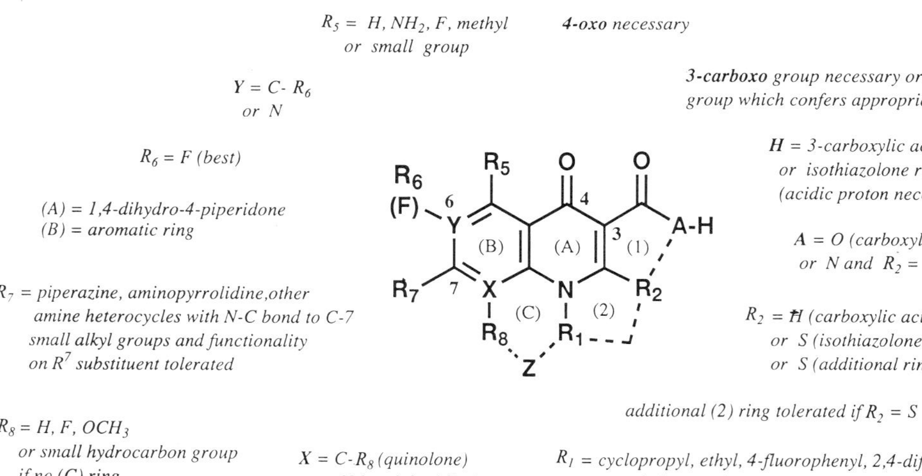

Figure 7.35 Summary of features which confer optimum antibacterial activity to the quinolone nucleus.

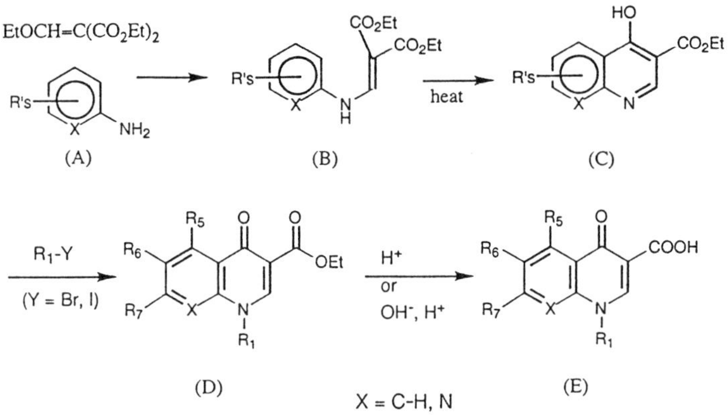

Figure 7.36 Gould–Jacobs cyclization to quinolones.

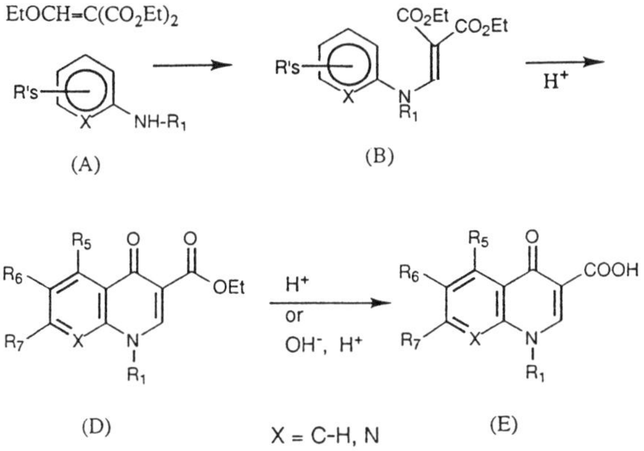

Figure 7.37 Modified Gould–Jacobs method to quinolones.

A modified approach (Figure 7.37) resorts to the use of a mono-substituted aniline (A) as starting material which avoids subsequent N-1 amine alkylation. A strong acid (such as polyphosphoric acid) is often needed to induce cyclization, directly resulting in the formation of an N-1-alkyl-4-oxoquinolone-3-carboxylate ester (D). In either case, the final manipulation is acidic or basic hydrolysis to cleave the ester generating the biologically active free carboxylic acid (E) (or carboxylate salt).

Appropriately substituted aminopyridines and aminopyrimidines can be elaborated using this methodology to yield naphthyridines (8-aza-quinolones) as well as pyridopyrimidines (6,8-diaza-quinolones). Indeed the Gould–Jacobs cyclization and its modifications constitute an extremely valuable process by which quinolone antibacterial agents have been constructed. Oxolinic acid, norfloxacin, pefloxacin, fleroxacin, ofloxacin and enoxacin are among the quinolones that have been synthesized using this approach.

A second synthetic strategy that has been extensively used to construct the quinolone or naphthyridine nucleus employs an intramolecular nucleophilic aromatic addition process, sometimes referred to as a 'cycloaracylation'. A halogenated benzoic acid derivative (A) is acylated to afford an unsymmetrical malonate (B) to which an aminomethylidene group is subsequently introduced (C). Treatment with a base induces cyclization to produce the quinolone (D) or naphthyridine (Figure 7.38). As in the case of the Gould–Jacobs method, the N-1 amino group can be introduced either prior to or after construction of the bicyclic nucleus. In most cases, the N-1 substituent is introduced prior to cyclization since the corresponding primary alkyl or aryl amines are readily available and undergo efficient reaction with the alkyloxymalonate intermediate. The problems encountered with the use of bulky alkylating agents described earlier are usually avoided.

This cycloacylation process is used to synthesize quinolones and naphthyridines that have hindered alkyl groups at N-1. The synthesis of ciprofloxacin and of a promising N-1 *t*-butyl naphthyridine illustrate the usefulness of this method (Figure 7.39).

Since electrophilic arylating agents are rare and often pose traditional problems (indiscriminate selectivity, inefficient reactivity with weakly nucleophilic species), the N-1 aryl quinolones (e.g. difloxacin **14** (Figure 7.6), temafloxacin **23** (Figure 7.8)) and naphthyridines (e.g. tosufloxacin **22** (Figure 7.8)) are prepared using the above method (Figure 7.40).

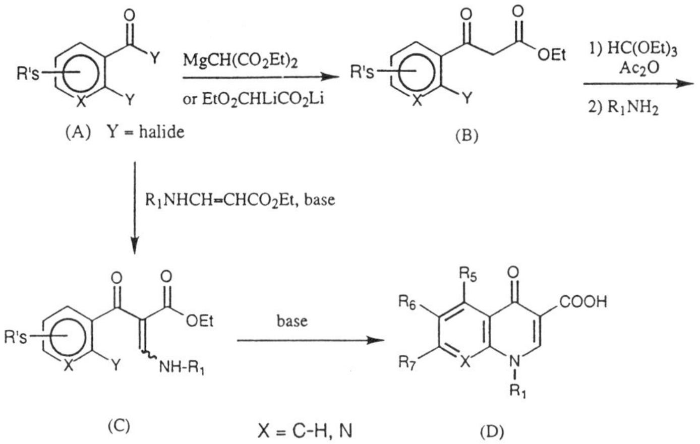

Figure 7.38 Intramolecular nucleophilic aromatic addition process (cycloacylation) to quinolones.

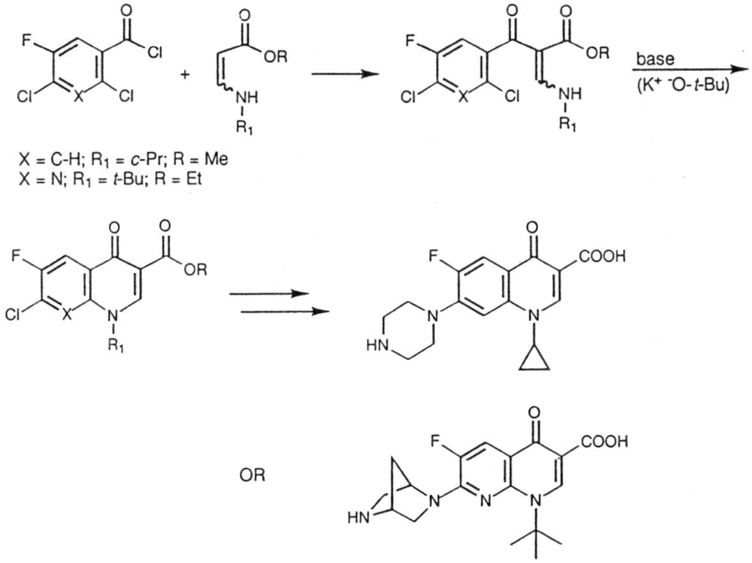
base
(K+ -O-t-Bu)
X = C-H; R1 = c-Pr; R = Me
X = N; R1 = t-Bu; R = Et
OR

Figure 7.39 Introduction of bulky N-1 substituents.

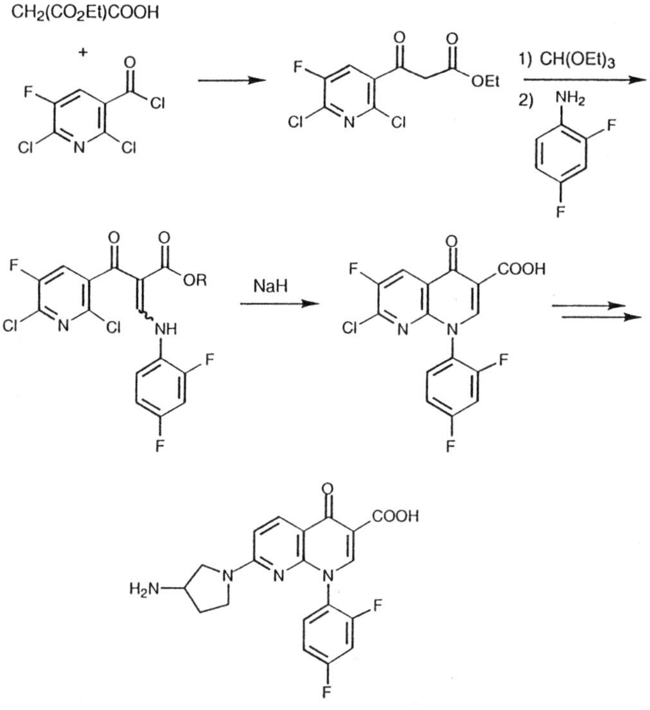
CH2(CO2Et)COOH
+
1) CH(OEt)3
2) NH2
NaH

Figure 7.40 Introduction of N-1 aryl substituents.

Halogenated anilines are common starting materials for both Gould–Jacobs cyclization procedures as well as cycloaracylations. The main reason is that a C-7 chloro or fluoro group can be placed prior to quinolone/naphthyridine formation. After cyclization, the halogen C-7 group is readily displaced upon reaction with nucleophilic amines, and therefore allows for the introduction of a variety of C-7 substituents at a late stage of the synthesis (Figure 7.41). This is significant since the C-7 groups are one of the most important moieties in influencing spectrum of overall potency and pharmacokinetic properties of the resultant quinolone antibacterial agent. The rather rare quinolone agent which contains a carbon bound to the C-7 position is not amenable to this approach (except for simple alkyl groups) and a more elaborate aniline starting material is required that usually necessitates its own separate synthesis.

The incorporation of a C-7 piperazine group is usually straightforward since the piperazine molecule is either symmetrical (and it does not matter which amine center undergoes bond formation) or has one amine 'blocked' (e.g. acylated, to allow for subsequent modification, or alkylated, such as a (C-7) 4-methylpiperazinyl group). More elaborate piperazines with alkyl or heteroatom appendages may have one amine center that is slow to react at C-7 due to adjacent steric bulk. However other groups may require the protection of one amine center in order to install the C-7 moiety properly. The introduction of pyrrolidinyl C-7 substituents can be somewhat more complicated. The most common variations encountered are amino-pyrrolidines or aminoalkylpyrrolidines. A number of protecting groups

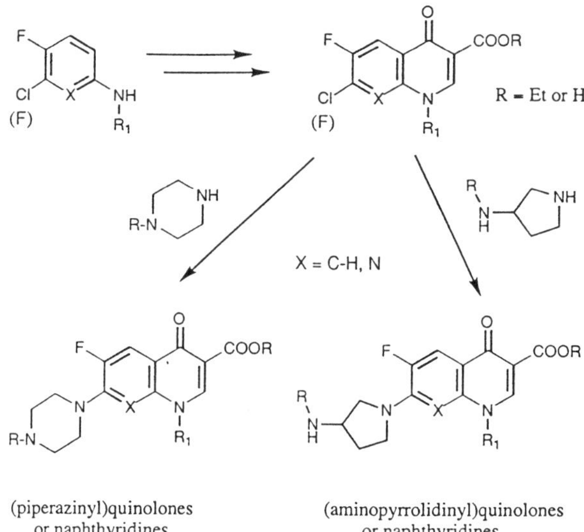

(piperazinyl)quinolones
or naphthyridines

(aminopyrrolidinyl)quinolones
or naphthyridines

Figure 7.41 Introduction of the C-7 cyclic amine substituent.

have been used to protect the exocyclic C-3 pyrrolidinylamine to allow the endocyclic secondary amine to undergo the desired bond formation at the C-7 center. Sometimes protection is not necessary; simple 3-aminopyrrolidine preferentially undergoes reaction at the secondary amine center since this position is more nucleophilic than the exocyclic primary amine. However, protecting groups are often utilized in carbon–nitrogen bond formation since selectivity is dependent upon rection conditions and the nature of the amine itself. Azetidinyl C-7 moieties are similar to the pyrrolidines although the azetidine nitrogen is even more nucleophilic. Bicyclic amines also follow these general considerations. Lastly, some valuable C-7 groups (e.g. C-7 morpholinyl) do not pose an issue of regioselectivity, and others (4-pyridinyl, imidazolyl, alkyl, etc.) are obtained in completely different ways.

An alternative to the introduction of C-7 amine groups is demonstrated in the synthesis of irloxacin **18** (Figure 7.7) using the Gould–Jacobs method (Figure 7.42): in this case, the C-7 amino group requires protection so as to prevent alkylation by ethyl bromide.

As alluded to earlier, quinolones and naphthyridines that possess a carbon–carbon bond linkage at the C-7 position have been prepared. The synthesis of these compounds is achieved using palladium coupling of stannanes with suitable aryl or alkyl halides or triflates (Figures 7.43 and 7.44). Simple C-7 alkyl groups (e.g. methyl as in the case of nalidixic acid) and others (e.g. methylenedioxy bridge of oxolinic acid) are in place prior to either mode of quinolone/naphthyridine nucleus formation (Gould–Jacobs cyclization or cycloaracylation).

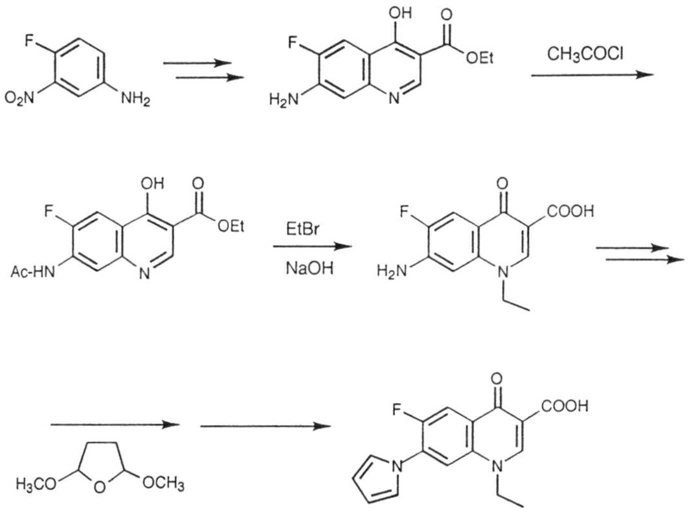

Figure 7.42 Synthesis of irloxacin: construction of the C-7 substituent from the C-7 amine.

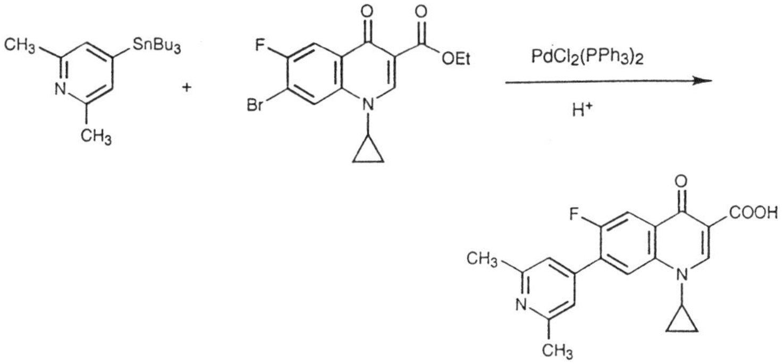

Figure 7.43 Introduction of a dimethylpyridinyl (DMP) C-7 substituent.

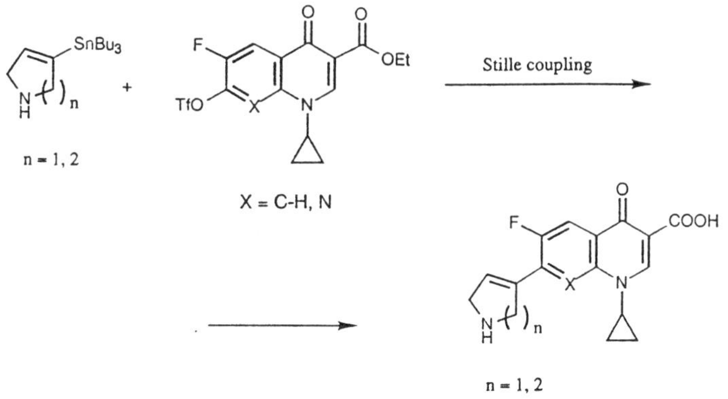

Figure 7.44 Synthesis of quinolones with a C-7 carbon center.

Interestingly C-1, C-8 bridged quinolones have been prepared using either the cycloaracylation or Gould–Jacobs method, as illustrated by the synthesis of ofloxacin (Figure 7.45) and of rufloxacin (Figure 7.46), respectively. In an approach to (racemic) ofloxacin, a tricyclic quinolone nucleus was assembled in one transformation from the ethanolamino-methylidene ketoester depending upon the reaction conditions (fluoride base). The rufloxacin precursors needed to be S-protected to avoid (C) ring cleavage; sulfide oxidation to the sulfoxide allows for the introduction of the C-7 piperazine group. Last, the sulfoxide was reduced via phosphorous(III)-mediated deoxygenation.

The incorporation of most C-5 substituents have been achieved by using suitable anilines as starting material (e.g. 5-methylated quinolones/naphthyridines) or by making use of a synthetic 'handle' such as a nitro (Figure 7.47) or fluoro (Figure 7.48) group. The 6,8-difluoro-N-1-cyclopropyl quinolone (PD 124816) (Figure 7.47) and sparfloxacin (**20** in Figure 7.8) (Figure 7.48) exemplify the latter strategy.

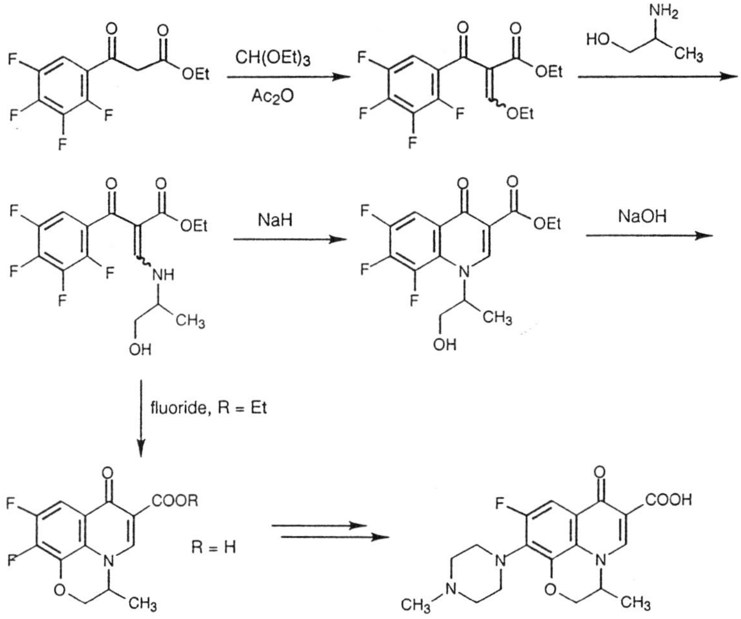

Figure 7.45 Synthesis of ofloxacin using the cycloaracylation method.

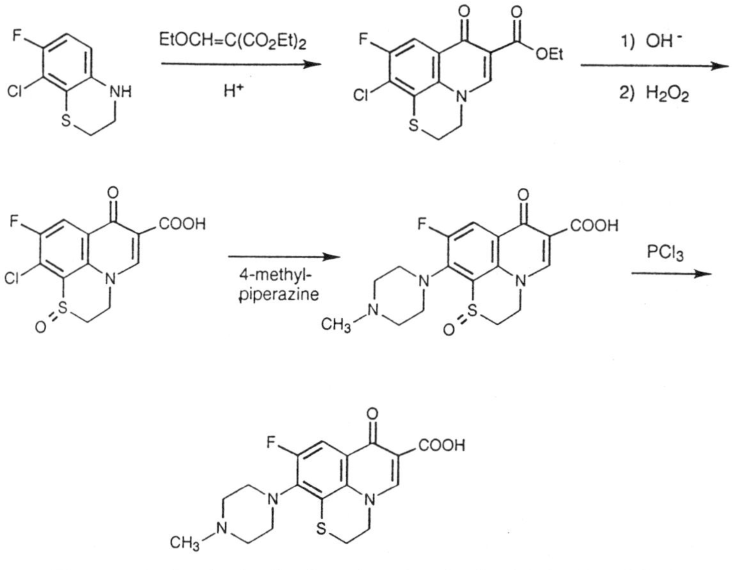

Figure 7.46 Synthesis of rufloxacin using the Gould–Jacobs method.

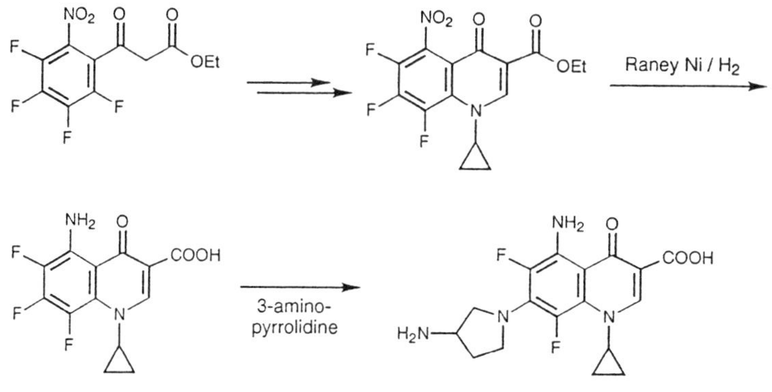

334 ANTIBACTERIAL CHEMOTHERAPEUTIC AGENTS

Figure 7.47 Introduction of a C-5 amino substituent from a C-5 nitro precursor.

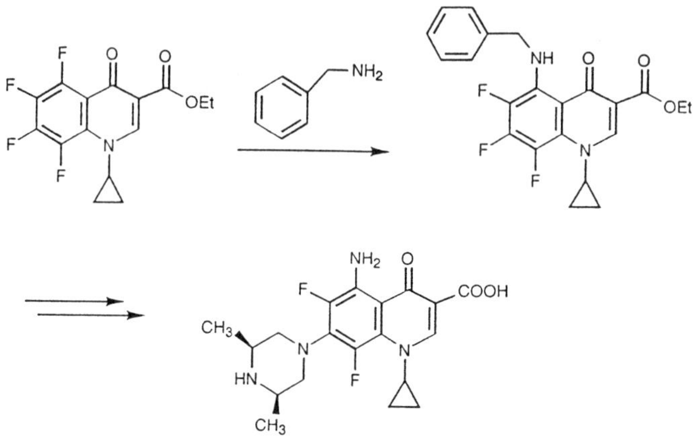

Figure 7.48 Introduction of a C-5 amino substituent from the C-5 fluoro precursor.

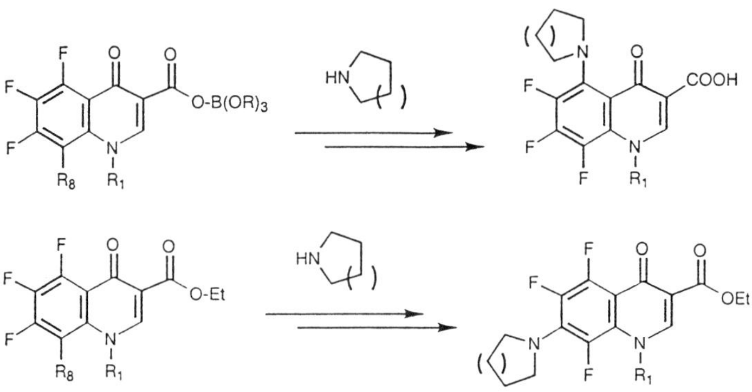

Figure 7.49 Regioselectivity of amine addition can be directed to C-5 or to C-7 positions.

Nucleophiles have been regioselectivity introduced at C-5 or at C-7 by the displacement of corresponding halogen groups. This selectivity is dependent upon substrate and reaction conditions, but in general the C-5 fluoride is displaced from quinolone borate esters by amine bases, whereas the C-7 position is more reactive in the case of quinolone alkyl esters (Figure 7.49).

Quinolone agents which contain a sulfur atom at C-2 require a more elaborate synthetic approach regardless of whether the Gould–Jacobs cyclization method or cycloaracylation technology is employed. In one example, the sulfur substituent destined for the C-2 quinolone position is introduced early in a synthesis of the tetracyclic quinolone (i.e. reaction of an aniline with carbon disulfide to produce an thiocyanate) (Figure 7.50).

Substituents at the C-8 position can be present in the starting material (e.g. alkyl, alkyloxy, trifluoromethyl) or introduced into the appropriate aniline (Gould–Jacobs cyclization) or benzoic acid (cycloaracylation) precursor using a number of aromatic manipulations. For example, a chlorination procedure has been used to prepare a C-8 chloro derivative but a number of subsequent alterations are needed in order to afford a benzoic acid substrate amenable to cyclization (Figure 7.51). In general, C-8 substitution is best handled case by case since a variety of benzene and aniline chemistry can often by applied.

A latent hydroxyl group equivalent has been used to access

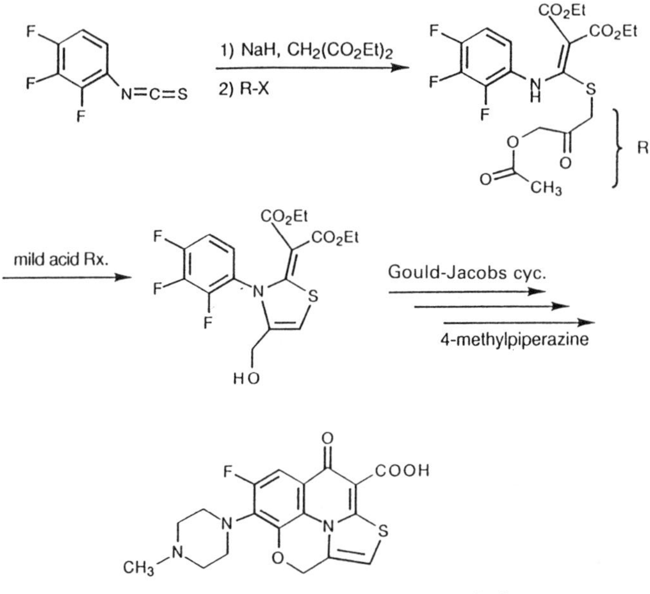

Figure 7.50 Introduction of a C-2 sulfur substituent.

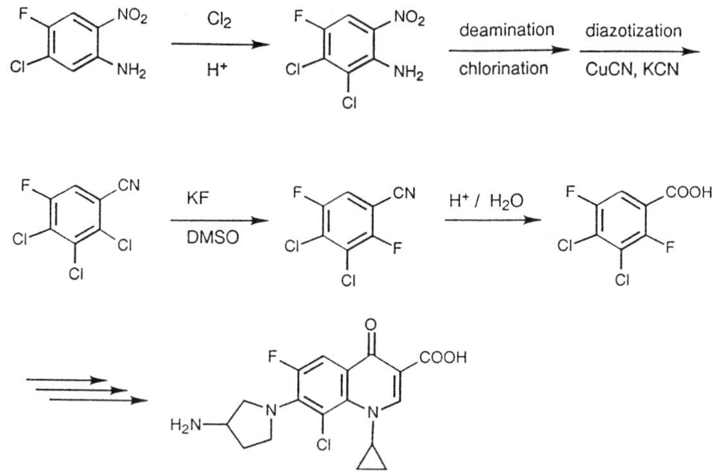

Figure 7.51 Introduction of a C-8 chloro substituent.

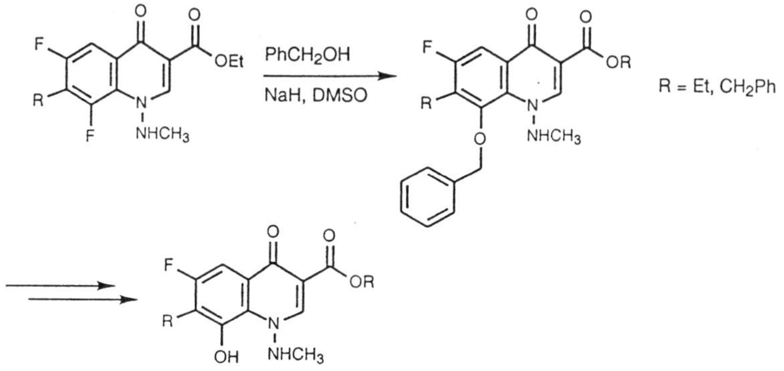

Figure 7.52 Introduction of a C-8 hydroxyl substituent.

8-hydroxyquinolones but this strategy is complicated by transesterification at C-3 (Figure 7.52). Later work described the formation of the desired oxydiazine ring which avoided this difficulty without the need for a hydroxyl group surrogate (Figure 7.53).

Carboxylic isosteres such as C-2,C-3-fused thiazoloquinolones/ naphthyridines are quite involved as summarized in Figure 7.54. The key steps are the introduction of the thiomethyl group prior to cyclization to either a quinolone or naphthyridine nucleus. The thiomethyl moiety is replaced by the free mercaptan via an oxidation to the corresponding sulfoxide followed by reaction with hydrosulfide. Treatment with hydroxylamine-*O*-sulfonic acid in the presence of a base affords the desired isothiazolo derivative.

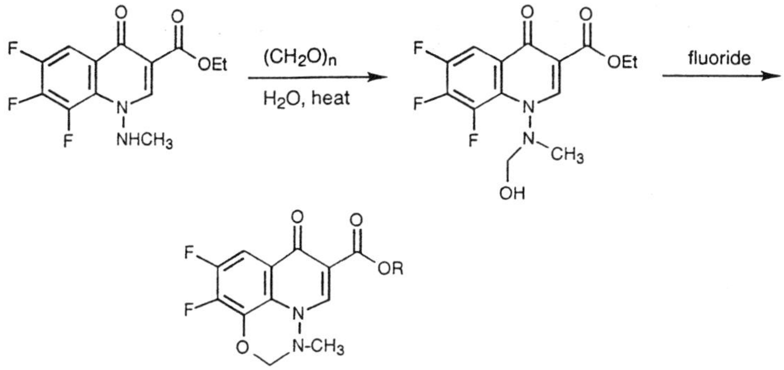

Figure 7.53 Introduction of a C-8 oxygen functionality: formation of a C-1,C-8 bridged quinolone.

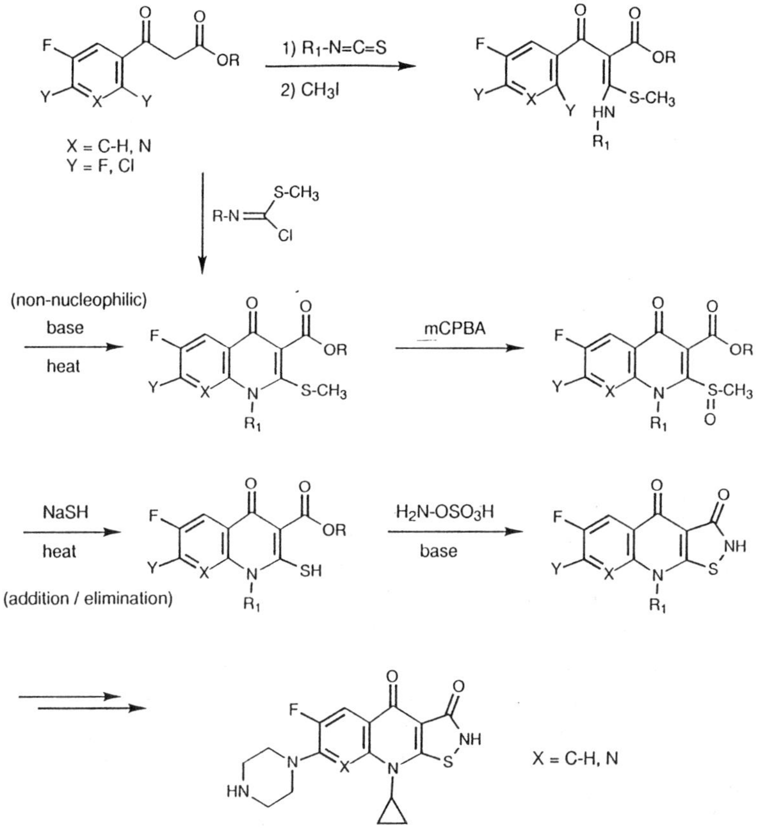

Figure 7.54 Synthesis of a C-2,C-3 isothiazoloquinolone.

This section has summarized the major synthetic approaches to the quinolone antibacterial agents; it is incomplete in scope and interested readers will find valuable references at the end of the chapter. However, with this information in mind, one can appreciate the current impetus that guides (or problems that hinder) the synthesis of new quinolone antibacterial agents. Unlike many classes of antibacterial agents, new quinolone congeners have been rapidly advanced due to incredibly productive synthetic protocols. If there is an impasse confronting the future development of the quinolones at this time, it is likely to exist in the lack of understanding of the details of quinolone interaction upon DNA gyrase, and certainly upon peripheral biological targets which infrequently produce troubling toxicological properties in the mammalian recipient. However, the future success of quinolone antibacterials will be governed by the ability to develop new agents that can tackle resistant bacterial pathogens.

7.5 Bacterial resistance to quinolone antibacterials

The onset of drug resistance threatens virtually all classes of antibacterial agents and the quinolones are not spared in this regard. Fortunately, the fluoroquinolones have been less prone to resistance compared to many other antibacterials. For example, most bacteria acquire quinolone resistance through a lone mechanism that is not transmitted on plasmids. In contrast, the sulfonamides for example, suffer from a variety of mechanisms that cause resistance, including plasmid-mediated phenomena. Furthermore, sulfonamide-resistant bacteria often have several mechanisms in operation that destroy the antibacterial activity of these drugs. In the case of the quinolones, the site of action, bacterial DNA gyrase, is a protein that mutates rather infrequently and this may be an inherent property of the enzyme. Of course, since the fluoroquinolones are relatively new chemotherapeutic agents, it may just be a matter of time before new mechanisms of quinolone resistance surface. Nevertheless, the emergence of quinolone-resistant pathogenic bacteria is alarming and has placed a constant pressure on researchers to develop novel potent derivatives that are less prone to the bacterial mechanisms of resistance. An increasing trend of ciprofloxacin-resistance is an indication that the entire class of quinolones could be in jeopardy; this agent is the best selling quinolone, but several bacterial species are already immune today. These findings are troubling when one considers that the quinolones are now widely used and are expected to play an even more dominant role in the marketplace for years, possibly decades, to come.

A wide variety of bacteria that are resistant to quinolones have been selected from laboratory strains or have been obtained from clinical isolates. In the former case, highly resistant strains can be harvested by exposing bacteria to increasing concentrations of quinolone and culturing viable organisms. This technique has allowed for the investigation and identification of specific resistant strains. Clinically, *Pseudomonas aeruginosa*, *Klebsiella pneumoniae*, *E. coli*, *Serratia* and *Staphylococcus* are most likely to be resistant.

One type of quinolone resistance, frequent among *P.-aeruginosa*, *S. marcescens*, *K. pneumoniae* and other *Enterobacteriaceae*, results from changes in permeability of the microorganism. In this context, many *E. coli* strains have been selected that are resistant to earlier non-fluorinated quinolone agents as well as norfloxacin and ciprofloxacin. In these cases, drug impermeability prevents sufficient intracellular accumulation of quinolone needed to inhibit bacterial DNA gyrase and thus no antibacterial effects are produced. Resistance of this type can be traced to chromosomal mutations that result in decreased expression of specific porin proteins (e.g. *Omp*F in *E. coli*) that normally allow passage of quinolone into the cytoplasm. In some cases, altered lipopolysaccharides may play a supporting role. (Recognized resistance mutations include *nfx*B, *nor*B and *nor*C (selected with nalidixic acid), and *cfx*B (selected with ciprofloxacin); all of which produce quinolone resistance in Gram negative species). In general, the phenomenon of permeation mutation usually confers low level resistance since (*in vitro*) concentrations need to be increased only several fold in order for an antibacterial effect to be established. Frequently, these organisms can still prove to be susceptible to a quinolone if adequate concentrations at the site of infection are achieved.

The most important acquired quinolone resistance arises from mutation of bacterial DNA gyrase enzyme. This phenomenon has been elucidated in a number of bacterial species as involving the change of a single nucleotide in the gyrase gene. The expression of mutated gyrase protein prevents the quinolone agent from undergoing binding to gyrase, and lethal effects on the organism as a whole are not produced. Similar mechanisms of resistance, namely mutation of the drug molecular target, occur with other classes of antibacterial agents but are often plasmid borne; plasmid-mediated resistance to quinolones is notably absent to date. Quinolone resistance caused by mutated DNA gyrase appears to be exquisitely chromosomal in nature although the reason for this is speculative at this time. One possibility is that wild-type gyrase genes are dominant and any hybrid gyrase enzymes which would result from plasmid conjugation are, for some reason, inherently less stable. In this context, perhaps any resistant strains that could be produced are unable to have functional gyrase activity and as a result are eliminated by cell death. It is also possible that plasmid conjugation is inhibited in the presence of quinolones, by some processes that have yet to be recognized. Since quinolones exert their activity on chromosomal DNA, there is perhaps some mechanism by which plasmids (which are non-chromosomal DNA molecules) are eliminated or rendered ineffective, at concentrations of quinolone that are below any static or killing threshold.

The frequency of resistance is somewhat quinolone specific; in general, the fluoroquinolones are less susceptible to gyrase mutation than the older, non-fluorinated congeners. As a class, the fluoroquinolones are two to three orders of magnitude less prone than prototypic agents such as nalidixic acid. It has been hypothesized that resistance to the fluoro-quinolones is restricted to specific single mutations within the gyrase gene, whereas nalidixic acid resistance, for example, can occur from any single change in a host (20) of nucleotides. There are notable exceptions depending upon the particular bacterial species. One of the most alarming species in this regard is *P. aeruginosa* strains that are only one order of magnitude less susceptible towards gyrase mutation when exposed to fluoroquinolones compared to nalidixic acid. Probably these pathogens can undergo any one of several mutations that are detrimental to both types of quinolone agent.

A number of quinolone-resistant mutations in gyrase-encoding genes have been identified. From a broad perspective, two commonly encountered variations are those to either the *gyr*A or the *gyr*B gene. Recall (section 7.2) that it is the *gyr*A gene product that is specifically targeted by quinolone antibacterials while other agents such as novobiocin exert their effects on the gyrase B subunit. The use of DNA-mapping techniques has clearly shown that quinolone resistance originates within the gene (*gyr*A) coding for the DNA gyrase A subunit. A single step mutation encodes for a gyrase protein that is less susceptible to the effects of quinolones. A single *gyr*A mutation can cause resistance to a variety of quinolones (aptly demonstrated in *E. coli*) but does not confer cross-resistance to other classes of antibacterial agents. Spontaneous single-step mutations of this type occur with a frequency of approximately 10^{-7} in many Gram negative bacteria such as many *Enterobacteriaceae* (including *E. coli*).

Nucleotide sequences have been determined for several *E. coli gyr*A mutants selected for spontaneous resistance to either nalidixic acid or pipemidic acid. In these mutants, a single base pair was changed that produced gyrase protein with defects (compared to the wild type) in a hydrophilic region near the amino terminus of the *gyr*A protein. These mutations occur near or within the codon responsible for the tyrosine-122 amino acid residue. (Recall (section 7.2)) that the same tyrosine amino acid residue was implicated in carrying out the transesterification reaction of the DNA phosphodiester linkage resulting in DNA strand cleavage). Apparently, the loss of this amino acid confers quinolone resistance; indeed the replacement of this specific tyrosine with phenylalanine or serine has been demonstrated to abolish gyrase activity (section 7.2). Presumably, the absence of the tyrosine residue renders the enzyme unable to carry the transesterification with the DNA substrate and DNA cleavage is not possible; the state of DNA supercoiling cannot be changed. (Resistance mutations of DNA gyrase gene include *nal*A, *nfx*A, *ofx*A, *nor*A and *cfx*A products).

Other DNA gyrase mutations that map away from the A subunit (*nal*D and *nal*C) can cause quinolone resistance. These mutations can confer resistance to nalidixic acid and other C-7 piperazinyl quinolones, even though the alterations are on the *gyr*B gene. These mutations have been characterized in terms of nucleotide sequence and compared to the *gyr*B gene; a single nucleotide change is responsible for resistance. Surprisingly, susceptibility to novobiocin, an antibacterial compound that exerts its effects on the gyrase B subunit, is not compromised. Lastly, although rare and not clinically important, there is at least one example of a quinolone-resistant bacterial strain that does affect the action of unrelated antibacterial agents. However, this nalidixic acid-resistant species was selected in the presence of other antibacterial agents and cross-resistance to tetracycline, for example, was low level. Resistance probably occurred from decreased accumulation of the tetracycline.

Two final observations that have recently emerged deserve comment before closing. First of all, decreased quinolone permeability along with mutated gyrase has been implicated in some *Pseudomonas* species that display exceptional high-level resistance. The presence of both mechanisms in a single organism would be frightening but more work needs to be done before a rational judgment can be formed. Also, such a phenomenon would be predicted to remain a rare occurrence in the near future.

Lastly, efflux of quinolone has been observed in some resistant *E. coli* and *S. aureus* species. In both cases, resistance can be traced to chromosomal mutations (a *nor*A gene product) that produce active efflux mechanisms selective for some quinolones (e.g. norfloxacin). Resistance of this type represents a third mechanism (in addition to mutated gyrase and altered permeability) by which bacteria are able to shun quinolones. This finding is of great significance and concern; the near future of the quinolone may be at stake and perhaps dismal if bacteria are able to acquire mutated gyrase along with an efflux mechanism.

7.6 Recent advances

7.6.1 New quinolones

There are literally dozens of new quinolone agents that may be poised to enter clinical trials within the near future. Some of the more advanced of these compounds are shown below (Figure 7.55). In most cases, each compound has undergone at least preliminary profiling in humans. The

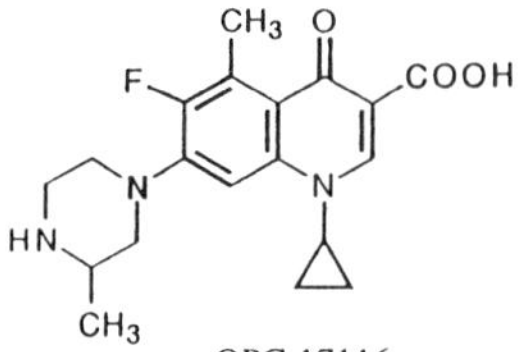

OPC-17116
(Grepafloxacin)

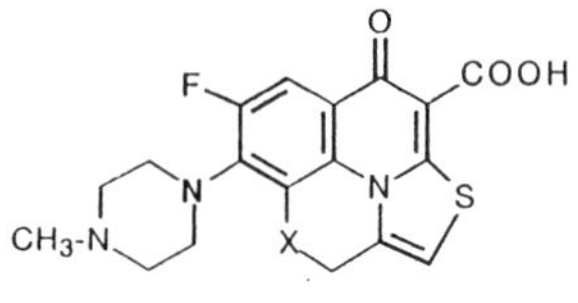

X = O (KB 5246)
X = NR, X = S, X =CO

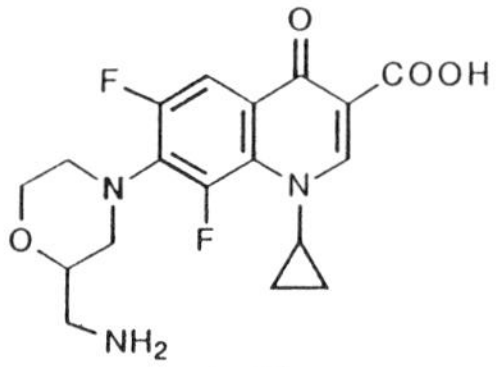

Y-26611

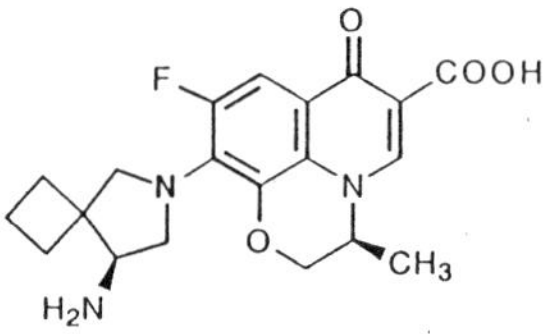

DV-7751a

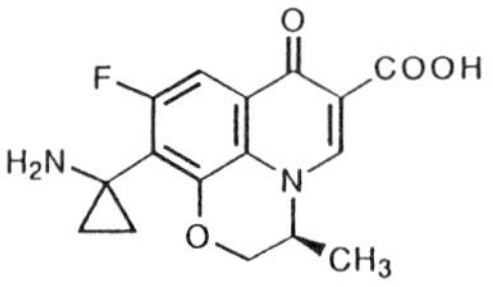

T-3761

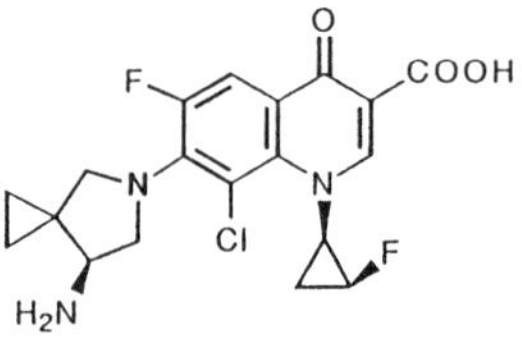

DU-6859

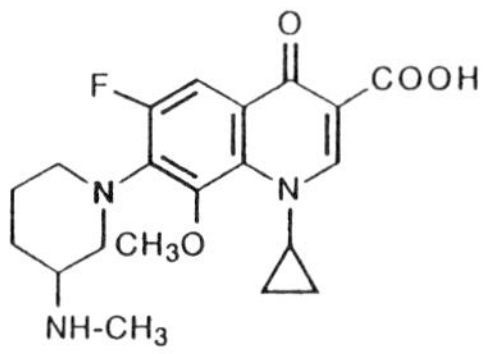

Q-35

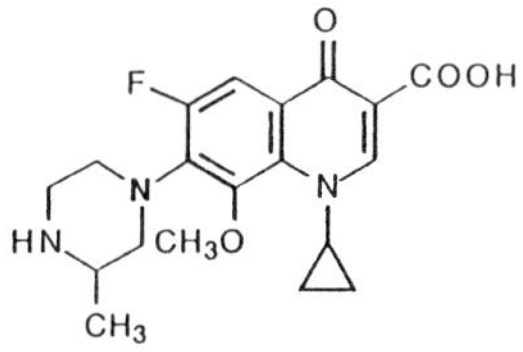

AM-1155

Figure 7.55 New quinolones.

selection of the C-7 substituent continues to dominate current efforts although the first C-5 substituted quinolone and the first N-1-fluorocyclopropyl derivatives are receiving attention. Also, there is a resurgence in the tricyclic quinolone nucleus following the development of the (*S*)-enantiomer of ofloxacin (levofloxacin) that is reflected in some new compounds. Tetracyclic-derived quinolones are also being investigated. As expected, many new congeners are C-6 fluorinated and an N-1-cyclopropyl substituent continues to be a dominant feature. Notably absent are N-1-fluoroaryl-derived quinolones, which is in response to the market withdrawal of temafloxacin.

In addition, various quinolones have been targeted for use in veterinary indications. Often, congeners that are exceptionally potent against common animal pathogens are selected for this purpose, although a marginal safety profile can relegate a given compound for development

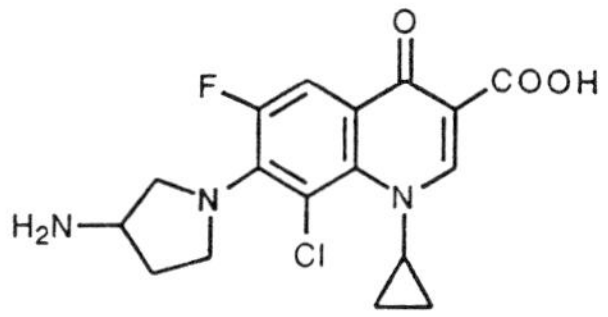

Clindafloxacin

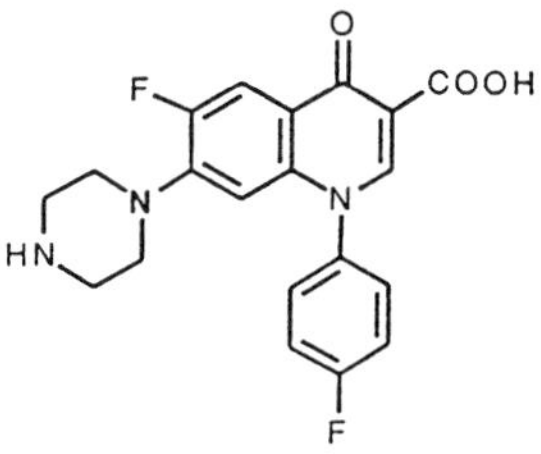

Sarafloxacin

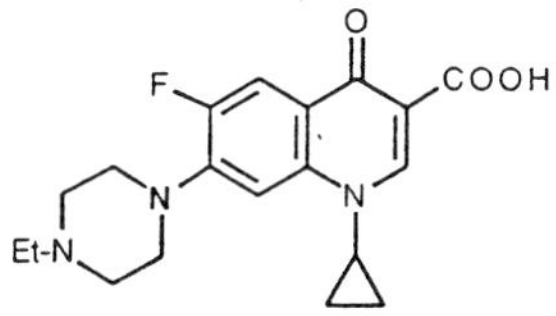

Enrofloxacin

Figure 7.56 Quinolones reserved for veterinary use.

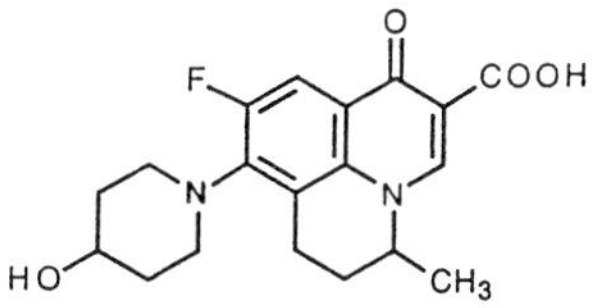

Nadifloxacin

Figure 7.57 New topical quinolone.

solely in animals. Some of the most recent additions to the veterinary market are shown below (Figure 7.56). A new quinolone, nadifloxacin, has been launched as a topical agent against bacteria that cause acne (Figure 7.57).

7.7 Summary

The quinolone antibacterial agents are valuable chemotherapeutics due to many factors.

1. Quinolones are potent, broad spectrum antibacterial agents. The early congeners (non-fluorinated at C-6) were limited to certain Gram negative infections such as urinary tract infections. However, the modern generation of fluoroquinolones containing a C-6 fluoro substituent and a cyclic basic amine moiety at C-7 surpass their

predecessors in terms of spectrum of activity and potency. This has allowed for their use against a variety of Gram negative as well as some Gram positive pathogens.

2. Quinolones are easily administered via parenteral and oral routes and are generally well tolerated. Once-a-day dosing is now available with some agents.

3. Quinolones are totally synthetic substances that are relatively easily prepared. As a therapeutic agent, their cost per dose and per therapy rivals any other class of therapeutic antibacterial agents.

4. Quinolones exert their antibacterial effects by inhibiting the bacterial DNA gyrase enzyme which is essential for bacterial survival and growth. This molecular target is absent in humans and therefore potential adverse effects are minimal.

5. The quinolones are powerful antibacterial agents and their use is on the increase but bacterial resistance, primarily caused by genetic mutations of the gyrase enzyme, threaten the future success of these agents.

Major quinolone antibacterials are shown in Table 7.1.

Table 7.1 Major antibacterial quinolone chemotherapeutic agents

Generic name	Trade names	Administration
Cinoxacin	Cinobac	PO
Ciprofloxacin	Cipro, Ciprobay, Ciproxin	PO, IV
Enoxacin	Penetrex	PO
Fleroxacin	Megalone, Quinodis	PO
Flumequine	Apurone	PO
Lomefloxacin	Maxaquin	PO
Levofloxacin ((S)-ofloxacin)	Floxacin, Cravit	PO
Nalidixic acid	Nogram, NegGram, Cybis	PO
Nadifloxacin	Acuatim, Aquatim	topical
Norfloxacin	Noroxin	PO
Ofloxacin	Floxacin	PO, IV
Oxolinic acid	Utibid, Uroxil, Uritrate	PO
Pefloxacin	Peflacine	PO
Pipemidic acid	Pipemid, Pipedac, Pipram	PO
Piromidic acid	Bactramyl, Panacid, Pirodal	PO
Rosoxacin	Eradacil, Winuron, Roxadil	PO
Rufloxacin	Monos, Qari	PO
Sparfloxacin	Spara, Zagam	PO
Temafloxacin*	Omniflox	PO
Veterinary use		
Danofloxacin (mesylate)	Advocin	
Difloxacin		
Enrofloxacin	Baytril	
Merafloxacin		
Sarafloxacin		

*withdrawn from US market in 1992.

Further reading

C. Siporin, C.L. Heifetz and J.M. Domagala (Eds) (1990) *The New Generation of Quinolones*, Marcel Dekker, New York.

J.S. Wolfson and D.C. Hooper (Eds) (1989) *The Quinolones*, American Society for Microbiology, Washington, D.C.

V.T. Andriole (Ed.) (1988) *The Quinolones*, Academic Press, San Diego, CA.

N.R. Cozzarelli and J.C. Wang (Eds) (1990) *DNA Topology and Its Biological Effects*, Cold Spring Harbor Laboratory Press, Cold Spring Harbor, NY.

H. Neu (1992) 'Quinolone antibacterial agents', *Annu. Rev. Med.*, **43**, 465.

D.C. Hooper and J.S. Wolfson (1991) 'Fluoroquinolone antimicrobial agents', *New England J. Med.*, **324**, 384.

D.T.W. Chu and P.B. Fernandes (1990) 'Quinolone antibacterial agents', in *Advances in Drug Research*, Vol. 21, B. Testa (Ed.) Academic Press, New York, pp. 42–144.

S. Radl (1990) 'Structure–Activity Relationships in DNA Gyrase Inhibitors', *Pharm. Ther.*, **48**, 1.

D.T.W. Chu and P.B. Fernandes (1989) 'Structure–activity relationships of the fluoroquinolones', *Antimicrob. Agents Chemother.*, **33**, 131.

S. Radl and D. Bouzard (1992) 'Recent Advances in the Synthesis of Antibacterial Quinolones', *Heterocycles*, **34**, 2143.

R. Albrecht (1977) 'Development of antibacterial agents of the nalidixic acid type', *Prog. Drug Res.*, **21**, 9.

J.C. Wang (1985) 'DNA topoisomerases', *Annu. Rev. Biochem.*, **54**, 665.

K. Drlica (1984) 'Biology of bacterial deoxyribonucleic acid topoisomerases', *Microbiol. Rev.*, **48**, 273.

N. Cozzarelli (1980) 'DNA gyrase and the supercoiling of DNA', *Science*, **207**, 953.

M. Gellert (1981) 'DNA topoisomerases', *Annu. Rev. Biochem.*, **50**, 879.

L.L. Shen, L.A. Mitscher, P.N. Sharma, J.J. O'Donnell, D.W.T. Chu, C.S. Copper, T. Rosen and A.G. Pernet (1989) 'Mechanism of inhibition of DNA gyrase by quinolone antibacterials: a cooperative drug–DNA binding model', *Biochemistry*, **28**, 3886.

L.A. Mitscher, B.D. Davis, T. Ichikawa and K. Maeda (1987) 'Horizons on antibiotic research' from *Proceedings of the Symposium Dedicated to the Late Professor Hamao Umezawa on the 25th and 40th Anniversary of the Institute of Microbial Chemistry and the Japan Antibiotics Research Association*, Nov. 25–26, Tokyo, Japan, 166–193.

D.C. Hooper and J.S. Wolfson (1988) 'Mode of action of the quinolone antimicrobial agents', *Rev. Infectious Diseases*, **10**(S1), S14.

J.S. Wolfson and D.C. Hooper (1985) 'The fluoroquinolones: structures, mechanisms of action and resistance, and spectra of activity *in vitro*', *Antimicrob. Agents Chemother.*, **28**, 581.

L.L. Shen and A.G. Pernet (1985) 'Mechanism of inhibition of DNA gyrase by analogues of nalidixic acid: the target of the drugs is DNA', *Proc. Natl. Acad. Sci. USA*, **82**, 307.

D.S. Horowitz and J.C. Wang (1987) 'Mapping the active site tyrosine of *Escherichia coli* DNA gyrase', *J. Biol. Chem.*, **262**, 5339.

C.J.R. Willmott and A. Maxwell (1993) 'A single point mutation in the DNA gyrase A protein greatly reduces binding of fluoroquinolones to the gyrase–DNA complex', *Antimicrob. Agents Chemother.*, **37**, 126.

G.W. Kaatz, S.M. Seo and C.A. Ruble (1993) 'Efflux-mediated fluoroquinolone resistance in *Staphylococcus aureus*', *Antimicrob. Agents Chemother.*, **37**, 1086.

D.B. Wigley, G.J. Davies, E.J. Dodson, A. Maxwell, and G. Dodson (1991) 'Crystal structure of an N-terminal fragment of the DNA gyrase B protein, *Nature*, **351**, 624.

C.D. Lima, J.C. Wang and A. Mondragon (1994) 'Three-dimensional structure of the 67K N-terminal fragment of *E. coli* DNA topoisomerase I', *Nature*, **367**, 138.

C. Hubschwerlen, P. Pflieger, J.-L. Specklin, K. Gubernator, H. Gmunder, P. Ahgehrn and I. Kompis (1992) 'Pyrimido[1,6a]benzimidazoles: a new class of DNA gyrase inhibitors', *J. Med. Chem.*, **35** 1385.

8 Peptidic antibacterial agents

8.1 Introduction

There are a variety of peptidic substances that have gained significant use in antibacterial chemotherapy. A definition of a peptidic family of antibacterial agents is arbitrary; some simple amino acid and small peptide derivatives, as well as various complex peptides, some even resembling small proteins, have been marketed. The criterion chosen here groups substances that are superficially or primarily, if not totally peptidic in nature. In many cases, D-amino acids, cyclic structures and fatty acid prosthetic groups are distinguishing features. Many peptidic compounds act upon surfaces and membranes of susceptible bacteria and often a single particular molecular target is lacking. However, in some cases, a similarity to other major classes of antibacterial agents can be drawn due to a specific mechanism of action. This is not surprising since other classes of antibacterials, such as the β-lactams (chapter 3), behave predominantly as peptide surrogates. Nevertheless, the peptidic antibacterial agents, as a whole, offer a variety of chemical properties and biological functions that distinguishes these compounds from other major classes of antibacterial agents.

In general, the peptidic antibacterials have not been advanced through chemical modification for a number of reasons. The most important antibacterial peptides are but one member of a family of structurally related congeners; often the agents that have been promoted to the market exist as mixtures. Separation of the individual components is usually difficult which has deterred semi-synthetic modifications. Very importantly, many peptidic agents are complex in structure and chemical alteration has proven to be extremely demanding. Therefore, most compounds are still best obtained from microbial sources and used as natural products. There are other shortcomings that are common to this family. The peptidic agents are seldom orally active and often poorly tolerated in systemic use. Even under optimal conditions, most antibacterial peptides are unable to contest the spectrum of activity and potency afforded by other well-established chemotherapeutic agents. Some peptidic agents are prone to degradation due to the action of peptidases and other endogenous mammalian enzymes that cleave peptidic linkages. Lastly, in many cases, sound structure–activity relationships have yet to be fully defined; the desire to tackle such a challenge is probably academic since superior unrelated agents are available today.

Regardless of these generalizations, a text on antibacterial chemotherapy would be incomplete without a discussion of the peptidic antibacterial agents. These compounds are important therapeutic agents although reserved for specific infections which can often be severe in nature but limited in scope. It is also important to acknowledge that there are many naturally occurring and synthetic peptides which do display some level of antibacterial activity. For example, some dipeptides exert antibacterial effects by merely competing with specific amino acids for uptake into the bacterial cell. However, in the case of the complex antibacterial peptides, a precise mode of action is often absent and activity results from some gross effect upon the bacterial cell.

8.2 Individual peptidic antibacterial compounds

8.2.1 D-Cycloserine

8.2.1.1 Introduction. D-Cycloserine **1** (Figure 8.1) is an antibacterial substance produced by a *Streptomycete* (*S. garyphalus* sive *orchidaceus*) and was first reported in 1955. While its spectrum of antibacterial activity includes both Gram negative and Gram positive bacteria, cycloserine is very effective against mycobacteria and has thus been used primarily as a tuberculostatic agent. Due to the occurrence of adverse side effects, cycloserine is today regarded as an alternative chemotherapeutic agent for the treatment of tuberculosis.

Cycloserine is an amino acid surrogate that behaves as a structural analog of D-alanine, an essential amino acid for bacterial growth and survival. Cycloserine is able to compete with D-alanine and exert primarily bactericidal effects upon susceptible bacteria by inhibiting cell wall biosynthesis. Due to its simple structure, cycloserine can be produced by synthetic means, as well as being available as a fermentation product. Cycloserine has been structurally modified in a variety of ways and a structure–activity relationship has been gathered; however no synthetic analog has replaced cycloserine nor has been marketed.

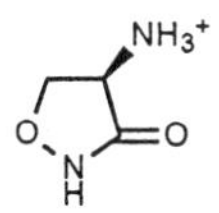

1

Figure 8.1 Cycloserine.

The mechanism of action of cycloserine has been elucidated and it remains somewhat unique among the major classes of antibacterial agents. Despite this, research interest in cycloserine today is likely to remain of historical and academic interest in the light of newer and more promising agents. However, the concept of targeting simple amino acid nutrients as a means of inducing an antibacterial action is noteworthy. Other simple amino acid surrogates (e.g. diaminopimelic acid analogs) are currently being explored due to their ability to disrupt enzymatic processes that are also necessary for bacterial cell wall construction.

8.2.1.2 Mode of action: inhibition of alanine racemase and D-Ala-D-Ala synthetase. Cycloserine is a potent inhibitor of cell wall biosynthesis in susceptible bacteria due to its ability to function as a D-alanine mimetic as recognized by two enzymatic processes. In both cases, cycloserine ultimately foils production of a major component that is needed to construct the peptidoglycan barrier. The peptidoglycan sheath is paramount to cell survival and growth; some pertinent issues regarding its biosynthesis have been described earlier (chapter 3). (The reader is strongly encouraged to review this material). The gross effects produced by cycloserine are the inability of the susceptible bacterial organism to maintain and complete construction of the peptidoglycan matrix, which leads to profound morphological changes and eventually bacterial cell death.

Cycloserine interferes with the formation of the nucleotide-bound pentapeptide normally destined for incorporation into the peptidoglycan matrix. Recall (chapter 3, section 3.2) that peptidoglycan biosynthesis entails two distinct cytoplasmic processes. Cycloserine has no effect upon the initial stage in which uridine diphosphate-*N*-acetylmuramic acid is synthesized. It is this nucleotide 'anchor' to which the Gram negative or Gram positive pentapeptide is progressively attached. During this process, the first three amino acids are individually added (to give UDP-MurNAc-tripeptide) in a specific and sequential fashion directed by a series of ligase enzymes. However, the last two amino acids that are added, D-Ala-D-Ala, are joined in a different manner in being attached as a single dipeptide unit (Figure 8.2).

This event is dependent upon the action of highly specialized enzymes which produce a suitable D-Ala-D-Ala conjugate and subsequently join this dipeptide to the terminus of the UDP-MurNAc-tripeptide. The D-Ala-D-Ala conjugate originates from L-alanine and so two distinct operations must occur; the configuration of the alanine chiral center must be formally inverted and the two amino acid units must be joined together. The inversion is catalysed by the enzyme alanine racemase. The joining of two molecules of D-alanine to one another is catalysed by the ligase, D-alanine:D-alanine synthetase. The attachment of D-Ala-D-Ala to the UDP-MurNAc-tripeptide substrate is carried out by a D-alanyl-D-alanine 'adding

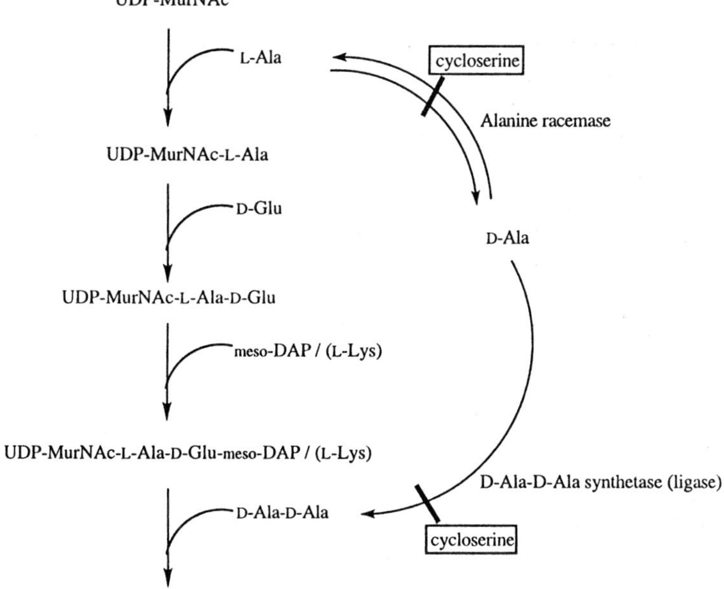

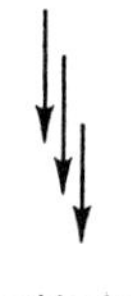

Figure 8.2 Sites of action of cycloserine.

enzyme' which is another ligase (and synthetase). The incorporation of the D-Ala-D-Ala segment completes the formation of the crucial peptidoglycan precursor, UDP-MurNAc-pentapeptide. The D-Ala-D-Ala terminus plays an exquisite role in subsequent transformations that lead to the completion of the peptidoglycan matrix. Cross-linking between strands of peptide–glycan polymer eventually occurs by peptide bond formation onto the D-alanyl-D-alanine amide linkage with subsequent loss of the terminal alanine residue. Conversely, the premature removal of the terminal alanine residue prevents this transpeptidation process, thus preventing extensive cross-linking. It is this interplay of opposing processes that guarantees the peptidoglycan matrix is sturdy yet yielding to repair and the influx of nutrients. The completion and regulation of peptidoglycan biosynthesis is jeopardized if the D-Ala-D-Ala species is not properly synthesized nor attached onto the UDP-MurNAc-tripeptide.

Cycloserine produces antibacterial effects by inhibiting the synthesis of the D-Ala-D-Ala conjugate, but does not prevent its incorporation into the UDP-MurNAc peptide. Cycloserine mimics alanine-derived substrates of both the alanine racemase and the D-alanine:D-alanine synthetase, and acts as competitive inhibitors to both enzymes (Figure 8.2). As a result, the conversion of L-alanine into D-alanine is inhibited as is the joining of two molecules of D-alanine to afford D-alanyl-D-alanine. This dual antibacterial action gives cycloserine potent antibacterial activity, although cycloserine is accepted by these enzymes as a competitive substrate without permanently abolishing their activity (alanine racemase and D-Ala-D-Ala synthetase). Thus the effects of cycloserine can be reversed if the natural substrate, alanine, is supplied.

8.2.1.3 Structural features and structure–activity relationships. Cycloserine is extremely simple in structure but few analogs have been shown to retain useful antibacterial activity. Since cycloserine is an amino acid surrogate of D-alanine (or D-Ala-D-Ala), it is important that the amine group be present for enzyme recognition. Accordingly removal of the C-4 amine functionality results in loss of antibacterial activity. Within this same context, the amine is best left unsubstituted; alkylation or acylation greatly diminish or eliminate activity. The D-configuration of this lone chiral center is required for useful antibacterial activity. The appropriate heteroatom substitution pattern is also important and the isoxazolidone ring system is crucial. Replacement of either the cyclic nitrogen or oxygen atoms is detrimental, probably since such modifications destroy the planarity of the ring needed for enzyme recognition. Likewise, cleavage of the ring is usually disadvantageous, regardless of the site of bond breakage. The notable exception is *O*-carbamyl-D-serine (not discussed) which is also an inhibitor of alanine racemase and does display notable antibacterial properties. The remaining position, that of the C-5 center, is unsubstituted in the case of cycloserine, and there have been few groups tolerated at this position in terms of antibacterial activity. The criteria at this site appears to be substituents that are small in size (e.g. methyl) and preferably without heteroatoms/functional groups.

No structural modifications have produced a developed agent superior to cycloserine. However, as previously suggested, research efforts are probably continuing in some laboratories in attempts to investigate novel amino acid surrogates, unrelated to serine/cycloserine, as potential chemotherapeutic compounds.

8.2.1.4 Resistance. The antibacterial effects of cycloserine can be reversed by simply supplying additional alanine to the bacterial culture. In doing so, alanine is often able successfully to compete with cycloserine and re-establish sufficient peptidoglycan biosynthesis. While such a phenomenon is interesting from a laboratory perspective, bacteria are unable to obtain excesses of alanine from a host. Realistically, resistance to cycloserine does not occur as a result of increased levels of alanine, however, some bacteria have evolved other ways to avoid the effects of cycloserine.

Some bacterial species are able to produce increased levels of either alanine racemase or D-Ala-D-Ala synthetase (or both enzymes). Increased supplies of these enzymes enhance the production of either D-alanine or of D-alanyl-D-alanine which can eventually translate into operational peptido-glycan synthesis and loss of antibacterial activity. Another mechanism of resistance to cycloserine occurs in bacterial species that are able to prevent cycloserine from either entering into the cytoplasm, or from accumulating to harmful levels within the cell. Simple amino acids such as alanine can be transported into the bacterial cells via an active transport system and cycloserine, being an alanine mimic, can be ushered into the cell in a similar fashion. Impairment of the alanine transport system can reduce intracellular levels of cycloserine. Some cycloserine-resistant bacteria are able to undergo a single genetic event which leads to the loss of an effective alanine transport system. Consequently, uptake of cycloserine is greatly diminished and the microorganism is resistant.

8.2.2   The tyrocidines and gramicidin S

The tyrocidines are a family of structurally related complex cyclic decapeptides that possess useful antibacterial properties, notably good potency against some Gram positive bacteria. Gramicidin S is often included in this same family although its name is certainly a misnomer since the 'true' gramicidins are larger linear peptides that contain modifications of both the N-terminal amino acid and the C-terminal amino acid (discussed later). The tyrocidines and gramicidin S **2** (Figure 8.3) were first isolated as a mixture from cultures of a soil bacterium (*Bacillus brevis*). There are three main tyrocidines in addition to gramicidin S and all of these compounds have similar structural features and mode of antibacterial action. Tyrocidine A **3** (Figure 8.4) is probably the most significant of the tyrocidines; the other members are analogs that contain either a single or double replacement of L-tryptophan for the phenylalanine units (sixth and seventh amino acid residues).

Despite useful antibacterial activity, there are a number of features that have greatly limited the use of these agents. In general, the tyrocidines and gramicidin S are considered today to be too toxic for systemic use since they demonstrate hemolytic properties and other adverse side effects. Furthermore, these agents do not display as potent nor as broad of a spectrum of antibacterial activity as many structurally unrelated agents currently available. In addition, the tyrocidines (and gramicidin S) are not orally absorbed and undergo degradation in the gut. As a result, the

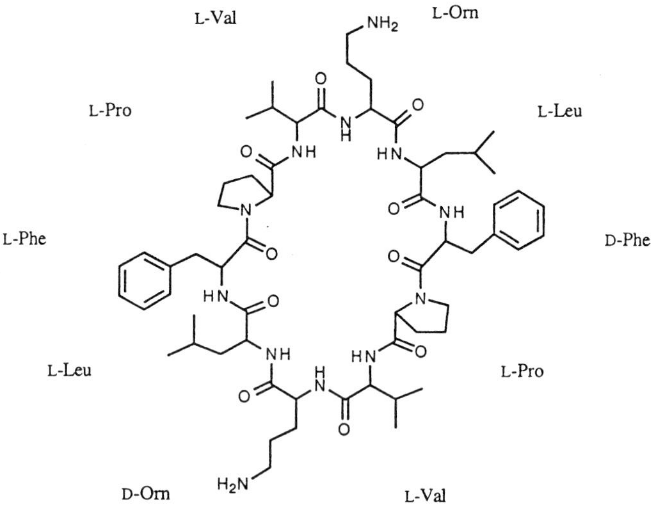

2

Figure 8.3 Gramicidin S.

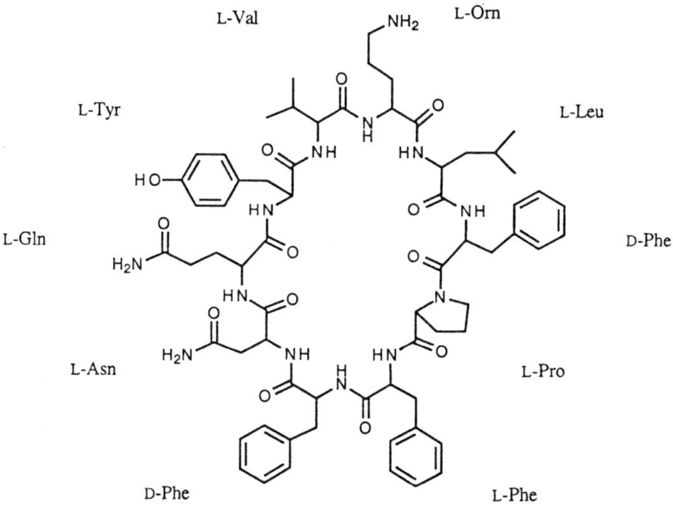

3

Figure 8.4 Tyrocidine A.

tyrocidines were once primarily utilized as topical agents. Today, these agents probably remain mostly of academic interest due to their effects upon both prokaryotic and eukaryotic cells.

8.2.2.1 Mode of action. The mode of antibacterial action of the tyrocidines, including gramicidin S, has been extensively explored. These antibiotics exert antibacterial properties due to their ability to act as surfactants which disrupt the integrity of the bacterial cell surface. The tyrocidines (and gramicidin S) impair the functions of the cell surface rendering the external cellular barrier(s) weak and leaky in nature. In essence, these compounds behave as bactericidal detergents towards many Gram positive bacteria as well as some Gram negative species. Their action results in the loss of amino acids, purines and pyrimidine bases, phosphate and other vital material from the susceptible bacterial organism. In addition, these substances can have profound effects upon cellular functions such as oxidative phosphorylation, ATPase activity, as well as inducing changes in the function of certain intracellular organelles such as the mitochondria. Unfortunately, many of these events are not necessarily specific towards the bacterial cell, which probably contributes to the toxic disposition of these compounds in the mammalian host.

8.2.2.2 Structural features and issues of structure–activity relationships. The characteristic feature of the tyrocidines, including gramicidin S, is a cyclic decapeptide consisting of basic amino acid residues in concert with hydrophobic amino acid units. The arrangement of these components within a cyclic constraint is also important. Linear analogs generally do not possess antibacterial properties. In addition, the optical configuration of each amino acid residue has also remained somewhat optimal for activity. The array of amino acid residues containing specific D- and L-configurations allows for bioactive conformations which place hydrophobic residues on one side of the macrocycle while disposing basic hydrophilic residues on the complementary side. Species which can readily adopt this general conformation appear to be responsible for a maximum effect as a molecular detergent and maximum antibacterial activity.

A notable characteristic of the tyrocidines and gramicidin S is the presence of an amino acid sequence of L-Val-L-Orn-L-Leu-D-Phe-L-Pro (designated as AA$_1$ through AA$_5$). In the case of the 'pure' tyrocidines (tyrocidines A, B and C), the remaining residues contain two hydrophobic amino acids which are joined onto the fifth amino acid (L-Pro) followed by two basic residues, namely L-Asn and L-Gln. The remaining position of the decapeptide is usually L-Tyr with the exception of gramicidin S which has L-Pro at this position. Gramicidin S has a slightly different construct in that a basic amino acid, L-Orn is the seventh amino acid in the sequence using this nomenclature, and L-Leu-D-Phe-L-Pro complete the peptidic macrocycle.

The tyrocidines can be structurally modified to an extent, so long as the composite analog properly disposes the hydrophobic and basic hydrophilic amino acid residues. Other members of the tyrocidine family can be regarded as analogs in which an amino acid residue is replaced with another of similar hydrophobicity.

8.2.2.3 Resistance. Bacteria resistant to the effects of the tyrocidines (including gramicidin S) are primarily Gram negative species which have a cell wall composition that is often impervious to the antibacterial detergent-like effects of these antibiotics. Some Gram positive organisms are also unaffected. The tyrocidines and gramicidin S produce rather indiscriminate and gross effects on susceptible bacteria, and although these agents have been implicated in targeting some sites, a discussion on specifics surrounding resistant microorganisms will not be presented here.

8.2.3 The gramicidins

The gramicidins are another family of complex polypeptidic substances that exert gross effects upon the surface and membranes of susceptible bacterial cells resulting in antibacterial activity. The true gramicidins are linear (non-cyclic) molecules and are not totally peptidic in composition, unlike their namesake gramicidin S (as well as the tyrocidines). The gramicidins are neutral peptides and do not contain the characteristic basic amino acid residues found in the tyrocidines. In addition, the mechanism of antibacterial action of the gramicidins is somewhat distinct from that of the tyrocidines although similar effects are produced.

The gramicidins are hydrophobic pentadecapeptides **4** (Figure 8.5) which are covalently modified at each terminal amino acid residue. Specifically the gramicidins are formylated at the N-terminus and contain an ethanolamine-derived amide at the C-terminus. Members of the gramicidin family vary in their composition at two specific amino acid residues. Gramicidins A, B and C are distinguished based upon their corresponding eleventh amino acid residue (the N-terminus amino acid is designated as the first residue by convention). Gramicidin A has tryptophan at this position whereas gramicidin B and C have phenylalanine and tyrosine, respectively (Figure 8.5). In addition, each gramicidin actually exists as a mixture of two components which differ in composition at the first amino acid residue. The majority of the gramicidins (nearly 90% of a given member) have a (formylated) valine as the first amino acid, but variants containing isoleucine at this position are also isolated from the fermentation mixture and are virtually inseparable from the major (valine-containing) components. These minor congeners are often referred to as Ile-gramicidins (gramicidins with valine as the first amino acid can either have no designation or can be referred to as Val-gramicidins). Commercially

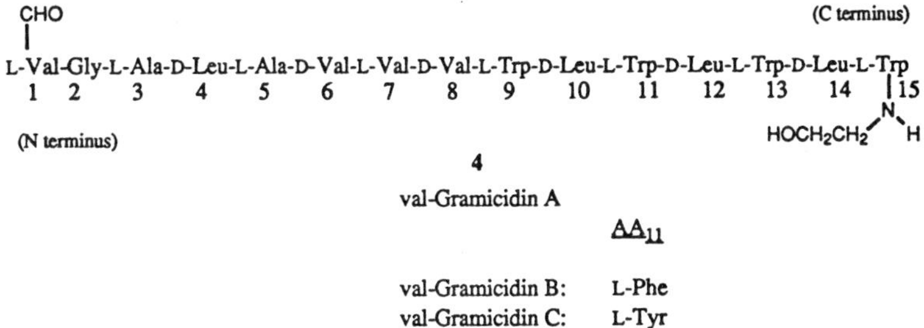

Figure 8.5 The gramicidins.

available gramicidin is actually a mixture of components with gramicidin A
(valine as the N-terminus amino acid) being the major compound.

8.2.3.1 Mode of action. Like the tyrocidines, the gramicidins target the
outer barrier(s) of the bacterial cell responsible for protecting the
intracellular machinery and genetic information needed for survival and
growth. The gramicidins exert profound effects upon the bacterial
cytoplasmic membrane in a somewhat more distinctive manner than the
grossly damaging effects caused by the tyrocidines. Since the gramicidins
are more precise in their mechanism of action, these compounds are
generally less toxic than the tyrocidines. As a result, the gramicidins have
found greater use in antibacterial chemotherapy. However, the gramicidins
cannot be used for systemic indications and have followed a similar fate as
the tyrocidines in being limited to topical administration.

The gramicidins behave as ionophoric substances and cause devastating
effects upon the bacterial cytoplasmic membrane. The gramicidins appear
to self associate and presumably form small pores or channels that cause
the leakage of essential cations from within the bacterial cytoplasm. A
variety of cations are efficiently effluxed from the susceptible bacterial cell,
including protons and ammonium cations, and to a lesser degree,
potassium and sodium cations. It is the ensuing loss of these cationic
materials that eventually leads to the disruption of normal bacterial
metabolic functions and ultimately cell death. The establishment of
intramolecular hydrogen bonds between local environments of a coiled
gramicidin peptide chain are essential for the formation of pore-like
structures capable of acting as ionophores. Gramicidin molecules are likely
to adopt a bioactive helical conformation in which hydrogen bonding
interactions between the polar amide portions of the peptidic linkages
stabilize this structure. The polar region probably occupies the interior of
the channel so as to escort the cation along the channel via electronic
associations. This conformation aligns the hydrophobic amino acid

residues along the periphery of the channel which favors placement of the molecular pore within the cytoplasmic membrane.

It has been suggested that at least two gramicidin molecules must associate with each other in order for a functional channel to form. The formyl group is likely to be a vital feature in such a process and the C-terminus ethanolamine substituent may also play a secondary role. It has been hypothesized that head-to-head aggregation (formyl groups associated with each other) of two gramicidin molecules occurs and that such a superstructure is stabilized by as many as 22 intramolecular and six intermolecular hydrogen bonds. However, several such dimeric units may need to be properly aligned so that the channel is able to cross the cytoplasmic membrane and thus behave as a true ionophore.

8.2.3.2 Structural features and issues of structure–activity relationships. The gramicidins are composed of a variety of neutral hydrophobic amino acids. Due to this structural complexity, it has been virtually impossible to explore extensively all imaginable modifications of the gramicidins, such as the incorporation of novel amino acids at any given position. In addition, since these compounds cannot rival the tolerability, potency and spectrum of activity of many other unrelated classes of chemotherapeutic agents, there has been little impetus directed at the synthesis of novel analogs. However, there have been some structural modifications that have provided insight into the features needed for antibacterial properties. In general, any change which disrupts or prevents the formation of the gramicidin channels within the cytoplasmic membrane is detrimental. Removal of the formyl groups leads to complete loss of antibacterial activity. Likewise, *N*-methylation of the amide linkages also results in the destruction of antibacterial properties. The ethanolamine-derived C-terminus is tolerant of some modification, but no variation has resulted in an agent with market potential. In summary, although the gramicidins possess some useful antibacterial activity, particularly against gram positive organisms, these compounds have remained more as curiosities than as leads for the construction of new analogs.

8.2.3.3 Resistance. Many gram negative bacteria are unaffected by the gramicidins due to an unyielding outer bacterial wall which prevents the compounds from exerting their properties upon the bacterial cytoplasmic membrane.

8.2.4 *Polymyxins*

The polymyxins are a family of antibacterial peptides that contain a characteristic cyclic heptapeptidic core from which extends a tripeptidic side chain containing an amide derived from a fatty acid. The polymyxins are quite distinctive from other peptidic antibiotics due to some rather notable components. The polymyxins contain a total of ten amino acid residues with at least half of these units being α,γ-diaminobutyric acid (DAB) (Figure 8.6). The fatty acid moiety is either 6-methylheptanoic or 6-methyloctanoic acid. The members of the polymyxin family differ either in the fatty acid substituent or at specific amino acid residues.

The polymyxins are elaborated by a bacterium (*Bacillus polymyxa*) and were first discovered in 1947. These compounds initially generated considerable interest following their discovery due to their spectrum of antibacterial activity. In general, the polymyxins display moderate to good potency (*in vitro*) against Gram negative bacteria, but are inactive against Gram positive species and mycobacteria. In addition, there are a number of structurally related peptidic antibiotics such as the octopeptins (not

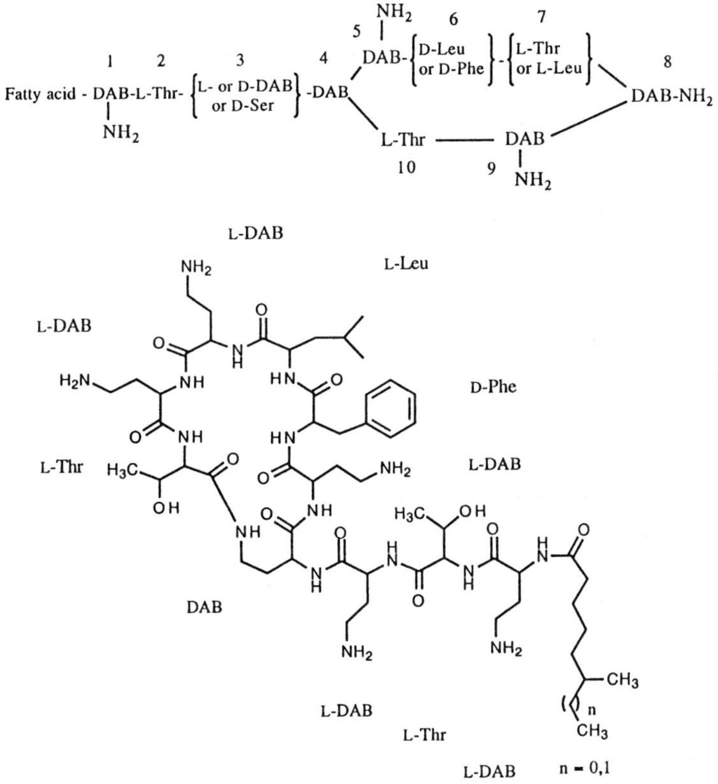

Figure 8.6 Polymyxin B.

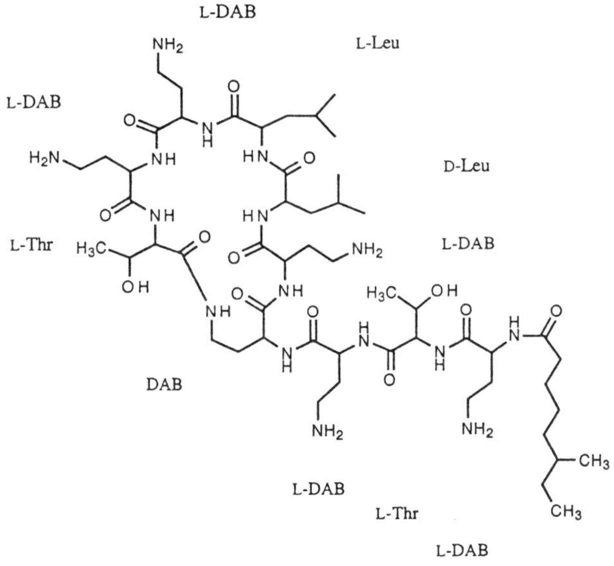

Figure 8.7 Polymyxin E (colistin).

discussed) that are generally inferior in terms of antibacterial activity but contain the same general type of cyclic peptidic core. Polymyxin B **5** (Figure 8.6) and polymyxin E **6** (Figure 8.7), which is also called colistin (A), are among the best tolerated of all polymyxins and display useful antibacterial properties. The issue of tolerability has traditionally been a key factor in the development of the polymyxins since these compounds can produce serious adverse side effects in mammalian hosts such as nephrotoxicity. Unfortunately, even these landmark polymyxins are not readily absorbed and so their use has been limited to either topical administration or indications in which accumulation within the urinary tract or digestive tract is appropriate.

8.2.4.1 Mode of action. The polymyxins exert antibacterial properties as a consequence of their association and interaction with the bacterial cytoplasmic membrane. Susceptible organisms undergo loss of purines, pyrimidines and various nucleosides, as well as phosphate and some simple saccharides. These events are a direct result of membrane damage that produces non-discriminate leakage of small molecular precursors to DNA synthesis and the inhibition of normal cellular respiration. Depending upon the degree of membrane damage, the polymyxins can be bactericidal or merely temporarily halt the growth of bacterial organisms. There appears to be no receptor specific to the polymyxins; these compounds appear to be randomly accepted by the cytoplasmic membrane and able to occupy an environment among the lipid and protein layers of this structure. This event ultimately weakens the membrane to the point at which osmotic regulation is hampered, thus leading to the efflux of the vital materials (as described above). The lodging of polymyxins within the membrane, in essence, produces holes in the phospholipid layer of the membrane which allows the passage of small molecules but not large proteins. Thus while nucleosides, purines and pyrimidines are readily lost, cytosolic proteins are not.

8.2.4.2 Structural features and issues of structure–activity relationships. The structural features of the polymyxins include a cyclic heptapeptidic core with a high content of α,γ-diaminobutyric acid and an amino acid side chain that terminates with a fatty acid-derived amide (as mentioned earlier). The intimate interaction of the polymyxins with the phospholipid layer is perhaps the single most important event since all other effects primarily arise as a consequence of this phenomena. The polymyxins are able to mimic 'head and tail' features of phospholipids and this allows the polymyxins to become incorporated within the membrane. The heptapeptide core of the polymyxins serves as a polar head to complement the polar glycerol-derived phosphate moiety of the phospholipid unit. The fatty acid tail which extends from the polymyxin cyclic core serves as a hydrophobic tail similar to that of the unsaturated and saturated long chain aliphatic fatty acids which comprise the tail of the phospholipids (Figure 8.8).

The polymyxins have been produced synthetically and a number of interesting structural features have been realized. The cyclic heptapeptidic core is an absolute requirement and modification via ring cleavage and ring expansion is detrimental to antibacterial activity. Smaller peptidic cores appear to be less favorable in studies both related and unrelated to the polymyxins; the heptapeptide appears to be optimal in the case of the polymyxins, and there has been little reason to explore additional variations. The peptide sequence of the polymyxins has also been modified. The replacement of specific amino acid residues by other amino acids that contain similar hydrophobicity and size has been a common theme. Members of the polymyxin family vary at the two hydrophobic residues within the macrocycle that are flanked by DAB (i.e. the sixth and seventh amino acid residues, see Figure 8.6) as well as at the residue adjacent to the macrocycle (i.e. the third amino acid residue). However, the α,γ-diaminobutyric acid (DAB) residues are crucial for good antibacterial activity. Acylation of the terminal amino group of any DAB residue decreases activity compared to the parent compound, although antibacterial potency can still be appreciable. For example, acetylation can afford active derivatives providing that acetylation is not exhaustive.

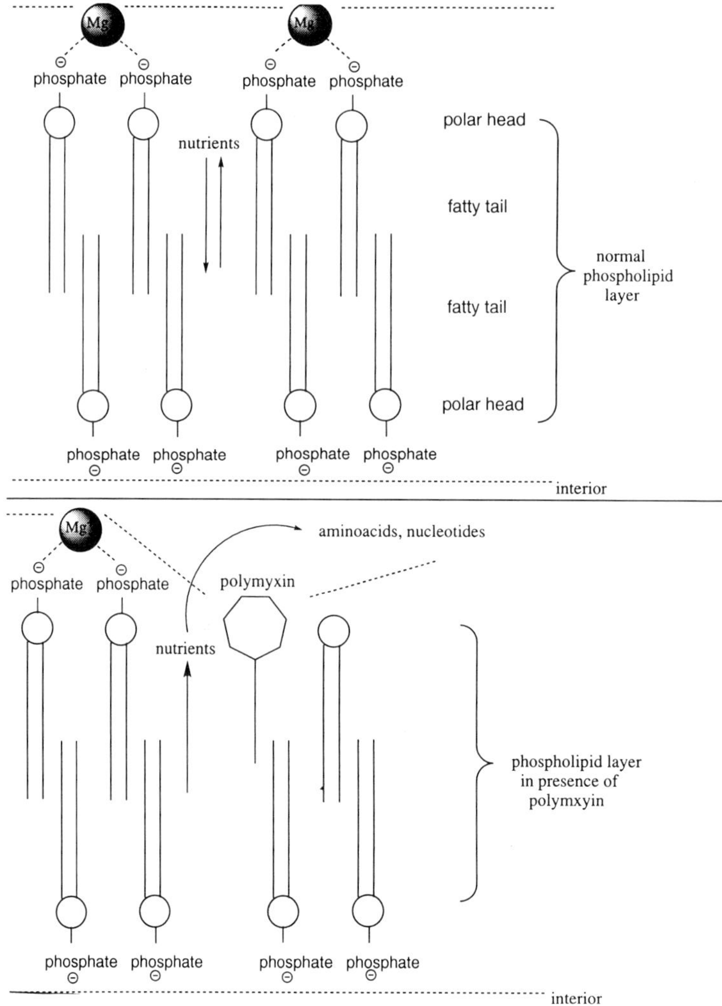

Figure 8.8 Effects of the polymyxins.

Monoacetylated congeners are more active than bis-acetylated derivatives, and the incorporation of additional acetyl moieties yields progressively worse compounds in terms of antibacterial activity.

The fatty acid group is also necessary for antibacterial activity; its removal produces essentially inactive derivatives. However, variation of the fatty acid component itself has afforded some compounds with improved potency against certain bacteria. The length of the fatty acid has an effect in this regard, and in general, longer chain acids are better than those present in conventional polymyxins. The optimum length appears to the C_{12} to C_{14} but shorter chain analogs can display enhanced potency against some Gram negative pathogens such as *E. coli* and pseudomonads. Fatty acids at the upper end of this range can display activity against pathogens which are usually not susceptible to the polymyxins, such as certain Gram positive species. However, no (semi-)synthetic polymyxin derivative has made an impact on antibacterial chemotherapy to date. Despite a few promising leads, the polymyxins have changed little since their introduction into chemotherapy decades ago, and there is scant work underway today in any attempt to re-explore these antibiotics since more potent and better tolerated unrelated agents are available.

8.2.4.3 Resistance. Resistance to the polymyxins is displayed by certain Gram negative species (*Proteus*, *Serratia*) and is a consequence of outer membrane composition which prevents the drug from acting on the cytoplasmic membrane. Most Gram positive bacteria are inherently not susceptible to the polymyxins.

8.2.5 *Bacitracin*

Bacitracin **7** (Figure 8.9) is a complex antibiotic that is primarily peptidic in structure but also contains an unusual thiazoline-based terminus. Bacitracin has a peculiar origin; this substance was first isolated from wound material contaminated with a bacillus extracted from a girl named Tracy (hence its name similarity). It possesses notable activity against Gram positive

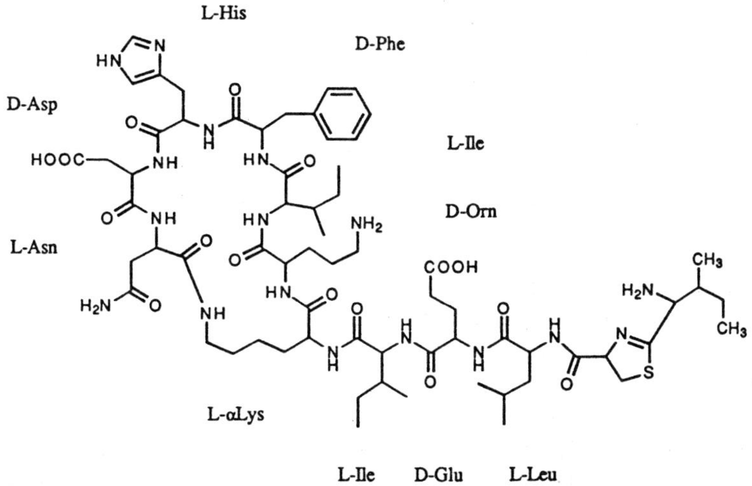

Figure 8.9 Bacitracin.

bacteria, but is generally inactive against many common Gram negative species such as the *Enterobacteriaceae* and pseudomonads. Bacitracin is unacceptable as a systemic agent due to host intolerability, poor absorption and stability, and has found use exclusively as a topical antibacterial agent. It is available as a mixture of structurally similar compounds; bacitracin A is the major component.

8.2.5.1 Mode of action.   Bacitracin exerts its antibacterial properties as a consequence of its ability to halt bacterial cell wall biosynthesis. Specifically, bacitracin inhibits the synthesis of the peptidoglycan polymer. Recall that the β-lactam antibacterials, as well as cycloserine, target this same protective sheath. However, the β-lactam antibacterials act 'downstream' from bacitracin by inhibiting the final step that cross-links the peptide–glycan strands (transpeptidation). In contrast, cycloserine acts 'upstream' from bacitracin and inhibits the production of alanine-derived peptidoglycan precursors in the cytoplasmic stage. Bacitracin, on the other hand, does not affect the cytoplasmic events, nor cell wall functions directly, but rather acts during the membrane-associated stage that links the operations of the cytoplasm with those of the cell wall.

Although peptidoglycan biosynthesis commences within the cytoplasm, a mechanism is operative that allows key precursors to cross the cytoplasmic membrane; the peptidoglycan matrix is eventually completed outside on the opposite side of the cytoplasmic membrane. The nucleotide-bound pentapeptide is transported from the relatively polar environment of the cytoplasm across the non-polar cytoplasmic membrane, and finally to its destination in the cell wall. This journey, the second stage of peptidoglycan biosynthesis, is made possible by the action of the lipid carrier, undecaprenyl phosphate (C_{55}-P). This lipid can exist in two distinct phosphorylated states, both of which are crucial for the regulation of the translocation process. The monophosphorylated lipid is an acceptor of the UDP-pentapeptide and the lipid adduct, MurNAc-pentapeptide-P-P-lipid, is formed (Figure 8.10). This species undergoes glycosylation; the resultant peptidic-sugar lipid (GlcNAc-MurNAc-pentapeptide-P-P-lipid) is further elaborated, delivered to the outside surface of the cytoplasmic membrane and cleaved from the lipid. The lipid by-product is undecaprenyl pyrophosphate (C_{55}-P-P) which is unable to accept another UDP-pentapeptide unit. However, a phosphatase catalyses cleavage of phosphate from undecaprenyl pyrophosphate to regenerate monophosphorylated lipid (C_{55}-P). This recycling of the lipid carrier continues the transport process and structural elaboration of new UDP-pentapeptide across the cytoplasmic membrane (Figure 8.11).

The recycling of the undecaprenyl lipid is crucial for the continuation of peptidoglycan biosynthesis, a process that is necessary for the survival of the bacterial organism. This matrix is susceptible to damage from external

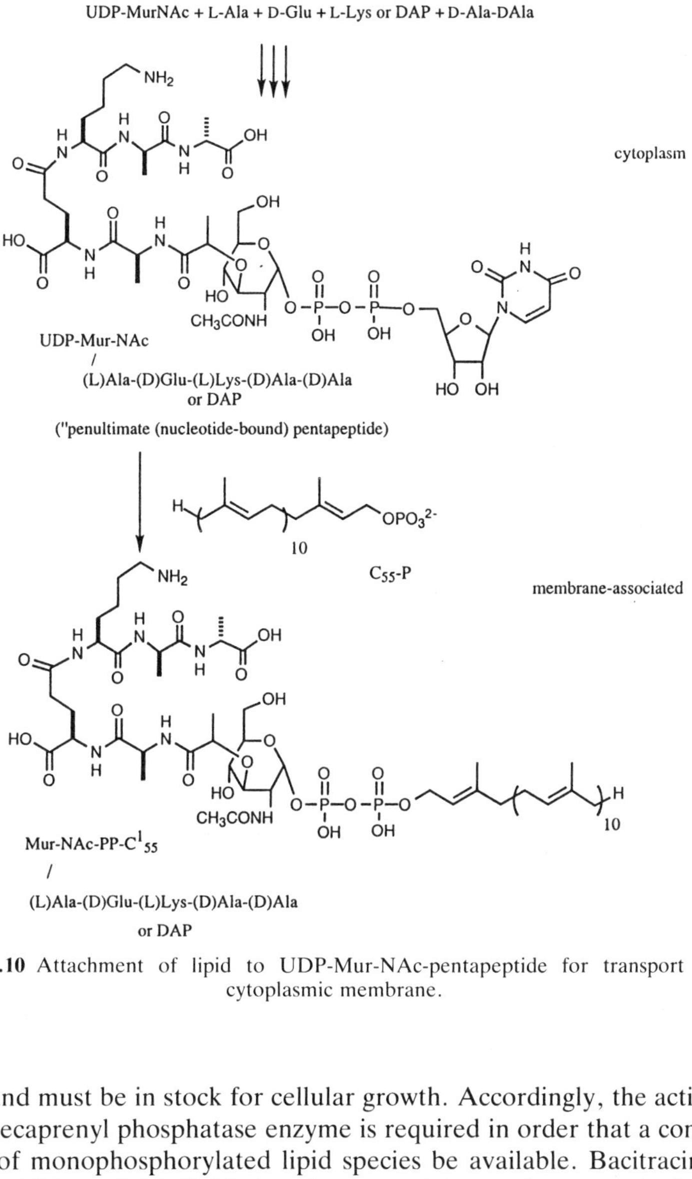

Figure 8.10 Attachment of lipid to UDP-Mur-NAc-pentapeptide for transport across cytoplasmic membrane.

forces and must be in stock for cellular growth. Accordingly, the action of the undecaprenyl phosphatase enzyme is required in order that a constant supply of monophosphorylated lipid species be available. Bacitracin is a potent inhibitor of this lipid phosphatase activity, and as a result displays antibacterial activity in susceptible bacteria. Bacitracin inhibits the dephosphorylation of undecaprenyl pyrophospate but does not inactivate the corresponding phosphatase enzyme (undecaprenyl pyrophosphatase) (Figure 8.11). This is accomplished via complexation of the drug to the undecaprenyl pyrophosphate substrate in such a way that pyrophosphate unit is protected from the hydrolytic action of the phosphatase enzyme.

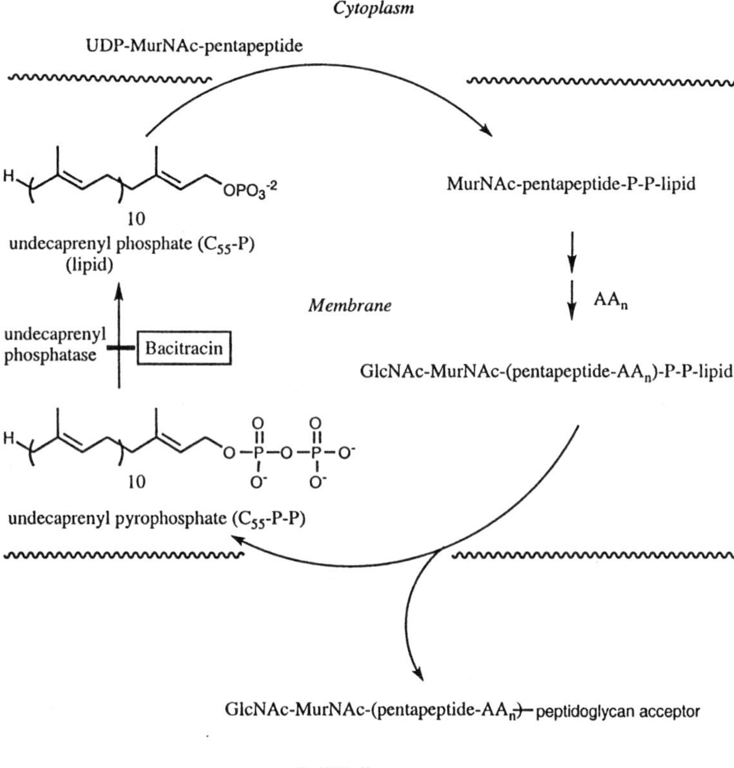

Figure 8.11 Site of action of bacitracin.

Divalent cations such as zinc are needed for binding to occur on a magnitude that confers useful antibacterial properties. The histidine residue and the thiazoline moiety of bacitracin are vital in this regard. Bacitracin has a considerably weaker affinity for other lipid-like pyrophosphates such as farnesyl pyrophosphate, a key intermediate in the biosynthesis of steroidal materials; this selectivity contributes to a favorable therapeutic effect.

8.2.5.2 Structural features. Bacitracin consists of ten amino acid units and the terminal leucine amino acid residue is joined to a thiazoline ring via a (ketone) carbonyl center. Extending from the thiazoline ring is an aminoalkyl side chain. Unlike the other peptidic antibacterials, bacitracin contains a histidine vital for antibacterial activity since chelation to a divalent cation such as zinc is necessary for optimum activity. Because of these exquisite requirements and its structural complexity, bacitracin has not been significantly investigated in terms of structure–activity relationships.

8.3   General comments

The peptidic antibacterial agents have contributed significantly to current chemotherapy although in general these agents are often regarded as alternative agents and/or are limited to indications that tolerate topical treatment. In some ways, peptidic agents are at an inherent disadvantage with regard to systemic use since mammalian hosts contain a host of enzymes, as well as numerous organ environments that can often induce degradation of peptide substrates. Indeed, this very problem is hampering the development of systemic peptidic agents in other areas such as antiviral (e.g. HIV protease inhibitors) and cardiovascular (e.g. renin inhibitors) therapies. Despite this, many new peptidic substances are being routinely isolated from naturally occurring sources and as these new materials are being investigated, new modes of antibacterial action are being elucidated (e.g. the mureidomycins as bacterial translocase inhibitors). It seems that interest in petitdic substances will continue, based upon the therapeutic potential generated by each specific class of compounds. In the face of increasing resistance, novel mechanisms of antibacterial action are not only welcomed but are often directing new strategies in infectious disease therapy. In the case of peptidic substances, advances using peptidomimetic strategies may provide a means to avoid the problems traditionally associated with peptide-based drugs.

Additional reading

General

R. Walker and J. Meienhofer (Eds) (1975) *Peptides: Chemistry, Structure and Biology*, Proceedings of the Fourth American Peptide Symposium, Ann Arbor Science, Ann Arbor, MI.

Cycloserine

F.C. Neuhaus (1967) 'D-Cycloserine and *O*-Carbamyl-D-serine', in *Antibiotics I (Mode of Action)*, D. Gottlieb and P.D. Shaw (Eds), Springer-Verlag, New York, pp. 40–83.

Tyrocidines

N. Izumiya, T. Kato, H. Aoyagi, M. Waki and M. Kondo (1979) *Synthetic Aspects of Biologically Active Cyclic Peptides*, Halsted Press, a Division of John Wiley, New York.

F.E. Hunter, Jr. and L.S. Schwartz (1967) 'Tyrocidines and gramicidin S (J_1, J_2)', in *Antibiotics I (Mode of Action)*, D. Gottlieb and P.D. Shaw (Eds), Springer-Verlag, New York, pp. 636–641.

Polymyxins

D.R. Storm, K.S. Rosenthal and P.E. Swanson (1977) 'Polymyxin and related peptide antibiotics', *Ann. Rev. Biochem.*, **46**, 723.

O.K. Sebek (1967) 'Polymyxins and circulin', in *Antibiotics I (Mode of Action)*, D. Gottlieb and P.D. Shaw (Eds), Springer-Verlag, New York, pp. 142–152.

Gramicidins

F.E. Hunter, Jr. and L.S. Schwartz (1967) 'Gramicidins', in *Antibiotics I (Mode of Action)*, D. Gottlieb and P.D. Shaw (Eds), Springer-Verlag, New York, pp. 642–648.

Bacitracin

E.D. Weinberg (1967) 'Bacitracin', in *Antibiotics I (Mode of Action)*, D. Gottlieb, P.D. Shaw (Eds), Springer-Verlag, New York, pp. 90–101.

K.J. Stone and J.L. Strominger (1971) 'Mechanism of action of bacitracin complexation with metal ion and C_{55}-isoprenyl pyrophosphate', *Proc. Natl. Acad. Sci.*, **68**, 3223.

G. Siewert and J.L. Strominger (1967) 'Bacitracin: an inhibitor of the dephosphorylation of lipid pyrophosphate, an intermediate in biosynthesis of the peptidoglycan of bacterial cell walls', *Proc. Natl. Acad. Sci.*, **57**, 767.

9 Miscellaneous antibacterial agents

9.1 Chloramphenicol

9.1.1 Introduction

Chloramphenicol 1 (Figure 9.1) has been a useful antibacterial drug in the past but has since been replaced by newer agents which offer comparable or enhanced potency without particular adverse side effects (potentially fatal aplastic anemia). Nonetheless, chloramphenicol was the first orally active broad spectrum antibiotic to reach the market. It is also a landmark agent since it was the first systemic agent to be produced by total chemical synthesis in an inexpensive fashion and abundant quantity. As a result, chloramphenicol is a historically important chemotherapeutic agent and its use was prevalent for decades from the 1950s. Chloramphenicol was first reported in 1947 as an isolable natural product from composted material containing a *Streptomyces* microorganism and its broad spectrum of primarily bacteriostatic action was duly noted.

Chloramphenicol is active against a host of microbes including a variety of Gram negative and Gram positive bacteria, as well as anaerobic strains. In addition, it is potent against intracellular Rickettsial infections (e.g. Rocky Mountain spotted fever) and mycoplasmas. Chloramphenicol is well absorbed from the gastrointestinal tract and often blood levels can be achieved that are sufficient to thwart many common bacterial pathogens. However, in practice, high serum levels may pose a danger to the patient and so other antibacterials, such as the β-lactams or tetracyclines, are usually preferred. A valuable property of chloramphenicol is that it readily crosses the blood-brain barrier and can therefore be administered for infections of the brain and spinal tissues. Today, chloramphenicol remains an alternative therapy for the treatment of meningitis caused by *Neisseria*

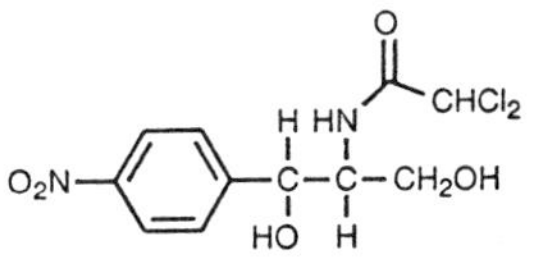

1

Figure 9.1 Chloramphenicol.

meningitidis, *Streptococcus pneumoniae* and *Haemophilus influenzae* in penicillin-allergic patients and is also used for the treatment of bacterial brain abscesses. Traditionally, chloramphenicol has been a premier agent against typhoid fever (*Salmonella typhi*) although resistance is quite widespread at this time.

9.1.2 Mode of action

Chloramphenicol exerts its bacteriostatic action by inhibiting bacterial protein biosynthesis. (The reader is encouraged to refer to chapter 4.2 for a discussion on ribosome-directed bacterial protein biosynthesis). However, a number of events that prime the bacterial 'machinery' for protein synthesis are not affected. Specifically, the activation of individual amino acids as t-RNA conjugates is not hindered; the action of aminoacyl-t-RNA synthetases is unimpeded. Chloramphenicol does not directly compete with any amino acid for incorporation into a t-RNA conjugate, nor is the availability of any amino acid compromised by the presence of this agent. In addition, the m-RNA molecule encoding for a given protein is not altered by chloramphenicol, nor is the assemblage of the ribosome–m-RNA complex. Lastly, the various protein factors that govern initiation, elongation and termination of the ribosome-bound peptidic substrate also remain functional, although the release of the peptide from the ribosome can be partially inhibited at high concentration of chloramphenicol.

Chloramphenicol targets the bacterial ribosome, as do some other classes of antibacterial agents (e.g. the macrolides). In essence, chloramphenicol undergoes binding to the bacterial ribosome by adopting a conformation which mimics that of a peptidyl adenylyl terminus of a t-RNA molecule (Figure 4.11, chapter 4). The end result is that amino acid units, available as t-RNA conjugates, are unable to be incorporated upon the ribosome-bound peptide chain, and thus protein production is halted. Interestingly, protein biosynthesis is not stopped immediately upon ribosomal binding of the drug. In fact, ribosomal-bound peptidic substrates are formed in the presence of chloramphenicol, however functional proteins are not elaborated. This tolerance may reflect a length or size requirement which must be fulfilled before protein biosynthesis can be completely shut down; the same phenomenon occurs with the erythromycins and some other macrolides (chapter 6.2).

Chloramphenicol binding occurs on the 50S ribosomal subunit, but can be 'washed out' (diluted away from the bacterial cell) indicating that this association is weak in nature and reversible. There are actually two sites on the bacterial ribosome that can accommodate chloramphenicol. One site exerts a rather high affinity for chloramphenicol while the other site is weaker in binding the drug. As a result, more than one molecule of chloramphenicol can bind to the bacterial ribosome under optimal conditions. The effcts of ribosome binding of chloramphenicol have been demonstrated in several ways.

In cell-free systems which contain all of the components (ribosomes, m-RNA aminoacylated t-RNA molecules and various protein factors) required for peptide synthesis, the polymerization of amino acids is inhibited. For example, poly-A nucleotide substrates are not fully elaborated to the encoded poly-Lys peptide. However, lysine-derived dipeptides and tripeptides products are formed, demonstrating that protein synthesis is not immediately halted upon chloramphenicol binding to the bacterial ribosome; there is a tolerance for 'early' peptide chain formation. Interestingly, the composition of the polynucleotide template is a factor since peptide synthesis directed by poly-U and poly-UA templates is less effected than that directed by poly-A and poly-UC substrates. These features are similar in gross action to those of the erythromycins (chapter 6, section 6.2). Therefore, it is not surprising that erythromycins, as well as some other macrolide antibacterials (e.g. oleandomycin, spiramycin) can inhibit ribosomal binding of chloramphenicol, indicating that these agents compete for the same or overlapping site on the bacterial ribosome. (For comparison, aminoglycosides (chapter 5) and tetracyclines (chapter 4) do not effect chloramphenicol binding to the ribosome, but both of these classes of antibacterial agents do affect the bacteria ribosome).

Chloramphenicol binding to the 50S bacterial ribosome subunit requires potassium or ammonium cation, another feature shared in common with the macrolides. The ribosomal 50S subunit is known to provide the peptidyl transferase component needed for elaboration of peptides (chapter 4, section 4.2); catalytic peptidyl transferase activity requires potassium and ammonium cations. Therefore, chloramphenicol binding to the bacterial ribosome probably compromises catalytic peptidyl transferase activity as a primary mode of antibacterial activity.

Chloramphenicol inhibition of peptide bond formation has been demonstrated in a number of ways. As described above, cell-free ribosomal complexes of polynucleotides do not elaborate the encoded polyamino acid. Chloramphenicol suppresses peptidyl transferase activity in these instances although movement of the ribosome along the m-RNA molecule is not affected (this distinguishes chloramphenicol from some macrolides which can inhibit this process of translocation). Another revealing example involves the treatment of bacteria with puromycin in the presence of chloramphenicol. Puromycin is a well-recognized structural analog of the aminoacyl terminus of an aminoacylated-t-RNA (AA-t-RNA). If puromycin alone (no chloramphenicol) is added to bacteria actively engaged in protein biosynthesis, puromycin will be incorporated in the growing peptide chain instead of the incoming amino acid. Protein biosynthesis is halted since the peptide chain is transferred to puromycin; this so-called 'puromycin reaction' is a simpler version of peptide bond formation since puromycin does not need to interact with t-RNA. However, in the presence of chloramphenicol, peptidyl-puromycin formation is inhibited indicating that peptide bond formation is inhibited and hence chloramphenicol is an inhibitor of peptidyl transferase. This phenomenon has been demonstrated in whole bacteria as well as cell-free ribosomal preparations supplied with polynucleotide substrate.

9.1.3 *Structural features and structure–activity relationships*

Synthetic chloramphenicol exists as a mixture of all four possible diastereomers which are a consequence of two asymmetric carbon centers that are indiscriminately produced. The D-(−)-threo-diastereomer, produced originally by fermentation, is responsible for the antibacterial activity of chloramphenicol. Most structural modifications have been to the aromatic ring system, the dichloroacetamido side chain and the propanediol backbone. The most promising alterations have included replacement of the aromatic nitro group substituent, alteration of the amide moiety and (covalent) conjugation of the free hydroxyl groups. Although it is impossible to address all of the analogs which have been reported to date (more than 500), several obvious trends have been recognized.

The presence of an electron-withdrawing group at the *para* position of the aromatic ring is a requirement. Thiamphenicol 2 (Figure 9.2), a methylsulfonyl analog, is the most significant chloramphenicol analog and has been marketed outside of the United States. In general, other electron-withdrawing moieties can be tolerated at the *para* position although most offer no advantage compared to the prototypes (chloramphenicol, thiamphenicol). Even amine-derived moieties that are electron-withdrawing in nature (e.g. amine, nitroso) are inferior. Monosubstitution at the *para* position with a nitro or sulfonyl group remains optimal with regard to antibacterial potency. Unfortunately, the nitro group is regarded with suspicion as the cause of aplastic anemia, and thiamphenicol is a marginal chemotherapeutic agent for human use since it is also toxic.

Both chloramphenicol and thiamphenicol contain a characteristic dichloroacetamido side chain that extends from the central carbon atom of

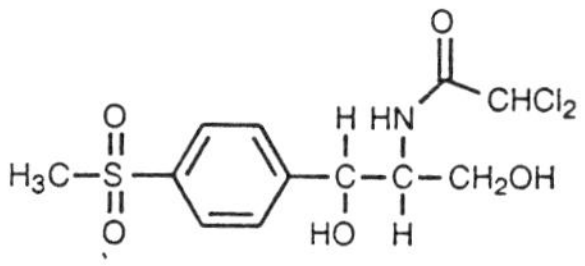

2

Figure 9.2 Thiamphenicol.

the propanediol chain. An amide substituent must be in place at the C-2 position for useful antibacterial activity and few substituent alterations at this site are tolerated. In general, the carboxamido moiety must have electron-withdrawing substituents for potent activity, and the dichloroacetamido group remain most favorable. However, the iodoacetyl- and bromoacetyl-analogs (also called monoiodoamphenicol and monobromoamphenicol, although such nomenclature is a misnomer) behave similarly to chloramphenicol with respect to binding to the bacterial ribosome. None of these variations have offered an advantage over chloramphenicol and thiamphenicol.

The hydroxyl moieties of the propanediol chain are also a requirement for ribosome binding and antibacterial activity. These sites have been structurally modified and again, few variations are permitted. Virtually any alterations which disrupts the hydrogen bonding capabilities of the hydroxyl group(s), or results in the replacement of the oxygen atom(s) abolishes activity. For example, alkylation of the diol group is detrimental but esterification can afford congeners that retain activity although this is due to hydrolytic release of the free parent substrate *in vivo*. The palmitate and hemisuccinate esters have been commercially significant prodrugs of chloramphenicol. Lastly, the length of the propanediol chain is also vital and extension of the propane backbone results in loss of activity.

9.1.4 Resistance

Bacteria can be intrinsically resistant to chloramphenicol and thiamphenicol, or resistance can be acquired. Intrinsic resistance usually results from impermeability, a phenomenon which prevents the entry of the agent into the cell and prevents access to the bacterial ribosome. Cell-free ribosomal preparations of such bacterial strains remain sensitive and so resistance can be traced to the nature of the bacterial cell barriers.

Resistance to chloramphenicol can be acquired. In some cases, the chloramphenicol resistance arises from the ribosomal mutations and although the drug reaches the bacterial ribosome, it is unable to undergo binding and so antibacterial activity is lost. Mutations of certain ribosomal proteins have been identified and resistance of this type appears to result from multi-step mutations rather than by a single event.

More commonly, resistance to chloramphenicol occurs from the action of bacterial enzymes that deactivate the drug. These enzymes, chloramphenicol acetyl transferases (CAT), attach acetyl groups onto the chloramphenicol hydroxyl functionality. Both the resultant mono- and bis-acetylated conjugates are inactive. Chloramphenicol-deactivating enzymes have been identified from many bacterial species. One of the most troublesome CAT-mediated resistances is that frequently found in strains of *Salmonella typhi*, the pathogen of typhoid fever.

9.2   Lincosamides

The lincosamides are a small group of antibiotics that originated from the discovery of lincomycin **3** (Figure 9.3) as an antibacterial fermentation product of a *Streptomyces* species in 1963. Lincomycin was recognized for both its unique chemical composition, as well as for its oral antibacterial properties. Lincomycin contains a thiolated-D-erythro-α-D-galacto-octopyranoside that is joined to a proline residue through an amide linkage. It is active against a number of clinically important Gram positive bacteria (some streptococci and staphylococci) and its spectrum includes the important anaerobic Bacteroides species and the malarial parasite *Plasmodium falciparium*. The lincosamides are generally impotent against most Gram negative pathogens (*E. coli*, *Klebsiella*, *Proteus*, *Shigella*, *Serratia*, *Pseudomonas* sp.) but are synergistic with some aminoglycosides against these bacteria. In addition, clindamycin is moderately against *Haemophilus influenzae* and exquisitely potent against Bacteroides.

A number of lincomycin analogs have been prepared in a semi-synthetic fashion, or by supplying various additives to the *Streptomyces linconensis* fermentation. Only one such derivative has advanced on to the market. Clindamycin **4** (Figure 9.4) is a chlorinated analog in which the (C-7) hydroxyl group is displaced with inversion of stereochemistry. Clindamycin

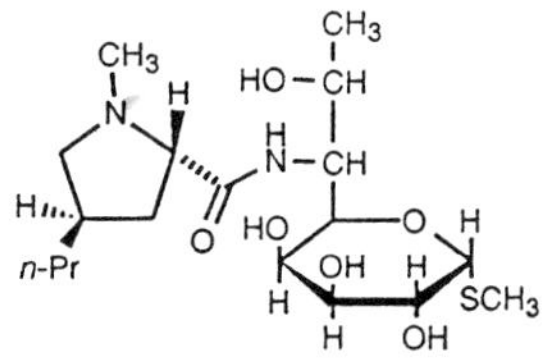

3

Figure 9.3 Lincomycin.

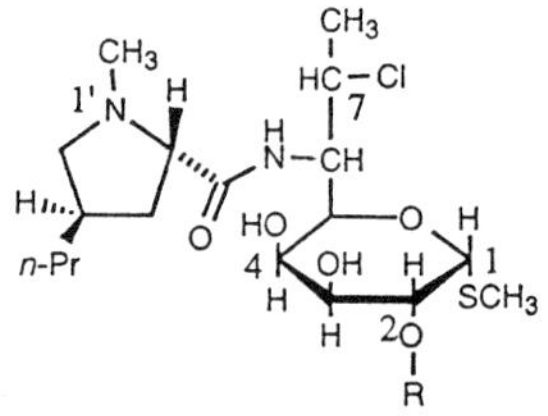

4: R = H
5: R = P(O)(OH)$_2$
6: R = C(O)(CH$_2$)$_{14}$CH$_3$

Figure 9.4 Clindamycin, its C-2 phosphate and C-2 palmitate esters.

is generally more potent than lincomycin and is better absorbed from the human gastrointestinal tract. These properties have made clindamycin a valuable agent for the treatment of various anaerobic bacterial infections. Unfortunately, the lincosamides can cause severe colitis (pseudo-membranous colitis) due to *Clostridium difficile* superinfection. (*Clostridium difficile* is an opportunistic anaerobic pathogen that is not inhibited by the lincosamides. A toxin produced by *C. difficile* is responsible for the potential life-threatening diarrhea and pseudomembranous colitis that have been associated with a number of other broad spectrum antibiotics.) For this reason, clindamycin is usually administered for confined anaerobic infections. Typical examples included vaginal infections (bacterial vaginosis) involving *Bacteroides* and *Peptostreptococcus* species.

A variety of prodrugs of both lincomycin and clindamycin have been prepared. Most of these prodrugs are simple ester derivatives of various hydroxyl group functionality; several esters can be present in a given molecule. The ester prodrug derivatives undergo cleavage *in vivo* and it is the free lincosamide that is responsible for the antibacterial effect. The most significant lincosamide prodrugs are the C-2 phosphate derivative of clindamycin **5** (Figure 9.4) which is administered intravaginally, and the C-2 palmitate ester of clindamycin **6** (Figure 9.4) which has been used for certain pediatric indications.

9.2.1 Mode of action

The lincosamides are inhibitors of bacterial protein biosynthesis as a result of interaction with the bacterial ribosome. The lincosamides undergo binding to the 50S ribosomal subunit of susceptible bacteria and this has several effects. In cell-free ribosomal preparations derived from several bacterial species, the lincosamides prevent incorporation of certain amino acids into a polypeptide. For example, polyuridine-directed polyphenyl-alanine synthesis is inhibited in extracts of *B. stearothermophilus*. Whole bacteria are similarly affected assuming that the lincosamide penetrates into the cell and reaches susceptible ribosomes. In *Staphylococcus aureus*, the incorporation of lysine into protein is halted in the presence of lincomycin. In general, Gram positive bacteria are more susceptible to the effects of the lincosamides; streptococci, staphylococci and *Corynebacterium diphtheriae* are particularly sensitive.

Erythromycin antagonizes lincosamide–ribosome binding and so both compounds probably associate with overlapping regions of the 50S subunit. However, erythromycin allows for the construction of small peptidic substrates which remain attached to the ribosome. In fact, erythromycin (and some other macrolides) can even enhance the formation of some small di- and tripeptides before protein synthesis is completely shut down. In contrast, the lincosamides inhibit the initiation of ribosomal protein biosynthesis; there is no induction of peptide formation and even small peptide substrates can be absent.

9.2.2 Structural modifications and structure–activity relationships

The lincosamides have been obtained via chemical synthesis; clindamycin was first produced by semi-synthetic means in 1966 from lincomycin. The total chemical synthesis of any lincosamide is not trivial and so new compounds nearly always originate from a naturally occurring product, usually lincomycin itself. This has ben accomplished in two ways. Many derivatives have been prepared by chemically modifying a lincosamide through a series of specific organic transformations. Alternatively, some lincosamides can be obtained by using additives during the fermentation process. This latter approach, that of 'precursoring' or 'directed bio-synthesis', has afforded some interesting analogs. For example, ethionine added to a *Streptomyces lincolnensis var. lincolnensis* fermentation produced 1'-demethyl-1-demethylthio-1'-ethyl-1-ethylthiolincomycin, the congener in which an ethyl substituent is located on both the proline nitrogen atom as well as the anomeric sulfur center (see numbering scheme as shown in Figure 9.4).

Demethylated lincosamide derivatives have been obtained in a similar fashion. The addition of certain sulfonamides inhibits biogenic methylation of the proline (N-1') nitrogen center. N-1'Demethylclindamycin has been isolated from fermentation using this protocol (it has also been identified as a clindamycin metabolite) and retains excellent antibacterial activity. In addition, the free (N-1'-) amine moiety of desmethyllincosamides has been further elaborated by acylation or alkylation, and although no congener of this type has advanced onto market, some interesting compounds have been obtained in this way (see below).

Treatment of lincomycin with hydrazine results in cleavage of the proline ring amide linkage (as a hydrazide) from the octopyranoside. This has allowed for modification of the nitrogen heterocycle. A variety of 'unnatural' prolines have been attached to the sugar portion to afford novel congeners. The hydroxyl group functionality of the pyranoside unit has also been altered through esterification and etherification, as well as through the formation of carbamates and phosphate esters. In general, useful modifications to the hydroxyl group functionality have been limited to esterifications provided that the ester group can be cleaved enzymatically or metabolically following administration. Most lincosamide esters are devoid of useful antibacterial activity or only weakly active depending upon the size of the ester group and, to a degree, which particular hydroxyl groups are modified. Removal of hydroxyl group functionality has been effected in some cases; however C-2 deoxylincosamides are useless as antibacterials.

Certain lincosamide ester prodrugs have been successfully developed. In general, the C-2 esters are superior to C-3 and C-7 ester analogs; estrification at the C-4 hydroxyl center is detrimental. Monoesterification is superior and progressive esterification diminishes antibacterial activity; however, lincosamide C-2–C-7 diesters can retain notable antibacterial activity. Lincosamide triesters and tetraesters are only weakly active. Selective esterification at the C-2 hydroxyl center is best carried out by masking the C-3 and C-4 hydroxyl groups in the form of a cyclic acetal. Under appropriate experimental conditions, it is possible to acetylate the C-3 hydroxyl group directly since it is the least sterically hindered of the equatorial hydroxyl groups. The C-4 group has been esterified by inducing migration of the acyl group from the corresponding C-3 analog under acidic or basic conditions.

In addition to clindamycin phosphate, simple alkyl esters at C-2 are the most significant since often these compounds are better absorbed than the bioactive free lincosamides. Esters containing alkyl chains of between four to eight carbon units (unbranched) are optimal, although larger esters can retain appreciable activity. Interestingly, carbamate derivatives, whether at C-2 or joined to another hydroxyl group, are inferior. The C-7 position alcohol can be esterified in keeping with the requirement highlighted above.

Halogenation at the C-7 position is well tolerated as is evident by the properties of clindamycin. The corresponding bromo and iodo analogs are more potent *in vitro* than lincomycin or clindamycin against strains of staphylococcus and streptococcus. The treatment of these compounds with nucleophiles results in displacement of the halide with inversion of the stereocenter. Azido and cyano derivatives have been prepared using this methodology and both retain notable activity; the amino analog has been obtained by subsequent reduction but is essentially inactive. The C-7 alcohol itself has been directly converted to ethers or into a leaving group that can be displaced to afford ethers, thioethers and mercaptans. The C-7(S)-O-methylated analog is more potent than lincomycin *in vivo* and nearly as potent as clindamycin in some animal models. In general, the introduction of larger alkyloxy substituents produces compounds inferior to either lincomycin or clindamycin. In addition, epimerization of the C-7 alcohol (or ether) or oxidation to the corresponding ketone is detrimental.

The C-1 anomeric center is tolerant of a number of structural alterations and, in general, the α-anomers produce better activity than the β-counterparts. For example, the α-C-1 thioether substituent can incorporate larger alkyl chains; the ethyl and butyl homologs are active. Several lincosamide-related antibiotics have been discoverd that possess an α-2-hydroxyethylthio C-1 substituent (celesticetin) or a salicyclic acid-derived ester of this moiety (desalicetin).

The proline unit has also been explored in terms of structural modifications. As described earlier, there are several ways in which N-1'-desmethyllincoamides have been synthesized. These compounds are generally less active than the N-methylated parents, but the placement of an ethyl substituent at N-1' affords derivatives which are nearly as potent as the methylated homologs. In all of these cases, the substituent at the C-4' position can play a contributing role. Lincomycin and clindamycin both have a *n*-propyl substituent at the C-4' center. In the case of N-1'-methyl congeners, groups larger than *n*-propyl are well tolerated; straight chains of seven and eight carbon units in length are quite potent. However, the presence of an N-1 ethyl substituent requires that smaller groups be placed at C-4', such as alkyl groups of five carbon units or less.

Lastly, the amide linkage that joins the proline unit to the octopyranoside is absolutely necessary; the thioamide derivatives are considerably less potent as antibacterials.

9.2.3 Resistance

Bacteria can develop resistance to the lincosamides in a variety of ways. Resistance determinants can be carried on plasmids or arise from chromosomal mutation. A common and particularly interesting form of resistance is that of self modification of the bacterial ribosome, the target of the lincosamides. Specifically the 23S r-RNA molecule is subject to adenine methylation as a result of the so-called MLS (macrolide-lincosamide-streptogramin B) determinant. (The reader is encouraged to review the material on resistance to the macrolides in chapter 6).

Certain staphylococci, such as some strains of *S. aureus*, are able to resist the antibacterial effects of the lincosamides by directly altering the structure of the agent. Recall that this phenomenon is a common mode of resistance to the aminoglycosides (see chapter 5, section 5.5), and involves the action of various enzymes which catalyse attachment of small organic moieties onto a reactive portion of the antibacterial agent. In the case of the lincosamides, the C-4 hydroxyl group is prone to conjugation caused by bacterial adenylyl transferases. For example, adenylylation of clindamycin results in loss of antibacterial activity.

9.3 Nitrofurans and nitroimidazoles

A number of totally synthetic nitrofurans and the nitroimidazoles have found use as antibacterials; the first agents were introduced nearly 50 years ago. More recently, nitrofurantoin **7**, nitrofurazone **8** and furazolidine **9** (Figure 9.5) have been the most widely used of the nitrofurans developed for human purposes; a variety of structurally related congeners have found use in animals, fish or as food additives. The nitrofurans are medium spectrum antibactrial agents which are potent against a variety of Gram positive (e.g. *Staphylococcus aureus*) and Gram negative bacteria (e.g. *Escherichia coli*) assuming that sufficient concentrations are achieved at the site of infection. Metronidazole **10** (Figure 9.6) is a marketed antibacterial nitroimidazole that has found widespread use primarily against anaerobic bacteria (*Bacteroides* and *Clostridium* species) and several protozoa (*Trichomonas* species). Metronidazole is particularly well absorbed orally and widely distributed throughout the human body.

These agents, being nitroheterocycles, are also generally toxic to hypoxic tumor cells and in some cases, to fungi and nematodes as well. Unfortunately, these compounds frequently display mutagenic and/or carcinogenic activity in a variety of assays, properties that have been attributed to the presence of the aromatic nitro substituent. However, it is not clear how relevant these assays are with respect to short term use. For example, metronidazole is positive in the Ames test (for mutagenicity) but has been used safely for over three decades. Nonetheless, due to these potential toxicities the nitrofurans and nitroimidazoles are often used for urinary tract and vaginal infections respectively, rather than systemic use. However, metronidazole is valuable for the treatment of bone and joint, spinal, intra abdominal and lung infections caused by *Bacteroides*, *Clostridium* and *Peptostreptococcus* species, as well as endocarditis and septicemia. Some nitrofurans (e.g. nitrofurazone) are also available for

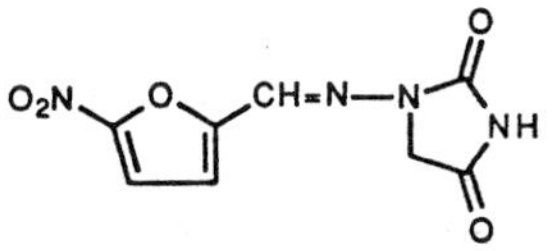

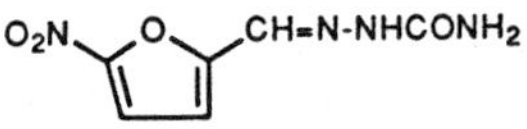

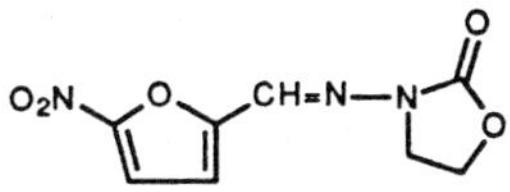

Figure 9.5 Nitrofurantoin, nitrofurazone and furazolidine.

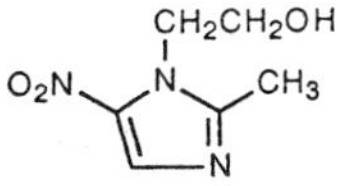

Figure 9.6 Metronidazole.

topical administration to prevent skin surface infections; this use is commonly encountered with burn victim and skin-graft patients.

9.3.1 Mode of action

The nitrofurans and nitroimidazoles exert a variety of actions upon prokaryotic and eukaryotic cells. This relative lack of selective action is responsible for the toxicities associated with these agents. It is believed that noxious nitro-derived radicals (nitrosoamine and hydroxylamine intermediates) are formed *in vivo* by metabolic reduction and that these species are responsible for the lethal effects exerted upon the cell. These intermediates are chemically reactive and so it has been difficult to gather definitive evidence on which is the most dangerous moiety. Nevertheless, a variety of cytoplasmic bacterial enzymes and proteins that are normally involved in respiration are targeted by the action of the nitrofurans and nitroimidazoles. Normal pyruvate and glucose metabolism is disrupted; pyruvate dehydrogenase, glutathione reductase and citrate synthetase activity is usually involved. Even genetic material (DNA) is likely to undergo damage and ribosomal function may be impaired. The nitro-imidazoles, specifically metronidazole, undergo reduction to their respective toxic intermediates through the action of highly reducing ferroproteins or perhaps electron-transferring proteins. Since anaerobes perform respiration at lower redox potentials (the end product of their metabolism is not oxygen or carbon dioxide) than aerobes, anaerobic bacteria are especially susceptible to metronidazole.

9.3.2 Structural features

The aromatic nitro group is absolutely necessary for the nitrofurans and nitroimidazoles to possess antibacterial activity. A host of substituents are tolerated on the hetrocyclic ring; the agents highlighted are illustrative. Many other compounds have been prepared but have been judged to be less acceptable for purposes of antibacterial chemotherapy.

The nitrofurans are readily synthesized from commercially available 5-nitro-2-furancarboxaldehyde; the nitroimidazoles are often obtained from nitration of imidazoles.

9.3.3 Resistance

The nitrofurans and nitroimidazoles target several bacterial targets and perhaps because of this, resistance is not a problem with the use of these agents.

9.4 Vancomycin and teicoplanin

Vancomycin **11** (Figure 9.7) is a very important glycopeptide antibiotic with a narrow spectrum of antibacterial activity essentially limited to a few Gram positive species as well as some spirochetes. The recent emergence of multi-drug resistant enterococci (*E. faecium*) has renewed a strong interest in vancomycin and its structural congener teicoplanin **12** (Figure 9.8), which is more active *in vitro*. Vancomycin is the preferred agent for the treatment of methicillin-resistant (*β*-lactam) *Staphylococcus aureus* (MRSA) infections including septicemia, endocarditis and tissue and bone infections. It can be given to patients who are allergic to the penicillins. Vancomycin is also active against streptococci, particularly when given with an aminoglycoside (synergism). Oral administration of vancomycin is effective against toxigenic *Clostridium difficile* and therefore can be used to combat antibiotic-induced pseudomembranous colitis. Vancomycin is a very valuable antibacterial for these reasons; unfortunately, it is often the last resort when *β*-lactum agents and others have failed miserably.

Vancomycin was isolated in 1956 from a *Streptomyces* species (*Streptomyces orientalis*) present in soil samples obtained from Indonesia, and has been used clincially for decades with success. Teicoplanin is the most significant glycopeptide since vancomycin to be marketed although it remains an experimental drug in the United States at this time. The glycopeptide family today encompasses over 50 compounds, but vancomycin and teicoplanin remain the most important in antibacterial chemotherapy. However, actinoplanin and avoparcin have found use as animal feed additives, and interestingly ristocetin has been used in determining

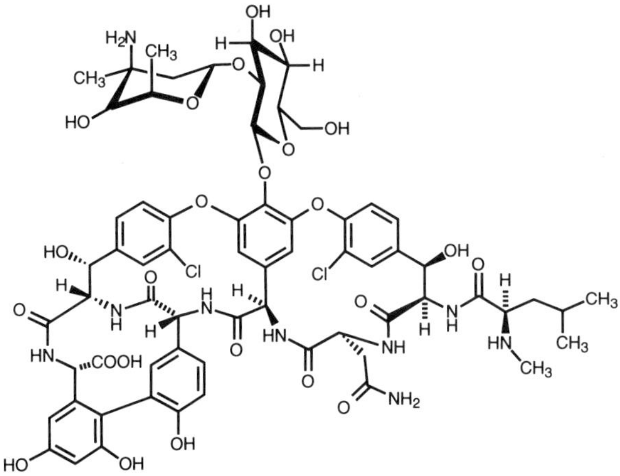

11

Figure 9.7 Vancomycin.

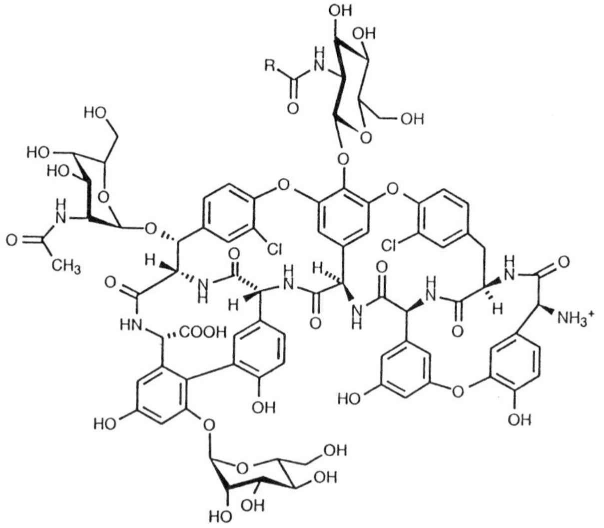

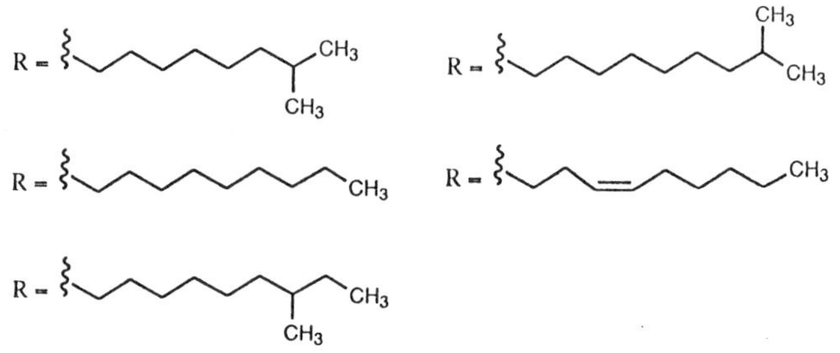

Figure 9.8 Teicoplanin.

certain blood-clotting disorders due to its ability to react with certain plasma proteins and cause platelet aggregation (not discussed).

9.4.1 Mode of action

The vancomycin group of glycopeptides exert antibacterial properties primarily as a result of disruption of normal peptidoglycan biosynthesis, but in a different manner than that caused by the *β*-lactam antibiotics, bacitracin and cycloserine. The site of action of vancomycin is 'upstream' from that of the *β*-lactams; vancomycin inhibits a key transformation in the second stage (lipid-bound) of the biosynthetic pathway. Vancomycin undergoes tight binding to the D-alanyl-D-alanine terminus at the free carboxyl end of the UDP-pentapeptides. This inhibits the final steps of transglycosylation and transpeptidation that are needed to afford functional

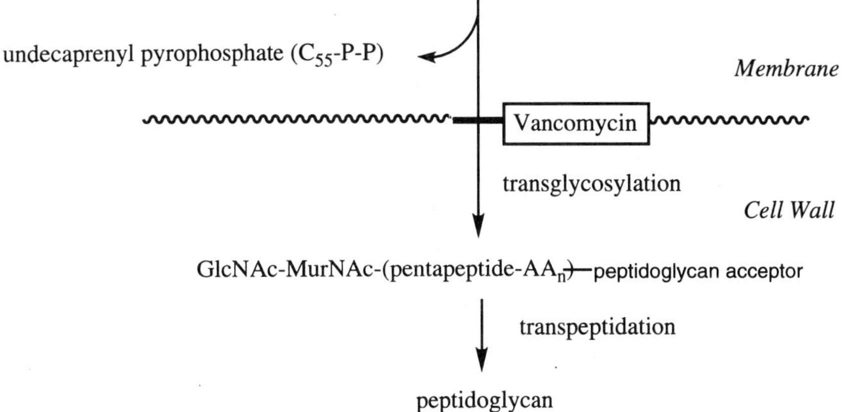

Figure 9.9 Mode of action of vancomycin.

peptidoglycan and usually results in bacterial cell lysis (Figure 9.9). This has the same effect upon susceptible bacteria as the β-lactams exert, but the molecular mode of action is very different. (The reader is encouraged to review the mode of action of the β-lactam agents (chapter 3, section 3.2) and of cylcoserine and bacitracin (chapter 8) for background information).

9.4.2 Basic structural features

The vancomycin group of antibiotics are very complex in stucture; the important members of this family are obtained directly from fermentation. The main structural feature is assumed to be a glycopeptidic configuration that forms a groove which accepts the peptidic hydrophilic D-Ala-D-Ala terminus and establishes hydrogen bonding interactions with this substrate. The aromatic residues likely form an 'outer wall' of the drug–substrate complex. In essence, the glycopeptides behave as (di)peptide receptors but accept the opposite optical series (D) to those found in mammals, which is a basis of some selectivity.

9.4.3 Resistance

Resistance to vancomycin has been rather rare until recently. This has been partially attributed to the secondary effects vancomycin exerts on the susceptible bacterial cell which can include alteration in the permeability of the bacterial cytoplasmic membrane and impairment of RNA synthesis. Nevertheless, vancomycin-resistant enterococci are becoming troublesome at this time.

Resistance can be endowed by the action of plasmids or transposons which express proteins that override the effects of vancomycin. For example, a specific plasmid-borne transposon (Tn1546) encodes for nine polypeptides and is responsible for *VanA*-mediated resistance. The inducible *VanA* phenotype is characterized by the presence of a ligase protein that allows for the construction of the bacterial cell wall even in the presence of the antibacterial glycopeptide. *VanA* proteins contain considerable amino acid homology to D-Ala-D-Ala ligases (*E. coli*) and appear capable of redirecting peptidoglycan assembly, by affording precursors with reduced affinity for binding to vancomycin. Six other genes encode for proteins that also circumvent the effects of vancomycin in some manner (two gene products are not involved). The *VanH* gene product supplies α-ketoacid reductase activity which affords a D-Ala-D-lactate peptidoglycan substrate, rather than the corresponding D-Ala-D-Ala congener, to the *VanA* ligase. The D-Ala-D-lactate substrate has been shown to be approximately one thousand-fold less susceptible to vancomycin. *VanS* and *VanR* gene products provide, respectively, a transmembrane protein which signals the presence of the glycopeptide and a DNA-binding protein which regulates DNA transcription for the expression of other gene products.

9.5 Isoniazid and other antituberculosis agents

Isoniazid **13**, ethonamide **14**, pyrazinamide **15**, *para*-aminosalicyclic acid **16** and ethambutol **17** are compounds that have been used for the treatment of tuberculosis (caused by *Mycobacterium tuberculosis*) (Figure 9.10). In general, these compounds rarely constitute a sole first line of therapy today. However, many of these agents are given in conjunction with a rifamycin (a preferred agent, or streptomycin (chapter 5)), especially in light of the emergence of multi-drug resistant *M. tuberculosis*. Isoniazid is

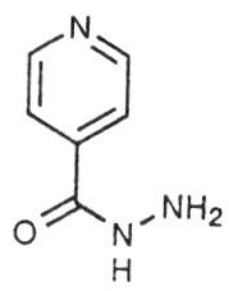

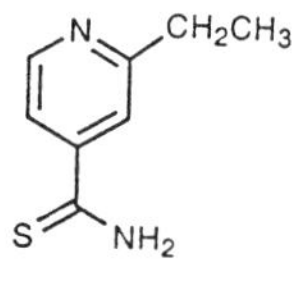

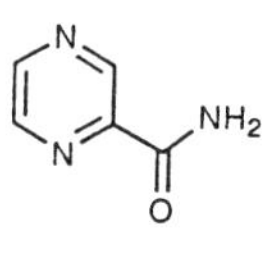

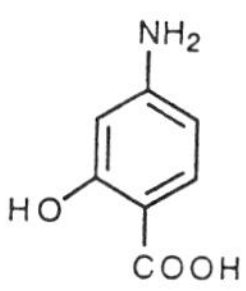

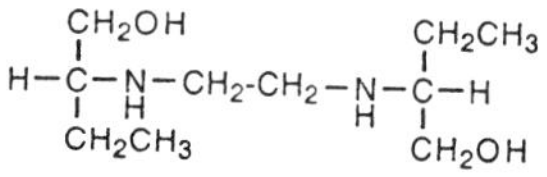

Figure 9.10 Antituberculosis agents: isoniazid, ethonamide, pyrazinamide, *p*-aminosalicyclic acid and ethambutol.

Table 9.1 Miscellaneous antibacterial agents

Generic name	Common trade names	Administration
Chloramphenicol	Chloromycetin	PO, IM, topical
Lincomycin	Linocin	PO, IM, IV
Clindamycin	Cleocin	PO
Clindamycin Phosphate	Cleocin Phosphate,	IV
	Cleocin T,	Topical
	Cleocin Vaginal	Vaginal
Clindamycin Palmitate	Cleocin Pediatric	PO
Furazolidone	Furoxone	PO
Nitrofurantoin	Macrodantin, Macrobid,	PO
	Furadantin	
Nitrofurazone	Furacin	Topical
Metronidazole (HCl)	Flagyl	PO, IV
	MetroGel	Topical, vaginal
Vancomycin	Vancocin, Vancoled	IV
	Vancomycin HCl	PO
Teicoplanin	Targocid, Targosid	IM, IV
Isoniazid	INH, Hyzyd, Niadox, Nydrazid	PO, IM
Pyrazinamide		PO

exceptionally useful against the slowing-growing mycobacteria, but unfortunately this agent tests positive for carcinogenic activity in some assays. The mechanism of action of isoniazid appears to involve the disruption of mycolic acid biosynthesis but details remain obscure. Mycolic acids are characteristic and essential components of the cell wall and are a rather unique target of mycobacteria compared to other bacterial species. Ethonamide and pyrazinamide are also (iso)nicotinamide mimetics. *para*-Aminosalicyclic acid is an inhibitor of bacterial folate metabolism in a manner similar to the sulfonamide antibacterials. The mechanism of ethambutol is essentially unknown.

Table 9.1 shows miscellaneous antibacterial agents.

Suggested reading

Chloramphenicol

F.E. Hahn (1967) 'Chloramphenicol', in *Antibiotics I (Mechanism of Action)*, D. Gottlieb and P.D. Shaw (Eds), Springer-Verlag, New York, pp. 308–330.

Hahn, F.E. (1983) 'Chloramphenicol', in *Antibiotics VI (Modes and Mechanism of Microbial Growth Inhibitors)*, F.E. Hahn (Ed.), Springer-Verlag, Berlin, pp. 34–45.

S. Pestka (1967) 'Chloramphenicol', in *Antibiotics III (Mechanism of Action of Antimicrobial and Antitumor Agents)*, J.W. Corcoran and F.E. Hahn (Eds), Springer-Verlag, New York, pp. 370–395.

The lincosamides

V.K. Dhawan and H. Thadepalli (1982) 'Clindamycin', *Rev. Infectious Diseases*, **4**, 1133.

F.N. Chang and B. Weisblum (1967) 'Lincomycin', in *Antibiotics I (Mechanism of Action)*, D. Gottlieb and P.D. Shaw (Eds), Springer-Verlag, New York, pp. 440–445.

B.J. Magerlein (1977) 'Modification of lincomycin', in *Structure–Activity Relationships Among the Semisynthetic Antibiotics*, D. Perlman (Ed.), Academic Press, New York, pp. 601–651.

The nitrofurans and nitroimidazoles

D.I. Edwards (1983) 'Metronidazole', in *Antibiotics VI (Modes and Mechanism of Microbial Growth Inhibitors)*, F.E. Hahn (Ed.), Springer-Verlag, Berlin, pp. 121–135.

T.J. Schwan and F.H. Ebetino (1992) 'Nitrofurans' (antibacterial agents), in *Kirk-Othmer Encyclopedia of Chemical Technology*, 4th edn, Vol. 2, John Wiley, New York, pp. 870–876.

Vancomycins and related glycopeptides

F. Parenti (1988) 'Glycopeptide antibiotics', *J. Clin. Pharmacol.*, **28**, 136.

J.C.J. Barna and D.H. Williams (1984) 'The structure and mode of action of glycopeptide antibiotics of the vancomycin group', *Annu. Rev. Microbiol.*, **38**, 339.

R. Nagarajan (1991) 'Antibacterial activities and modes of action of vancomycin and related glycopeptides', *Antimicrob. Agents Chemother.*, **35**, 605.

R. Nagarajan (Ed.) (1994) *Glycopeptide Antibiotics*, Marcel Dekker, New York.

J.C. Jordan and P.E. Reynolds (1967) 'Vancomycin', in *Antibiotics III (Mechanism of Action of Antimicrobial and Antitumor Agents)*, J.W. Corcoran and F.E. Hahn (Eds), Springer-Verlag, New York, pp. 704–718.

J.C. Jordan and P.E. Reynolds (1967) 'Vancomycin', in *Antibiotics I (Mechanism of Action)*, D. Gottlieb and P.D. Shaw (Eds), Springer-Verlag, New York, pp. 102–116.

M. Arthur and P. Courvalin (1993) 'Genetics and mechanisms of glycopeptide resistance in enterococci', *Antimicrob. Agents Chemother.*, **37**, 1563–1571.

C.T. Walsh (1993) 'Vancomycin resistance: decoding the molecular logic', *Science*, **261**, 308–309.

Index